AF322759

Polymer electrolyte membrane and direct methanol fuel cell technology

Related titles:

Polymer electrolyte membrane and direct methanol fuel cell technology Volume 1: Fundamentals and performance of low temperature fuel cells
(ISBN 978-1-84569-773-0)
Polymer electrolyte membrane fuel cells (PEMFCs) and direct methanol fuel cells (DMFCs) represent promising low-temperature electrochemical power generation technologies that operate on hydrogen and methanol respectively. These technologies have attracted intense worldwide commercialisation research and development efforts with a large element of these efforts directed at materials developments for fuel cell durability and long term operation. This two volume set presents a comprehensive and detailed review of the fundamentals, performance, and *in situ* characterisation of PEMFCs and DMFCs. Volume 1 covers the fundamental science and engineering of these low temperature fuel cells, focusing on understanding and improving performance and operation.

Functional materials for energy applications
(ISBN 978-0-85709-059-1)
Functional materials are a class of advanced energy conversion materials of use in photoelectric, thermoelectric, electrochemical, piezoelectric or electromagnetic applications, such as photovoltaics (PV), hydrogen production and storage, fuel cell systems, and demand-side energy management systems. Global demands for lower cost, higher efficiency, mass-production and, of course, sustainably sourced materials have coupled with advances in nanotechnology to enable an increasingly important role for functional materials in the sustainable energy mix. This book presents a comprehensive review of the issues, science and development of functional materials in renewable and sustainable energy production and management applications.

Advanced membrane technology for sustainable energy and environmental applications
(ISBN 978-1-84569-969-7)
Progress in membrane materials, selective membrane design, and computer modelling and simulation have contributed greatly to the application of advanced membranes in conventional and alternative power sectors, as well as to clean industry applications. Research and development of membrane systems continues apace towards the goal of simple, efficient, and easily integrated systems that offer low-cost, reliable processing and operation. This book presents a comprehensive review of membrane science and technology, focusing on developments and applications in the areas of sustainable energy and clean-industry.

Details of these and other Woodhead Publishing materials books can be obtained by:

- visiting our web site at www.woodheadpublishing.com
- contacting Customer Services (e-mail: sales@woodheadpublishing.com; fax: +44 (0) 1223 832819; tel.: +44 (0) 1223 499140 ext. 130; address: Woodhead Publishing Limited, 80 High Street, Sawston, Cambridge CB22 3HJ, UK)
- contacting our US office (e-mail: usmarketing@woodheadpublishing.com; tel. (215) 928 9112; address: Woodhead Publishing, 1518 Walnut Street, Suite 1100, Philadelphia, PA 19102-3406, USA)

If you would like e-versions of our content, please visit our online platform: www.woodheadpublishingonline.com. Please recommend it to your librarian so that everyone in your institution can benefit from the wealth of content on the site.

Woodhead Publishing Series in Energy: Number 31

Polymer electrolyte membrane and direct methanol fuel cell technology

Volume 2: *In situ* characterization techniques for low temperature fuel cells

Edited by
Christoph Hartnig and Christina Roth

WOODHEAD
PUBLISHING

Oxford   Cambridge   Philadelphia   New Delhi

Published by Woodhead Publishing Limited,
80 High Street, Sawston, Cambridge CB22 3HJ, UK
www.woodheadpublishing.com
www.woodheadpublishingonline.com

Woodhead Publishing, 1518 Walnut Street, Suite 1100, Philadelphia,
PA 19102-3406, USA

Woodhead Publishing India Private Limited, G-2, Vardaan House,
7/28 Ansari Road, Daryaganj, New Delhi – 110002, India
www.woodheadpublishingindia.com

First published 2012, Woodhead Publishing Limited
© Woodhead Publishing Limited, 2012
The authors have asserted their moral rights.

British Library Cataloguing in Publication Data
A catalogue record for this book is available from the British Library.

Library of Congress Control Number: 2011945946

ISBN 978-1-84569-774-7 (print)
ISBN 978-0-85709-548-0 (online)
ISSN 2044-9364 Woodhead Publishing Series in Energy (print)
ISSN 2044-9372 Woodhead Publishing Series in Energy (online)

The publisher's policy is to use permanent paper from mills that operate a sustainable forestry policy, and which has been manufactured from pulp which is processed using acid-free and elemental chlorine-free practices. Furthermore, the publisher ensures that the text paper and cover board used have met acceptable environmental accreditation standards.

Typeset by RefineCatch Limited, Bungay, Suffolk, UK
Printed by TJI Digital, Padstow, Cornwall, UK

Contents

3   Differential electrochemical mass spectrometry (DEMS) technique for direct alcohol fuel cell characterization   65

C. Cremers and D. Bayer, Fraunhofer Institute for Chemical Technology, Germany

4   Small angle X-ray scattering (SAXS) techniques for polymer electrolyte membrane fuel cell characterization   87

X. Tuaev and P. Strasser, Technical University Berlin, Germany

5   X-ray absorption near edge structure ($\Delta\mu$ XANES) techniques for low temperature fuel cell characterization   120

D. E. Ramaker, George Washington University, USA and C. Roth, Technische Universität Darmstadt, Germany and Karlsruhe Institute of Technology, Germany

Contributor contact details

(* = main contact)

Editors

C. Hartnig*
Chemetall GmbH
Trakehner Str. 3
60487 Frankfurt
Germany
E mail: chartnig@gmx.net

C. Roth
IAM-ESS
Karlsruhe Institute of Technology
Hermann-von-Helmholtz-Platz 1
76344 Eggenstein-Leopoldshafen
Germany
E mail: christina.roth@kit.edu

Chapter 1

E. Principi
FERMI@Elettra Free
    Electron Laser
Sincrotrone Trieste S.C.p.A.
S.S. 14 – Km 163.5, Area Science
    Park
34149 Basovizza (TS)
Italy
E-mail: emiliano.principi@elettra.
    trieste.it

Chapter 2

H. Schulenburg*
Electrochemistry Laboratory
Paul Scherrer Institut (PSI)
5232 Villigen
Switzerland
E-mail: Hendrik.schulenburg@web.de

C. Roth
IAM-ESS
Karlsruhe Institute of Technology
Hermann-von-Helmholtz-Platz 1
76344 Eggenstein-Leopoldshafen
Germany
E-mail: christina.roth@kit.edu

F. Scheiba
IFW Dresden
Institute for Complex Materials
    (IKM)
Helmholtzstrasse 20
01069 Dresden
Germany

Chapter 3

C. Cremers* and D. Bayer
Fraunhofer Institute for Chemical
    Technology
Joseph-von-Fraunhofer Str. 7
76327 Pfinztal
Germany
E-mail: Carsten.Cremers@ict.
    fraunhofer.de

Chapter 4

X. Tuaev and P. Strasser*
The Electrochemical Energy,
    Catalysis and Materials Science
    Laboratory
Department of Chemistry, Chemical
    Engineering Division
Technical University Berlin
Strasse des 17 Juni 124
10623 Berlin
Germany
E-mail: xenia.tuaev@tu-berlin.de;
    pstrasser@tu-berlin.de

Chapter 5

D. E. Ramaker*
Chemistry Department
George Washington University
725 21st Street NW
Washington, DC 20052
USA
E-mail: ramaker@gwu.edu

C. Roth
IAM-ESS
Karlsruhe Institute of Technology
Hermann-von-Helmholtz-Platz 1
76344 Eggenstein-Leopoldshafen
Germany
E-mail: christina.roth@kit.edu

Chapter 6

T. Swamy
Fuel Cell Dynamics and Diagnostics
    Laboratory
Department of Mechanical and
    Nuclear Engineering
The Pennsylvania State University
University Park
Pennsylvania 16802
USA
E-mail: tswamy@a123systems.com

E. C. Kumbur*
Electrochemical Energy Systems
    Laboratory
Department of Mechanical
    Engineering and Mechanics
3104 Chestnut Street
Drexel University
Philadelphia
Pennsylvania 19104
USA
E-mail: eck32@drexel.edu

Chapter 7

D. S. Hussey* and D. L. Jacobson
Physical Measurement Laboratory
National Institute of Standards and
    Technology
100 Bureau Dr., Mail Stop 8461
Gaithersburg
Maryland 20899
USA
E-mail: daniel.hussey@nist.gov

Chapter 8

K. Wippermann* and A. Schröder
Institute of Energy and Climate
    Research
IEK-3: Fuel Cells
Forschungszentrum Jülich GmbH
52425 Jülich
Germany
E-mail: k.wippermann@fz-juelich.de

Chapter 9

I. Manke* and N. Kardjilov
Helmholtz Centre Berlin for
    Materials and Energy
Institute of Applied Materials
Hahn-Meitner-Platz 1
14109 Berlin
Germany
E-mail: manke@helmholtz-berlin.de;
    kardjilov@helmholtz-berlin.de

C. Hartnig
Chemetall GmbH
Trakehner Str. 3
60487 Frankfurt
Germany
E-mail: chartnig@gmx.net

Chapter 10

K. W. Feindel
National Research Council Canada
Institute for Biodiagnostics
    (Atlantic)
Suite 3900, 1796 Summer Street
Halifax
Nova Scotia
B3H 3A7
Canada
E-mail: kirk.feindel@nrc.ca

Chapter 11

H. Bettermann* and P. Fischer
Institute of Physical Chemistry
Heinrich-Heine-University of
    Düsseldorf
Geb. 26.33
Universitätsstr. 1
40225 Düsseldorf
Germany
E-mail: hb@ak-bettermann.de;
    Peter.Fischer@ict.fraunhofer.de

Chapter 12

I. A. Schneider*, M. H. Bayer and
    S. von Dahlen
Fuel Cell Diagnostics Activities
Electrochemistry Laboratory
Paul Scherrer Institut (PSI)
5232 Villigen
Switzerland
E-mail: ingo.schneider@psi.ch

Chapter 13

W. Schuhmann
Analytische Chemie
Elektroanalytik & Sensorik
Ruhr-Universität Bochum
Universitätsstr. 150
44780 Bochum
Germany
E-mail: Wolfgang.schuhmann@rub.de

M. Bron*
Naturwissenschaftliche Fakultät II
Chemie, Physik und Mathematik
Institut für Chemie – Technische
    Chemie I
Martin-Luther-Universität Halle-
    Wittenberg
Von-Danckelmann-Platz 4
06120 Halle (Saale)
Germany
E-mail: michael.bron@chemie.uni-halle.de

Chapter 14

R. Lindken* and S. Burgmann
Zentrum für
    BrennstoffzellenTechnik ZBT
    GmbH
Carl-Benz-Str. 201
47057 Duisburg
Germany
E-mail: R.Lindken@zbt-duisburg.de

Chapter 15

C. Hartnig*
Chemetall GmbH
Trakehner Str. 3
60487 Frankfurt
Germany
E-mail: chartnig@gmx.net

I. Manke
Helmholtz Centre Berlin for
    Materials and Energy
Institute of Applied Materials
Hahn-Meitner-Platz 1
14109 Berlin
Germany
E-mail: manke@helmholtz-berlin.de

Woodhead Publishing Series in Energy

xv

Preface

Low temperature fuel cells constitute an important piece of the sustainable and secure energy puzzle. They are one of the most promising energy conversion technologies and can be used in a broad variety of applications, ranging from stationary and portable to automotive applications, each posing different challenges and design requirements. Depending on the required power and energy density the most appropriate technology may be polymer electrolyte membrane fuel cells (PEMFCs) or direct methanol fuel cells (DMFCs). Methanol fuel can be either directly fed to the DMFC or reformed and purified to provide low-impurity hydrogen for use in PEMFCs.

Both types of fuel cell run at low operating temperatures in the region of 50–90 °C, which results in the formation of product water in a liquid state. However, non-optimized fuel cell water management not only leads to performance losses but also to accelerated degradation effects in the different cell components. These correlated effects are not easy to solve, as the respective investigations require a multidisciplinary approach and a range of expertise, from materials science and catalysis research up to in-depth field experience in mechanical engineering. Successful fuel cell design depends on combining different highly spatially and time-resolved analytical and spectroscopic methods with simulation and modeling over many orders of magnitude. This two-volume reference work provides coverage of the essential topics and techniques for improving low-temperature fuel cell design and operation.

Volume 1 opens with a comprehensive introduction to low-temperature fuel cell technology, reviewing performance issues, materials and systems developments and issues surrounding the storage and processing technologies of fuels. The second part of Volume 1 provides detailed coverage necessary to understand the characterization and degradation of fuel cell materials and their properties. Transport issues are addressed and modeling approaches detailed to provide a sound basis for a phenomenological understanding of the processes taking place in an operating fuel cell. Emphasis is put here on the transport of reactants and products and effects resulting from non-optimal conditions, with multi-scale modeling approaches offering new insights into materials properties and the correlations that can be drawn from these. Accelerated aging protocols are also described as they help to discover weak spots in fuel cell design, improve the development times of new membrane and catalyst materials, and thereby help to ensure more reliable long-term operation of low-temperature fuel cells.

Based on these fundamental insights into nano- to macro-scale processes, Volume 2 is dedicated to the investigation and observation of processes taking place under operating conditions, with a particular attempt made in the book's coverage to bridge the resolution gap, starting with enhanced spectroscopic and microscopic methods for online monitoring of catalyst activities and processes. Each individual characterisation method is described and assessed, in a way that elucidates its strengths and weaknesses. The second part of this volume reviews water and fuel management, which can be directly observed via imaging methods based on neutron and X-ray radiation as well as magnetic fields allowing for the description of almost the complete electrochemically active area. The final part focuses on locally resolved processes, where the typical methods applied are based on synchrotron X-ray radiation, laser optical methods and scanning electron microscopy.

We are convinced that the readership will benefit from the broad scope of this two-volume work, covering as it does a wealth of detailed coverage from fundamental issues to advanced *in situ* characterisation techniques. With methods ranging from the nano- to the macro-scale to investigate effects that are observed in lab-size fuel cells, complete fuel cell stacks and even systems, this work is hoped to be of great use to anyone involved in low-temperature fuel cell research and development.

The editors would like to thank the authors for their dedication and hard work and the publishers for their supporting role in realizing this ambitious project.

Christoph Hartnig and Christina Roth

Part I
Advanced characterization techniques for polymer electrolyte membrane and direct methanol fuel cells

1

Extended X-ray absorption fine structure (EXAFS) technique for low temperature fuel cell catalysts characterization

E. PRINCIPI, Sincrotrone Trieste S.C.p.A, Italy

Abstract: Extended X-ray absorption fine structure (EXAFS) is a unique tool for investigating the atomic structure of a large class of metallic electro-catalysts under selected working conditions inside polymer electrolyte membrane fuel cells (PEMFCs) and direct methanol fuel cells (DMFCs). The chapter first provides an overview of the theoretical basis of the EXAFS technique and discusses how to exploit its characteristics for *in situ* measurements on PEMFCs and DMFCs. It then describes how to adapt a conventional fuel cell to improve the quality of the EXAFS signal of the catalyst while retaining realistic working conditions.

Key words: extended X-ray absorption fine structure (EXAFS), *in situ* EXAFS on polymer electrolyte membrane fuel cells (PEMFCs) and direct methanol fuel cells (DMFCs), atomic structure of electro-catalysts, EXAFS data analysis.

1 1　Introduction

The role of the atomic structure of fuel cells nano-catalysts in the catalysis process is crucial. In particular, the atomic arrangement on the surface of nano-catalyst particles, where catalysis takes place, appears to critically affect the electrochemical performance of the fuel cell. Furthermore, the atomic structure of nano-catalysts has been found to undergo significant changes upon alteration of working conditions, exhibiting a fascinating flexible behaviour (Principi *et al.*, 2009a). However, it is still an open question how those changes can be exploited to accelerate the catalysis process. Therefore, future promising fuel cell electro-catalysts have to be the subject of fine structural analysis aimed at elucidating the role of the atomic structure during fuel cell operation. This challenging task can be achieved by developing effective and reliable techniques able to monitor the atomic structure of electro-catalysts under *in situ* conditions, i.e. inside a working fuel cell. Extended X-ray absorption fine structure (EXAFS) is a unique tool for investigating the atomic structure of a large class of metallic electro-catalysts under selected working conditions inside polymer electrolyte membrane fuel cells (PEMFCs) and direct methanol fuel cells (DMFCs), thus being a natural candidate for this research.

The aim of this chapter is to provide a compact but fairly comprehensive compendium for those researchers not expert in EXAFS, who intend to use such a technique to investigate the structure of electro-catalysts for PEMFCs and DMFCs, possibly under working conditions. The chapter first provides an overview on the

3

theoretical basis of the EXAFS technique and discusses how to exploit its characteristics for *in situ* measurements on PEMFCs and DMFCs. It then describes how to adapt a conventional fuel cell to improve the quality of the EXAFS signal of the catalyst while still retaining realistic working conditions. Effective experimental strategies are suggested to carry out *in situ* EXAFS measurements, focusing attention on possible drawbacks that can affect the quality of the results. Finally, advantages as well as limits of the EXAFS technique are discussed in detail.

1.2   Basic principles and methods

Electromagnetic radiation within the X-ray energy range (10^2–10^5 eV) is absorbed by matter mainly through the photoelectric effect. In this process, X-ray photons yield their energy to electrons lying in an atomic quantum core level (such as 1s or 2p electrons), thus promoting one or more electrons to discrete or continuum states. The probability that such a process takes place can be related to a physical quantity termed 'atomic photoabsorption cross-section' (σ_{ph}) that approximates the 'total atomic cross-section' (σ_{tot}) in the X-ray range. X-ray absorption spectroscopy (XAS) is the experimental technique that provides σ_{ph} as a function of the photon energy for a sample material. For a monoatomic substance, σ_{ph} can be roughly approximated in terms of the atomic number Z, the photon energy E (eV), the Avogadro's number N_A and the density ρ (g cm^{-3}): $\sigma_{ph} \approx \rho Z^4 N_A^{-1} E^{-3}$. The latter equation points out that the probability of a material to absorb X-ray photons increases rapidly as a function of the atomic number and decreases considerably for high-energy photons.

The smooth atomic photoabsorption cross-section exhibits sharp rises at particular energies as an effect of the quantization of the atomic energy levels. These discontinuities, called 'absorption edges', manifest themselves when the energy yielded by the X-ray photon is equal to the binding energy of a specific electron. Fig. 1.1 shows the theoretical photoabsorption cross-sections of Pt, Pd and Ni within the energy range 2–30 keV (XDB, 2011). The absorption edges of K electrons (1s), L electrons (2s, $2p_{1/2}$, $2p_{3/2}$) and M electrons (3s, $3p_{1/2}$, $3p_{3/2}$, $3d_{3/2}$, $3d_{5/2}$) can be easily identified in Fig. 1.1.

When the energy of the absorbed X-ray photon exceeds the binding energy, an excited electron (termed 'photo-electron') can be promoted to the continuum with kinetic energy equal to the energy surplus, leaving the original atom (termed 'photo-absorber') in an excited metastable state. A fraction of photo-electrons interacts elastically with the surrounding neutral atoms and comes back to the photo-absorber thus describing well-defined closed paths. In what follows, we will take into consideration only this specific class of photo-electron. The average length of the photo-electrons' path (i.e. the so called 'mean-free path') is typically limited to a few angstroms as an effect of the short lifetime of the excited state of the photo-absorber and of the inelastic losses experienced by the photo-electron. The photo-electrons' paths can be characterized by one or more 'scattering events' involving different surrounding neutral atoms as well as the photo-absorber itself.

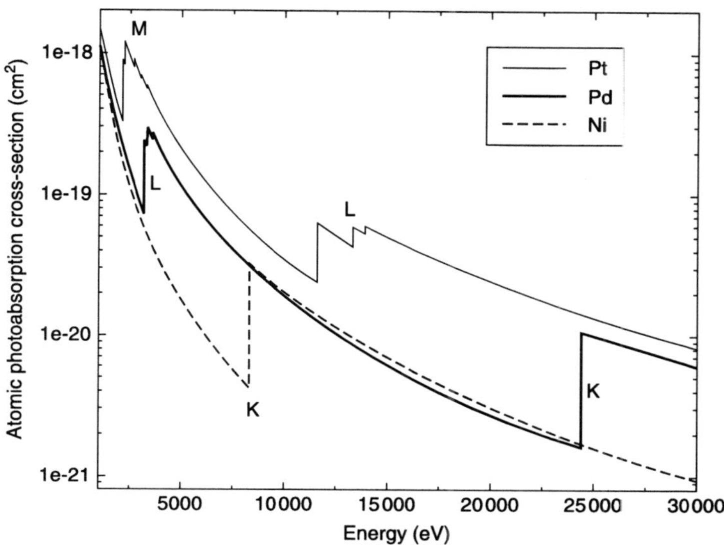

1.1 Theoretical atomic photoabsorption cross-sections of Pt, Pd and Ni. The labels 'M', 'L', 'K' refer to the absorption edges.

While a single scattering path is definitely determined by a couple of atoms, multiple scattering paths can include two or more atoms.

It is common to describe the photo-electron in terms of a spherical wave invoking its wavelike nature. The wave number associated with a photo-electron is $k = (2m\hbar^{-2}(E - E_e))^{1/2}$, where E_e is the absorption edge energy (usually defined by the inflection point at the absorption edge) and m is the electron mass. Such waves originate from the photo-absorber, propagate in the atomic lattice and are partially reflected by the atoms around the photo-absorber. The interaction between the outgoing and back-scattered waves leads to interference effects that alter the absorption cross-section as a function of the photo-electron energy or wave number. These effects result in a small oscillation of the absorption cross-section that usually extends over an energy range of hundreds of eV, known as 'extended X-ray absorption fine structure' (EXAFS). Therefore, the EXAFS is pronounced in condensed matter where the inter-atomic distances are smaller than the photo-electron mean free path. In Fig. 1.2 the EXAFS can be easily identified above the absorption edges L_3 (11 564 eV) and L_2 (13 273 eV) of Pt. The so called 'EXAFS signal' can be defined (Filipponi *et al.*, 1995b) as

$$\chi(E) = \frac{\sigma_{ph}(E) - \sigma_{ph}^0(E)}{\sigma_{ph}^0(E)} \qquad [1.1]$$

where $\sigma_{ph}^0(E)$ is the photoabsorption cross-section of the isolated atom ('bare atom absorption').

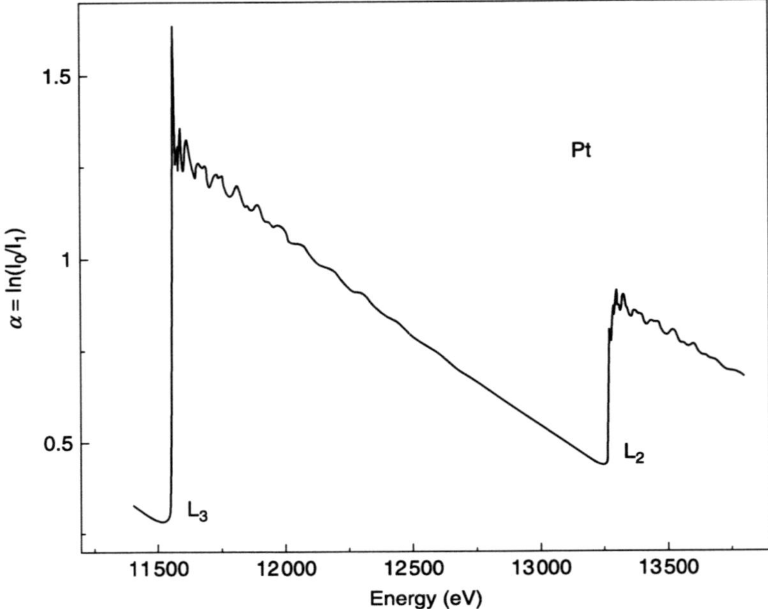

1.2 Measured X-ray absorption ($\alpha(E) \propto \sigma_{ph}(E)$) of a Pt foil within an energy range including the absorption edges L_3 (11564 eV) and L_2 (13273 eV). The oscillating extended X-ray absorption fine structure (EXAFS) is noticeable in the post-edge regions.

Above the main absorption edge, $\sigma^0_{ph}(E)$ is not a smooth function of the energy, due to the presence of additional absorption edges associated with the opening of double-electron excitation channels (Li *et al.*, 1992; Soldatov *et al.*, 1993; Filipponi and Di Cicco, 1995a; Kodre *et al.*, 1997). Currently, available theories for theoretical EXAFS signal calculation only account for single-electron absorption processes, thus severely hampering a reliable calculation of double-electron excitation edges. However, the absorption edges related to double-electron excitations can alter the EXAFS analysis and should be included in the atomic background. The lack of a theoretical approach can be tackled exploiting a semi-empirical method based on the use of properly defined step-like functions that mimic the shape of the additional absorption edges (Di Cicco *et al.*, 1996). These additional features can be suitably shaped and added to the background polynomial splines by means of a fitting procedure. Their energy position can be roughly defined invoking the so called '$Z + 1$' approximation and then optimized upon the same fitting procedure. According to this method, the excitations of further electrons for an atom with atomic number Z, occur approximately around the energies $E^i_{e2} = E_e + E^i_{Z+1}$, where E^i_{Z+1} is the binding energy of the *i*-th electron in the neutral atom owning one proton more than the photo-absorber.

Concerning the case of Pt, the binding energies of 5s, $4p_{3/2}$ and $4p_{1/2}$ electrons of the 'Z + 1' atom (Au) are respectively 107.2 eV, 546.3 eV and 642.7 eV. The expected double-electron excitations 1s5s (E_e + 107.2 eV), $1s4p_{3/2}$(E_e + 546.3 eV) and $1s4p_{1/2}$ (E_e + 642.7 eV) are indicated by three vertical arrows in Fig. 1.3. The double-electron edges found after background fitting procedure are in fair agreement with the energies defined above by the 'Z + 1' method (Fig. 1.3). The magnified EXAFS signal obtained removing the additional discontinuities and a polynomial smooth background is shown in Fig. 1.3.

Being strictly related to the local atomic arrangement, the EXAFS signal conceals detailed and precious structural information about the photo-absorber neighbourhood. Since the early 1970s more and more accurate theoretical frameworks were developed in order to retrieve such a structural information from the EXAFS (Sayers *et al.*, 1971; Lee and Pendry, 1975; Lee *et al.*, 1981; Filipponi *et al.*, 1995b; Rehr and Albers, 2000). Four typical approximations have been used historically for the development of those theoretical models. Briefly, one assumes that:

- the photo-absorption process involves only one electron per neutral atom
- the remaining *N*-1 electrons stay 'frozen' in their state ('sudden approximation')

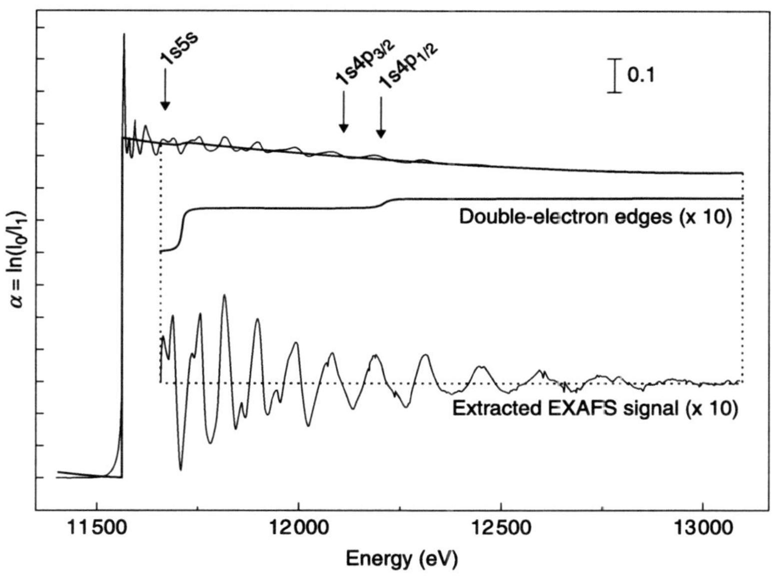

1.3 Magnified EXAFS signal of a Pt foil (L_3-edge) after atomic background removal. The background consists of a set of smooth polynomial splines and two additional edges (magnified in the figure) associated with double-electron excitations. The arrows indicate the energies of possible additional absorption edges as obtained using the 'Z + 1' approximation.

- the incident electric field is spatially constant during the absorption process ('dipole approximation', realistic for core electrons)
- the atomic potential has spherical symmetry and it is null above a cut-off radius ('muffin-tin approximation', acceptable for photo-electrons kinetic energy greater than 20–30 eV).

Under these assumptions, the EXAFS signals can be expressed analytically in terms of a set of structural and non-structural parameters.

A possible simplified analytical form of the EXAFS signal, valid for crystalline matter, including only single-scattering contributions from the j-th atomic coordination shells, is given by (see for example Lee and Pendry (1975)):

$$\chi(k) = \sum_j \left(N_j S_0^2 f_j(k) e^{-2k^2 \sigma_j^2} e^{-2R_j/\lambda(k)} / kR_j^2 \right) \cdot sin(2kR_j + \delta_j(k)) =$$

$$= \sum_j A_j(k, R_j) \cdot sin(2kR_j + \delta_j(k)) \quad [1.2]$$

The structural parameters of the j-th coordination shell are R_j (the distance of the j-th coordination shell), N_j (number of atoms in the j-th coordination shell, known as 'coordination number' or 'degeneracy'), σ_j^2 (the structural and thermal variance affecting the parameter R_j). The amplitude $A(k, R_j)$ of the signal $\chi(k)$ rapidly decays as an effect of the exponentials $e^{-2k^2\sigma_j^2}$ and $e^{-2R_j/\lambda(k)}$, depending respectively on the atomic disorder and on the lifetime of the excited photo-absorber (strictly related to the photo-electron mean free path λ). The parameter S_0^2, resulting from the 'sudden approximation', accounts for the relaxation of the excited photo-absorber and further attenuates the amplitude of the signal. Typical values for S_0^2 are restricted within the range 0.6–1.0. The terms $f_j(k)$ and $\delta_j(k)$ belong to the diffusion theory and they are respectively the backscattering and the phase-shift functions. From Eq. 1.2 it emerges that the EXAFS signal of the first coordination shell (i.e. the first neighbour atoms) exhibits the largest amplitude and the lowest frequency. Eq. 1.2 provides a theoretical description of the EXAFS signal and allows retrieval of the structural and non-structural parameters by a fitting procedure of the model theoretical signal on the experimental EXAFS signal.

A more sophisticated analytical expression for the EXAFS signal, applicable both to crystals and disordered matter (amorphous or liquid), shows in a more elegant fashion the structural insight of the EXAFS technique by including explicitly both 2-body and 3-body configurations in the definition of the total signal. It is given by Filipponi and Di Cicco (1995c) and Filipponi (2001):

$$\langle \chi(k) \rangle = \int_0^\infty dr\ 4\pi\rho r^2 g_2(r)\gamma^{(2)}(r,k) +$$

$$+ \int dr_1\, dr_2\, d\theta\ 8\pi^2 r_1^2 r_2^2 \sin(\theta)\ \rho^2 g_3(r_1, r_2, \theta)\gamma^{(3)}(r_1, r_2, \theta, k) + \dots \quad [1.3]$$

$$= \int_0^\infty dr\ N^{(2)}(r)\gamma^{(2)}(r,k) + \int dr_1\, dr_2 N^{(3)}(r_1, r_2 \theta)\gamma^{(3)}(r_1, r_2, \theta, k) + \dots$$

In Eq. 1.3 the kernel functions $\gamma^{(n)}$ can be expressed as the sum of single and multiple-scattering signals associated with the n-body configurations around the photo-absorber (see Fig. 1.4a). $\gamma^{(2)}$ refers to 2-body configurations and consists of a single-scattering signal χ_2 (the most intense) and multiple-scattering signals (χ_4, χ_6, χ_8 ...). $\gamma^{(3)}$ refers to 3-body configurations, and therefore consists of multiple-scattering signals (χ_3, χ_4, χ_5, ...). The functions $N^{(n)}$ represent the number densities associated with the specific atomic arrangement around the photoabsorber and contain additional structural information such as configuration degeneracies and atomic disorder. The distribution functions g_n and consequently the number densities $N^{(n)}$ extend formally to the whole coordinate space but they can be truncated as the $\gamma^{(n)}$ signals in Eq. 1.3 tend rapidly to zero when the photo-electron path-length increases. The short-range $N^{(n)}$ functions can be modelled as the sum of independent peaks ($p_i^{(n)}$) properly weighted by a configuration degeneracy factor ($\eta_i^{(n)}$): $N^{(n)} = \Sigma_i \eta_i^{(n)} p_i^{(n)}$ (Filipponi and Di Cicco, 1995c). Note that $\eta_i^{(2)} = N_i$, where N_i are the coordination numbers defined in Eq. 1.2. Such peaks are not necessarily Gaussian. For example, the peak $p_1^{(2)}$, proportional to the first peak of pair distribution function $g_2(r)$, is modelled more realistically by asymmetric functions such as the Euler Γ-like distribution (Minicucci and Di Cicco, 1997). The asymmetry level of that distribution can be easily tuned by varying a single additional parameter β (skewness). For β tending to 0 the Euler Γ-like distribution reduces to a symmetric Gaussian distribution.

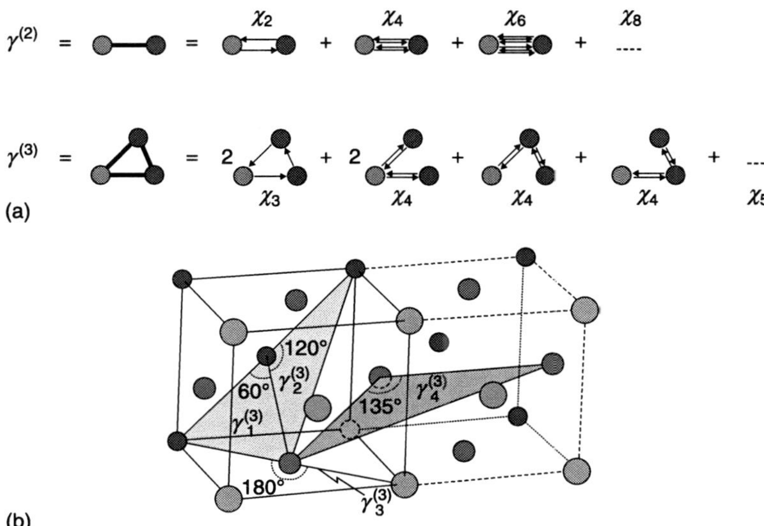

1.4 (a) Pictorial view of the principal single- and multiple-scattering signals χ_i that compose the signal functions $\gamma^{(n)}$. Adapted from Filipponi et al. (1995b). (b) Three-body configurations present in the Pt crystalline fcc lattice associated with the $\gamma^{(3)}$ signals used to fit the EXAFS signal of Pt (see Fig. 1.5).

Several computer programs are available developed to analyse the XAS raw data and get the structural information of the EXAFS signal. An exhaustive list is available on the web (XAFS, 2011). Here, we mention two robust packages that calculate *ab initio* the theoretical EXAFS signal (generating the phase shifts and the amplitude functions) and perform a fit procedure on the experimental EXAFS signal to extrapolate the structural quantitative information. They are FEFF (http://leonardo.phys.washington.edu/feff/) and GNXAS (http://gnxas.unicam.it). At the time of writing the GNXAS code can be freely run as a Web 2.0 server-side application on http://gnxas2.unicam.it.

In Fig. 1.5 the analysis of the EXAFS signal at the L_3-edge of Pt (foil) is shown, carried out using the GNXAS code (Witkowska *et al.*, 2007). The atomic background is subtracted as shown in Fig. 1.3. The total EXAFS signal is successfully modelled with six signals, of which two are associated with 2-body configurations and four with 3-body configurations. Fig. 1.4b illustrates the 3-body configurations related to the $\gamma^{(3)}$ signals used in the fitting procedure. The 3-body configurations are defined by the angle between the two shortest atomic bonds. The multiple scattering signal associated with the collinear three-body configuration (180°) is particularly intense as an effect of the 'focusing' of the central atom.

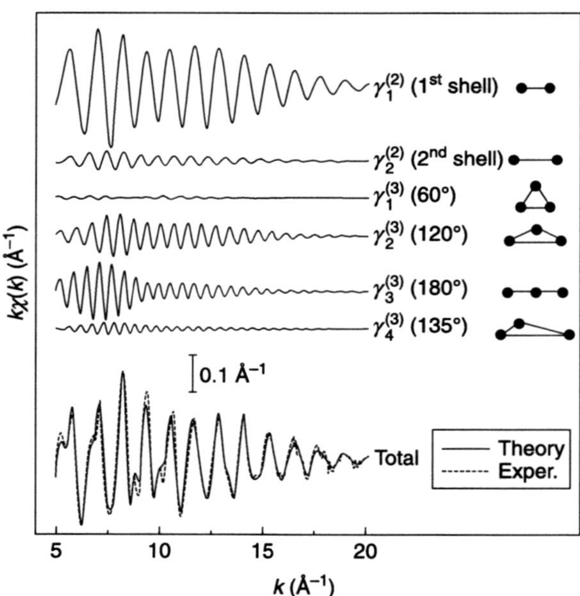

1.5 Analysis of the EXAFS signal (GNXAS package) of a Pt foil by using a suitable set of signal functions $\gamma^{(n)}$. The signals $\gamma^{(3)}$ refer to the 3-body configurations illustrated in Fig. 1.4b. (Source: adapted from Witkowska *et al.* (2007).)

1.3   Development of techniques

XAS measurements are mainly carried out at synchrotron radiation facilities in dedicated beamlines equipped with an X-ray spectrometer. A complete list of synchrotrons is provided by the International X-ray Absorption Society (IXAS, 2011). Conventional X-ray spectrometers are equipped with a double crystal monochromator and guarantee a very large operational energy range (approximately 5–80 keV), high energy resolution (down to below 1 eV), high spectral signal-to-noise ratio and high beam stability. The beam size has a rectangular shape as an effect of the more pronounced horizontal divergence of the beam of the X-ray source (often a bending magnet), with typical sizes at the sample around 10 mm^2. The size of the beam can be significantly reduced by using vertical and horizontal slits or possibly by using additional focusing mirrors that can generate spot diameters smaller than 100 μm. Typical time required to collect a conventional XAS measurement ranges from 30 to hundreds of minutes, depending on the experimental set-up and conditions. When smaller X-ray spots (~10 μm) and/or shorter acquisition times (1 ms or less) are desirable, in principle one can use an alternative kind of X-ray spectrometer based on the energy dispersive geometry (polychromator coupled with a CCD detector). However, in this chapter we focus on conventional X-ray spectrometers that represent the state-of-the-art devices for *in situ* EXAFS measurements on PEMFCs and DMFCs.

In practice, X-ray spectrometers allow for two main experimental approaches, both suitable for *in situ* EXAFS measurements on PEMFCs and DMFCs: the transmission and the (energy dispersive) fluorescence mode. An illustration of these set-ups is depicted in Fig. 1.6. Three ionization chambers (I_0, I_1, I_2) monitor the absorption of the sample and of a reference sample. An additional multi-channel X-ray detector (I_f) is required in the energy dispersive fluorescence geometry. The X-ray detector should be positioned on the horizontal plane at a right-angle to the incident beam, to maximize the suppression of the elastic and inelastic X-ray scattering of the sample as an effect of the polarization of the synchrotron radiation.

Typical acquisition times are approximately 1–3 s/point and 10–20 s/point respectively for transmission and fluorescence. Usually, several X-ray absorption spectra are collected subsequently and averaged to further minimize the noise. Typical noise level for *in situ* measurements is within the 10^{-4}–10^{-3} range in transmission and 10^{-3}–10^{-2} range in fluorescence mode. The transmission mode provides intrinsically better EXAFS signals, but its efficacy decreases drastically for diluted samples such as in a modern fuel cell catalyst layer. Therefore, the choice between the two experimental approaches should be carefully evaluated when preparing EXAFS measurements on a working PEMFC or DMFC.

The physical quantity measured in X-ray absorption experiments is the linear absorption coefficient μ (cm^{-1}) that, for a monoatomic substance, is simply related

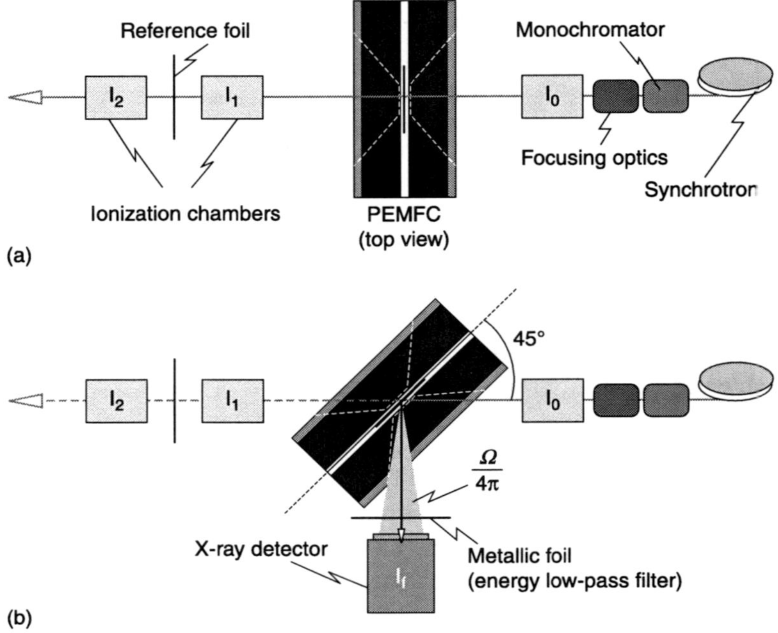

1.6 (a) Conventional transmission and (b) fluorescence set-up available at synchrotrons for *in situ* XAS measurements on PEMFCs and DMFCs. (Source: adapted from Principi *et al.* (2009a).)

to the atomic photoabsorption cross-section: $\mu = \rho\, N_A\, A^{-1}\, \sigma_{ph}$, where N_A is the Avogadro's number, ρ is the density (g/cm^3) and A is the atomic weight. In transmission geometry $d\cdot\mu(E) = \ln(I_0/I_1)$, according to the Lambert-Beer law ($I_1 = I_0\, e^{-\mu d}$), where d is the effective sample thickness. The quantity $\alpha(E) = d\cdot\mu(E)$ is commonly called 'absorption'. In the fluorescence geometry of Fig. 1.6, under the 'small thickness' approximation (Lee *et al.*, 1981) that is usually valid for diluted catalysts (Principi *et al.*, 2009a), one has $\xi\cdot d\cdot\mu(E) = \ln(I_f/I_0)$, where the constant $\xi = 2^{-1/2} 4\pi^{-1}\Omega\varepsilon$ depends on the solid angle ($(4\pi)^{-1}\Omega$) covered by the detector and on the fluorescence yield (ε) of the photo-absorbing atom. According to Eq. 1.1 the EXAFS signal characteristics and the related structural information definitely reside in the experimentally measured quantities I_0, I_1, I_2 and I_f. Other experimental parameters such as the sample amount or the acquisition time cannot affect the EXAFS. A possible mistake for beginners is to increase arbitrarily the absorption $\alpha(E)$ by sample overloading with the aim, for instance, of increasing the EXAFS signal amplitude.

Contrary to the transmission geometry, where the whole X-ray beam is efficiently absorbed by the ionization chambers, in fluorescence mode X-rays can be detected only partially due to the limited solid angle covered by the X-ray detector. The use of wide-angle detectors could be ineffective if it is not possible to make a large X-ray window on the fuel cell. It could be preferable to position a compact detector as close as possible to the X-ray window obtained in the cell. This approach also reduces the absorption due to air that is not negligible below 10 keV. This aspect should be taken into account because below 10 keV one observes the K_α fluorescence lines of 4th period metals (e.g. Ti, Cr, V, Fe, Co, Ni, . . .) and the L_α fluorescence lines of 6th period metals (e.g. W, Pt, Au . . .) that are often present in the electrodes of PEMFCs and DMFCs.

A limit of the fluorescence set-up shown in Fig. 1.6b is the poor energy resolution of solid-state X-ray detectors (about 150 eV at 10 keV for a cooled Ge detector) often used in this class of measurements. This should affect the EXAFS signal if further atomic elements present in the membrane electrolyte assembly (MEA) or in the fuel cell components generate X-ray fluorescence lines close to that of the catalyst. If those lines are more energetic than the sample lines, an energy low-pass filter, made by a thin metallic foil with a suitable absorption edge, can efficiently discriminate between the sample and the environmental fluorescence. Moreover, the filter is effective to suppress possible residual scatter peaks generated by the cell (Fig. 1.6b). Attention should be devoted to fluorescence lines possibly generated by the filter itself that can be eventually removed by Soller slits centred on the sample, although this increases the level of complexity of the set-up and attenuates the intensity of the sample fluorescence at the detector. As an alternative to the energy-dispersive fluorescence set-up, a high-energy resolution fluorescence detection (HERFD) geometry could be used, which is not affected by energy resolution limits and additional fluorescence peaks, but requires a dedicated beamline.

The attenuation of the EXAFS amplitude is an interesting effect noticeable when dealing with EXAFS of nano-catalysts. A direct example of such an effect is provided by Witkowska *et al.* (2007). The reason of this phenomenon can be attributed to the large surface-to-volume ratio of nano-particles. For instance, a particle of diameter 1.5 nm has approximately 80% of its constituent atoms on its surface (Principi *et al.*, 2009a). The presence of many partially unbounded atoms placed on the surface leads to an appreciable raise of the static (or structural) disorder (σ_s^2). Although the dynamic (or thermal) disorder (σ_d^2) is substantially unchanged, the total average bond-length variance ($\sigma^2 = \sigma_s^2 + \sigma_d^2$) increases, thus leading to the reduction of the exponential term $e^{-2k^2\sigma_j^2}$ in Eq. 1.2. Therefore, Eq. 1.2 points out that the atomic disorder attenuates the EXAFS signal amplitude especially at high values of k. Other structural parameters affected by large values of the surface-to-volume ratio are the coordination numbers or more generally the degeneracies of the n-body atomic configurations. Calculations of the configurations degeneracies for a quasi-spherical fcc cluster of Pt atoms as a function of the particle diameter are shown in Fig. 1.7 (Witkowska *et al.*, 2007; Frenkel *et al.*, 2001). A substantial decrease of the degeneracies is observable for Pt particles with diameters smaller than 4–5 nm. Such calculations can provide a

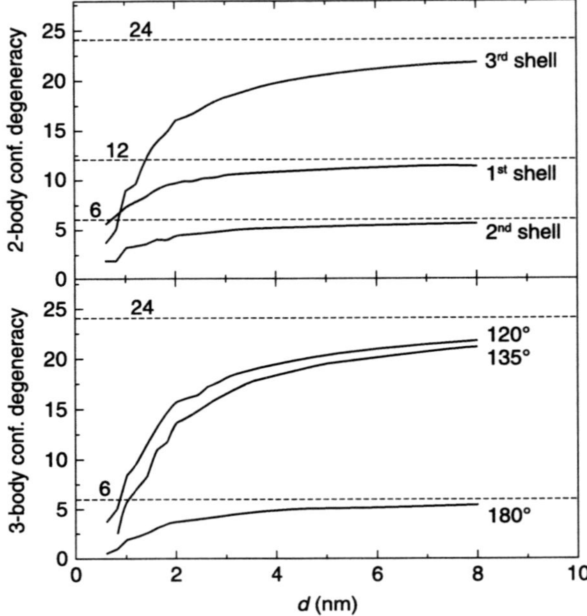

1.7 Calculated trend of the degeneracies of 2-body and 3-body configurations for a quasi-spherical particle of Pt as a function of the particle diameter. The horizontal dashed lines indicate the expected degeneracies for bulk Pt particles. (Source: dapted from Witkowska *et al.* (2007).)

realistic evaluation of nano-particle diameters if complemented by the coordination numbers obtained by a robust EXAFS analysis. Equation 1.2 emphasizes that the decrease of the coordination numbers N_j leads to a uniform attenuation of the EXAFS signal along the whole range of k values.

Structural and non-structural parameters that affect the EXAFS amplitude are often correlated, thus making their quantitative determination arduous. A possible approach to tackle this problem is to collect the X-ray absorption spectrum of a bulk sample (such as a foil, if available) of well-known structure (inter-atomic distances, signal degeneracies, thermal disorder at room temperature, etc.) made of the same material as the catalyst, using the same experimental set-up. The accurate EXAFS analysis of that sample should provide several important reference values such as σ_j^2 and S_0^2 to be used in successive analysis of the nano-catalyst. In particular, the value of the S_0^2 parameter should be carefully evaluated in order to get reasonable values of the structural parameters for the bulk sample (e.g. at room temperature). The obtained S_0^2 value for the bulk sample should not be changed when analysing the EXAFS signal of the nano-catalyst as it is strongly correlated with the signals degeneracies (parameters N_j in Eq. 1.2 or functions $N^{(n)}$ in Eq. 1.3) and with the atomic disorder parameters (σ^2).

A further effect that alters the EXAFS signal in nano-catalysts is oxygen acsorption. The EXAFS signal associated with oxygen in the vicinity of the ploto-absorber is characterised by very low frequency (according to the short inter-atomic distance between metals and oxygen) and rapidly decreasing amplitude as an effect of the oxygen backscattering function $f(k)$. In Fig. 1.8 the EXAFS signal of a Pt nano-catalyst at the cathode of a dry PEMFC is shown. The experimental signal consists of a Pt-Pt and a Pt-O signal. The presence of the strong Pt-O signal is indicated by a considerable peak in the Fourier transform presented in Fig. 1.9. By monitoring the length, variance and asymmetry of the Pt-O bond under different working conditions, one can obtain details on the role of oxygen in the oxygen reduction reaction (ORR) mechanism. In Fig. 1.9, for instance, the attenuation of the Pt-O peak reveals that a pronounced reduction occurs at the Pt nano-catalyst surface when the PEMFC is activated by fuels supply. This is an essential prerequisite to achieve an effective ORR, as the catalysis process is inhibited when the nano-catalysts surface is oxidised.

During conventional XAS measurements, the absorption of a reference sample can be measured in transmission geometry, simultaneously with the sample under examination, using the ionization chambers I_1 and I_2 (see Fig. 1.6). The reference sample should be carefully selected being crucial to monitor possible energy shifts of the monochromator as well as to corroborate the results of the EXAFS

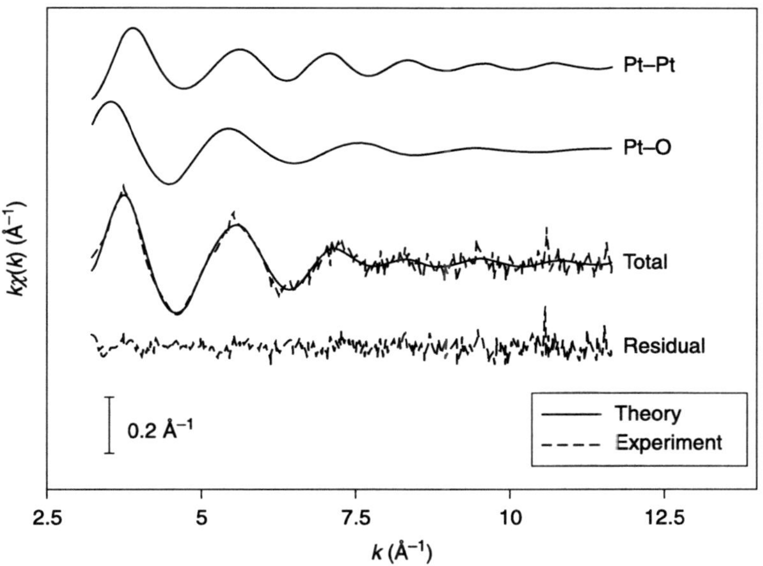

1.8 Analysis of the EXAFS signal (GNXAS package) of a Pt nano-catalyst inside a PEMFC (at the cathode) prior to supply of the gases (dry cell). The strong Pt-O oxygen signal indicates a high level of oxidation of the catalyst. (Source: adapted from Principi *et al.* (2009b).)

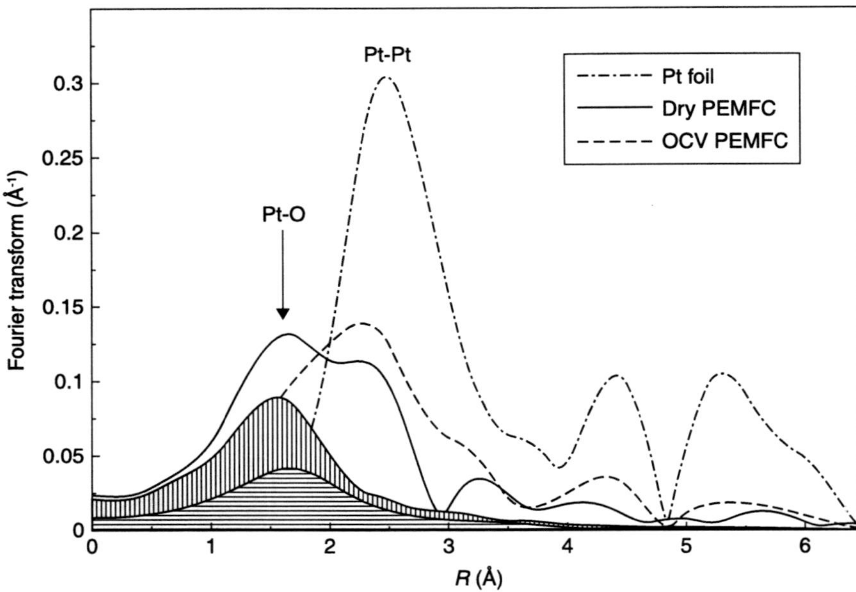

1.9 Fourier transform (FT) of the experimental EXAFS signal shown in Fig. 1.8 (dry cell) compared with the FT of the signal obtained under open circuit voltage (OCV) conditions. The FT of a Pt foil (negligible Pt-O signal) is shown as a reference. The FT of the theoretical Pt-O signal both for dry and OCV conditions gives two different peaks: dry PEMFC (peak filled in by vertical lines pattern), OCV PEMFC (peak filled in by horizontal lines pattern). (Source: adapted from Principi *et al.* (2009a).)

analysis on the sample. If available, a metallic foil of suitable absorption (typical absorption jump: $\Delta\alpha \sim 1$) and same atomic number as the catalyst should be used.

1.4   Application to fuel cell inspection

The EXAFS technique is particularly suitable for *in situ* measurements due to the capability of hard X-rays to penetrate through matter, therefore it is largely employed in various classes of experiments requiring confinement of the sample inside a specific device. Typical examples are the Paris-Edinburgh large-volume presses or the diamond anvil cells developed for studying matter under high pressure and temperature conditions (Eremets, 1996), as well as the L'Aquila-Camerino furnace specifically designed to carry out XAS experiments on very-high-temperature liquids (Filipponi and Di Cicco, 1994). Also, catalysts operating inside PEMFCs and DMFCs can be the subject of *in situ* EXAFS measurements. During the last ten years many research groups modified opportunely conventional fuel cells in order to carry out EXAFS measurements on catalysts under different working conditions (Principi *et al.*, 2007; Wiltshire *et al.*, 2005; Roth *et al.*, 2005;

Roth *et al.*, 2002; Viswanathan *et al.*, 2002a; Viswanathan *et al.*, 2002b; Maniguet *et al.*, 2000). This is possible providing a specific modification of the fuel cell aimed at maximizing the transmission of the constituent parts still retaining the electrochemical performance of the fuel cell.

The quality of the EXAFS signal is severely affected under *in situ* conditions. In order to obtain an acceptable EXAFS signal various strategies must be developed. The first aspect to take into consideration is the spatial homogeneity of the catalyst. It must be uniformly dispersed on the catalyst layer, thus leading to a smooth absorption on the area probed by the X-ray beam. This requirement should be typically fulfilled in fuel cell electrodes as the electrochemical performance can be affected by inhomogeneous catalyst layers. However, a careful visual inspection of the catalyst layer by optical microscope or scanning electron microscope (SEM) should be carried out to test its macroscopic quality. Pin-holes and cracks possibly occurring upon dehydration of the catalyst ink alter the absorption coefficient and the EXAFS signal when they cover more than 10% of the area probed by the X-ray beam (Witkowska *et al.*, 2007).

A serious problem is to limit as much as possible the absorption of the constituent parts of the fuel cell. The X-ray transmission of metallic foils thicker than 100 μm is negligible within the range 2–20 keV, including most of the core levels of metals used as electro-catalysts (e.g. Pt L_3-edge: 11564 eV). Therefore, thick metallic parts, such as metallic collector plates, must be completely removed. However, within the X-ray energy range under consideration, the absorption of light elements is *also* pronounced. Therefore, for instance, the use of a thinner Nafion® proton exchange membrane is desirable if crossover effects are limited. The estimated X-rays transmission of a Nafion® N-115 membrane (thickness 127 μm) around 10 keV is about 8% higher than the Nafion® N-117 membrane (thickness 183 μm). Similarly, the amount of carbon in the gas diffusion layer and in the catalyst layer should be reduced if this not to alter significantly the catalysis process. However, the major source of absorption from carbon is due to the graphite bipolar plates, which should be as thin as possible. It has been found that non-porous graphite, usually employed as a material for bipolar plates in PEMFCs (Herman *et al.*, 2005), can be thinned down to 250 μm, in the vicinity of the central serpentine flow channel, still being gas- and light-tight (Principi *et al.*, 2007). The required mechanical strength can be obtained providing that the thin region is limited to a few mm^2, which is the typical spot size of conventional X-ray spectrometers at synchrotrons.

In Fig. 1.10 the X-rays transmission curves of several light substances indispensable for the operation of PEMFCs or DMFCs are shown. These are carbon (C), water (H_2O), methanol (CH_3OH) and Nafion® ($C_7HF_{13}O_5S \cdot C_2F_4$). Fig. 1.10 emphasizes how the presence of such functional substances can alter severely the X-ray absorption level. This leads to a considerable background absorption in the XAS raw data and can decrease the signal-to-noise ratio of the EXAFS. For example, Fig. 1.10 shows that a modified PEMFC with graphite

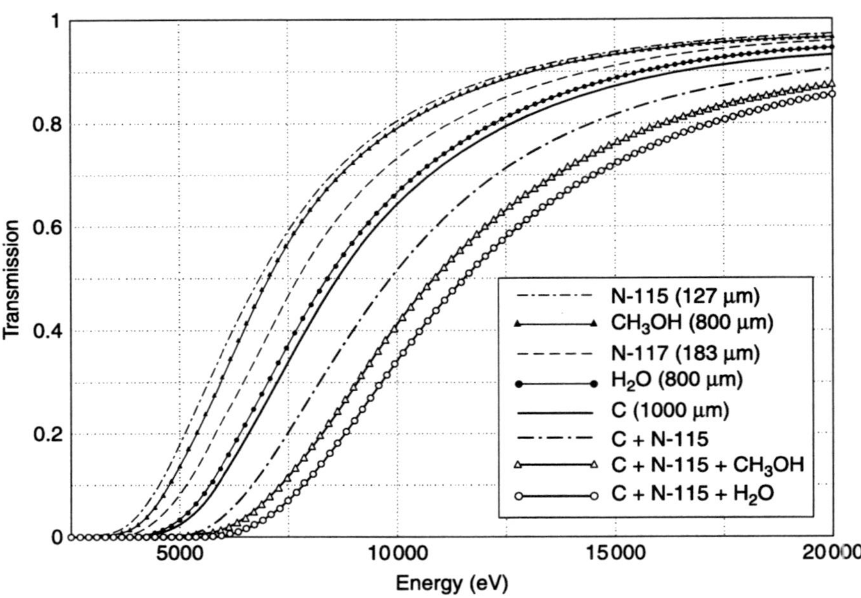

1.10 X-ray transmission of functional low-Z substances typically present in PEMFCs and DMFCs: Nafion® membranes N-115 and N-117, carbon (C), water (H_2O) and methanol (CH_3OH). (Source: CXRO (2011).)

plates thinned down to 500 μm (1 mm of overall thickness) and a Nafion® N-115 membrane is expected to transmit about the 60% of the incident X-rays at the Pt L_3-edge (excluding absorption attributed to the sample and electrodes). A similar DMFC under working conditions (flow channels filled with 800 μm of methanol) exhibits a transmission of 50%. Moreover, the considerable absorption of water below 10 keV can also suddenly affect the EXAFS measurements when water condenses inside the flow channels. This happens when the humidity inside the fuel cell is too high. The problem can be limited by decreasing the level of humidification of the gases or alternatively, if the latter is not possible, by positioning the X-ray beam between two parallel channels as shown in Fig. 1.11 (position 'B'). In that position the transmission is attenuated by the graphite plate, but the undesired rapid changes of absorption due to movements of water droplets under the X-ray beam cannot occur.

In order to guarantee realistic working conditions, the MEA parts should be preserved although they limit X-ray transmission. A common problem is to limit or exclude the X-ray absorption of the electrode not under investigation. It is practically impossible to exclude completely the EXAFS signal associated with the catalyst present in the other side of the cell, as the impinging X-rays are unavoidably absorbed by both the electrodes. A crude solution is to remove a

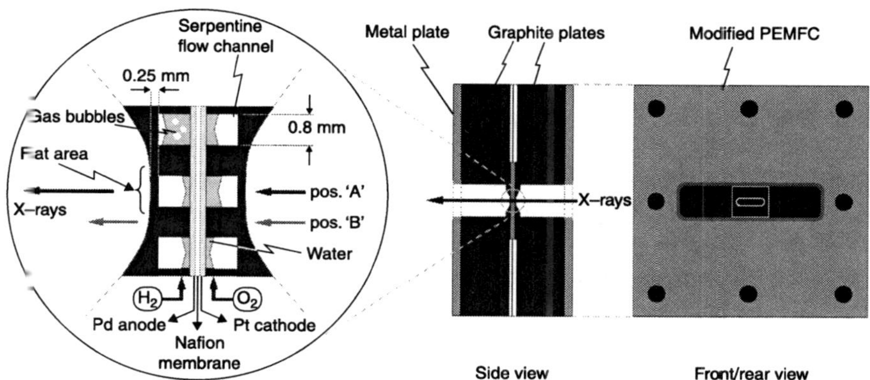

1.11 Detail of the flow channels of a PEMFC modified for *in situ* transmission and fluorescence EXAFS measurements. Position A provides higher X-ray transmission. Position B can be used when water condensation or gas bubbles occur in the flow channels as schematically illustrated in the figure. (Source: adapted from Principi *et al.* (2009a).)

limited region of the undesired electrode. However, the presence of holes of whatever size in the electrode not under investigation should be avoided, as the electric field in the region probed by the X-ray beam will be affected thus altering the catalysis process. Alternatively, the drawback can be limited by using a different catalyst at the anode. For example, if one is working at the Pt L_3-edge at the cathode, a Pd catalyst can be used at the anode preserving the overall electrochemical performance of the fuel cell. The Pd photoabsorption cross-section around the Pt L_3-edge is smooth and no Pt fluorescence lines are present around the Pt lines.

The catalyst load is another important aspect to evaluate as it can affect both the transmission and the functionality of the fuel cell. In transmission geometry, absorption jumps ($\Delta\alpha$) between 0.1 and 0.5 are reasonable for *in situ* measurements. Higher jumps should be carefully evaluated as the absorption of the sample adds to the absorption of the fuel cell. Moreover, an excessive amount of catalyst can lead to a partial under-exposure to the fuel, thus generating two classes of samples, i.e. the exposed catalyst and the under-exposed one, which are in different chemical states, and that cannot be discriminated by the XAS technique as they contribute simultaneously to the X-ray absorption. For these reasons the overload of the catalyst can be counterproductive and should be avoided. Jumps smaller than a value of 0.1 are typical of advanced electro-catalysts characterized by high dilutions (less than 0.1 mg/cm^2). For such a class of samples, which usually fulfils the 'small thickness' approximation, the fluorescence mode is preferable. Again, catalyst overload is not advisable in fluorescence mode as it can lead to undesired self-absorption effects that must be corrected through a cumbersome data analysis (Tröger *et al.*, 1992; Pfalzer *et al.*, 1999).

In Fig. 1.12 a commercial PEMFC by Electrochem® (EFC-05-02) is shown after modification as described by Principi *et al.* (2007). The graphite plates have been drilled obtaining truncated prism-shaped cavities, thus allowing for both transmission and fluorescence measurements. The flat area of about 1 mm × 7 mm in the middle of the graphite plate cavity has a minimum thickness of 250 μm (Fig. 1.11). A specific fuel cell holder shown in Fig. 1.13 positions the cell at the desired angle with respect to the X-ray beam, allowing connection of the gas pipes and easy removal and opening of the cell. The latter aspect is crucial when many catalysts have to be subsequently measured *in situ*.

In order to operate a PEMFC or a DMFC at a synchrotron beamline, an additional simplified set-up has to be installed that includes humidifiers, flow-meters, fuels, control devices (i.e. a galvanostat to monitor and tune the cell voltage) and one or more power supplies combined with resistive devices to heat the fuel cell and the fuel pipes if desired. Examples of these set-ups combined with X-ray spectrometers are described in the literature (Principi *et al.*, 2007; Wiltshire *et al.*, 2005; Viswanathan *et al.*, 2002b). Although additional equipment and devices can be safely installed by the user, it would be proper to inform the beamline responsible prior to their handling, in particular when O_2 and H_2 cylinders are to be used.

1.12 Commercial PEMFC modified for *in situ* transmission and fluorescence EXAFS measurements.

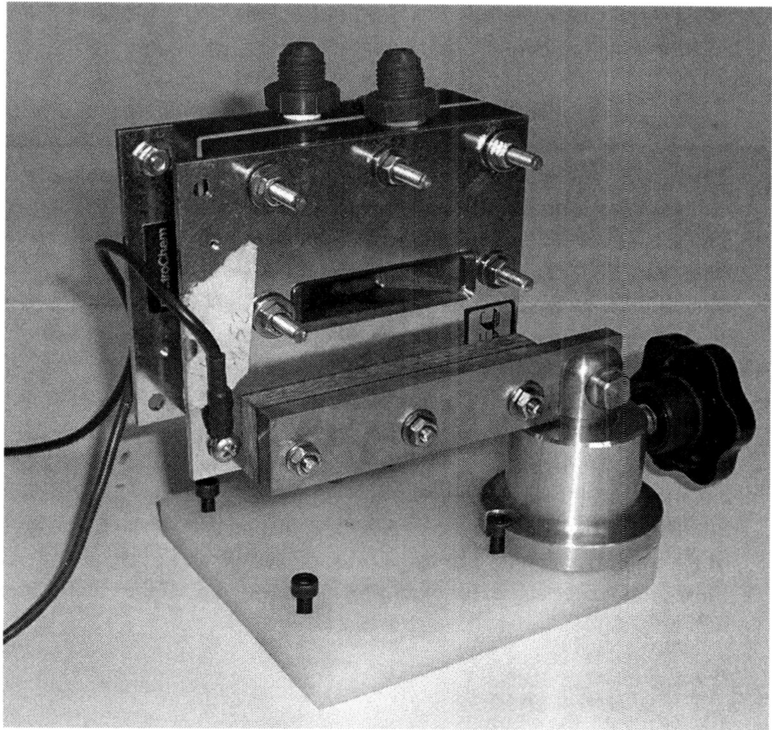

1.13 PEMFC (Fig. 1.12) connected to a special cell holder allowing connection of gas pipes and positioning of the cell at the desired angle with respect to the X-ray beam.

1.5   Advantages and limitations

Advantages are:

- X-rays exhibit a good transmittance through matter that allows for *in situ* measurements on properly modified PEMFCs and DMFCs.
- EXAFS has outstanding sensitivity to the local atomic structure. When used on nano-catalysts particles the technique is even more sensitive to the structure of the surface where the catalysis takes place.
- EXAFS is site-selective, therefore it is suitable to study poly-atomic catalysts as well as to exclude undesired effects related to other high-Z substances present in the electrodes. This peculiarity is crucial when one wants to focus the structural investigation on a specific photo-absorber.
- Very diluted innovative catalysts (less than 0.1 mg/cm^2) can be the object of EXAFS investigation by using the fluorescence mode. Moreover, diluted

catalysts are likely to fulfil the 'small thickness' approximation that simplifies considerably the EXAFS data analysis.

Limitations are:

- EXAFS measurements cannot be carried out at local facilities such as X-ray diffraction laboratories. Furthermore, beam-time at large-scale synchrotron radiation facilities has to be gained and is limited.
- The EXAFS quality is affected by spatial inhomogeneities of the catalyst layer, especially in transmission geometry.
- *In situ* XAS measurements can be severely altered by water condensation or air bubbles in the serpentine channels, which can occur during fuel cell operation.
- XAS does not permit discrimination between superficial and bulk atoms in catalyst particles.
- XAS does not permit discrimination between photo-absorbers under different chemical or physical conditions. For instance, if the catalyst is partially oxidised, the EXAFS signal of that catalyst consists of the linear combination of the signals belonging to the oxidised catalyst and to the metallic one. The same drawback occurs, for example, if the structure of the catalysts consists of two well-defined crystalline structures.

1.6   Future trends

EXAFS is a state-of-the-art technique; however, future technological advances can further extend the class of experiments accessible by this technique. In this section we limit our considerations to possible future trends of EXAFS *in situ* measurements on fuel cells.

The high-energy resolution fluorescence detection (HERFD) is a promising innovative technique that can be used possibly to collect high-quality *in situ* EXAFS measurements on diluted electro-catalysts. The HERFD allows selection of the desired fluorescence line by means of one or more analyser crystals thus avoiding the limitations caused by the poor energy resolution of solid-state detectors. Moreover, the technique allows improvement of the energy resolution above the value set by the lifetime broadening of the core hole according to Heisenberg's indeterminacy principle (Singh *et al.*, 2010). Energy resolutions around 1 eV have been already obtained. The gain in energy resolution may lead to more accurate EXAFS signals and more reliable structural analysis. Another important peculiarity of HERFD is the capability to discriminate, under specific conditions, the EXAFS signals of the same photo-absorber under different chemical states. This aspect should be crucial when investigating electro-catalysts under working conditions. A pioneering HERFD-XAS experiment has been recently carried out on a working electrochemical cell (Friebel *et al.*, 2011).

Conventional X-ray spectrometers can be successfully improved by installing focusing mirrors. Such additional devices can effectively generate spots of linear dimensions well below 100 µm still retaining the original X-ray beam intensity. This advance offers the opportunity to select the desired region of the catalyst layer to investigate. For instance, the catalyst may be under a specific chemical condition only in well-defined places within the MEA (e.g. regions where the fuel is supplied with lower efficiency as an effect of the shape of the flow channels, or regions at the margin of the MEA).

Advances in the so-called 'quick EXAFS' monochromators have reduced drastically the acquisition times characteristic of the fluorescence mode. Modern X-ray spectrometers conceived to carry out 'quick EXAFS' measurements, such as the new SuperXAS beamline at Swiss Light Source (SLS), can collect an X-ray absorption spectrum in a few seconds. The future development of this class of X-ray spectrometers will possibly allow fast *in situ* EXAFS measurements in PEMFCs and DMFCs, with the advantages of a very small spot size as discussed above. This technical progress will give the unique opportunity of monitoring the dynamics of the structural effects occurring at the catalyst upon changing the working conditions of the fuel cell on a time-scale of seconds.

1.7   Sources of further information

For a theoretical comprehensive overview of the XAS technique including EXAFS the reader should refer to Koningsberger and Prins (1988). A more recent practical guide to EXAFS has been written by Bunker (2010). A very useful primer released by Matt Newville is freely available on the web (Newville, 2004). A review specifically focused on *in situ* XAS for fuel cell catalysts has been written by Russell and Rose (2004). Further information on XAS applied to catalysis can be found in *Topics in Catalysis* (2000) **10** 3–4 where several interesting papers are collected.

1.8   References

Bunker G (2010), *Introduction to XAFS*, Cambridge University Press.

CXRO (2011), 'X-Ray interactions with matter'. Available from: http://henke.lbl.gov/optical_constants/ [Accessed 21 Mar 2011].

Di Cicco A, Filipponi A, Itié J P and Polian A (1996), 'High-pressure EXAFS measurements of solid and liquid Kr', *Phys Rev B* **54**, 9086–9098.

Eremets M I (1996), *High Pressure Experimental Methods*, Oxford University Press.

Filipponi A and Di Cicco A (1994), 'Development of an oven for X-ray absorption measurements under extremely high temperature conditions', *Nucl Instrum Meth B*, **93**, 302–310.

Filipponi A and Di Cicco A (1995a), 'Atomic background in x-ray absorption spectra of fifth-period elements: Evidence for double-electron excitation edges', *Phys Rev A*, **52**, 1072–1078.

Filipponi A, Di Cicco A and Natoli C R (1995b), 'X-ray-absorption spectroscopy and *n*-body distribution functions in condensed matter. I. Theory', *Phys Rev B*, **52**, 15122–15134.

Filipponi A and Di Cicco A (1995c), 'X-ray-absorption spectroscopy and *n*-body distribution functions in condensed matter. II. Data analysis and applications', *Phys Rev B*, **52**, 15135–15149.

Filipponi A (2001), 'EXAFS for liquids', *J Phys Cond Mat*, 13, R23–R60.

Frenkel A, Hills C W and Nuzzo R G (2001), 'A view from the inside: complexity in the atomic scale ordering of supported metal nanoparticles', *J Phys Chem B*, **105**, 12689–12703.

Friebel D, Miller D J, O'Grady P C, Anniyev T, Bargar J *et al.* (2011), '*In situ* X-ray probing reveals fingerprints of surface platinum oxide', *Phys Chem Chem Phys*, **13**, 262–266.

Hermann A, Chaudhuri T and Spagnol P (2005), 'Bipolar plates for PEM fuel cells: A review', *Int J Hydrogen Energ* 30, 1297–1302.

IXAS (2011), List of synchrotron radiation facilities. Available from: http://www.ixasportal.net/ixas/index.php?option = com_content&view = article&id = 102&Itemid = 145 [Accessed 21 Mar 2011].

Kodre A, Arčon I and Frahm R (1997), 'Exact atomic absorption background for Rb K-edge EXAFS', *J Phys IV*, **7**, 195–197.

Koningsberger D C and Prins R (1988), *X-ray Absorption: Principles, Applications, Techniques of EXAFS, SEXAFS, and XANES*, New York, Wiley.

Lee P A and Pendry J B (1975) 'Theory of the extended x-ray absorption fine structure', *Phys Rev B*, **11**, 2795–2811.

Lee P A, Citrin P H, Eisenberger P and Kincaid B M (1981), 'Extended x-ray absorption fine structure – its strengths and limitations as a structural tool', *Rev Mod Phys*, **53**, 769–804.

Li G, Bridges F and Brown G S (1992), 'Multielectron x-ray photoexcitation observations in x-ray-absorption fine-structure background', *Phys Rev Lett*, **68**, 1609–1612.

Maniguet S, Mathew R and Russell A E (2000), 'EXAFS of carbon monoxide oxidation on supported Pt fuel cell electrocatalysts', *J Phys Chem B* 104, 1998–2004.

Minicucci M and Di Cicco A (1997), 'Short-range structure in solid and liquid CuBr probed by multiple-edge x-ray absorption spectroscopy', *Phys Rev B*, **56**, 11456–11464.

Newville M (2004), Fundamentals of XAFS. Available from: http://xafs.org/Tutorials?action = AttachFile&do = get&target = Newville_xas_fundamentals.pdf [Accessed 21 Mar 2011].

Pfalzer P, Urbach J-P, Klemm M, Horn S, denBoer M L *et al.* (1999), 'Elimination of self-absorption in fluorescence hard-x-ray absorption spectra', *Phys Rev B*, **60**, 9335–9339.

Principi E, Di Cicco A, Witkowska A and Marassi R (2007), 'Performance of a fuel cell optimized for *in situ* X-ray absorption experiments', *J Synchrotron Radiat*, **14**, 276–281.

Principi E, Witkowska A, Dsoke S, Marassi R and Di Cicco A (2009a), 'An XAS experimental approach to study low Pt content electrocatalysts operating in PEM fuel cells', *Phys Chem Chem Phys*, **11**, 9987–9995.

Principi E, Witkowska A and Di Cicco A (2009b), 'Local atomic order in low Pt-content nanocatalysts investigated *in situ* by XAS', *J Phys Conf Ser*, **190**, 12173–12176.

Rehr J J and Albers R C (2000) 'Theoretical approaches to x-ray absorption fine structure', *Rev Mod Phys*, **72**, 622–654.

Roth C, Benker N, Mazurek M, Scheiba F and Fuess H (2005),'Development of an *in-situ* cell for X-ray absorption measurements during fuel cell operation', *Adv Eng Mater*, **7**, 952–956.

Roth C, Martz N, Buhrmester T, Scherer J and Fuess H (2002), '*In-situ* XAFS fuel cell measurements of a carbon-supported Pt-Ru anode electrocatalyst in hydrogen and direct methanol operation', *Phys Chem Chem Phys*, **4**, 3555–3557.

Russell A E and Rose A (2004), 'X-ray absorption spectroscopy of low temperature fuel cell catalysts', *Chem Rev*, **104**, 4613–4635.

Sayers D E, Stern E A and Lytle F W (1971) 'New technique for investigating noncrystalline structures: Fourier analysis of the extended x-ray-absorption fine structure', *Phys Rev Lett*, **27**, 1204–1207.

Singh J, Lamberti C and van Bokhoven J A (2010), 'Advanced X-ray absorption and emission spectroscopy: *in situ* catalytic studies', *Chem Soc Rev*, **39**, 4754–4766.

Soldatov A V, Ivanchenko T S, Della Longa S and Bianconi A (1993), 'Two-electron excitations and one-electron multiple-scattering resonances in the x-ray absorption of solid neon', *Phys Rev B*, **47**, 16155–16161.

Tröger L, Arvanitis D, Baberschke K, Michaelis H, Grimm U *et al.* (1992), 'Full correction of the self-absorption in soft-fluorescence extended x-ray-absorption fine structure', *Phys Rev B*, **46**, 3283–3289.

Viswanathan R, Hou G, Liu R, Bare S, Modica F *et al.* (2002a), '*In-situ* XANES of carbon-supported Pt-Ru anode electrocatalyst for reformate-air polymer electrolyte fuel cells', *J Phys Chem B*, **106**, 3458–3465.

Viswanathan R, Liu R and Smotkin E S (2002b), '*In situ* x-ray absorption fuel cell', *Rev Sci Instrum*, **73**, 2124–2127.

Wiltshire R, King C, Rose A, Wells P, Hogarth M *et al.* (2005), *Electrochim Acta*, **50**, 5208–5217.

Witkowska A, Di Cicco A and Principi E (2007), 'Local ordering of nanostructured Pt probed by multiple-scattering XAFS', *Phys Rev B*, **76**, 104110.

XAFS (2011), List of data analysis codes for EXAFS. Available from: http://www.xafs.org/Software [Accessed 21 Mar 2011].

XDB (2011), *X-Ray Data Booklet*. Available from: http://xdb.lbl.gov/ [Accessed 21 Mar 2011].

2

Advanced microscopy techniques for the characterization of polymer electrolyte membrane fuel cell components

H. SCHULENBURG, Paul Scherrer Institut (PSI), Switzerland,
C. ROTH, Technische Universität Darmstadt, Germany and
Karlsruhe Institute of Technology, Germany and
F. SCHEIBA, IFW Dresden, Germany

Abstract: High and low resolution electron microscopy is a powerful tool for detailed structure characterization of novel materials. In fuel cell research the catalyst, electrode and membrane electrode assembly (MEA) can be studied though the tool is currently underused. The chapter reviews the use of advanced electron microscopy techniques for the detailed characterization of fuel cell components emphasizing recent experimental and methodical developments. Progress in the analysis of core-shell nanoparticles, the application of identical-location (IL) transmission electron microscopy (TEM), 3D-TEM, and TEM lamellae preparation will be highlighted and future perspectives discussed.

Key words: electron microscopy, SEM, TEM, fuel cells, ionomer distribution, electrode porosity, electron tomography.

2.1 Analytical challenges in fuel cell research

Polymer electrolyte fuel cell (PEFC) components can be analyzed using a variety of electrochemical, spectroscopic and microscopic methods. Microscopic investigations of fuel cell components rely predominantly on imaging techniques, e.g. scanning electron microscopy (SEM), transmission electron microscopy (TEM) and energy-dispersive X-ray spectroscopy (EDX) mapping. Until recently, the application of electron microscopy in fuel cell technology mostly focused on the characterization of the catalyst. Particle size, particle size distribution, and dispersion on a support material were analyzed by default; but also catalyst degradation by leaching of the less noble alloy components as well as by nanoparticle growth via either Ostwald ripening or coalescence (a question not yet resolved) were prominent topics of microscopy studies. A first step towards the analysis of complete electrodes, however, was made, when sample preparation progressed and embedding and ultra-microtomy were applied to stabilize the fragile structure of the membrane-electrode assembly (MEA). This development created increasing attention from industry, as it allowed the study of MEAs before and after their operational life. Among the addressed questions were the electrode's porosity as well as changes to the distribution of the ionomer in pristine and aged electrodes providing the proton-conducting pathways in the electrode.

26

While the standard TEM/SEM characterization techniques become increasingly routine, though still valuable, new developments are on their way, which could help to significantly broaden the impact of advanced microscopy on fuel cell research. One main limitation of electron microscopy is the need of ultra-high vacuum (UHV) conditions, which so far prevent an analysis in an environment as close as possible to real conditions. A first step in this direction was the design of an *in situ* TEM sample holder, where a powder sample can be exposed to low pressures of gases, and the reaction of the catalyst studied in great detail. By this technique, changes in the surface facets of a supported nanoparticle depending on the surrounding atmosphere were probed with high resolution. Along the same lines, the so-called environmental scanning electron microscope (ESEM) was designed using sequentially pumped vacuum chambers to introduce small volumes of gases, as well as water vapour into the sample chamber. This appeared to be an important development for fuel cell technology, as it allows imaging of complete MEAs in quasi-*in situ* humid conditions. The surface wettability of differently prepared electrodes with water was studied, as well as the water transport behaviour of gas diffusion layers (GDL), and even ice was grown on the samples. Recent efforts concern the freeze-thaw behaviour of the MEA with respect to sub-zero start-up in automotive applications.

In the future, several analytical questions with respect to ionomer morphology in the electrode, carbon corrosion, electrode delamination and water transport need to be asked and answered, aiming for a more detailed understanding of fuel cell performance.

The answers to these questions are still an analytical challenge, and further progress is needed to satisfy the fuel cell community. In this chapter we will discuss advanced microscopic techniques, which may be capable of solving these problems. The visualization of the ionomer using staining and elemental mapping techniques is discussed in Section 2.2. Imaging electrode porosity before and after carbon support corrosion using two-dimensional (2D) and three-dimensional (3D) microscopy is discussed in Section 2.3. Visualization of delamination effects observed between flexible membrane and rigid electrode will be discussed thereafter (Section 2.4.3). Finally, we review other innovative microscopic techniques applied to fuel cell research only recently (Section 2.5), namely aberration corrected TEM, 3D-TEM, identical location TEM (IL-TEM) and *in situ* X-Ray tomography.

2.2   Imaging of the ionomer

Transport processes play a significant role in the proper operation of polymer electrolyte membrane fuel cells (PEFC). In PEFCs the reactant gases must have access to the catalytically active sites. Protons and electrons must be conducted through the electrode and the reaction product water must be removed from the pore system to avoid blocking of the gas diffusion paths. One disadvantage in the current standard electrode design is that each transport process is realized by a different component. For instance, gas transport is accomplished by a network of pores in the electrode structure. Electrons are conducted either by the catalyst particles themselves or by a conductive support, whereas proton diffusion in the electrode is realized by the addition of a polymer electrolyte, in most cases a perfluorosulfonated ionomer (PFSI), e.g. Nafion®. Furthermore, polytetrafluoroethylene (PTFE) is often added to the electrode to increase its hydrophobicity in order to enhance the removal of reaction water from the pore system.[1–3] Since the various components influence each other and therefore the electrode properties in a non-constructive manner, optimization of the electrode structure is far from being trivial.

One of the key components in electrode design is the ionomer, because it influences proton conductivity as well as the catalyst's electrochemically active surface, mass transport, electronic resistivity and electrode porosity. Several publications have focused on the effect of the polymer electrolyte concentration in the electrode.[4–8] In general, intermediate contents of 30–40 wt.% are reported as optimum concentrations for perfluorosulfonated ionomers such as Nafion® with carbon black supported catalysts.[4–7] At low and high electrolyte contents a much poorer electrode performance is observed. For low electrolyte contents the reduced electrode performance is explained by a limited proton conductivity and incomplete wetting of the catalyst particles by the ionomer, thus decreasing the electrochemically active surface.[4,5] In contrast, at high electrolyte contents exceeding 40 wt.% the pore size and average pore diameter decrease dramatically, and the electrode performance becomes diffusion-controlled.[5, 7]

Besides the absolute amount of the polymer electrolyte in the electrode, its distribution has a significant effect on the performance of the cell. The distribution of the electrolyte is mainly influenced by the preparation technique of the electrode, e.g. the catalyst ink preparation and coating procedure. Furthermore, significant changes in the ionomer distribution of the electrodes are expected after operation. Numerous publications deal with the degradation of the catalyst nanoparticles by growth, coalescence, and Ostwald ripening. Further work discusses corrosion of the support material in the harsh conditions at the fuel cell cathode as well as damage of the membrane due to peroxide attack.[8–12] However, comparatively little is known about the ionomer distribution and degradation in different operation conditions, and only recently this topic attracted more attention.[13] Electron microscopy studies of the membrane electrode assembly might be one key to understanding electrode degradation, helping to improve the current electrode materials to withstand it.

2.2.1 Comparison between scanning electron microscopy (SEM) and transmission electron microscopy (TEM) approach

Scanning electron microscopy (SEM) may in principle be used to obtain direct information on the polymer electrolyte distribution if it is combined with energy-dispersive X-ray spectroscopy (EDX).[14]

However, SEM on bulk electrodes has a number of limitations. Due to the high porosity and low density of the carbon support material the electron beam penetrates deep into the sample. This limits the obtainable resolution for Z-contrast (BSE) or element resolved (X-ray) imaging significantly, because – unlike the secondary electrons (SE), which are collected from the sample surface and used for normal imaging – back scattered electrons (BSE) and X-rays are collected from the excitation volume (up to 1 μm depth for the X-rays). Due to the higher penetration depth of X-rays this is especially severe for elemental mapping by X-rays.

Another drawback of the technique is due to electrode porosity; as the density of the material varies throughout the electrode, the excitation volume and therefore the amount of material contributing to the signal varies, too. Furthermore, the porosity induces a high surface roughness of the sample, which modulates the X-ray signal. Therefore, the elemental information obtained by the X-ray signal cannot be quantified correctly, and elemental maps obtained by this method may not correspond to the real elemental distribution in the sample.

A further limitation, which does not only apply to SEM, but to all electron microscopy techniques and therefore also to TEM, is the beam sensitivity of the commonly used PFSA polymer electrolyte, which is extremely prone to beam damage.[14–16] However, long recording times are necessary for mapping, since the sensitivity of EDX for light elements such as fluorine, which is the predominant element in PFSA polymer electrolytes, is rather low, and hence beam damage of the sample becomes an important issue. Some strategies, which may get us closer to the solution, will be presented below.

Although TEM in principle suffers from the same beam damage problems as SEM, TEM has a number of advantages over SEM for the characterization of the polymer electrolyte distribution in fuel cell electrodes. First of all, TEM allows imaging of all parts of the electrode structure, including the nanometer-sized catalyst particles. Secondly, TEM can also provide information about the pores. These will appear as voids in the catalyst structure, since TEM sample preparation requires the samples to be thinned to less than 100 nm so that they become transparent for the electron beam. Moreover, with energy-filtered transmission electron microscopy (EFTEM) another powerful technique for elemental mapping is at our disposal. In contrast to EDX, EFTEM is more sensitive towards light elements and allows acquisition of the complete image at the same time. This reduces the acquisition time from several hours to a few minutes and therefore helps to significantly reduce the beam damage to the sample.

2.2.2 Sample preparation techniques

Depending on the material, sample preparation for TEM can be a very tedious process. For complete MEAs, however, it is even a major obstacle due to their highly porous structure. In 2003, Blom *et al.* proposed the adoption of a sample preparation technique that is generally used for biological samples.[17] To stabilize the structure of the porous electrode the MEA is embedded in an epoxy resin, which either partially or completely infiltrates the pores. Thereafter, it is cut with an ultramicrotome using a diamond knife. As the epoxy resin infiltrates the pore space, the porous electrode structure is preserved during the sectioning process. Thus, sections can be obtained, which are sufficiently thin (~70–100 nm) for TEM analysis. The sectioning process furthermore yields sections of rather homogeneous thickness. This allows imaging of the whole cross-sectional area, including also the membrane, which is important for detailed degradation studies. Since the electrode's structural integrity is preserved by this preparation method and characterization of the membrane is possible, too, it is ideally suited for studying catalyst degradation processes after long-term operation.[18–20]

Sample preparation for TEM was carried out as follows: A small piece was removed from the MEA and embedded in Araldite 502© resin (SPI Supplies Inc.). Subsequently, the resin was cured at 60 °C for at least 16 hours. Sectioning of the embedded samples was carried out with a Reichert-Jung Ultracut E microtome at room temperature using a diamond knife (DDK). The sections were then collected from the surface of a water basin and transferred to copper grids. To reverse the deformation of the samples by the compressive cutting force, the obtained thin cuts were subjected to xylol vapor at 80 °C and dried at room temperature for 1–2 minutes. Sections of 200–500 nm thickness and ultra-thin sections of 70–100 nm were prepared for SEM and TEM microscopy, respectively. Thin sections were analyzed with a FEI Quanta 200 FEG environmental scanning electron microscope equipped with an energy dispersive X-ray detector for elemental analysis and mapping. To minimize interaction of the electron beam with the sample holder a specially designed holder was used, which supports the copper grids only on their outer rim (Fig. 2.1). Ultra-thin sections were examined with a Jeol JEM-3010 transmission electron microscope operating at 300 kV acceleration voltage with an LaB_6 cathode. The instrument is equipped with a Gatan Imaging Filter (GIF) for energy-filtered imaging and electron energy loss spectroscopy (EELS).

2.1 Sample holder for SEM analysis of thin sections.

2.2.3   Analysis of polymer electrolyte distribution in the electrode

For the analysis of the polymer electrolyte distribution in the electrode, infiltration with an epoxy resin has a significant drawback. Since the polymer electrolyte and the embedding resin have almost identical scattering contrast, the polymer electrolyte cannot be distinguished directly. As pores in the electrode structure may be filled by the polymer electrolyte or the embedding resin, it is also not possible to distinguish between open and closed pores. There exist two different approaches to tackle this problem, which will be discussed in more detail below:

- Enhancement of the scattering contrast by selective insertion of heavy metal ions in either the polymer electrolyte or the embedding resin (referred to as *staining techniques*).
- Elemental mapping of an element that is characteristic for either the polymer electrolyte or the embedding resin (referred to as *mapping techniques*).

Staining techniques

One main requirement for the polymer backbone of the polymer electrolyte is a high chemical stability in order to withstand the harsh chemical and electrochemical conditions in the electrode during fuel cell operation. A high chemical stability, however, is almost always linked to a low reactivity, which is extremely unfavourable for the incorporation of staining agents into the polymer backbone itself. Thus, staining can only be realized via the ionic groups of the polymer electrolyte. The polymer electrolyte can be easily transferred to an ion exchanged form by exposing it to a metal salt-containing solution. It has been shown in the literature that a wide variety of metal salts and even rather bulky organic cations, such as tetrabutylammonium (TBA^+), can be incorporated into PFSI polymer electrolytes.[21, 22]

Due to its high electron density and low charge allowing for a maximum number of ions to be incorporated into the polymer matrix, cesium appears as an ideal staining agent to enhance the scattering contrast of the polymer electrolyte. Rieberer *et al.*[22] reported the successful application of cesium ions as a marker for sulfonic clusters in Nafion® in a TEM study. Fig. 2.2 shows cross-sections of an unstained and a Cs^+-stained MEA in comparison. In both images the interface between the membrane and the electrode appears well-defined indicating that penetration of the membrane into the catalyst layer is low. A clear effect of the staining can be seen in the membrane of the Cs-treated sample: it has a speckled appearance with stripes of more or less strongly contrasted regions close to the electrode interface. The electrode of the stained sample also appears spotted and darker, when compared to that of the unstained sample. The staining in the electrode layer is most pronounced close to the catalyst support particles indicating that the Nafion® ionomer mainly covers the catalyst support particles, but does not

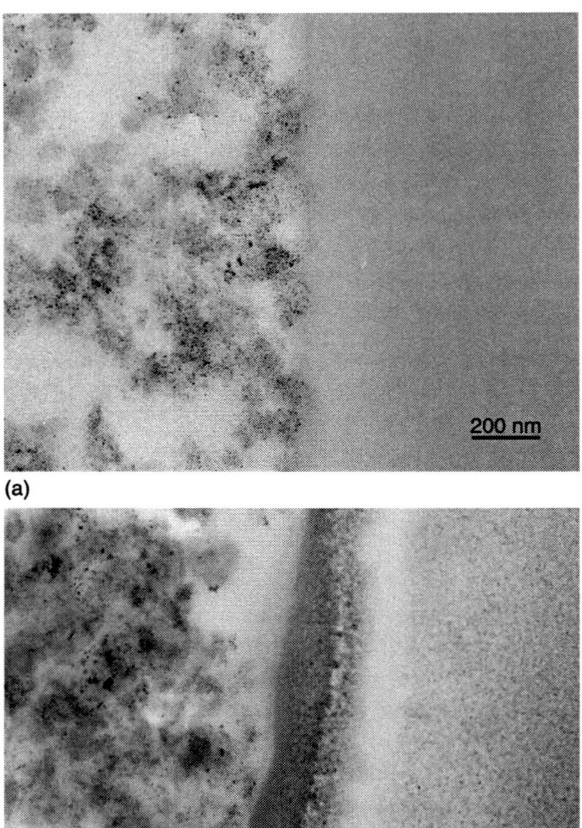

2.2 TEM bright field images of (a) an unstained and (b) Cs⁺ stained membrane electrode cross-section. Both images were taken at the membrane electrode interface with the membrane on the right hand side of the image.

flood larger pores. This is in good agreement with results of Gode *et al.*,[5] who assumed that the catalyst agglomerates become infiltrated by the polymer electrolyte only when the electrodes were prepared from inks containing Nafion® in solution. In contrast, inks containing colloidal Nafion® should not clog larger pores.[23]

The differently-contrasted regions in the membrane close to the membrane-electrode interface are most likely an artefact of the staining process and not due to the inherent structure of the membrane. Immersion of the sample in the staining solution causes swelling of the membrane. Since the water uptake in the membrane and the embedding resin is different, the membrane accommodates its increased lateral dimensions by a 'wavelike' deformation. Inhomogeneous drying of the stained sample, followed by a redistribution of the staining agent could be the reason for the contrast changes observed in Fig. 2.2. However, it has also been reported that the membrane pre-treatment process, which involves boiling of the membrane in H_2O_2 and H_2SO_4, causes an increase of the sulfonate group concentration at the membrane surface.[15] Thus, the observed contrast changes could also be linked to differences in the sulfonate group concentration at the membrane surface.

One measure to avoid these staining artefacts is to carry out the ion exchange prior to the infiltration with resin. This approach was not successful, since cesium is leached almost completely from pre-stained samples during the sectioning process. In a subsequent experiment, cesium was therefore replaced by barium, which is known to bond more strongly to sulfate or sulfonate groups. A sample, which was pre-stained with barium, is shown in Fig. 2.3. It can be observed directly from the image that the staining effect on the membrane and the electrode layer is weaker for the barium than for the cesium. This is mainly due to the higher charge of the barium ion (Ba^{2+}) compared to cesium (Cs^+), resulting in a lower amount of barium that can be incorporated into the polymer electrolyte. Due to the stronger bond, however, the staining appears to be more homogeneous. In

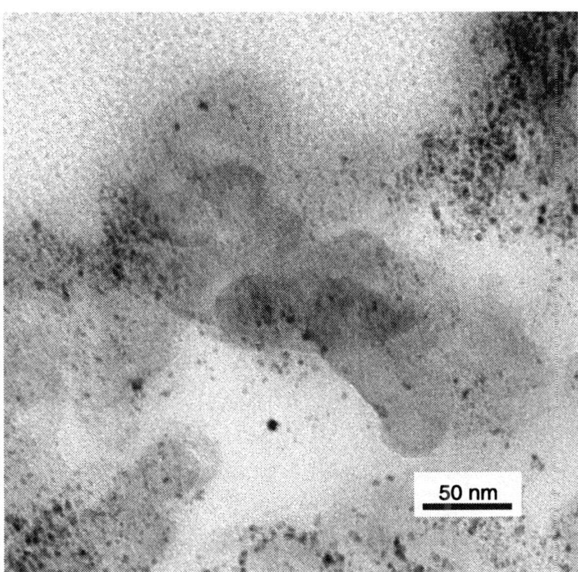

2.3 TEM bright field images of an MEA pre-stained with Ba(OH)$_2$ solution before embedding. The image shows the membrane-electrode interface with the membrane in the upper left corner.

addition, no contrast variations were observed in the membrane indicating that the staining artefacts were significantly reduced.

One interpretation for the 'speckling' is that it is due to stained ionic clusters, which are formed by the segregation of the sulfonate groups-containing side chains in the fluorocarbon matrix of the polymer backbone.[22] However, the high-resolution image shown in Fig. 2.4 reveals that the speckled appearance is caused by small nanocrystals deposited in the membrane. Evaluation of the lattice fringes by Fourier transformation (FT) indicates that these nanocrystals are most probably barium fluoride. As both the sulfonic groups and the fluorinated polymer backbone are known to decompose under ionizing radiation,[15] it is likely that the electron beam is the reason for the growth of these crystals. Barium ions, which are released from 'destroyed' sulfonic groups, may react with fluorine while the side chains and the polymer backbone degrade in the electron beam. Crystalline structures have also been observed in the darker regions of the Cs-stained samples, indicating that this phenomenon is not only limited to barium staining.

The size of the barium fluoride nanocrystals may be used to calculate the average number of sulfonate side chains contributing to their formation. The mean barium fluoride crystal size determined from Fig. 2.4 is about 2.9 nm. Assuming spherical symmetry for the barium fluoride nanocrystals the average number of barium atoms in each crystal is 215. As each side chain in Nafion®

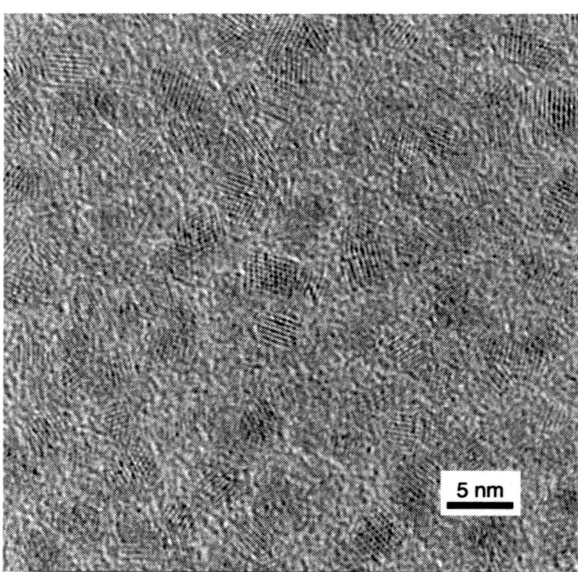

2.4 High resolution TEM bright field image of the Nafion® membrane doped with Barium ions. Small crystalline regions were found throughout the membrane with lattice spacings, which could be matched to BaF_2.

contains only one sulfonate group and one barium ion bonds to two sulfonate groups to maintain charge balance, the average number of side chains contributing to the formation of the barium fluoride crystals is about 430. According to the results of Gierke *et al.*[24] the number of sulfonate groups per ionic cluster is about 27 for a completely dried sample. This means that on average at least 16 ionic clusters contribute to the formation of one barium fluoride nanocrystal.

This result is certainly a rough approximation, since it is based on several simplifications regarding the structure of Nafion© and the structure of the barium fluoride crystals. However, it demonstrates that either significant structural re-arrangement has to take place when the sample is exposed to the electron beam or that significant percolation of ionic clusters does already exist in the non-humidified sample. Nevertheless, although significant structural re-arrangement of the polymer electrolyte may occur in the TEM on an atomic and molecular level, micro-structural changes of the electrode and membrane were not observed unless very high beam intensities were used.

Elemental mapping techniques

Elemental mapping techniques certainly are an elegant alternative to staining experiments, as they do not require the insertion of a staining agent and are therefore less prone to artefacts. Instead, differences in the chemical composition can be used to distinguish the polymer electrolyte from the resin. For PFSI polymer electrolytes, the fluorine signal can be used to visualize the electrolyte distribution. The sulfur signal stemming from the sulfonate groups is less suitable for elemental mapping, as it overlaps with the platinum $M\alpha_1$ line,[14] and sulfur impurities of the carbon support may be mapped as well.

Electron microscopy allows the use of two different elemental mapping techniques: energy dispersive X-ray analysis (EDX) and energy-filtered imaging (EFI). A major advantage of the EDX method is its wide distribution and its availability. Both methods have advantages and limitations that are rather complementary to each other and therefore they are best used in combination. As already pointed out in the introduction, the resolution of the EDX method is rather limited due to the large excitation volume for X-ray generation, which can reach a few micrometers for materials of low density. Although the use of thin sections instead of bulk samples reduces the excitation volume and therefore improves the resolution, it is difficult to resolve fine details of the polymer distribution by EDX. As the sensitivity of EDX for light elements such as fluorine is rather low and the elemental information is recorded sequentially for each image point, recording times of several hours are necessary to obtain an elemental map, thereby introducing problems of sample drift and sample stability. This is partly compensated by the fact that EDX allows parallel recording of several elements. In contrast, recording times for energy-filtered images are only in the order of a few minutes, but only one element can be mapped at a time.

Furthermore, EFTEM fails to record elemental maps of heavy elements such as platinum and hence cannot be used for mapping of the catalyst. And while it offers a much higher resolution than EDX, it is difficult to obtain survey images at low magnification.

Fluorine mapping by energy dispersive X-ray analysis (EDX)

Figure 2.5 shows the SE and BSE image of a thin section of a fuel cell electrode as well as elemental maps for carbon (C), fluorine (F), platinum (Pt) and ruthenium (Ru). In the lower half of the image, part of the membrane is visible. Since the fluorine signal recorded for the membrane was much more intense than that of the electrode, a log-transform of the image intensity was necessary to visualize the fluorine content in the electrode. The fluorine map of the electrode shows a similar intensity distribution as the carbon map, indicating that the polymer electrolyte is homogeneously mixed with the catalyst. However, especially close to the interface with the membrane, a slightly higher fluorine concentration can be observed. This could be caused to some degree by smearing of some polymer electrolyte of the membrane into the electrode during sectioning of the sample. It seems, however, more likely that the higher fluorine signal is caused by segregation of the polymer electrolyte during the MEA preparation. An airbrush was used to paint the electrode layer on the heated membrane spraying thin layers of a catalyst and ionomer-containing ink. Capillary forces in the already sprayed and drying layers may force polymer electrolyte, which is still in the liquid phase of the ink, towards the membrane leading to a concentration gradient perpendicular to the electrode layer.

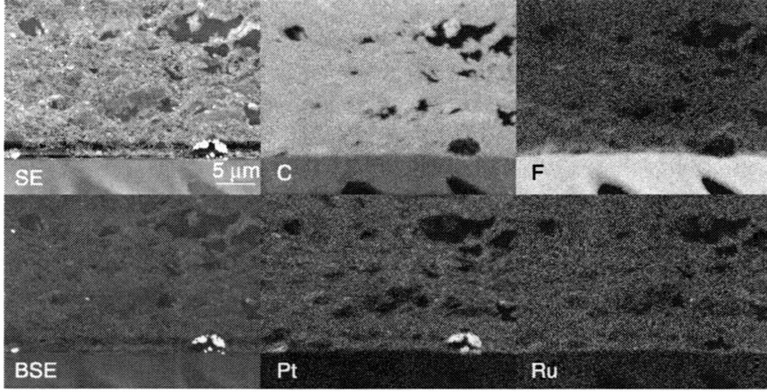

2.5 SE and BSE micrographs of a DMFC anode catalyst layer and elemental maps for carbon (C), fluorine (F), platinum (Pt) and ruthenium (Ru) for the same region obtained by EDX mapping.

Energy-filtered imaging (EFI)

Elemental information can be obtained from energy-filtered image data either by the three- or the two-window-method.[25] For the former, three images need to be acquired: one at an energy corresponding to the characteristic loss feature of the element investigated (post-edge image) and two at energy losses slightly lower than the characteristic energy loss feature (pre-edge image). The latter images are used to construct a background image, which is subtracted from the post-edge image to obtain the elemental distribution.

EFTEM uses only a relatively small fraction of the electrons leaving the sample to form an energy-filtered image. As a consequence, high electron doses are necessary during the exposure. Since high electron doses accelerate the degradation of the polymer electrolyte, it is necessary to know the degradation rate to reliably interpret the obtained fluorine concentration maps. For a first estimation, an EELS sequence was acquired from the membrane at electron doses similar to that used for the acquisition of the energy-filtered images. Fluorine losses of about 2.5% per minute were determined by integration of the intensity of the fluorine edge from the sequence of the EEL spectra. Since a typical acquisition of three energy-filtered images takes about 3 min, the fluorine loss during acquisition is less than 10%. The effect of fluorine loss can be further minimized, when the post-edge image is acquired first, since the influence of the fluorine loss on the pre-edge images should be negligible.

Unfiltered transmission electron micrographs and the corresponding fluorine maps of a PEFC electrode at two different locations are shown in Plate I in the colour section between pages 252 and 253. The fluorine maps were obtained using the three-window method with pre-edge images at 643 and 670 eV and a post-edge image at 698 eV. All energy-filtered images were recorded with a slit width of 25 eV and an exposure time of 30 s. Images Plate I(a–c) were acquired directly at the membrane electrode interface, which is clearly visible in the fluorine map (Plate I(b)) as a stripe of low fluorine concentration (dark colour). On the right side of the interface some pore space is visible, which is completely filled by the polymer electrolyte, as indicated by bright contrast or blue colour in Plate I(b, c). At those image positions, where catalyst material is present, the signal of the fluorine concentration essentially drops to zero, because the volume of the cross-section at these sample positions is almost completely occupied by the catalyst and the support particles. However, in the lower right corner of the image a fluorine signal can also be detected in between two support particles, which lie in close proximity to each other, indicating that Nafion® can penetrate into the void space of the catalyst agglomerates.

Plate I(d–f) shows a part of the electrode structure further apart from the membrane-electrode interface. In contrast to the images taken at the membrane-electrode interface the pore space in the electrode layer is not completely filled by the polymer electrolyte. In some parts of the image the fluorine distribution can be seen to closely track the outlines of the catalyst support particles, indicating that the polymer electrolyte coats the catalyst and the catalyst support particles. The catalyst agglomerates appear rather dense with no apparent void space. Therefore, the void space inside the catalyst agglomerates must be significantly smaller than the thickness of the ultra-thin section (i.e. << 100 nm). This is consistent with the agglomerate model described by Uchida et al.,[23] who assigned a pore range from 20–40 nm to pores inside the catalyst agglomerates. Due to the small volume of these ionomer-filled pores, the fluorine signal originating from these parts is very low and cannot be clearly distinguished from experimental noise. So far it was not possible to clearly determine the extent to which the polymer electrolyte penetrates into the catalyst agglomerates. In the future, optimization of the sample preparation and the acquisition parameters might further improve the elemental sensitivity, so that information about the fluorine distribution from inside the catalyst agglomerates may be obtained.

2.3   Imaging of electrode porosity

Besides the ionomer distribution, the electrode porosity is also of critical importance for fuel cell performance. However, the correct determination of the electrode porosity is not such an easy and straightforward task. For the characterization of the pore structure in PEFC electrodes, mercury intrusion porosimetry has been applied frequently,[26–30] but the application of gas adsorption[5, 31, 32] and thermoporometry[33] has also been reported in the literature. These studies showed that the range of pore sizes present in an MEA covers several orders of magnitude. The largest pores are found in the gas diffusion layer, but the electrode itself may contain cracks and holes with characteristic sizes of 5–200 μm.[32] However, the main fraction of pores in the catalyst layer is made up of pores smaller than about 1 μm.[29, 32] The pores in this size range are either pores inside the catalyst agglomerate particles, i.e. pores between the individual agglomerated carbon support particles, or pores between the larger agglomerates. The lower end of the pore range with pore sizes below 5 nm is due to pores in the primary particles of the carbon black support. These pores are generally not considered for reactant transport, since they do not form an interconnected pore network. However, noble catalyst particles deposited in these pores may be lost for the electrochemical reaction, since they are difficult to access by both the reactants and the polymer electrolyte.[7, 34]

2.3.1 2D techniques

In recent work by Roth's group, an effort has been made to determine the electrode porosity from electron microscopy information.[35] One main drawback of microscopy compared to other analytical techniques, however, is that it is generally rather time-consuming. Moreover, it might not be possible to obtain quantifiable information from images due to limitations in both imaging and analysis. For a standard MEA, the contrast between pores and other parts of the electrode is rather well defined in SE images, thus making the pores accessible for image analysis. A quantifiable image of the electrode (Fig. 2.6a) can be obtained by thresholding. If the threshold value is chosen correctly, a binary map of the pores in the image is obtained, as is shown in Fig. 2.6b. The thresholded image can be further processed to yield the relative pore volume by computing the ratio of pore pixels to the total number of pixels in the image, and image analysis software provides even more advanced analysis options. For instance, particle detection algorithms[36, 37] can be used to identify single pores; pore area, perimeter, length, width and other geometrical information can be evaluated separately for each pore allowing calculation of pore size distribution curves (Fig. 2.7).

The availability of geometric pore information, such as the pore aspect ratio or perimeter, is a clear benefit of image analysis when compared to other pore characterization techniques, such as mercury porosimetry or gas adsorption. However, the image-based approach has other limitations. The pore size distribution shown in Fig. 2.7 rapidly falls off for pores smaller than 200 nm. This does not reflect the real situation in the sample, but is a finite size effect of the specimen thickness. Only pores larger than the thickness of the thin cut provide sufficient contrast to be detected by the thresholding procedure. Also, pores with diameters close to the section thickness will be under-evaluated, as they may not be sectioned through the pore centre. Moreover, large inter-agglomerate pores, which can be easily observed by SEM or TEM imaging of thin sections, make only a minor contribution to the total electrode porosity.

Plate II(a) in the colour section between pages 252 and 253 shows a TEM micrograph, as it was obtained after digitization of the photographic slide and slight contrast adjustments. The unprocessed image does not provide much contrast, which makes identification of particle boundaries or the exact location of

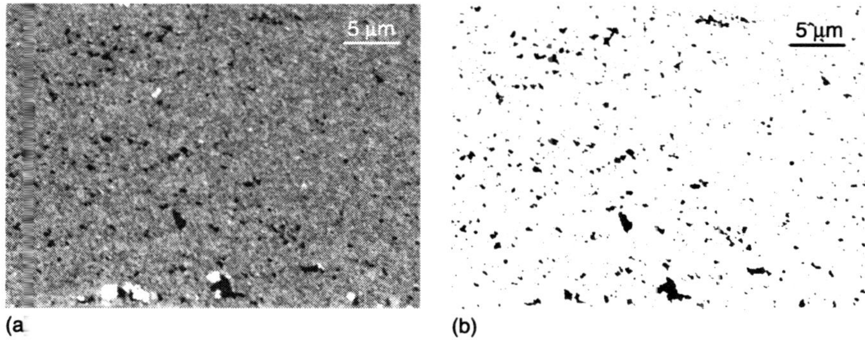

(a)   (b)

2.6 (a) SEM image of an electrode made of Pt/C catalyst and 40 vol.% Nafion. (b)Thresholded version of the image highlighting the pores in the structure.

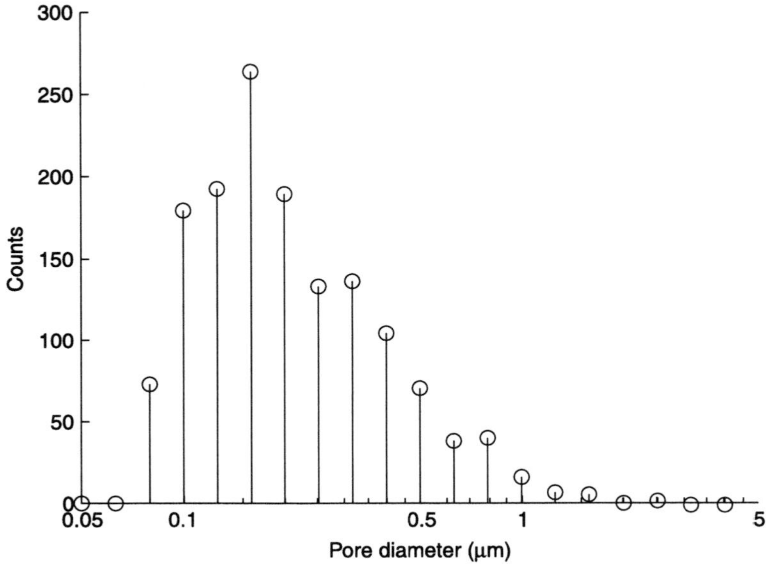

2.7 Pore size distribution obtained by analysis of the thresholded image data from Fig. 2.6.

catalyst agglomerates, the polymer electrolyte or even pores almost impossible. The visibility of weak contrast changes, however, can be improved significantly by applying colour maps to grey-level images. Plate II(b) in the colour section between pages 252 and 253 displays the same image as shown in Plate II(a) with the grey-level intensities re-mapped to a rainbow colour palette. In this case, the holes in the specimen were reproduced in white and the core of the agglomerate particles in black. The intermediate intensity range was distributed on colours from red to magenta following a rainbow colour scheme.

Although a clear separation of agglomerates and the polymer electrolyte is not feasible due to overlapping contrast ranges, additional information can be gathered from these false-colour images. Comparison with the original image shows that the colours red, yellow and green can be attributed to image parts containing mainly polymer electrolyte, whereas regions coloured in blue and magenta correspond to less electron-dense parts of the catalyst agglomerates. The assignment of different colours to the intensity range of the polymer electrolyte reveals that it is not uniformly filling or coating the pores in the electrode layer. The thickness variations within the polymer electrolyte may be due to pores in the electrolyte; however, the decreasing thickness of the polymer electrolyte found around pore walls rather suggests that the polymer electrolyte forms thin polymer films spanning across pores instead of coating layers on the catalyst. Apart from aiding the structural analysis of the polymer electrolyte, the colouring also helps to identify individual agglomerate particles. Since the core of the catalyst agglomerates has a slightly higher density than the boundary region between two aggregated agglomerates, the magenta and blue colours may be interpreted as agglomerate boundaries. The colours reveal that catalyst agglomerates have rather different sizes and shapes, ranging from 1 μm (upper right corner) to agglomerates with diameters even less than 100 nm.

Another approach to determine the average porosity from TEM information is shown in Fig. 2.8 using a method adapted from metallurgy. By this technique, the image in question is covered with a set of parallel and equidistant lines. The intersection with either high-contrast (electrode: carbon or Pt) or low-contrast (pores or ionomer) regions is then measured. By using the low-contrast to total length ratio, an average value for the porosity can be obtained.

2 3.2   3D techniques

Microscopic investigations of fuel cells and fuel cell components predominantly rely on 2D imaging techniques such as SEM, TEM or EDX mapping. These well-proven techniques allow the quantification of parameters such as (spherical) particle size distributions, active layer and membrane thickness, elemental distribution and others. Although most 3D imaging techniques require complex experimental work and data analysis they provide valuable additional information that is not accessible by 2D imaging techniques. A variety of 3D imaging

2.8 Quantification of TEM image using a method adapted from physical metallurgy.

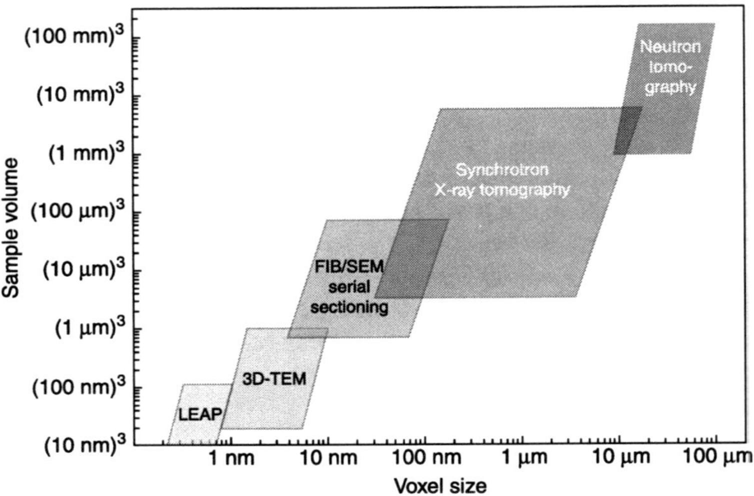

2.9 Comparison of 3D imaging techniques applicable for analysis of fuel cell components. Typical analyzed sample volumes are plotted against the voxel size of the resulting 3D image. Larger sample volumes induce lower image resolution/voxel size.

techniques can be used in fuel cell research (Fig. 2.9). The resolution of these techniques varies from about 1 nm to 100 µm at investigated sample volumes of about $(30\ nm)^3$ to $(150\ mm)$.[3] The main advantage is that volume images yield information about the connectivity of the investigated features such as pore structures or water channels. For instance the imaging of the liquid water distribution in running PEFCs is possible by neutron and synchrotron tomography and radiography techniques (see Chapters 7–11) whereby drying and flooding phenomena in the gas diffusion layers and gas channels can be followed. The catalyst dispersion on the carbon support and the shape of the catalyst particles can be investigated by 3D-TEM. However, none of these techniques seems appropriate to investigate the pore structure of the catalyst support because the investigated probe volume in 3D-TEM is too small and the resolution of X-ray tomography is not sufficient. This gap can be filled by focused ion beam/scanning electron microscopy (FIB/SEM) serial sectioning.[38–40] With state of the art instruments voxel sizes as small as 4 nm are accessible. This opens up the possibility of studying, for instance, carbon corrosion processes due to exposure of the MEA to damaging operation conditions.

FIB systems were developed in the 1970s and have been extensively used in the semiconductor industry for site-specific defect analysis, cutting unwanted electrical connections, depositing conductive material and sample preparation for TEM.[41, 42] Most FIB systems use gallium metal as an ion source. Heated liquid gallium metal wets a tungsten needle and an electric field causes ionization and field emission. The Ga^+ ions are accelerated to energies of about 5–50 keV and focused by electrostatic lenses. When the accelerated Ga^+ ions reach a surface they ablate material by a sputter process. In this manner the FIB can be used as a 'knife' that cuts slices of an adjustable thickness from a specimen. In early FIB systems the freshly cut surfaces could be analyzed by tilting the sample and recording FIB images, either by using secondary electrons or ions, both produced by the Ga^+ ion beam. A drawback of this method is that the sample has to be tilted for recording images, which results in extended experimental time and sample drift. Even more importantly, FIB imaging causes appreciable material ablation and therefore modification of the investigated surfaces.

To reduce sample damage and artefacts, so-called 'dual beam' or 'cross beam' systems have been developed and have been commercially available for several years. These systems combine a FIB with an SEM column, allowing FIB/SEM serial sectioning. The FIB cuts slices from the specimen followed by non-destructive SEM imaging. Repositioning of the sample is not necessary because the FIB and SEM columns are arranged at an angle, typically 52–54°. This serial sectioning process can be repeated several hundred times, allowing volume imaging of the investigated sample, but requiring rather long sampling times.

In fuel cell research FIB/SEM serial sectioning was applied to study the 3D microstructure of solid oxide fuel cell anodes.[43] Segmentation of porous nickel-, and Y_2O_3-stabilized ZrO_2-phases was successful. Volume images with voxel sizes of 44 nm^3 were achieved, which were sufficient to characterize the three-phase boundary, pore structure and connectivity of the Ni- and Y_2O_3-network. For volume imaging of the porosity of PEFC electrodes, however, even smaller voxel sizes are desirable. Nitrogen adsorption measurements of commonly used carbon support powders such as Vulcan XC-72 and Ketjenblack suggest that micropores (<2 nm), mesopores (2–50 nm) and macropores (>50 nm) are present.[44] Imaging micropores by FIB/SEM is unfeasible, but larger mesopores and macropores are accessible. The benefit of FIB/SEM serial sectioning over N_2 absorption measurements is that very small sample volumes, starting from ~1 μm^3, are sufficient, and these sample volumes can be chosen at different locations of the PEFC catalyst layers. Furthermore, the shape and interconnectivity of meso- and macropores becomes visible, which helps to improve modelling of mass transport in catalyst layers before and after degradation (Fig. 2.10).

Using FIB/SEM serial sectioning a catalyst coated membrane (CCM) supplied by W.L. Gore & Associates (PRIMEA MEA Series 5710) was investigated[45, 46] which had a cathode loading of 0.4 mg_{Pt}/cm^2 and an anode loading of 0.1 mg_{Pt}/cm^2. Electrochemical degradation experiments were carried out in a 16 cm^2 PEFC. To simulate varying load operation the cell voltage was cycled between open circuit and 0.6 V 24 000 times. In another set of experiments the PEFC anode was purged repeatedly with hydrogen and air to simulate start-up and shut-down of the fuel cell.

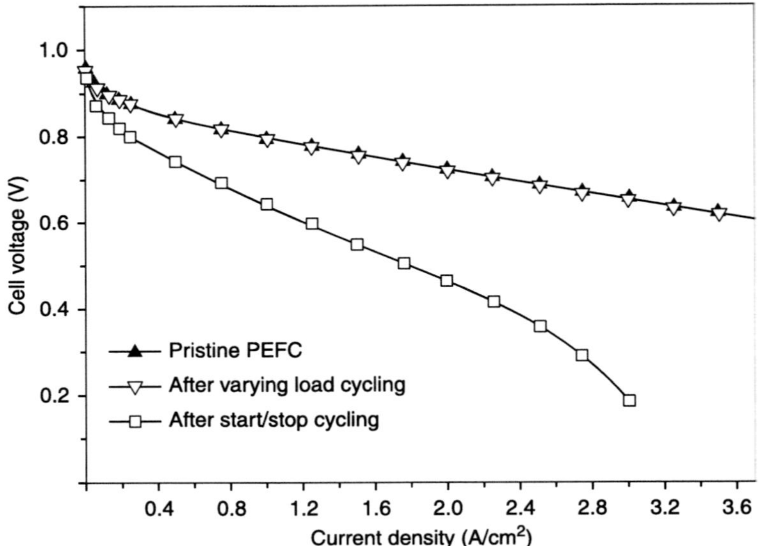

2.10 PEFC polarization curves before and after varying load or start/stop degradation. Degradation protocol for varying load test: 24 000 square wave cycles between open circuit voltage (OCV) and 0.6 V, time step 10 s. Protocol for start/stop degradation: 1000 purging steps of the anode side with hydrogen and air, time step 30 s. Polarization curves were recorded at 80°C, H_2/O_2, 2.5/2.5 bar, 1.5/1.5 stoichiometric flows, 100% humidified gases.

A piece (~0.5 cm^2) of the pristine and degraded CCMs was cut out using a scalpel. The gas diffusion layers of the degraded CCMs were gently removed with two tweezers. Afterwards the samples were glued to a sample holder with the cathode layer on top. After drying, the samples were inserted into the microscope (Zeiss NVision 40). To facilitate the subsequent segmentation a cross-section of the cathode catalyst layer was investigated by SEM. Parameters such as magnification, detector type, aperture and operating voltage were optimized to obtain images with the highest viable resolution and a minimum depth of focus. For the pre-cut, the cathode catalyst layer was machined using the FIB. In this manner a cube of material was exposed in the region of interest. Typically the cube had edge lengths of 2 μm. Prior to serial sectioning the surface of the cube was covered with a protective carbon layer.

Serial sectioning begins with cross-section FIB milling. Slices with thicknesses between 5–10 nm were removed and SEM images recorded. For later image processing it is important to minimize sample drifts in order to be able to consistently remove a known thickness of the material. FIB imaging of the protective carbon layer helps to estimate drift effects and check slice thickness. Software-controlled drift correction is possible and allows automated serial sectioning with a minimum of supervision.

In general, FIB milling leads to implantation of gallium in the surfaces. It may also modify the surface. For example, amorphization of crystalline samples has been reported. To minimize these unwanted surface modifications, FIB beam currents should be as small as possible. Under these conditions ion milling is a relatively damage-free process, especially when compared to mechanical sectioning methods. Features such as the porosity of PEFC catalyst layers are preserved during low-current FIB milling.

FIB/SEM serial sectioning of the catalyst cubes results in a stack of several hundred SEM images. If any lateral misalignment of succeeding images is visible, alignment software corrects the distortions. The next step is segmentation of SEM images, which is by far the most time-consuming step in image processing. The goal is to assign pore pixels with one colour and material pixels with another colour. Due to resolution limits of the SEM, further discrimination of the material in Pt particles, ionomer or carbon support is not possible unless very large Pt particles are present after degradation. Segmentation is carried out by selecting an appropriate range of grey scales, which is representative for the material and subsequent colouring of these pixels. After appropriate configuration of the SEM parameters pore pixels appear almost black. They are marked with another colour. Even after optimizing the SEM parameters, the images have a finite depth of focus and therefore it is not obvious if the material is located at the machined surface or deeper in the z axis (Fig. 2.11). This effect results in too-low porosity values, when segmentation is carried out from single SEM images.

To correct for this effect, succeeding images are inspected and it is checked whether the shape of the material region changes or not. If it changes, we may assume that the respective region is located at the machined surface and therefore the assignment 'material' is justified. If the shape of the inspected material does not change in succeeding images the material is obviously not in the machined surface but deeper in the z axis. Therefore those regions are assigned as pores. The whole stack of images is segmented in this way. Finally, a volume image is constructed from the segmented images. Calculation of the porosity is the last step in image processing.

For the investigated CCM one obtains a porosity of 38–40% for the pristine cathode catalyst layer. After conditioning of the PEFC a polarization curve of the pristine MEA was recorded (Fig. 2.10). In the first degradation experiment, potential cycling was conducted by subjecting the cathode side of the cell to a potential square wave between 0.6 V and open circuit with a time step of 10 s. In total 24 000 cycles were accumulated. This degradation test had little effect on fuel cell performance. The iE-curves before and after the test were almost identical, although the electrochemically active Pt surface area of the cathode shrinks to 25% of the initial value. In the second degradation experiment start/stop processes were simulated by purging the anode alternately with hydrogen and air. After 1000 start/stop cycles, the fuel cell performance is considerably

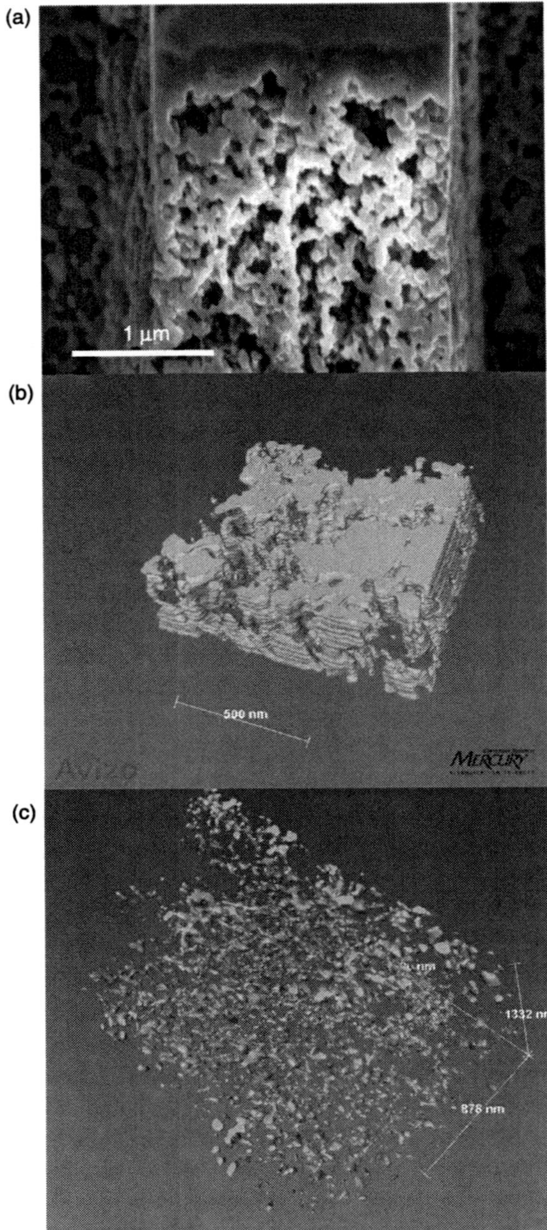

2.11 FIB/SEM serial sectioning. (a) SEM image of an exposed cathode catalyst cube adjacent to the membrane (below) and covered with carbon (top) before sectioning. (b) 3D image of catalyst pore structure reconstructed from 455 segmented SEM images. (c) 3D image after start/stop cycling.

lower, especially in the high current regime. The electrochemically active surface area decreases to 15% of the initial value.

Several cubes of degraded cathodes were analysed using FIB/SEM serial sectioning. After potential cycling the porosity is almost unchanged compared to the pristine material. A clearly lower porosity of 5% is obtained after start/stop cycling due to carbon corrosion and the collapse of the support structure (Fig. 2.11c). As a result of the hindered mass transport in the electrode, this leads to the poor performance especially at high current densities. This example illustrates how FIB/SEM serial sectioning can be used to separate different degradation mechanisms. Both applied degradation protocols lead to substantial loss of active Pt surface area, but only after start/stop degradation does this lead to performance loss. The intact electrode morphology after potential cycling ensures a sufficient mass transport of oxygen and water, whereas the collapsed electrode structure after start/stop cycling obviously leads to additional mass transport limitations, which limit PEFC performance.

2.4 Imaging of the interface between electrode and gas diffusion layer

A low-temperature fuel cell consists of a hybrid structure, where several functional layers are brought into contact in a sandwich-like manner. This is due to the fact that each component offers a different functionality, e.g. the membrane and the ionomer in the electrode serve as the proton conductor, whereas the carbon support and the catalytically active metallic components provide the electron conductivity. The functions of the gas diffusion layer are manifold; it not only serves as a gas distributor, but also helps in water and heat management of the fuel cell.[47, 48] Therefore it needs to be chemically, but also mechanically stable, inert, electron-conducting and inexpensive. Despite its many functions, the GDL has mostly been underestimated, and only recently fuel cell research has focused on this specific part of the fuel cell.

During the last years, it has been understood that the fuel cell's performance is not only dependent on the chosen catalyst and the electrode materials and structure, but also on the structure of the interface between the gas diffusion layer (GDL) and the electrode. At this interface gas and water need to be transferred between the pore space of the GDL and the pores of the electrode, which requires matching pore size regimes. Moreover, for enhanced transfer a tight contact between the GDL and the electrode layer in the final fuel cell design is desirable. This is achieved by either hot-pressing the GDL with the MEA (catalyst-coated membrane, CCM[49]) or by direct coating of the electrode onto the GDL and pressing of this so-called gas diffusion electrode (GDE) onto the membrane.[50, 51] However, while such a tight connection between the GDL and electrode is beneficial for gas and water transport as well as for the reduction of the Ohmic contact resistance, it significantly complicates the structural characterization of the whole assembly.

Although numerous methods exist for the analysis of either the GDL or the MEA, not many techniques render themselves suitable for the simultaneous imaging of both fuel cell components. Electron microscopy would be a promising option, but the preparation of cross-sections across the GDL-electrode interface is extremely demanding due to the materials' very different mechanical properties. Using intrusion with low-melting Wood's metal, the porous GDL-electrode interface can be stabilized and subsequently cut into thin sections.

2.4.1　Applicability of Wood's metal intrusion for specimen stabilization

Conventional sample preparation using the standard embedding resin and a subsequent cutting procedure often leads to delamination of the porous, but predominantly brittle, electrode structure from the smooth and flexible polymer membrane. The presence of a stiff gas diffusion layer (as in a GDE approach) only adds to the problem. This is why a novel sample preparation method was adopted, described by Cody and Davis,[52] for the analysis of pore space in coals. This technique was adapted to the examination of MEAs by replacing the epoxy resin commonly used to embed a piece of the MEA by an alloy with a low melting point.

Using a metal instead of the common epoxy resin has several advantages for the characterization of the fragile structure of the GDL-electrode interface by electron microscopy:

- Due to the large difference in atomic number between the metallic embedding media and the mostly carbonaceous materials of the MEA and GDL, a high Z-contrast is obtained between the embedding media and parts of the sample.
- The high thermal and electrical conductivity of the embedding medium reduces charging of the specimen, thus reducing specimen damage during imaging.
- The use of metals with a high atomic number reduces the excitation volume for the collection of X-ray emission from the sample, which improves the lateral resolution obtainable for back-scattered electron (BSE) images and elemental maps in the SEM.

Due to the presence of polymer materials such as the polymer electrolyte in the electrode or PTFE in the GDL, the melting point of the metallic embedding media should not exceed temperatures of about 120 °C. This limits the number of alloy systems, which are applicable as embedding materials. However, a number of bismuth-based alloys offer melting temperatures low enough so as not to destroy MEAs and backings during sample preparation. Wood's metal, also known by the commercial names cerrobend, bendalloy and pewtalloy, is the most prominent. It is a eutectic alloy with a melting temperature of 70 °C, consisting of bismuth (50%), lead (26.7%), tin (13.3%) and cadmium (10%).[53]

2 4.2   Sample preparation

The sample investigated was a commercial MEA from E-TEK inc., consisting of a carbon-supported platinum catalyst on both the anode and the cathode side, a carbon black-based microporous layer (MPL) and a woven carbon fibre backing attached to both electrodes. Thin sections of the GDL embedded in epoxy resin were obtained with a Reichert-Jung Ultracut E microtome operated at room temperature using a DDK diamond knife equipped with a 3 mm diamond blade polished at an angle of 45° ± 2° comparable to the procedure described in Section 2.2.

For the infiltration with Wood's metal, a 5 mm × 5 mm piece of the MEA was cut and placed between two disk-shaped pieces of Wood's metal. The sample was evacuated in a sealed container and heated to about 80 °C to encapsulate the MEA in the metal. The sample was then exposed to a hydrostatic pressure of 200 bar in a heated, pressurised cell to infiltrate the pore space of the sample with the metal. After cooling to room temperature, the sample was cut and polished to obtain a sample suitable for microscopy.

The samples were analyzed with either a ZEISS scanning electron microscope or a FEI Quanta 200 FEG environmental scanning electron microscope equipped with an energy dispersive X-ray detector for elemental analysis and mapping.

2 4.3   Effect of the gas diffusion layer on electrode morphology

Figure 2.12 shows scanning electron micrographs of an air-brushed MEA prepared by conventionally cutting a resin-embedded sample with a fresh razor blade. The image is of the top view of an electrode, which appears homogeneous on the

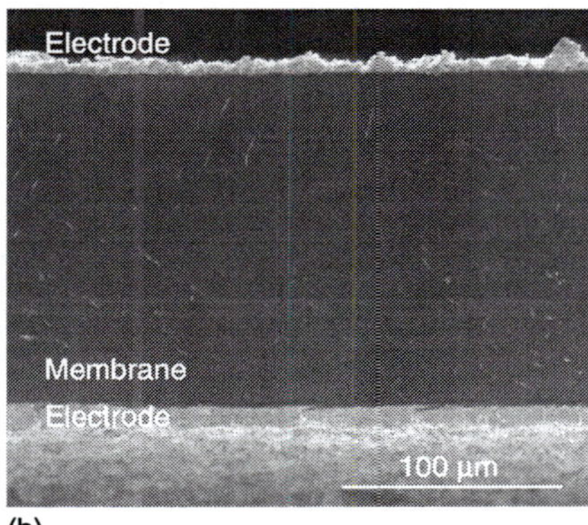

(a)   (b)

2.12 (a) Typical example of scanning electron micrograph of an electrode. (b) MEA cross-section.

micro scale but exhibits a rather porous and rough surface, and the corresponding cross-sectional view. Both the porous structure of the electrodes and the roughness of the electrode surfaces can be observed in this micrograph.

Figure 2.13 demonstrates a rather extreme case, in which the very different mechanical properties of soft and flexible polymer membrane and brittle carbonaceous electrode led to the severe delamination of the electrode during the sample preparation process. In similar cases with degraded MEA samples, it would be impossible to distinguish between specimen damage by sample preparation and damage experienced during operation. For instance, it would not be possible to conclude that any serious electrode delamination was caused by the operation conditions (important, e.g. for freeze-thaw experiments). Consequently, the specimen preparation needs to be modified to allow for meaningful results.

In Fig. 2.14, a thin cross-section of a standard GDL made from a PTFE-impregnated carbon fibre fabric is shown. In such a gas diffusion backing embedded in epoxy resin, the sectioning procedure leads to uncontrolled fracture and pull-out of the carbon fibres from the epoxy matrix. As a consequence, the standard sample preparation procedure is already not suitable for the imaging of the GDL itself and certainly could not be applied to image the electrode-GDL interface. The pull-out and brittle fracture of the carbon fibres as well as the

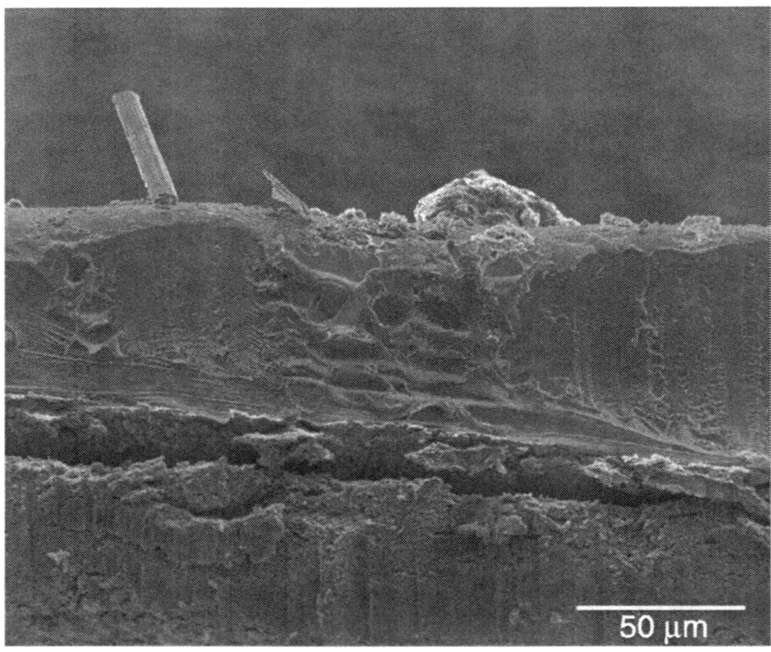

2.13 SEM image of a case of severe sample damage during sample preparation caused by the different brittleness of the materials.

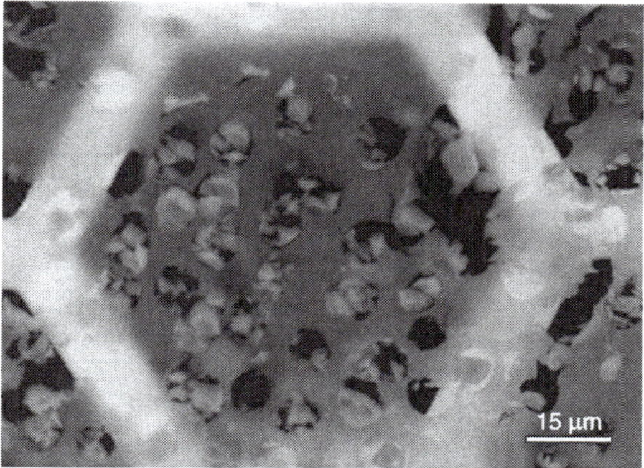

2.14 GDL-electrode interface. The image demonstrates that fibres break in a brittle manner and can be pulled out of the matrix when cut into ultrathin sections. The bright hexagon is the sample grid.

largely different hardness of the electrode and backing layer would result in severe mechanical deterioration of the electrode layer.

Consequently, Wood's metal intrusion was applied to a commercial E-TEK MEA with GDL and backing in order to image the undisturbed electrode-GDL interface for the first time. Figure 2.15 displays SEM images of the polished section recorded at different magnifications. The cut carbon fibres of the diffusion backing are clearly visible as black disks of circular or elliptical shape, depending on their orientation to the sectioning plane. Moreover, the woven structure of the fibre bundles is well-preserved in the thin cut. The fibre bundles approach the membrane and recede from it according to the texture of the fabric. In Fig. 2.15a

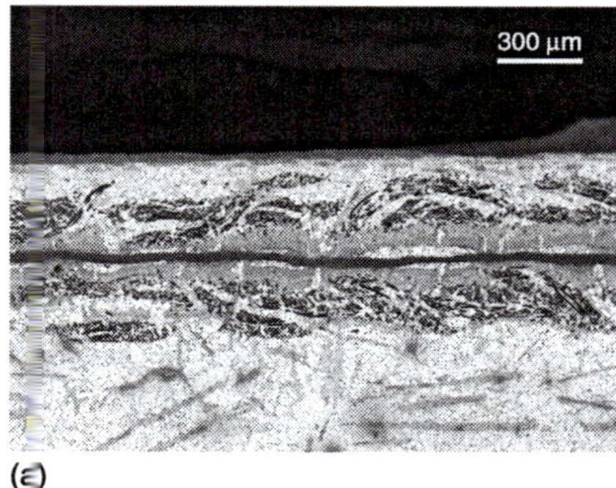

(a)

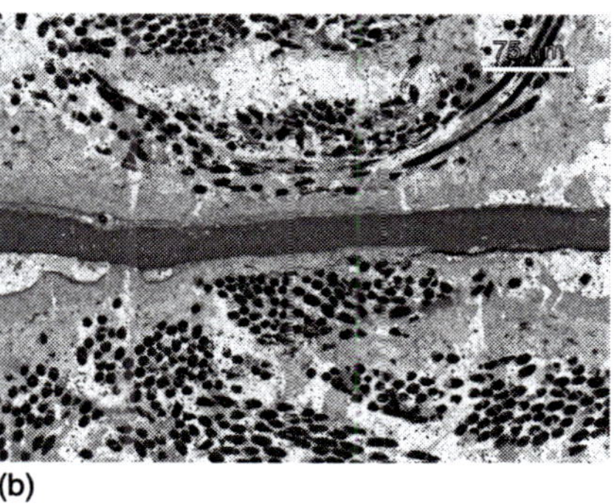

(b)

2.15 (a, b) SEM images of a commercial E-TEK MEA with GDL and backing embedded in Wood's metal. The carbon fibres of the diffusion backing can be seen as black disks, while the membrane is visible as a dark grey line running almost horizontally through the centre of the images.

it can be clearly seen that the thickness of the electrode layer, which is sandwiched between the two gas diffusion backings and the membrane, is lowest, where the carbon fibres approach the membrane. The structure of the fabric apparently has a significant influence on the morphology of the electrode. In regions where the fibre bundles from the backing recede from the membrane, delamination appears to occur more frequently. Furthermore, penetration of the microporous layer (MPL) into the carbon fibre backing was observed.

In some places, the electrode layer becomes so thin that some of the carbon fibres can be seen in direct proximity to the membrane, separated from it by only a few micrometres. In these areas, the MPL appears to be completely absent. Instead, fragments of the MPL are located in the void space between the carbon fibre bundles. From these results it is assumed that the woven structure of the GDL leads to a highly non-uniform contact pressure resulting in a poor contact between the electrode and membrane in considerably large parts of the electrode area. In fact, this effect could become even more severe when the fuel cell is in operation, i.e. the MEA is water-swollen and pressed into a rigid flow-field geometry.

Rapid removal of product water is one of the key functions of the GDL, and that is why it is often mixed or coated with PTFE to render it water-repellent.[54, 55] In order to characterize the water-repellent properties of the GDL and to understand their effect on water management in the entire fuel cell, it is essential to learn more about PTFE distribution in the GDL. Furthermore, recent studies suggest that PTFE distribution may be altered during long-term operation, leading to significant lifetime limitations.[56, 57] Consequently, EDX was used to probe PTFE distribution in the gas diffusion layer. Although it is rather sensitive to radiation losses and makes extra precautions for very short acquisition times necessary, the fluorine signal was chosen for this purpose.

In Plate III(a) in the colour section between pages 252 and 253, a BSE image of the respective sample area is shown. EDX mappings of the carbon and fluorine signal (Plate III(b)) shows that in the pristine GDL, the fluorine (i.e. the PTFE) is not homogeneously distributed, but appears in the form of large globules, most pronounced at the GDL-electrode interface. Most of the observed PTFE particles are found either in the MPL or in parts of the backing next to the MPL. We assume that the GDL attached to the commercial E-TEK MEA is either non-hydrophobized or that the PTFE is rather uniformly distributed. Consequently, the large PTFE particles found in the backing are most probably due to parts of the MPL penetrating into the GDL.

In addition to the fluorine and carbon data, maps of the Wood's metal constituents tin and cadmium were also acquired (Plate III(c)). They showed that not all the medium grey parts in the BSE image correspond to fragments of the MPL, but that some of them might be artefacts having their origin in tin- or cadmium-rich precipitations caused by the chosen sample preparation. This observation slightly complicates the direct interpretation of the BSE images; however, superposition of the element maps of cadmium and tin with the BSE image can be used to distinguish the precipitations from MPL fragments and thus prevent misinterpretation of the images and sloppy conclusions.

2.5   The future of advanced microscopy in fuel cell research

In operating fuel cells many components such as electrocatalysts and membranes are subject to complex degradation phenomena. These degradation processes may include carbon support corrosion, Pt dissolution, sintering, radical-induced membrane thinning and others. The ultimate analytical goal would be to monitor these processes *in situ*, 3D and at different length scales ranging from centimetres to atomic resolution. While this immodest goal is not within reach, several new microscopic techniques were recently developed to approach this aim. Of particular interest for fuel cell research are aberration corrected (S)TEM, 3D-TEM, identical-location TEM and *in situ* tomographic techniques. These techniques will be briefly discussed in the following.

2.5.1   Aberration corrected (scanning) transmission electron microscopy

Investigating metal or alloy nanoparticles with atomic resolution is highly attractive for structure determination in materials science and heterogeneous catalysis. In case of fuel cell catalysis it may allow segregation and percolation effects to be followed in bimetallic particles. Lattice parameters could be extracted, particle facets indexed and core-shell structures identified. Future studies may include topics such as surface roughness calculations, metal-support-interactions or determination of atomic numbers in nanoparticles.

The point-to-point resolution of a TEM is given in Williams and Carter[58] (see Eq. 2.1). One way to improve the resolution is to increase the kinetic energy of the electrons and thus to reduce their De Broglie wavelength λ. Unfortunately, this strategy leads to increased beam damage for the TEM samples of many materials investigated at high voltage.

$$r \approx 0.91(C_s\lambda^3)^{\frac{1}{4}} \qquad\qquad [2.1]$$

The other approach is to minimize the value of C_s, the spherical aberration coefficient. Spherical aberration is caused by intrinsic imperfections of rotationally symmetric electron lenses. Direct atomic-resolution imaging without C_s-correction was possible only for rather heavy atoms with relatively large spacings between them, i.e. for a limited number of special materials. Atomic resolution imaging of more common materials, i.e. especially lighter atoms with smaller spacings, has become possible, mainly since Haider and co-workers have technically realized[59] a corrector system for the spherical aberration of the objective lens in 1998, as suggested by Rose.[60] With multipole-based aberration correctors that act as auxiliary lenses the resolution limit can be pushed to less than 1 Å. At the time of this review a resolution of 0.5 Å has been reached.[61] Aberration corrected (S)TEMs have been commercially available for several years and a few research groups used these new microscopes for the study of fuel cell catalysts[62, 63] and other heterogeneous catalysts.[64] STEM imaging is usually preferred, due to the straightforward contrast interpretation. In high-angle annular dark field (HAADF) STEM images the contrast is approximately proportional to the thickness and to the mean square of the atomic number, Z^2.

In a study of commercial Pt_3Co/C catalysts at the Massachusetts Institute of Technology, aberration-corrected STEM images were used to study surface atomic structures and chemical compositions after annealing or acid leaching.[62] Figure 2.16 displays acid-treated Pt_3Co particles. The intensity variation within single nanoparticles can be due to thickness changes or changes of the Pt/Co ratio in individual nanoparticles. After acid leaching and subsequent heat treatment at 1000 K one observes a surface segregation (Fig. 2.16). The second atomic layer beneath the surface is much darker than the outermost (110) plane. Obviously, this second lattice plane is very rich in Co ($Z_{Co} \ll Z_{Pt}$). This segregation was also reported after annealing of extended sputtered Pt_3Co surfaces. With aberration-corrected HAADF images it could be shown for the first time that the same or a similar segregation process occurs in Pt_3Co particles after heat-treatment.

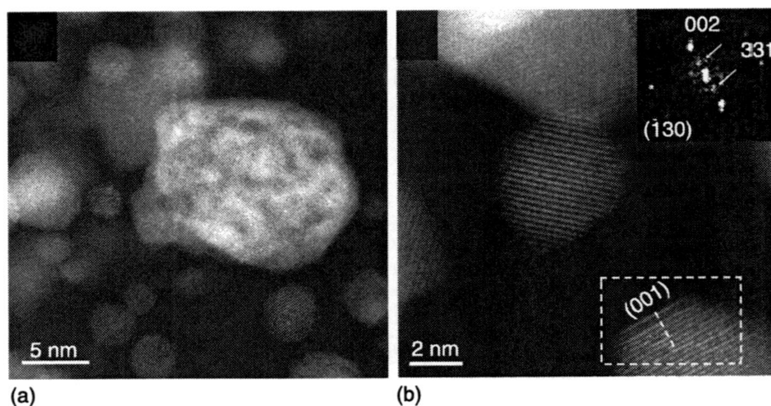

2.16 Aberration-corrected HAADF images of Pt_3Co particles.
(a) Large intensity variations are noted. (b) Image of heat-treated Pt_3Co nanoparticles. Surface segregation of the particle outlined by the dotted box is visible.

2 5.2   Electron tomography

E-en if we are able to investigate supported metal nanoparticles at atomic resolution, some morphological features of the catalyst and catalyst support cannot be analyzed using 2D (S)TEM images. The determination of, for example, the particle shape or dispersion of the particles on the support requires 3D images with a resolution of about 1 nm^3. This is possible using electron tomography, also referred to as transmission electron microtomography (3D-TEM). Using this method, a series of 2D-TEM-images images is collected by tilting the sample in the electron beam and recording an image at each tilt step. The tilt series is then aligned and the 3D image constructed.

Ito[65] investigated Pt/C catalyst powders using electron tomography. A modified TEM grid (Fig. 2.17) was made by cutting a regular Mo grid. A focused ion beam was employed to prepare a molybdenum pyramid on top of the exposed grid bar. A drop of a Pt/C catalyst suspension was deposited on top of the pyramid. This specimen preparation avoids shadowing effects from adjacent grid bars at high tilt angles which limit the rotation angle of common grids to typically ±70°. Since shadowing effects are absent in this modified holder the ideal tilt range of ±90° is possible. Therefore the 'missing wedge' problem[66] could be avoided. The constructed 3D images of the Pt/C catalyst suggest that the Pt nanoparticles are partially embedded in the carbon support. A 'degree of embeddedness' was defined for this effect.

In future studies, electron tomography may be applicable for Pt surface area determinations of non-spherical catalysts,[67, 68] such as Pt-prisms, multipods or 3M's nanostructured thin film electrodes.[69, 70] New possibilities open up by using FIB/SEM for sample preparation in electron tomography. Especially attractive are

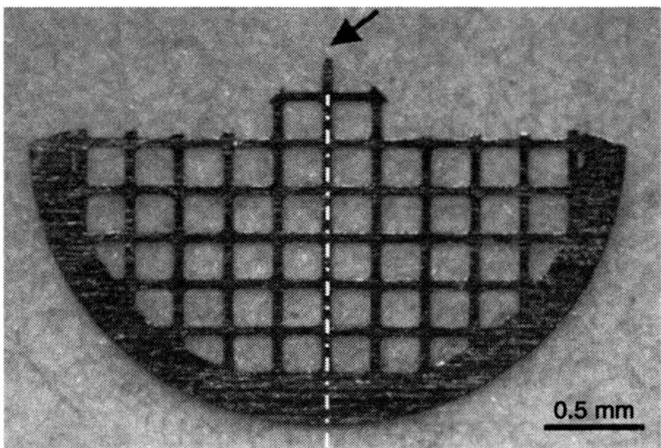

2.17 A modified TEM grid for electron tomography. The arrow indicates the position where the Pt particles on the carbon substrate are attached. The dotted line shows the rotational axis during the tilt series.[66]

degradation studies of catalyst layers, which include changes of the Pt particle shape and of the carbon support. Imaging the porosity of catalyst layers by FIB/SEM serial sectioning is limited to a resolution of about 10 nm. With electron tomography it is possible to resolve small mesopores and even micropores of the support. This will improve our understanding of carbon corrosion and may be useful for modelling the mass transport in catalyst layers. For this reason TEM lamellas have been prepared from catalyst layers of commercial catalyst coated membranes (Fig. 2.18), which may enable the imaging of meso- and micropores of pristine and degraded catalyst layers. Even catalyst pillars could be prepared from catalyst-coated membranes. In principle these pillars can be mounted to 360° sample holders and investigated without the missing wedge problem. In the case of catalyst pillars especially, bending and beam damage is still challenging.

2.5.3   Identical location (IL)-TEM

Microscopic studies of catalyst degradation usually focus on changes of the particle shape and particle size distribution. For that purpose, TEM images of

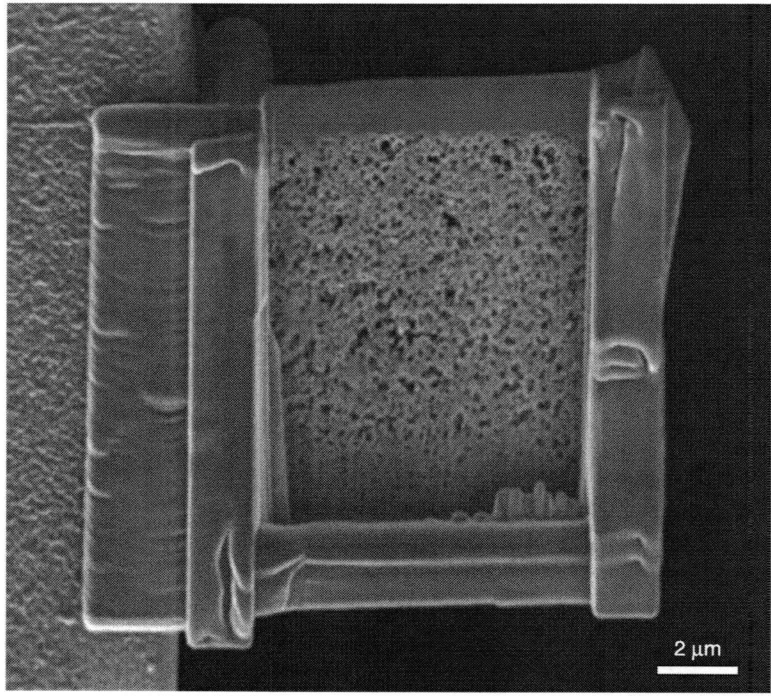

2.18 TEM lamella for use in electron tomography, prepared from the cathode catalyst layer of a catalyst-coated membrane (courtesy of J. Krbanjevic, EPFL).

pristine and degraded catalysts are compared, and changes in the particle size and distribution discussed. However, due to the randomness of the sampling and the destructiveness of the TEM method, the same regions would not be compared in their pristine and their post-mortem states. From a fundamental point of view, it would be interesting to follow the fate of a specific catalyst region during a degradation test. If the exact positions of the metal nanoparticles are known before and after degradation, this could help to distinguish between different ageing mechanisms, for example sintering and Ostwald ripening.

This wish came true, when in 2008 Mayrhofer *et al.*[71, 72] introduced a new analytical technique called identical location TEM (IL-TEM), which provides insight into the ageing mechanisms of specific catalyst sites. The underlying idea of IL-TEM is to use catalyst-covered TEM finder grids as working electrodes and collect TEM images before and after degradation experiments. By means of an alphabetical index, the identical quadrants on the grid can be retrieved in the microscope. Within these quadrants, it is indeed possible to investigate the same catalyst agglomerates viewed before treatment.

An illustrative example of an IL-TEM investigation is displayed in Fig. 2.19,[71] where a pristine Pt/C catalyst is deposited onto the TEM finder grid (Fig. 2.19a). After 3600 cyclic voltammograms between 0.4–1.4 V with a scan rate of 1 V/s the TEM grid was removed from the electrochemical cell, dried and investigated by TEM. Surprisingly the particle size distribution and the position of the nanoparticles on the support were almost unchanged; however, several

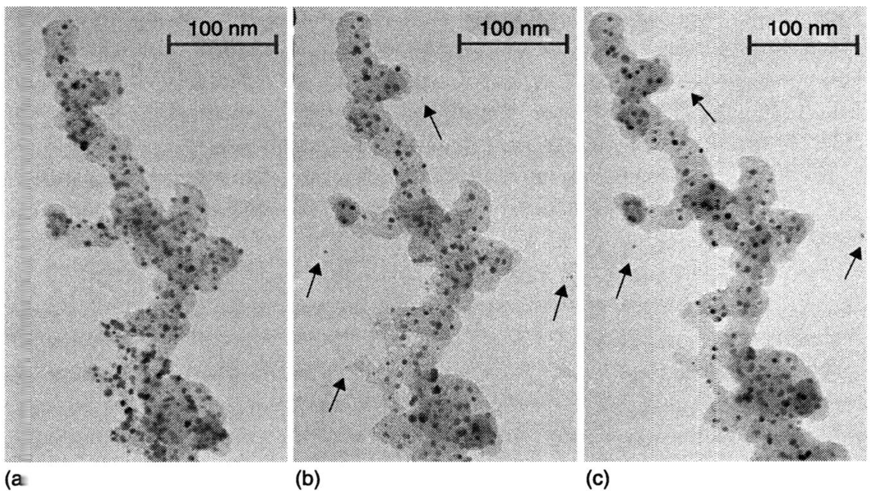

(a)   (b)   (c)

2.19 (a) IL-TEM micrographs of a catalyst region before 3600 potential cycles between 0.4 and 1.4 V reversible hydrogen electrode (RHE) with 1 V/s. (b) Afterwards. (c) After another 3600 cycles in 0.1 M $HClO_4$; arrows show the detachment of Pt particles from the carbon support.

nanoparticles were detached from the support (Fig. 2.19b) and were deposited on the carbon film of the grid (arrows in Fig. 2.19b and 2.19c). After additional cycling in the same potential range, even more particles appear to be detached (Fig. 2.19c). The number of particles decreased from 100% to 70% in Fig. 2.19b and 50% in Fig. 2.19c. Complementary ICP analysis of the electrolyte demonstrated that the amount of detached platinum is similar to the amount of platinum that was present in the electrolyte after the degradation experiments. The detachment of Pt particles was introduced as a new degradation mechanism.[72] More recently the Arenz group applied the IL-TEM method to study the effect of CO annealing on Pt/C[73] and Pt$_3$Co/C catalysts.[74]

2.5.4 *In situ* X-ray tomographic microscopy

At high current densities the water management of PEFCs becomes crucial for the power density. The gas diffusion layers are supposed to remove the product water quickly without hampering the transport of reactant gases to the catalyst layers. Analyzing *in situ* liquid saturation and water transport in the GDL provides fundamental understanding for the correlation of water management and the cell performance at high current densities and may explain differences in local degradation.

Computerized tomography uses a series of 2D projections at different sample rotation angles to reconstruct the 3D spatial distribution of a physical property. First results have been published by Sinha *et al.* with a 10 μm pixel size not resolving single fibres in the GDL.[75] X-ray radiography with a pixel size of 3 μm[76, 77] has then been used to study water in GDLs. However, radiography offers an inferior contrast and is also averaging along the beam. With the high flux available at synchrotron-based installations, exposure times for a radiogram of considerably below 1 s are achieved, and therefore synchrotron based X-ray tomographic microscopy (SRXTM) is feasible with relatively short measurement times.

Based on experience with *ex situ* SRXTM of GDLs,[78] first measurements with *in situ* SRXTM with a pixel size in the range of 0.7–1.8 μm have been made at PSI in 2008.[79] The measurement technology was further improved with the development of special cells (Fig. 2.20) containing small flow fields well adapted to the requirements of the field of view and rotation on the stage of the beamline endstation including all gas and electric (cell load and heating) connections.[80] Results obtained at the Tomcat beamline[81] at the Swiss Light Source (SLS) with a cell with an active area of 4.9 mm^2 (Fig. 2.20) operated at 200 mA/cm^2 show that SRXTM produces *in situ* data with unprecedented resolution and contrast for solid fibres and liquid water (see Fig. 2.21). *In situ* SRXTM is rapidly becoming a powerful *in situ* tool for analyzing liquid saturation on the scale of GDL pore sizes.

The future cannot be predicted. However, this overview shows the great potential of advanced electron microscopy for fuel cell research. Based on the steep development of new techniques and applications in recent years, the authors

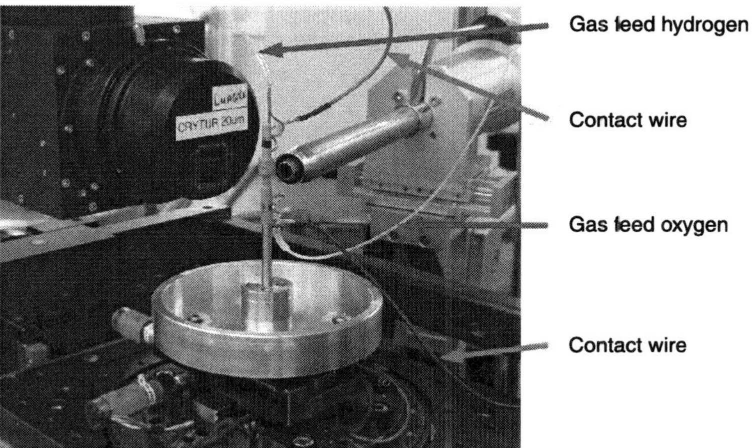

2.20 PEFC with cylindrically and horizontally aligned MEA on the rotation stage of the Tomcat beamline of the Swiss light source (courtesy of J. Eller, PSI).

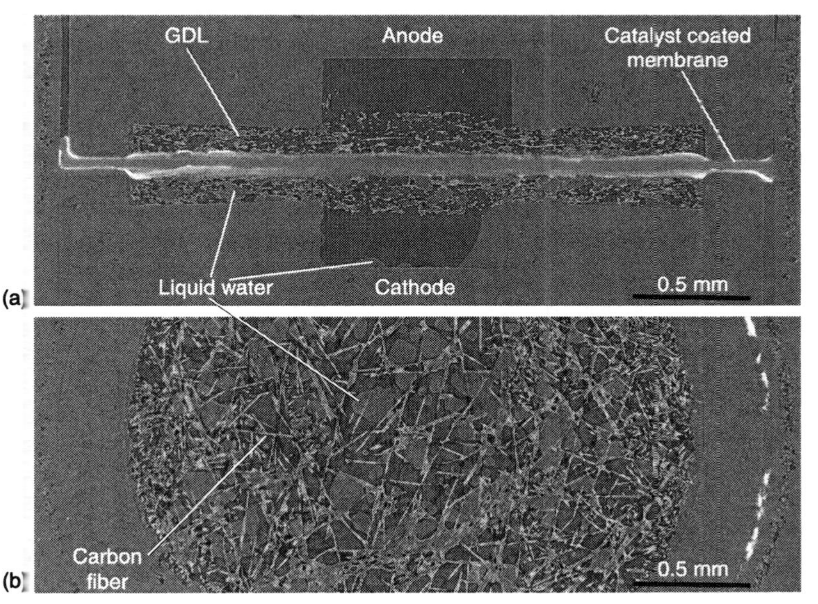

2.21 Reconstructed slices of cylindrical cell with 2.5 mm-long horizontal gas channels. Scan at 0.2 A/cm^2, 24 °C. (a) Through-plane cut. (b) Cropped in-plane cut through cathode GDL (courtesy of J. Eller, PSI).

believe electron microscopy to be a promising tool, which will help in solving the open questions raised in Part 1 in due course.

2.6   Acknowledgements

The authors gratefully acknowledge their present and former team members and colleagues Susanne Zils, Hanno Butsch, Ulrike Kunz (TU Darmstadt), Bernhard Schwanitz (PSI), Julijana Krbanjevic (EPFL-PSI), Jens Eller (PSI) and Nicolas Linse (PSI) for their contributions.

2.7   References

1. V. Gogel, T. Frey, Y.S. Zhu, K.A. Friedrich, L. Jorissen, J. Garche, 'Performance and methanol permeation of direct methanol fuel cells: dependence on operating conditions and on electrode structure', *J. Power Sources* **127** (2004) 172.
2. J. Nordlund, A. Roessler, G. Lindbergh, 'The influence of electrode morphology on the performance of a DMFC anode', *J. Appl. Electrochem.* **32** (2002) 259.
3. Z.B. Wei, S.L. Wang, B.L. Yi, J.G. Liu, L.K. Chen *et al.*, 'Influence of electrode structure on the performance of a direct methanol fuel cell', *J. Power Sources*, **106** (2002) 364.
4. E. Antolini, L. Giorgi, A. Pozio, E. Passalacqua, 'Influence of Nafion loading in the catalyst layer of gas-diffusion electrodes for PEFC', *J. Power Sources* **77** (1999) 136.
5. P. Gode, F. Jaouen, G. Lindbergh, A. Lundblad, G. Sundholm, 'Influence of the composition on the structure and electrochemical characteristics of the PEFC cathode', *Electrochim. Acta* **48** (2003) 4175.
6. E. Passalacqua, F. Lufrano, G. Squadrito, A. Patti, L. Giorgi, 'Nafion content in the catalyst layer of polymer electrolyte fuel cells: effects on structure and performance', *Electrochim. Acta* **46** (2001) 799.
7. M. Uchida, Y. Fukuoka, Y. Sugawara, N. Eda, A. Ohta, 'Effects of microstructure of carbon support in the catalyst layer on the performance of polymer-electrolyte fuel cells', *J. Electrochem. Soc.* **143** (1996) 2245.
8. R. Borup, J. Meyers, B. Pivovar, Y.S. Kim, R. Mukundan *et al.*, 'Scientific aspects of polymer electrolyte fuel cell durability and degradation', *Chem. Rev.* **107** (2007) 3904.
9. F.A. de Bruijn, V.A.T. Dam, G.J.M. Janssen, 'Review: Durability and degradation issues of PEM fuel cell components', *Fuel Cells* **8** (2008) 3.
10. S. Zhang, X.-Z. Yuan, J. Chen Hin, H. Wang, K.A. Friedrich *et al.*, 'A review of platinum-based catalyst layer degradation in proton exchange membrane fuel cells', *J. Power Sources* **194** (2009) 588.
11. T. Madden, M. Perry, L. Protsailo, M. Gummalla, S. Burlatsky *et al.*, 'Proton exchange membrane fuel cell degradation: mechanisms and recent progress', in *Handbook of Fuel Cells*, W. Vielstich, H. Yokokawa, H.A. Gasteiger (Editors), Vol. 5, p. 861, John Wiley & Sons Ltd, 2009.
12. J. Li, 'Catalyst layer degradation, diagnosis and failure mitigation', in *PEM Fuel Cell Electrocatalysts and Catalyst Layers*, J. Zhang (Editor), p. 1041, Springer, 2008.
13. F. Scheiba, N. Benker, U. Kunz, C. Roth, H. Fuess, 'Electron microscopy techniques for the analysis of the polymer electrolyte distribution in proton exchange membrane fuel cells', *J. Power Sources* **177** (2008) 273.

14 M. Schulze, M. von Bradke, R. Reissner, M. Lorenz, E. Gulzow, *J. Anal. Chem.* **365** (1999) 123.

15 M. Schulze, M. Lorenz, N. Wagner, E. Gulzow, *Fresen. J. Anal. Chem.* **365** (1999) 106.

16 Z. Porat, J.R. Fryer, M. Huxham, I. Rubinstein, 'Electron microscopy investigation of the microstructure of Nafion films', *J. Phys. Chem.* **99** (1995) 4667.

17 D.A. Blom, J.R. Dunlap, T.A. Nolan, L.F. Allard, 'Preparation of cross-sectional samples of proton exchange membrane fuel cells by ultramicrotomy for TEM', *J. Electrochem. Soc.* **150** (2003) A414.

18 T. Akita, A. Taniguchi, J. Maekawa, Z. Sirorna, K. Tanaka *et al.*, 'Analytical TEM study of Pt particle deposition in the proton-exchange membrane of a membrane-electrode-assembly', *J. Power Sources* **159** (2006) 461.

19 E. Guilminot, A. Corcella, F. Charlot, F. Maillard, M. Chatenet, 'Detection of Ptz^+ ions and Pt nanoparticles inside the membrane of a used PEMFC', *J. Electrochem. Soc.* **154** (2007) B96.

20 J. Xie, D.L. Wood, K.L. More, P. Atanassov, R.L. Borup, 'Microstructural changes of membrane electrode assemblies during PEFC durability testing at high humidity conditions', *J. Electrochem. Soc.* **152** (2005) A1011.

21 S.T. Iyer, D. Nandan, B. Venkataramani, 'Alkaline earth metal ion-proton-exchange equilibria on nafion-117 and dowex 50W X8 in aqueous solutions at 298 ± 1 K', *React. Funct. Polym.* **29** (1996) 51.

22 S. Rieberer, K.H. Norian, 'Analytical electron microscopy of Nafion ion exchange membranes', *Ultramicroscopy* **41** (1992) 225.

23 M. Uchida, Y. Aoyama, N. Eda and A. Ohta, 'New preparation method for polymer-electrolyte fuel cells', *J. Electrochem. Soc.* **142** (1995) 463.

24 T.D. Gierke, G.E. Munn and F.C. Wilson, 'The morphology in Nafion perfluorinated membrane products, as determined by wide- and small-angle x-ray studies', *J. Polym. Sci. Part B: Polym. Phys.* **19** (1981) 1687.

25 F. Hofer, P. Warbichler 'Elemental mapping using energy filtered imaging', in *Transmission electron energy loss spectrometry in material science and the EELS Atlas*, C. C. Ahn (Editor), 2nd ed. p. 159, Wiley-VCH 2004.

26 J. Kim, H.Y. Ha, I.H. Oh, S.A. Hong, H.I. Lee, 'Influence of the solvent in anode catalyst ink on the performance of a direct methanol fuel cell', *J. Power Sources* **135** (2004) 29.

27 M. Chisaka, H. Daiguji, 'Effect of glycerol on micro/nano structures of catalyst layers in polymer electrolyte membrane fuel cells', *Electrochim. Acta* **51** (2006) 4828.

28 J. Song, S. Suzuki, H. Uchida, M. Watanabe, 'Preparation of high catalyst utilization electrodes for polymer electrolyte fuel cells', *Langmuir* 22 (2006) 6422.

29 M. Uchida, Y. Aoyama, N. Eda, A. Ohta, 'Investigation of the microstructure in the catalyst layer and effects of both perfluorosulfonate ionomer and PTFE-loaded carbon on the catalyst layer of polymer electrolyte fuel cells', *J. Electrochem. Soc.* **142** (1995) 4143.

30 J. Xie, K. More, T. Zawodzinski, W. Smith, 'Porosimetry of MEAs made by "thin film decal" method and its effect on performance of PEFCs', *J. Electrochem. Soc.* **151** (2004) A1841.

31 J. Ihonen, F. Jaouen, G. Lindbergh, A. Lundblad, G. Sundholm, 'Investigation of mass-transport limitations in the solid polymer fuel cell cathode', *J. Electrochem. Soc.* **149** (2002) A448.

32 M. Tucker, M. Odgaard, P. Lund, S. Yde-Andersen, J.O. Thomas, 'The pore structure of direct methanol fuel cell electrodes', *J. Electrochem. Soc.* **152** (2005) A1844.

33. S. Escribano, P. Aldebert, M. Pineri, 'Volumic electrodes of fuel cells with polymer electrolyte membranes: electrochemical performances and structural analysis by thermoporometry', *Electrochim. Acta* **43** (1998) 2195.

34. H. Liu, C. Song, L. Zhang, J. Zhang, H. Wang *et al.*, 'A review of anode catalysis in the direct methanol fuel cell', *J. Power Sources* **155** (2006) 95.

35. Frieder Scheiba, 'Electrode structures of polymer electrolyte fuel cells (PEFC) – An electron microscopy approach to the characterization of the electrode structure of polymer electrolyte fuel cells', PhD Thesis, TU Darmstadt, 2009.

36. M.T. Pareschi, M. Pompilio, F. Innocenti, 'Automated evaluation of volumetric grain-size distribution from thin-section images', *Computers & Geosciences* **16** (1990) 1067.

37. C.F. Mora, A.K.H. Kwan, H.C. Chan, 'Particle size distribution analysis of coarse aggregate using digital image processing', *Cement Concrete Res.* **28** (1998) 921.

38. L. Holzer, F. Indutnyi, P. Gasser, B. Münch, M. Wegmann, 'Three-dimensional analysis of porous BaTiO3 ceramics using FIB nanotomography', *J. Microscopy* **216** (2004) 84.

39. M.D. Uchic, M.A. Groeber, D.M. Dimiduk, J.P. Simmons, '3D microstructural characterization of nickel superalloys via serial-sectioning using a dual beam FIB-SEM', *Scripta Mater.* **55** (2006) 23.

40. P.G. Kotula, M.R. Keenan, J.R. Michael, 'Tomographic spectral imaging with multivariate statistical analysis: comprehensive 3D microanalysis', *Microsc. Microanal.* **12** (2006) 36.

41. S. Reyntjens, R. Puers, 'A review of focused ion beam applications in microsystem technology', *J. Micromech. Microeng.* **11** (2001) 287.

42. J. Melngailis, 'Focused ion beam technology and applications', *J. Vac. Sci. Technol. B* **5** (1987) 469.

43. J.R. Wilson, W. Kobsiriphat, R. Mendoza, H.-Y. Chen, J.M. Hiller *et al.*, 'Three-dimensional reconstruction of a solid-oxide fuel-cell anode', *Nat. Mater.* **5** (2006) 541.

44. T. Soboleva, X. Zhao, K. Malek, Z. Xie, T. Navessin *et al.*, 'On the micro-, meso-, and macroporous structures of polymer electrolyte membrane fuel cell catalyst layers', *ACS Appl. Mater. Interfaces* **2** (2010) 375.

45. H. Schulenburg, B. Schwanitz, J. Krbanjevic, N. Linse, R. Mokso *et al.*, '3D Imaging of Polymer Electrolyte Fuel Cell Electrodes', *ECS Transactions* **33** (2010) 1471.

46. H. Schulenburg, B. Schwanitz, J. Krbanjevic, R. Mokso, M. Stampanoni, A. Wokaun, G.G. Scherer, '3D imaging of PEFC electrodes', *PSI Electrochemistry Laboratory-Annual Report 2009*, ISSN 1661–5379, http://ecl.web.psi.ch.

47. L. Cindrella, A.M. Kannan, J.F. Lin, K. Saminathan, Y. Ho *et al.*, 'Gas diffusion layer for proton exchange membrane fuel cells—A review', *J. Power Sources* **194** (2009) 146.

48. M. Mathias, J. Roth, J. Fleming, W. Lehnert, in: W. Vielstich, H. Gasteiger, A. Lamm (Eds.), *Handbook of Fuel Cells – Fundamentals, Technology and Applications*, Vol. 3, John Wiley & Sons Ltd., 2003

49. T. Frey, M. Linardi, 'Effects of membrane electrode assembly preparation on the polymer electrolyte membrane fuel cell performance', *Electrochim. Acta* **50** (2004) 99.

50. T.V. Reshetenko, H.-T. Kim, U. Krewer, H.-J. Kweon, 'The effect of the anode loading and method of MEA fabrication on DMFC performance', *Fuel Cells* **7** (2006) 238.

51. S. Gamburzev, A.J. Appleby, 'Recent progress in performance improvement of the proton exchange membrane fuel cell (PEMFC)', *J. Power Sources* **107** (2002) 5.

52. G. Cody, A. Davis, 'Direct imaging of coal pore space accessible to liquid metal', *Energy & Fuels*, 5 776 (1991).

53. B.F. Swanson, *J. Petrol. Technol.* **31** (1979) 10.

54. S. Zhang, H. Yu, H. Zhu, J. Hou, B. Yi *et al.*, 'Effects of freeze/thaw cycles and gas purging method on polymer electrolyte membrane fuel cells', *Chinese J. Chem. Eng.* **14** (2006) 802.

55. C. Lim, C.Y. Wang, 'Effects of hydrophobic polymer content in GDL on power performance of a PEM fuel cell', *Electrochim. Acta* **49** (2004) 4149.

56. I. Manke, C. Hartnig, M. Grünerbel, W. Lehnert, N. Kardjilov *et al.*, 'Investigation of water evolution and transport in fuel cells with high resolution synchrotron x-ray radiography', *Appl. Phys. Lett.* **90** (2007) 174105.

57. C. Hartnig, I. Manke, N. Karjilov, A. Hilger, M. Grünerbel *et al.*, 'Combined neutron radiography and locally resolved current density measurements of operating PEM fuel cells', *J. Power Sources* **176** (2008) 452.

58. D.B. Williams, C.B. Carter, *Transmission Electron Microscopy – A Textbook for Materials Science*, 2nd ed., p. 108, Springer 2009.

59. M. Haider, S. Uhlemann, E. Schwan, B. Kabius, K. Urban, 'Electron microscopy image enhanced', *Nature* **392** (1998) 768.

60. H. Rose, *Optik* **85** (1990) 19.

61. http://ncem.lbl.gov/TEAM-project/

62. S. Chen, W. Sheng, N. Yabuuchi, P.J. Ferreira, L.F. Allard *et al.*, 'Origin of oxygen reduction reaction activity on "Pt_3Co" nanoparticles: Atomically resolved chemical compositions and structures', *J. Phys. Chem. C* **113** (2009) 1109.

63. M.A. Asoro, D. Kovar, Y. Shao-Horn, L.F. Allard, P.J. Ferreira, 'Coalescence and sintering of Pt nanoparticles: *in situ* observation by aberration-corrected HAADF STEM', *Nanotechnolog.* **21** (2010) 025701.

64. L.C. Gontard, L.-Y. Chang, C.J.D. Hetherington, A.I. Kirkland, D. Ozkaya *et al.*, 'Aberration-corrected imaging of active sites on industrial catalyst nanoparticles', *Angew. Chem. Int. Ed.* **46** (2007) 3683.

65. T. Ito, U. Matsuwaki, Y. Otsuka, G. Katagir, M. Kato, K. Matsubara, Y. Aoyama, H. Jinnai, in: W. Vielstich, H. Yokokawa, H.A. Gasteiger, (Eds.), *Handbook of Fuel Cells-Fundamentals, Technology and Applications*, Vol. 6, John Wiley & Sons Ltd., 2009.

66. K.A. Taylor, J. Liu, H. Winkler, in *Electron Tomography*, J. Frank (Ed.), 2nd ed., p. 421, Springer 2006.

67. T.K. Sau, A.L. Rogach, F. Jäckel, T.A. Klar, J. Feldmann, 'Properties and applications of colloidal nonspherical noble metal nanoparticles', *Adv. Mater.* **22** (2010) 1805.

68. M. Subhramannia, K. Ramaiyan, V.K. Pillai, 'Comparative study of the shape-dependent electrocatalytic activity of platinum multipods, discs, and hexagons: Applications for fuel cells', *Langmuir*, **24** (2008) 3576.

69. A. Bonakdarpour, K. Stevens, G.D. Vernstrom, R. Atanasoski, A.K. Schmoeckel *et al.*, 'Oxygen reduction activity of Pt and Pt–Mn–Co electrocatalysts sputtered on nano-structured thin film support', *Electrochim. Acta* **53** (2007) 688.

70. A. Bonakdarpour, R. Löbel, R.T. Atanasoski, G.D. Vernstrom, A.K. Schmoeckel *et al.*, 'Dissolution of transition metals in combinatorially sputtered $Pt_{1-x-y}M_xM'_y$ (M, M'=Co, Ni, Mn, Fe) PEMFC electrocatalysts', *J. Electrochem. Soc.* **153** (2006) A1835.

71. K.J.J. Mayrhofer, S.J. Ashton, J.C. Meier, G.K.H. Wiberg, M. Hanzlik *et al.* 'Non-destructive transmission electron microscopy study of catalyst degradation under electrochemical treatment', *J. Power Sources* **185** (2008) 734.

72. K.J.J. Mayrhofer, J.C. Meier, S.J. Ashton, G.K.H. Wiberg, F. Kraus *et al.*, 'Fuel cell catalyst degradation on the nanoscale', *Electrochem. Commun.* **10** (2008) 1144.

73. K.J.J. Mayrhofer, M. Hanzlik, M. Arenz, 'The influence of electrochemical annealing in CO saturated solution on the catalytic activity of Pt nanoparticles', *Electrochim. Acta* **54** (2009) 5018.
74. K.J.J. Mayrhofer, V. Juhart, K. Hartl, M. Hanzlik, M. Arenz, 'Adsorbate-induced surface segregation for core-shell nanocatalysts', *Angew. Chem. Int. Ed.* **48** (2009) 3529.
75. P.K. Sinha, P. Halleck, C.-Y. Wang, 'Quantification of liquid water saturation in a PEM fuel cell diffusion medium using X-ray microtomography', *Electrochem. Solid-State Lett.* **9** (2006) A344.
76. I. Manke, Ch. Hartnig, M. Grünerbel, W. Lehnert, N. Kardjilov *et al.*, 'Investigation of water evolution and transport in fuel cells with high resolution synchrotron x-ray radiography', *Appl. Phys. Lett.* **90** (2007) 174105-3.
77. C. Hartnig, I. Manke, R. Kuhn, N. Kardjilov, J. Banhart *et al.*, 'Cross-sectional insight in the water evolution and transport in polymer electrolyte fuel cells', *Appl. Phys. Lett.* **92** (2008) 134106-3.
78. J. Becker, R. Flückiger, M. Reum, F.N. Büchi, F. Marone *et al.*, 'Determination of material properties of gas diffusion layers: experiments and simulations using phase contrast tomographic microscopy', *J. Electrochem. Soc.* **156** (2009) B1175.
79. F.N. Büchi, R. Flückiger, D. Tehlar, F. Marone, M. Stampanoni, 'Determination of liquid water distribution in porous transport layers', *ECS Transactions*, **16** (2008) 587.
80. F.N. Büchi, J. Eller, F. Marone, and M. Stampanoni, *ECS Transactions*, in press, (2010).
81. M. Stampanoni, A. Groso, A. Isenegger, G. Mikuljan, Q. Chen *et al.*, 'Trends in synchrotron-based tomographic imaging: the SLS experience', *Proc. of SPIE* 6318 (2006) 63180M.

3

Differential electrochemical mass spectrometry (DEMS) technique for direct alcohol fuel cell characterization

C. CREMERS and D. BAYER, Fraunhofer Institute for Chemical Technology, Germany

Abstract: In order to get an impression of the reactions taking place at an electro-catalyst, an appropriate analysis needs to combine electrochemical methods with a sensitive analysis technique capable of resolving transient processes. The detection of volatile products with differential electrochemical mass spectrometry (DEMS) matches these prerequisites. The evolution of DEMS from its beginnings into a sophisticated analysis tool is described, including basic principles, common cell designs, experimental techniques and the combination of DEMS with other analytical methods. Potential applications of DEMS with respect to fuel cells are highlighted and the advantages and limitations of the method are depicted.

Key words: differential electrochemical mass spectrometry (DEMS), electro-catalysis, monitoring of volatile reaction products, reaction pathway elucidation, fuel cell degradation studies.

3.1 Introduction

The membrane electrode assembly with its membrane and catalyst layers is one of the most important but also one of the most expensive parts of the fuel cell. Developments to increase the activity of the catalysts, their selectivity towards the desired reactions and the durability of both catalysts and membrane materials are still ongoing. However, such developments require a thorough understanding of the reactions going on in the fuel cell or at the catalyst, respectively, including such processes that only occur under a transient regime. In most cases electrochemical methods alone do not allow distinguishing between different products formed during an electrochemical reaction. Thus, combinations of electrochemical techniques with analytical ones are required. This is particularly true for the development of direct alcohol fuel cell electrocatalysts. Here, the detection of electrochemical reaction products right after their formation is essential for the investigation of the reactions of alcohols or other organic species at the catalyst surface. However, this is a difficult task as the product quantity formed is usually very low. From heterogeneous catalysis it is known that mass spectrometry is sensitive enough to detect desorption products in the submonolayer range (Baltruschat, 1999, Falconer and Schwarz, 1983, Cassuto and King, 1981). Consequently, this technique seemed to be suitable for the detection of intermediates and products of electrochemical reactions as well.

The first approach to detect gaseous products of an electrochemical reaction was started by Bruckenstein *et al.* in the early 1970s (Bruckenstein and Gadde, 1971). They attached the working electrode to a microporous non-wetting membrane, which was used as interface between the electrochemical cell and the vacuum system. However, Bruckenstein and Gadde employed a rather simple vacuum system with a single pumping stage resulting in a high time constant for the system in the range of tens of seconds.

A significant improvement was made by Wolter and Heitbaum using a more elaborate differentially pumped vacuum system including two pumping stages with turbomolecular pumps (Wolter and Heitbaum, 1984a, 1984b). This setup allows fast transport of products into the ionization chamber with a response time of approximately 0.2 s and a rapid elimination of the gas collected (Bittins-Cattaneo *et al.*, 1991). That way the mass intensities follow the current profile of a potentiodynamic measurement without major distortions. In order to distinguish this on-line detection technique from product sampling techniques, Wolter and Heitbaum called it differential (time resolved) electrochemical mass spectroscopy (DEMS).

From this starting point the DEMS technique was further developed: While the cell design changed from glass cells with stagnant electrolyte to flow-through cells made from inert materials, the possibility of combining DEMS with other detection techniques for instance quartz crystal microbalance (QCMB) and infra red spectroscopy in attenuated total reflection configuration (ATR-IR) also emerged. While the original DEMS setup was limited to metal lacquer-painted or metal-sputtered working electrodes, techniques were developed in order to investigate electrochemical processes at compact catalyst materials, single crystals and supported catalysts. Integration of DEMS sensors into actual fuel cells has also been reported (Seiler *et al.*, 2004). The high sensitivity and the variety of applications render DEMS a valuable tool for the investigation of gaseous products not only from electrochemical reactions. However, as no commercially available DEMS systems exist on the market, this technique is still restricted to research laboratories.

3.2   Basic principles, cell design and applications

A crucial feature common to all DEMS systems is the separation of the vacuum system from the electrolyte phase by a membrane. This membrane should hold back the electrolyte from entering the vacuum system while simultaneously being permeable allowing the reaction products to be detected. This prerequisite implies that DEMS is only sensitive for reaction products which are volatile enough to enter the vacuum system. For the most often used aqueous electrolytes and some other solutions, porous Teflon® membranes supported by steel or glass frits are appropriate.

The electrochemical cell is usually situated on top of the membrane inlet. In order to realize response times short enough to establish 'differential' conditions in the sense of DEMS, the working electrode has to be situated near the membrane inlet or at the membrane itself. The application of the working electrode directly onto the membrane by lacquer painting or sputter deposition has the advantage of the shortest response times. However, the formed catalyst layers have a comparatively high roughness factor, i.e. 50–100 for lacquer-painted electrodes and 5–20 for sputtered electrodes. In addition, the high thickness of lacquer-painted electrodes leads to depletion of the electroactive species in the catalyst layer due to slow diffusion (Baltruschat, 2004). Furthermore, these kinds of porous electrodes exhibit a mechanical instability against gas evolution resulting in the degradation of the catalyst layer (Bittins-Cattaneo *et al.*, 1991). On the other hand, when the working electrode is situated near the membrane inlet that enables also the use of massive metal electrodes or single crystals, response time rises and product transport by diffusion might not be sufficiently fast. This issue can be overcome by applying convection to the electrochemical cell.

When appropriate conditions for DEMS are established, product formation rates can be monitored by the respective ion intensity. As the mass spectrometrically detected ion intensity is directly proportional to the incoming flow of the species to be detected, and, furthermore, if this flow originates from an electrochemical reaction, the ion intensity can be expressed as follows:

$$I_i = (K^*/z)I_F \qquad [3.1]$$

$$K^* = K^0 N / F \qquad [3.2]$$

where I_i indicates the ion intensity of the respective mass fragment, K^* represents a calibration constant obtained from an appropriate calibration experiment, z equals the number of electrons transferred during the calibration reaction and I_F is the Faradaic current. The calibration constant K^* incorporates the constant K^0, which contains all settings of the mass spectrometer and the ionization probability of the corresponding species, N refers to the ratio of mass spectrometrically detected species to the total amount of species formed by the electrochemical reaction (transfer efficiency) and F is the Faraday constant. A detailed derivation of Eqs 3.1 and 3.2 can be found in Baltruschat (2004).

In order to obtain K^* from a calibration experiment, this experiment has to be designed in a way that the calibration reaction is a known electrochemical reaction with a single product. In this case, taking into account the number of electrons transferred (z), the Faradaic current (I_F) can directly be related to the ion intensity of the respective mass fragment (I_i) as shown in Eq. 3.3.

$$K^* = zI_i/I_F \qquad [3.3]$$

A commonly used calibration reaction to calibrate the mass spectrometer for carbon dioxide is the oxidation of pre-adsorbed carbon monoxide in an anodic potential sweep (CO-stripping). The quantification of reaction products turns DEMS from qualitative differential electrochemical mass spectroscopy into quantitative differential electrochemical mass spectrometry.

Depending on the task to fulfil, different electrochemical cell designs have evolved. Principally, one can differentiate between cells with stagnant electrolyte (Fig. 3.1) and those with a flow-through condition. While the traditional DEMS cells with stagnant electrolyte are often optimized with respect to a small cell volume allowing for the use of expensive electrolytes such as ionic liquids or isotopically labelled compounds, more flow cells are currently in use. With flow cells the internal electrolyte can easily be exchanged, maintaining potential control, while this is rather difficult in cells with stagnant electrolyte. Furthermore, the convective conditions in a flow cell allow the investigation of reactions at massive electrodes separated from the membrane inlet by an electrolyte layer of about 100 µm thickness. In the case of stagnant electrolyte the small volume between the electrode and membrane inlet would lead to a quick depletion of the electro-active species. The flow of electrolyte through the thin-layer flow cell can solve this problem; however, as the electrode is placed at a certain distance from the membrane inlet, the response time is limited to some seconds due to the transport of the products from the electrode into the membrane inlet. Furthermore, depending on the flow rate of the convective flow through the cell, a part of the products formed at the electrode will not reach the membrane inlet as they are transported out of the cell by the electrolyte flow. This lowers the transfer efficiency, which makes the analytical method less sensitive. More information on the application

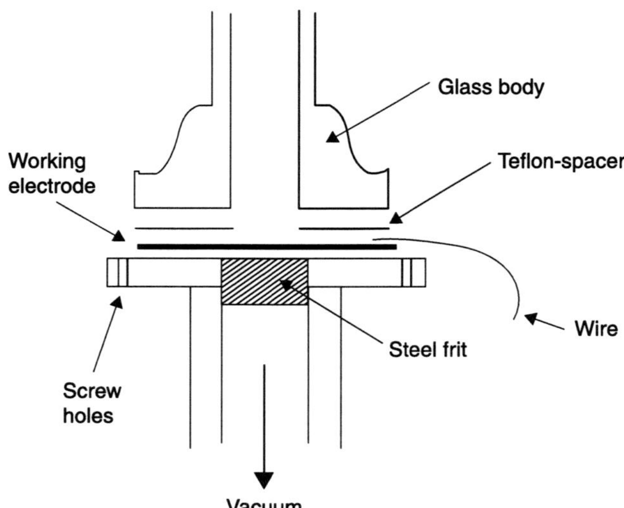

3.1 Conventional cell for DEMS with stagnant electrolyte. (Source: Baltruschat (2004) with permission from Springer.)

and the characterization of thin-layer flow cells can be found in Hartung and Baltruschat (1990), Baltruschat and Schmiemann (1993) and Smith *et al.* (2003).

Another type of flow cell is the 'dual thin-layer cell for continuous flow through of electrolyte' developed by Jusys *et al.* in the late 1990s (Jusys *et al.*, 1999, Baltruschat *et al.*, 1999). In this cell, the electrochemical reaction compartment is separated from the compartment where the species to be detected enters the vacuum system of the mass spectrometer (Fig. 3.2). Therefore, all species being formed in the electrochemical compartment enter the cell compartment in which they can be transferred to the mass spectrometer. Thus, compared to the thin-layer flow cell, the efficiency of product detection under flow conditions is increased. As the electrolyte enters the reaction compartment of the dual thin-layer flow cell in front of the electrode (wall-jet geometry) a uniform distribution of the electrolyte over the working electrode is obtained. Although the electrolyte needs to be transferred through six capillaries into the mass spectrometer compartment in order to be detected, the 2 s time constant of the dual thin-layer flow cell is comparable to the massive metal electrode capable thin-layer flow cell (Jusys *et al.*, 1999).

Another approach allowing the use of massive metal electrodes is the pinhole inlet (Gao *et al.*, 1994). Here, the mass spectrometric gas inlet is a pinhole located at the hemispherical end of a glass tube covered with a Teflon® film of 50 µm thickness. Gao *et al.* reported on the use of a pinhole inlet system to investigate electrochemical reactions at single crystals in the hanging meniscus configuration (Gao *et al.*, 1994). The advantage of this system is the comparatively small amount of gas sampled. Therefore, the vacuum system can be simplified using only one

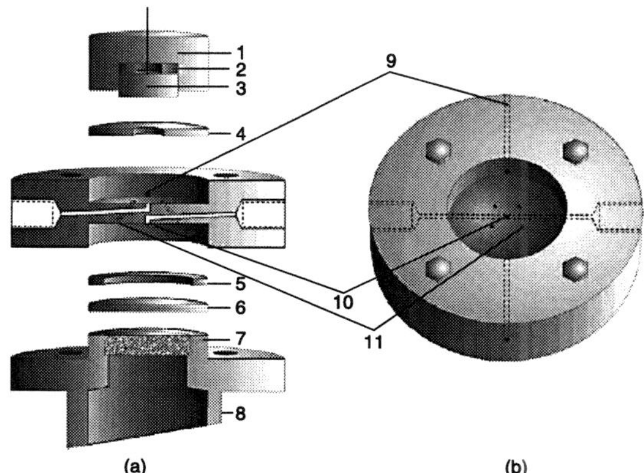

3.2 Dual thin layer cell for experiments under constant flow through. (a) Side view of Kel-F body of the cell; (b) top view of the cell. 1. Kel-F support; 2. Kalrez gasket; 3. Single crystal electrode; 4, 5. Teflon® spacer, 6. Porous Teflon® membrane, 7. Stainless steel frit, 8. Stainless steel connection to MS, 9. Capillaries for flushing with inert gas, 10. Inlet and outlet capillaries, 11. Connecting capillaries. (Source: Baltruschat (2004) with permission from Springer.)

turbomolecular pump. A disadvantage seems to be that volatile products are not only sampled from the small cylindrical volume between the pinhole and the electrode but from a larger volume. Furthermore, the complex diffusion behaviour between electrode and pinhole inlet should lead to complicated time dependence and larger response times (Baltruschat, 2004).

The pinhole inlet approach was adopted by Hillier *et al.*, who used it as the basis for scanning DEMS by scanning the probe some 100 µm above the surface of a multi-electrode array (Jambunathan and Hillier, 2003, Jambunathan *et al.*, 2004). This technique allows a lateral resolution in the millimetre range; however, time constants in the range of 10 s were rather high. The pinhole inlet system was further enhanced by Koper *et al.* by improving the inlet tip construction and adding a micrometer positioning system including a video camera (Wonders *et al.*, 2006 Housmans *et al.*, 2006). Thus, a gap of 10–20 µm between the tip of the probe and the electrode surface could reproducibly be formed. Interestingly, the presence of the probe in close proximity to the Pt(111) electrode in hanging meniscus configuration has no influence on the quality of the cyclic voltammograms. However, the response times of the system are still in the range of 10–15 s. As their technique does not require differential pumping, Koper *et al.* call it on-line electrochemical mass spectrometry (OLEMS) rather than DEMS.

Another application of DEMS which is even more relevant for fuel cell-related processes is the investigation of carbon-supported metal catalysts under defined mass transport conditions. In Jusys *et al.* (2001, 2002) a dual thin-layer flow-through cell with a thin-film electrode is described. A commercial Nafion®-bound carbon-supported Pt/Vulcan catalyst is applied to a polished glassy carbon disc resulting in a catalyst film of approximately 0.1 µm thickness. This catalyst film electrode is mounted in the electrochemical compartment of the dual thin-layer flow cell. This setup was used for a variety of studies concerning kinetic and mechanistic studies of, for example, the electro-oxidation of ethanol (Wang *et al.*, 2004a, 2004b). Pt alloy catalysts and other alcohols were also investigated (Wang *et al.*, 2006, Colmenares *et al.*, 2006, Wang *et al.*, 2009).

More recently, in contrast to all the aforementioned setups, the described approach was even extended to higher temperatures and higher pressures (Jusys and Behm, 2008, Sun *et al.*, 2009 and Chojak Halseid *et al.*, 2010). Heating up the electrolyte in a normal DEMS cell would lead to evaporation of the electrolyte or volatile ingredients resulting in bubble formation, which can block the various interconnections, leading to a loss of potential control by interrupting the current flow. Therefore, DEMS measurements are normally limited to temperatures well below 80 °C (Wang *et al.*, 2004b). In the high-temperature and high-pressure DEMS approach, the pressure is raised to 2–3 bar overpressure, which increases the boiling point of water and organic liquids in order to prevent their evaporation. Thus, temperatures of 100 °C and above can be achieved. Additionally, when using a pressurized cell with hot aqueous electrolyte, the membrane interface to the vacuum system of the mass spectrometer needs to fulfil strict requirements with respect to chemical and mechanical stability. Moreover, a porous membrane interface like the ones typically used in DEMS would result in extensive evaporation of water and organic molecules into the vacuum chamber. In Jusys and Behm (2008) a non-porous 10 μm-thick Teflon® membrane was used to overcome these issues. The membrane is sufficiently permeable for gases while it efficiently holds back water and organic species from entering the vacuum system. However, the non-porous membrane is not permeable for larger molecules, which therefore cannot be detected with this setup.

So far, all investigations referred to the acidic medium. However, as shown in recent studies, DEMS can also be applied in alkaline medium (Bayer *et al.*, 2010a, 2010b and Cremers *et al.*, 2010b). Carbon dioxide is one of the key components produced during the oxidation of organic molecules. However, in alkaline environment carbon dioxide is rapidly converted to carbonate or bicarbonate. Both are hardly volatile and thus not detectable with DEMS. When using a dual thin-layer flow cell, the time constant of approximately 2 s for transferring the product molecules from the electrode in the electrochemical compartment to the vacuum inlet in the detection compartment would certainly be enough to quantitatively convert all carbon dioxide formed at the electrode into carbonate or bicarbonate. In Bayer *et al.* (2010a) a single compartment thin-layer flow cell with a platinum sputter-deposited Teflon® membrane is used. The membrane serves simultaneously as working electrode and inlet to the vacuum system. Thus, the response time of the system is minimized to approximately 0.2 s, which enables the detection of carbon dioxide formed right at the electrode and before its conversion to carbonate or bicarbonate. The results of calibration experiments prove the suitability and reliability of this setup.

For all above setups the catalyst loading, i.e. the amount of catalyst in contact with the reactant solution, is rather low compared to the situation in actual fuel cells. Also, the catalyst layer is much thinner and any gas diffusion or microporous layers are entirely missing. Thus, the contact time of the reactants with the catalyst is rather short. In particular, in flow-through cells the measurements are therefore conducted in a differential regime with marginal conversion of the reaction. As it is known from heterogeneous catalysis, such measurements are very suitable for determining the initial processes in a multistep reaction. However, if subsequent secondary reactions are of importance, they will not be detected by this method. Therefore, there is some interest in using mass spectroscopy to determine products formed in actual fuel cells. Currently, two different setups are in use. The simpler setup is to have either a capillary inlet or a membrane inlet into the mass spectrometer situated in the gaseous or liquid exhaust stream of the fuel cell, respectively. This in-line setup is well suited to measure product distributions under steady-state conditions in the fuel cell or to monitor transient occurrences of products following load changes. A setup more similar to the common DEMS setup uses a membrane inlet into the mass spectrometer situated in the flow field of the fuel cell. Such a setup has been proposed by the Stimming group at Technische Universität München (Seiler *et al.*, 2004) (Fig. 3.3). Its advantage is

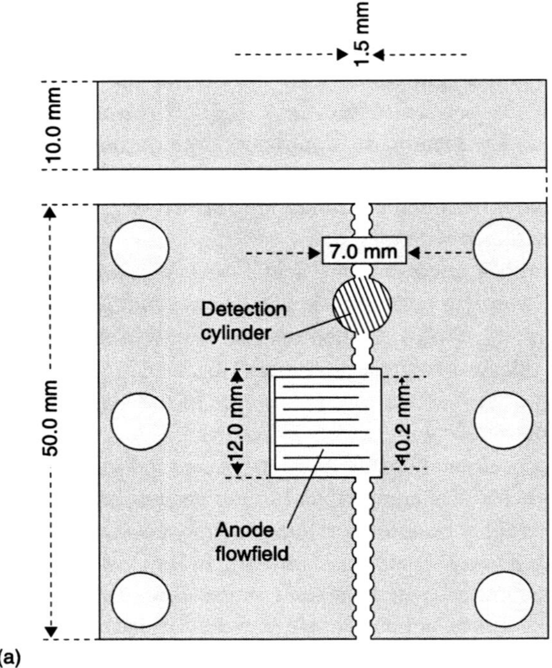

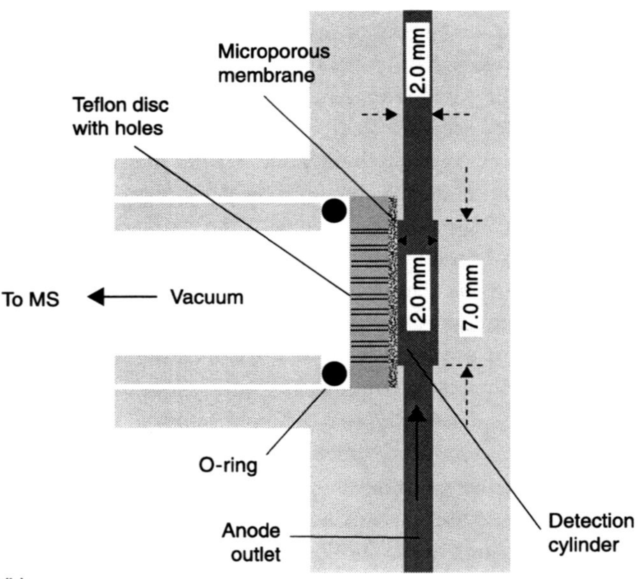

3.3 Scheme of the experimental setup used by Seiler *et al.* (a) fuel cell anode (b) DEMS sensor at the outlet of the fuel cell. (Source: Seiler *et al.* (2004) with permission from Elsevier.)

the possibility to perform potentiodynamic measurements that assign products to certain potentials at slow potential scan rates. For the aforementioned in-line mass spectrometry setup, strong backmixing in the piping would prohibit such an assignment, even for slow potential scan rates.

3 3   Experimental techniques

In principle, DEMS allows the same experimental working techniques which can be carried out in a standard three-electrode assembly. However, DEMS is much more versatile as it adds mass spectrometric analysis to the electrochemical experiment, and it can be combined with further analytical techniques.

3 3.1   Basic electrochemical working techniques

Provided that the response times for the ion intensities are not too long, basically all commonly used electrochemical working techniques can be carried out. In cyclic voltammetry (CV), the potential of the working electrode is scanned in a triangular manner at a constant scan rate between an anodic and a cathodic vertex potential. The resulting experiment from the combination of cyclic voltammetry (electrochemistry) and DEMS (mass spectrometry) is called mass spectrometric cyclic voltammetry (MSCV). Using this technique, an assignment of an electrochemical product to a certain potential is possible. In chronoamperometry (CA), the working electrode is held at a constant potential and the current–time transient is recorded. Both working techniques can be combined realizing so-called adsorption-stripping experiments in order to gain information on the adsorbate species from e.g. organic molecules. In an adsorption step, the electrode is held at a desired adsorption potential and the electrolyte containing the organic species is introduced into the cell. In a second step, the electrolyte containing the organic species is replaced by the supporting electrolyte maintaining potential control of the working electrode. In the stripping step, adsorbates are desorbed by scanning the electrode potential in either the anodic or cathodic direction beginning at the adsorption potential.

When using a thin-layer flow cell, the large IR drop in the thin layer of electrolyte can distort the shape of the cyclic voltammogram or lead to potential oscillations. As described in Baltruschat (1999), this can be addressed by installing two counter electrodes, one in the cell outlet and the other in the inlet. By connecting both counter electrodes to the potentiostat via different resistors, the total current is divided into two parts. After choosing the right value resistors, the current in the capillary to the second counter electrode and the reference electrode is much less than 50% of the total current. Hereby, the ohmic potential drop is reduced and higher electronic stability is obtained.

3.3.2 Combination of differential electrochemical mass spectrometry (DEMS) with other techniques

The combination of DEMS with a rotating disc electrode (RDE) was developed in the early stages of DEMS development (Tegtmeyer *et al.*, 1989, Wasmus *et al.*, 1990). This system allowed a defined convection at the working electrode and fast reactant transport to the electrode. The radial component of the mass flow at the rotating electrode transports the reaction products from the electrode to the nearby membrane inlet. It was shown that the sensitivity of this system was high enough to use smooth or even single crystal electrodes with a good signal-to-noise ratio.

Another approach was the combination of DEMS with electrochemical quartz crystal microbalance (EQCM) (Jusys *et al.*, 1999). This combination was made possible by integrating a gold-sputtered quartz crystal as working electrode into the electrochemical compartment of a dual thin-layer flow cell. Deposition or dissolution processes (adsorption, underpotential deposition (UPD), electropolymerization, corrosion, etc.) leading to a mass change can be detected at the quartz crystal working electrode through the change in the oscillation frequency of the quartz crystal, while volatile products can be simultaneously detected in the mass spectrometric compartment of the cell.

More recently, the combination of DEMS and attenuated total reflection Fourier-transformed infrared spectroscopy (ATR-FTIR) was established (Heinen *et al.*, 2007a, 2007b, 2009). In ATR configuration IR spectroscopy is a highly surface-sensitive analysis technique. Compared to conventional *in situ* IR spectroscopy performed in external reflection configuration, ATR-FTIR spectroscopy has the advantage of a significantly higher sensitivity. Moreover, as the reflection takes place internally, in the ATR prism, the IR beam does not have to pass through a liquid film as in conventional *in situ* IR spectroscopy. For the combination of DEMS and ATR-FTIR spectroscopy, once again, the dual thin-layer flow cell is modified. The ATR-FTIR compartment is formed by pressing an IR-transparent silicon prism via a circular gasket to the cell body. The mass spectrometric compartment remains unchanged. The working electrode is established by depositing a platinum film on the reflecting side on the silicon prism. In Heinen *et al.* (2009), the oxidation of a variety of small organic molecules is studied. Here, ATR-FTIR spectroscopy is employed for the detection of adsorbed reaction intermediates while product distribution is quantitatively determined by DEMS.

Figure 3.4 summarizes the evolution of DEMS and its combination with other analytical techniques.

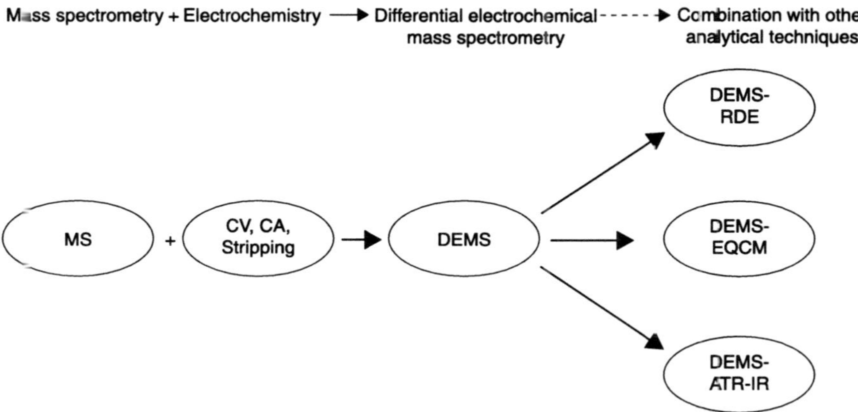

3.4 Schematic diagram of the evolution of DEMS and some combinations of DEMS with other analytical techniques.

3.4   Application with respect to fuel cell catalysis

Excepting hydrogen oxidation, most other fuel cell reactions could lead to the formation of side-products. The formation of these side-products is in most cases undesired as it causes efficiency losses. Also, some of the side-products are detrimental, e.g. because they are toxic, such as the aldehydes that can be formed as side-products of the alcohol oxidation. Other side-products, e.g. peroxides or super oxides from the oxygen reduction reaction, can damage the fuel cell. DEMS is a suitable tool to detect side-products as long as they are volatile and sufficiently stable. Thus, DEMS is often used to study alcohol oxidation reactions, but it cannot be employed to follow the peroxide formation during oxygen reduction. Hence, the development of anode catalysts for direct methanol fuel cells (DMFC) and other direct alcohol fuel cells is a classic field of application for DEMS. Questions that are regularly investigated using DEMS comprise:

- Product distribution for the oxidation of a certain alcohol.
- Effect of electrode composition.
- Effect of alcohol concentration.
- Reaction pathway elucidation.

An example is the investigation of factors influencing the methanol oxidation at catalysts relevant for DMFC. Generally it is accepted that methanol adsorbs via the abstraction of one carbon-bound hydrogen atom onto platinum surfaces. This is affirmed by DEMS findings of Krausa and Vielstich (1994) showing that the oxidative peak at 240 mV vs RHE observed during the first cycle of methanol oxidation CVs at platinum electrodes is not accompanied by the formation of any volatile product. In a subsequent step either all three remaining hydrogen atoms are abstracted, resulting in the formation of CO_{ads} or only two hydrogen atoms are

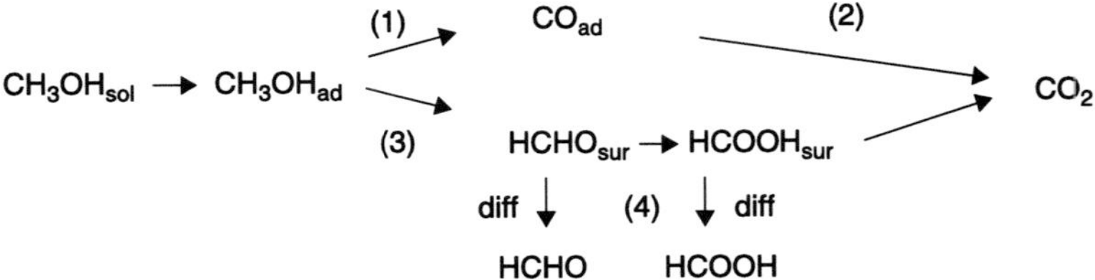

3.5 Reaction scheme for the oxidation of methanol. Reaction pathways, intermediates and products. (Source: Wang *et al.* (2001) with permission from Springer.)

abstracted resulting in the formation of H_2CO_{ads}. The former can react with an adsorbed hydroxyl radical to form carbon dioxide; the latter can either be released as formaldehyde or react with adsorbed OH to form formic acid. Also, a reaction to carbon dioxide is possible after either surface migration or readsorption (Fig. 3.5).

Using DEMS at thin-film electrodes it could be shown by Wang *et al.* (2001) that the carbon dioxide current efficiency is low at low potentials and high electrolyte flow rates; on the other hand it increased either at high potentials where CO oxidation occurs or at lower flow rates. The findings of Wang *et al.* are depicted in Fig. 3.6. Figure 3.6a shows results under potentiodynamic conditions (Table 1 in Wang *et al.* 2001) where the decrease of CO_2 current efficiency with increasing methanol concentration and flow rate is clearly apparent. Figure 3.6b shows results under potentiostatic conditions after 1 min of potential hold (Table 2 in Wang *et al.* 2001). For the solutions with 0.001 mol l^{-1} and 0.1 mol l^{-1} methanol concentration the flow rate was varied at a constant potential. The effect is less pronounced than that found under potentiodynamic conditions; however, the CO_2 current efficiency is still reduced for the highest tested flow rate. For the solution with 0.01 mol l^{-1} methanol concentration the potential was varied at a constant low flow rate. Here it becomes apparent that the methanol concentration is a dominant factor for the CO_2 current efficiency. In spite of the low flow rate and a comparatively higher potential, the maximum current efficiency is still lower than that obtained at the lower concentration of 0.001 mol l^{-1}.

Carbon dioxide current efficiency is also in general higher when supported catalysts forming thicker electrode layers are employed. This indicates that carbon dioxide formation proceeds via both pathways in parallel depending on the conditions at the electrode. It also indicates that readsorption of intermediates is of importance and must be considered in the design of the catalyst layer and the GDL, e.g. by making the catalyst layer sufficiently thick.

Possible side-products of the alcohol oxidation are acids, in the case of methanol oxidation, formic acid. Due to low volatility of the polar acids from aqueous solutions their detection in DEMS is already rather difficult in acidic solution and practically impossible in alkaline solutions. A usual way to measure acid formation has been to detect the alcohol esters as volatile products in the mass spectrum. The

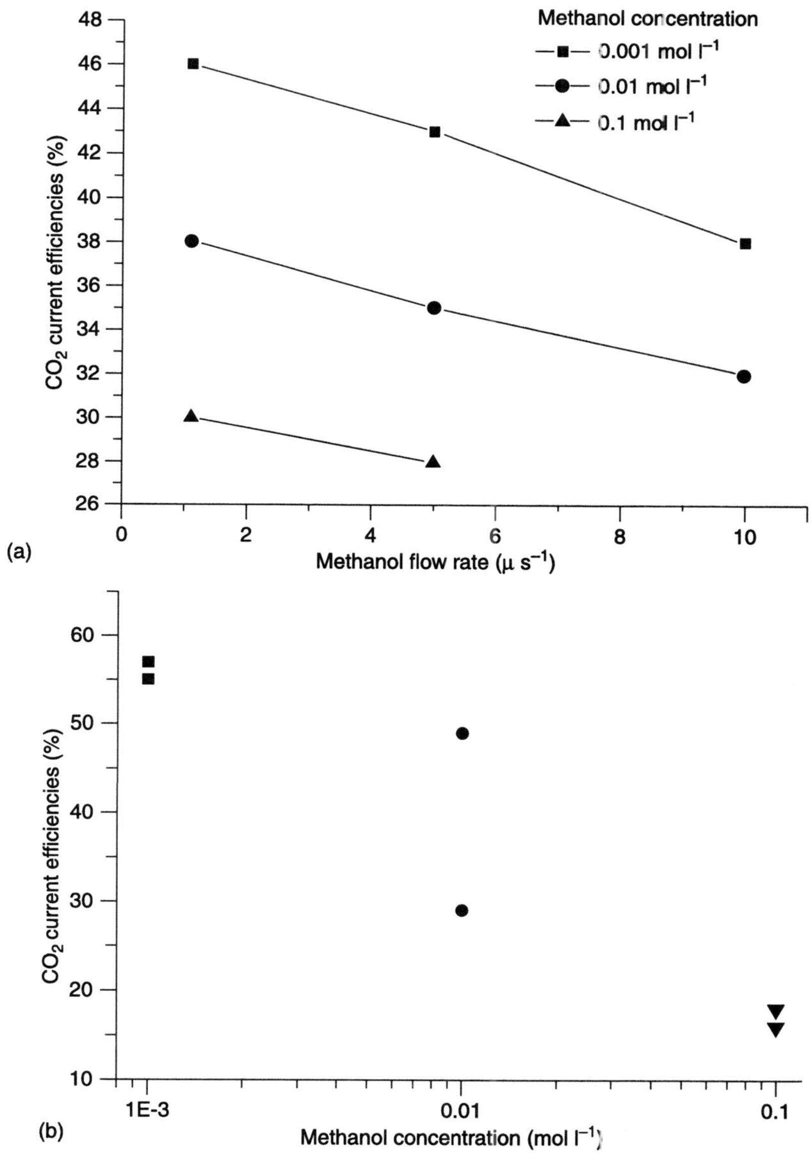

3.6 Depiction of the results by Wang *et al.*, 2001. (a) CO_2 current efficiency as a function of methanol flow rate for different methanol concentrations measured under potentiodynamic conditions. (b) CO_2 current efficiency as function of concentration and one additional parameter. ■: concentration 0.001 mol l⁻¹, potential 0.65 V vs RHE, flow rate 1.1, 5.0 and 10 μl s⁻¹ respectively; ●: concentration 0.01 mol l⁻¹, flow rate 1.1 μl s⁻¹, potential 0.65 V (lower) and 0.75 V (higher), respectively; ▼: concentration 0.1 mol l⁻¹, potential 0.6 V vs RHE, flow rate 1.9, 5.0 and 8.3 μl s⁻¹, respectively.

consideration was that the carboxylic acid resulting from the electrochemical reaction reacts with bulk alcohol in a secondary reaction. In this model this second reaction would neither depend on the electrode potential nor on any adsorption to the surface. Thus, the concentration of ester would only depend on the alcohol concentration that can be estimated to be constant and the acid concentration resulting from the electrochemical reaction (Krausa and Vielstich, 1994). However, recent results from Abd El Salehin *et al.* indicate that the formation of methyl formate proceeds via an electrocatalytic surface reaction and is thus itself dependent on the electrode potential in its formation rate (Abd El Salehin *et al.*, 2005). As shown in Fig. 3.7, similar results were found for the formation of ethyl acetate as a side-product of the ethanol oxidation by Cremers *et al.*, 2010a.

In general the mass spectrometric signal used to follow a certain reaction product needs to be carefully chosen. Due to short response time of the mass spectrometer as well as due to sensitivity, electron impact is used in most DEMS setups to generate charged ions for mass separation and detection. This technique has the advantages of high efficiency in ion generation and a high throughput; however, molecules get strongly fragmented. In Table 3.1 typical fragments for methanol and ethanol and their oxidation products are listed, also indicating those fragments that can be recommended for use as tracers for a specific product. An

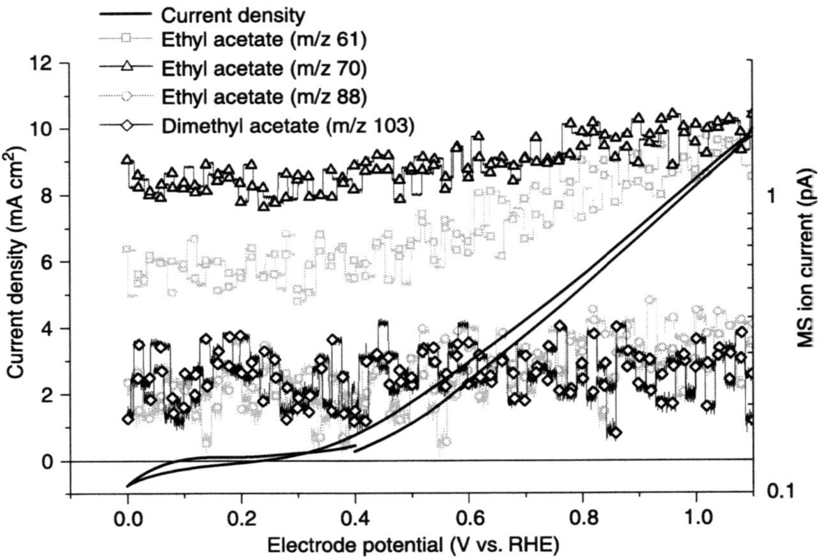

3.7 MSCV of a platinum sputtered electrode in 0.1 M ethanol/0.1 M KOH solution. Assignment of the mass fragments as indicated in the graph. Scan rate 10 mV s⁻¹, flow rate 7 ml min⁻¹. For experimental detail refer to Bayer *et al.* (2010a).

Table 3.1 Fragments of methanol and ethanol and their oxidation products. Fragments printed in italic have only a weak intensity. Fragments printed in bold are recommended for tracing the given substance. If several fragments are printed in bold for a substance then simultaneous tracing of at least two of these fragments is recommended to monitor this substance

m/z	15	22	28	29	30	31	32	43	44	45	46	60	61	88
Methanol	CH_3^+			CHO^+	CH_2O^+	$\mathbf{CH_3O^+}$	CH_4O^+							
Formaldehyde			CO^+	CHO^+	$\mathbf{CH_2O^+}$									
Formic Acid				CHO^+					CO_2^+	HCO_2^+	$\mathbf{HCOOH^+}$			
Methyl formate	CH_3^+			CHO^+		CH_3O^+	CH_4O^+					$\mathbf{HCOOCH_3^+}$		
Carbon dioxide		$\mathbf{CO_2^{(2+)}}$	CO^+						$\mathbf{CO_2^+}$					
Ethanol	CH_3^+			CHO^+		$\mathbf{CH_3O^+}$				$C_2H_5O^+$	$\mathbf{C_2H_6O^+}$			
Acetladehyde	$\mathbf{CH_3^+}$			$\mathbf{CHO^+}$				$C_2H_3O^+$	$\mathbf{CH_3CHO^+}$					
Acetic Acid	CH_3^+							$C_2H_3O^+$		$\mathbf{C_2H_5O^+}$		$\mathbf{CH_3COOH^+}$		
Ethyl Acetate				CHO^+				$C_2H_3O^+$					$\mathbf{C_2H_5O_2^+}$	$\mathbf{C_4H_8O_2^+}$
Carbon dioxide		$\mathbf{CO_2^{(2+)}}$	CO^+						CO_2^+					

alternative is the use of isotopic labelled substrates. This can help to separate certain peaks from those found in the environment, e.g. both CO and N_2 exhibit intensive molecular ion peaks at m/z = 28 so that a separation can be difficult. Using ^{13}C labelled educts would however result in ^{13}CO with molecular ion signal at m/z = 29 that can better be distinguished. For fuel cell application the use of partly labelled fuel is more important, e.g. to determine the carbon group of a fuel from which an observed product originates. An example is the differentiation of carbon dioxide produced from the α- or β-C atom of ethanol. This differentiation is feasible if one of both carbon atoms is ^{13}C labelled. Another possibility is the use of (partially) deuterated fuels, e.g. the use of ethanol-d_6 enables tracing the carbon dioxide formation with the primary m/z = 44 signal as the disturbing fragment from acetaldehyde at m/z = 44 (CH_3CHO^+) is shifted to m/z = 48 (CD_3CDO^+).

Recently, DEMS techniques have been successfully used to investigate the reaction mechanism of the ethanol oxidation reaction for direct ethanol fuel cells. An important part of the reaction is the adsorption of the fuel molecule onto the catalyst surface. A DEMS technique allowing for the determination of the adsorbed species that are formed during the alcohol adsorption is the recording of stripping MSCV. The technique is described, e.g. in Bayer *et al.* (2010a). In principle the investigated catalyst surface is first cleaned by repetitive potential cycling in the base electrolyte. Then the potential is set to the adsorption potential and the electrolyte is exchanged against a fuel containing electrolyte solution keeping the potential at the set value. After the desired adsorption time the electrolyte is changed back to neat base electrolyte still keeping the potential at the set adsorption value. After some time for rinsing the remaining fuel out of the bulk of the cell, a stripping cyclic voltammogram is recorded in the DEMS cell, also monitoring the products being released. Stripping cycles are either started in anodic or cathodic direction to determine redox-amphoter adsorbates.

Using this technique in an acidic solution it could be shown that the major adsorbates formed from ethanol adsorption are CO_{ads} and $CH_{x,ads}$ indicating facile C-C bond scission at clean platinum surfaces (Wang *et al.*, 2004a, Iwasita and Pastor, 1994 and Schmiemann *et al.*, 1995). As shown in Fig. 3.8, Bayer *et al.* could show that ethanol has similar adsorption behaviour in alkaline media (Bayer *et al.*, 2010a). For bulk oxidation a strong influence of the ethanol concentration on the product composition was found (Cremers *et al.*, 2010a). Carbon dioxide current efficiency drops when the ethanol concentration is increased. This is probably due to the poisoning of the surface by reaction intermediates which can only be removed at different potentials as it has been revealed by the DEMS investigation. A main problem thereby might be poisoning by $CH_{x,ads}$ species.

In total, DEMS is a well-established tool for the investigation of the electro-oxidation of small organic molecules and is therefore regularly employed in the development of electro-catalysts for direct alcohol fuel cells.

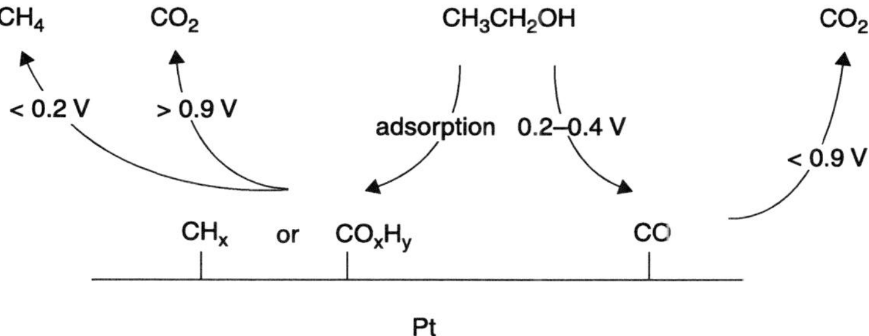

3.8 Reaction scheme for the adsorption and stripping of ethanol in 0.1 M KOH. (Source: adapted from Bayer *et al.* (2010a).)

3.5   Advantages and limitations of differential electrochemical mass spectrometry (DEMS)

Since its invention, DEMS has proven to be a valuable analytical tool for the on-line detection of electrochemical reaction products and adsorbed intermediates. The high sensitivity of the method was demonstrated by the detection benzene in a dual thin-layer flow cell at concentrations as low as 1 μM (Baltruschat, 1999). By applying the appropriate calibration procedures, reaction and desorption products can also be quantified. This can help to distinguish between main and side-products. As the dual thin-layer flow cell enables the use of various electrode materials, reactions at fuel cell relevant catalysts can also be studied. Separating the electrochemical compartment from the mass spectrometric compartment, the dual thin-layer flow cell provides an interface for coupling with other analytical methods. The main disadvantage of DEMS is, of course, that only volatile products can be detected. However, depending on the analytical task, coupling DEMS with another analytical method could help. In this respect, the combination of DEMS with ATR-FTIR has to be emphasized, as both methods are highly sensitive and yield complementary information.

3.6   Fuel cell DEMS and in-line mass spectrometry

Fuel cell electrodes differ particularly from typical DEMS electrodes in their dimensions. The surface areas, both the geometric surface area and the real surface area are much larger in a fuel cell than in a typical DEMS cell. Hence, chances for secondary reactions to occur are much higher than at the thin-film electrodes used in the DEMS setups described above. The monitoring of the anode product composition is therefore an important issue in particular for the development of direct alcohol fuel cells. The use of mass spectrometry is one possible approach to this goal. Other approaches use chromatographic methods such as gas chromatography or high-performance liquid chromatography (HPLC) to analyse the anode exhaust. The advantages of mass spectrometry are its rapidity, allowing for the online detection of volatile compounds, and the direct comparison to DEMS results obtained during the catalyst development.

The coupling of a fuel cell to the mass spectrometer can be performed in two different ways, either using a fuel cell DEMS setup or measuring online downstream of the fuel cell exhaust.

The first setup is mainly described by the Stimming group at the Technische Universität München. The setup developed there is described in Section 3.2. It was originally used to investigate processes in direct methanol fuel cells, in particular methanol adsorption on real-world fuel cell electrodes. Thereby it could be shown that in this case CO is the main stable product of the methanol adsorption (Seiler *et al.*, 2004). In some further studies ethanol oxidation in direct ethanol fuel cells was studied using the same setup. Here some interesting results were found. It could be shown that high carbon dioxide current efficiencies of the ethanol oxidation are possible if high-loading Pt/C catalysts are employed. The carbon dioxide current efficiency thereby correlates with the electrochemical active surface area determined by CO stripping experiments. This also holds true when platinum alloys such as PtRu/C or PtSn/C are used. The observation is assigned to the effect of readsorption of acetaldehyde that is known to be formed as the primary product. High catalyst loadings will increase the residence time of these intermediates, increasing the probability for readsorption and further oxidation. This hypothesis is supported by two additional findings reported in that paper (Rao *et al.*, 2007a). First, the carbon dioxide current efficiency can be reduced at a given catalyst loading when the ethanol feed rate is increased, which increases the mass transport and thus decreases the residence time of intermediately formed products in the electrode layer. Secondly, results for platinum black catalyst showed a lower carbon dioxide current efficiency than would be expected from the electrochemical active surface area that was provided. Here again, the thinner catalyst layer will reduce the residence time of any formed intermediates within the electrode layer. In another effort the DEMS cell was used to investigate the ethanol oxidation in an anion exchange membrane fuel cell (Rao *et al.*, 2007b). Here, a higher carbon dioxide conversion efficiency was found at low platinum loading compared to the PEM type fuel cell. However, these results were measured in aqueous ethanol solution without the addition of base electrolyte to the fuel. At the Teflon® electrodes only very small current densities were thus obtained. The measurement in the absence of a liquid base was required in order to be able to detect formed carbon dioxide. In presence of a base the carbon dioxide would be absorbed in the electrolyte as carbonate ions and thus could not be detected. Overall, fuel cell DEMS has thus the same deficiencies as the dual thin-layer DEMS cell described above.

The simpler alternative setup is to place a connection to the mass spectrometer gas inlet system downstream of the fuel cell. Such an inlet can either be a membrane inlet for liquid anode exhaust streams or a capillary for gaseous ones. Using a membrane inlet, Fraunhofer ICT investigated the ethanol conversion in

PEM type DEFC using a mixed catalyst anode which was developed there (Cremers and Krausa 2007). The basic idea was to exploit the findings by *Rao et al.* that carbon dioxide current efficiencies can be increased by promoting acetaldehyde readsorption. As very high catalyst loadings are not economically viable a catalyst which can oxidize acetaldehyde was looked for. It was found by standard DEMS measurements that platinum ruthenium alloys show some activity for this reaction (Cremers *et al.*, 2008). It could be shown that mixing PtSn/C catalyst and PtRu/C catalyst in the anode layer of a DEFC can increase both the electrical performance of the cell and the carbon dioxide current efficiency of the reaction (Cremers *et al.*, 2008).

There are further applications of mass spectrometry for the investigation of processes in fuel cells. One potential application is the monitoring of degradation processes caused, e.g. by the production of hydrogen peroxide yielding in the release of HF or CF_x species. Wu *et al.* found that after longer PEMFC run times, hydrogen peroxide and hydrofluoric acid were detectable in the cathode exhaust gas stream (Wu *et al.*, 2010). Considering the fast response time of the mass spectrometer and the transient nature of many processes relevant to fuel cell degradation, e.g. potential excursions encountered during hydrogen admittance at fuel cell start-up, mass spectrometry can become a very important tool for fuel cell degradation studies.

In more complex setups with several gas sample inlets distributed over the electrode area an investigation of gas distributions in the cell can also be realised as was reported by Schuler *et al.* (Schuler *et al.*, 2010).

Mass spectrometry can also be an important tool for investigation in actual fuel cells. As with the basic DEMS investigation, combination with other surface sensitive techniques is beneficial. However, as IR radiation will not be able to penetrate the catalyst layer to any significant distance, X-ray-based methods are more suitable. Recently, in particular, X-ray absorption spectroscopy (XAS) in the near edge region (X-ray adbsortion near edge structure) was employed using the $\Delta\mu$-XANES analysis method described in Chapter 5 and in the literature cited there. In a recent work Melke *et al.* studied ethanol oxidation in a DEFC (Melke *et al.*, 2010) using prior results from fuel cell DEMS measurements (Cremers *et al.*, 2008).

3.7   References

Abd El Salehin A A A, Lanova B and Baltruschat H (2009), 'Ethanol and methanol: Adsorption rates and rates of intermediate formation at Pt single crystal electrodes', *ECS Meeting Abstract*, 216th ECS Meeting, Abstract #1010.

Baltruschat H (1999), 'Differential electrochemical mass spectrometry as a tool for interfacial studies', in Wieckowski A, *Interfacial Electrochemistry–Theory, Experiment and Application*, New York, Marcel Dekker, 577–597.

Baltruschat H (2004), 'Differential electrochemical mass spectrometry'. *J Am Soc Mass Spectrom*, **15**, 1693–1706.

Baltruschat H and Schmiemann U (1993), 'The adsorption of unsaturated organic species at single-crystal electrodes studied by differential electrochemical mass-spectrometry', *Ber Bunsenges Phys Chem*, **97**, 452–460.

Baltruschat H, Jusys Z and Löffler T (1999), 'Differentielle elektrochemische Massenspektrometrie: Neue Entwicklungen', in *Gesellschaft Deutscher Chemiker, Elektrochemische Reaktionstechnik und Synthese – Von den Grundlagen bis zur industriellen Anwendung*, GDCh-Monographie, Bd. 14, Frankfurt am Main, 99–105.

Bayer D, Berenger S, Joos M, Cremers C and Tübke J (2010b), 'Electrochemical oxidation of C_2 alcohols at platinum electrodes in acidic and alkaline environment', *Int J Hyrdogen Energy*, **35–22**, 12660–12667.

Bayer D, Cremers C, Baltruschat H and Tübke J (2010a), 'Ethanol stripping in alkaline medium: A DEMS study', *ECS Trans* 25(13), 85–93.

Bittins-Cattaneo B, Cattaneo E, Königshoven P and Vielstich W (1991), 'New developments in electrochemical mass spectroscopy' in Bard A J, *Electroanalytical Chemistry: A Series of Advances*, Vol. 17, New York, Marcel Dekker, 181–220.

Bruckenstein R R and Gadde J (1971), 'Use of a porous electrode for *in situ* mass spectrometric determination of volatile electrode reaction products', *J Am Chem Soc*, **93**, 793–794.

Cassuto A and King D A (1981), 'Rate expressions for adsorption and desorption-kinetics with precursor states and lateral interactions', *Surf Sci*, **102**, 388–404.

Chojak Halseid M, Jusys Z and Behm R J (2010), 'Electrooxidation of ethylene glycol on a carbon-supported Pt catalyst at elevated temperatures and pressure: A high-temperature/high-pressure DEMS study', *J Electroanal Chem*, **644**, 103–109.

Colmenares L, Wang H, Jusys Z, Jiang L, Yan S *et al.* (2006), 'Ethanol oxidation on novel, carbon supported Pt alloy catalysts – Model studies under defined diffusion conditions', *Electrochim Acta*, **52**, 221–233.

Cremers C and Krausa M (Fraunhofer Gesellschaft zur Förderung der angewandten Forschung e.V.) 2007. *Verwendung einer Anode in einer Brennstoffzelle zur Oxidation von Ethanol und/oder zumindest eines C3 bis C10-haltigen Alkohols*. German Patent DE102007031526B4.

Cremers C, Bayer D, Kintzel B, Joos M, Jung F *et al.* (2008), 'Oxidation of alcohols in acidic and alkaline environments', *ECS Trans*, **16**(2), 1263–1273.

Cremers C, Bayer D, Meier J O, Berenger S, Kintzel B *et al.* (2010a), 'Development of electro-catalysts for the alcohol oxidation in anion exchange membrane direct alcohol fuel cells', *ECS Trans*, **25**(13), 27–37.

Cremers C, Kintzel B, Bayer D and Tübke J (2010b), 'Influence of the pH value and the carbonate ion concentration of the electrolyte solution on the ethanol oxidation at polycrystalline platinum', *ECS Trans*, **33**(1), 1681–1692.

Falconer J L and Schwarz J S (1983), 'Temperature-programmed desorption and reaction: Applications to supported catalysts', *Catal Rev Sci Eng*, **25**, 141–227.

Gao Y, Tsuji H, Hattori H and Kita H (1994), 'New on-line mass spectrometer system designed for platinum-single crystal electrode and electroreduction of acetylene', *J Electroanal Chem*, **372**, 195–200.

Hartung T and Baltruschat H (1990), 'Differential electrochemical mass spectrometry using smooth electrodes: Adsorption and H/D-exchange reactions of benzene on Pt', *Langmuir*, **6**, 953–957.

Heinen M, Chen Y X, Jusys Z and Behm R J (2007a), '*In situ* ATR-FTIRS coupled with on-line DEMS under controlled mass transport conditions – A novel tool for electrocatalytic reaction studies', *Electrochim Acta*, **52**, 5634–5643.

Heinen M, Chen Y X, Jusys Z and Behm R J (2007b), 'CO adsorption kinetics and adlayer build-up studied by combined ATR-FTIR spectroscopy and on-line DEMS under continuous flow conditions', *Electrochim Acta*, **53**, 1279–1289.

Heinen M, Jusys Z and Behm R J (2009), 'Surface species and product distribution in the electrooxidation of small organic molecules', *ECS Trans*, **25**(1), 259–269.

Housmans T H M, Wonders A H and Koper M T M (2006), 'Structure sensitivity of methanol electrooxidation pathways on platinum: An on-line electrochemical mass spectrometry study', *J Phys Chem B*, **110**, 10021–10031.

Iwasita T and Pastor E (1994), 'A DEMS and FTir spectroscopic investigation of adsorbed ethanol on polycrystalline platinum', *Electrochim Acta*, **39**, 531–537.

Jambunathan K and Hillier A (2003), 'Measuring electrocatalytic activity on a local scale with scanning differential electrochemical mass spectrometry', *J Electrochem Soc*, **150**, E312–E320.

Jambunathan K, Jayaraman S and Hillier A (2004), 'A multielectrodeelectrochemical and scanning differential electrochemical mass spectrometry study of methanol oxidation on electrodeposited PtxRuy', *Langmuir*, **20**, 1856–1863.

Jusys Z and Behm R J (2008), 'DEMS analysis of small organic molecule electrooxidation: A high-temperature high-pressure DEMS study', *ECS Trans*, **16–2**, 1243–1251.

Jusys Z, Kaiser J and Behm R J (2001), 'Electrooxidation of COH2/CO and mixtures on a carbon-supported Pt catalyst – A kinetic and mechanistic study by differential electrochemical mass spectrometry', *Phys Chem Chem Phys*, **3**, 4650–4660.

Jusys Z, Massong H and Baltruschat H (1999), 'A new approach for simultaneous DEMS and EQCM: Electro-oxidation of adsorbed CO on Pt and Pt-Ru', *J Electrochem Soc*, **146**, 1093–1098.

Jusys Z, Schmidt T J, Dubau L, Lasch K, Jörissen L *et al.* (2002), 'Activity of PtRuMeOx (Me = W, Mo or V) catalysts towards methanol oxidation and their characterization', *J Power Sources*, **105**, 297–304.

Krausa M and Vielstich W (1994), 'Study of the electrocatalytic influence of Pt/Ru and Ru on the oxidation of residues of small organic-molecules', *J Electroanal Chem*, **379**, 307–314.

Melke J, Schökel A, Dixon D, Cremers C, Ramaker D E *et al.* (2010), 'Ethanol oxidation on carbon-supported Pt, PtRu, and PtSn catalysts studied by operando x-ray absorption spectroscopy', *J Phys Chem C*, **114**(13), 5914–5925.

Rao V, Cremers C, Stimming U, Cao L, Sun S G *et al.* (2007a), 'Electro-oxidation of ethanol at gas diffusion electrodes: A DEMS study', *J Electrochem Soc*, **154**(11), B1138–B1147.

Rao V, Hariyanto, Cremers C and Stimming U (2007b), 'Investigation of the ethanol electro-oxidation in alkaline membrane electrode assembly by differential electrochemical mass spectrometry' *Fuel Cells*, **7**(5), 417–423.

Schmiemann U, Müller U and Baltruschat H (1995), 'The influence of the surface structure on the adsorption of ethene, ethanol and cyclohexene as studied by DEMS', *Electrochim Acta*, **40**, 99–107.

Schuler G A, Wokaun A and Büchi F N (2010), 'Local online gas analysis in PEFC using tracer gas concepts', *J Power Sources*, **195**(6), 1647–1656.

Seiler T, Savinova E R, Friedrich K A and Stimming U (2004), 'Poisoning of PtRu/C catalyst in the anode of a direct methanol fuel cell: A DEMS study', *Electrochim Acta*, **49**, 3927–3936.

Smith S P E, Casado-Rivera E and Abruna H D (2003), 'Application of differential electrochemical mass spectrometry to the electrocatalytic oxidation of formic acid at a modified Bi/Pt electrode surface', *J Solid State Electrochem*, **9**, 582–587.

Sun S, Chojak Halseid M, Heinen M, Jusys Z and Behm R J (2009), 'Ethanol electrooxidation on a carbon-supported Pt catalyst at elevated temperature and pressure: A high-temperature/high-pressure DEMS study', *J Power Sources*, **190**, 2–13.

Tegtmeyer D, Heindrichs A and Heitbaum J (1989), 'Electrochemical on line mass-spectrometry on a rotating electrode inlet system', *Ber Bunsenges Phys Chem*, **93**, 201–206.

Wang H, Jusys Z and Behm R J (2004a), 'Ethanol and acetaldehyde adsorption on a carbon-supported Pt catalyst: A comparative DEMS study', *Fuel Cells*, **4**, 113–125.

Wang H, Jusys Z and Behm R J (2004b), 'Ethanol electrooxidation on a carbon-supported Pt catalyst: Reaction kinetics and product yields', *J Phys Chem B*, **108**, 19413–19424.

Wang H, Jusys Z and Behm R J (2006), 'Ethanol electro-oxidation on carbon-supported Pt, PtRu and Pt3Sn catalysts: A quantitative DEMS study', *J Power Sources*, **154**, 351–359.

Wang H, Jusys Z and Behm R J (2009), 'Adsorption and electrooxidation of ethylene glycol and its C-2 oxidation products on a carbon-supported Pt catalyst: A quantitative DEMS study', *Electrochim Acta*, **54**, 6484–6498.

Wang H, Löffler T and Baltruschat H (2001), 'Formation of intermediates during methanol oxidation: A quantitative DEMS study', *J Appl Electrochem*, **31**, 759–765.

Wasmus S, Cattaneo E and Vielstich W (1990), 'Reduction of carbon-dioxide to methane and ethene – An online MS study with rotating electrodes', *Electrochim Acta*, **35**, 771–775.

Wolter O and Heitbaum J (1984a), 'Differentiel electrochemical mass spectroscopy (DEMS) – A new method for the study of electrode processes', *Ber Bunsenges Phys Chem*, **88**, 2–6.

Wolter O and Heitbaum J (1984b) 'The adsorption of CO on a porous Pt-electrode in sulfuric-acid studied by DEMS', *Ber Bunsenges Phys Chem*, **88**, 6–10.

Wonders A H, Housmans T H M, Rosca V and Koper M T M (2006), 'On-line mass spectrometry system for measurements at single-crystal electrodes in hanging meniscus configuration', *J Appl Electrochem*, **36**, 1215–1221.

Wu J, Yuan X-Z, Martin J J, Wang H, Yang D *et al.* (2010), 'Proton exchange membrane fuel cell degradation under close to open-circuit conditions: Part I: *In situ* diagnosis', *J Power Sources*, **195**(4), 1171–1176.

4

Small angle X-ray scattering (SAXS) techniques for polymer electrolyte membrane fuel cell characterization

X. TUAEV and P. STRASSER, Technical University
Berlin, Germany

Abstract: Small angle X-ray scattering (SAXS) is ideally suited for the non-invasive *in situ* analysis of structural changes of fuel cell components such as polymer membranes, porous supports and supported nanoparticle ensembles. This chapter first discusses some basic concepts of scattering and in particular SAXS and then addresses special methods such as *in situ* SAXS and anomalous SAXS. The chapter then reviews selected important reports where SAXS was applied to analyse the structure of fuel cell components, in particular supported monometallic and bimetallic nanoparticle electrocatalysts. The chapter concludes with an outlook on the important future role of SAXS as an analytical tool in fuel cell research.

Key words: supported Pt alloy nanoparticles, small angle X-ray scattering (SAXS), nanoparticle growth and coarsening, fuel cell electrode degradation mechanisms.

4.1 Introduction

Max von Laue's ground-breaking theoretical predictive work (Friedrich *et al.*, 1912) on the interference phenomena of X-rays – elegantly experimentally verified by Friedrich and Knipping and leading up to the Nobel Prize in Physics a mere two years later – has revolutionized our understanding and our investigative set of tools in regards to the structure of matter at the atomic scale forever. Max von Laue realized that the interaction of waves with wavelengths in the range of atomic diameters (10^{-10} m = 1 Å) has the power to probe the geometry of atomic arrangements (structures of size on the order of 0.1–1 nm). Applied to well-ordered solid crystals, von Laue showed that planar X-ray waves on their way through atomic lattices produced sharp interference patterns, which contained detailed structural information on the symmetry and bond distances of the atomic arrangement. Depending on the symmetry and lattice parameters of the crystals, the interference maxima were located at scattering angles between 3° and perhaps 120°.

During the late 1930s, researchers realized that X-rays with wavelengths of atomic dimensions also result in useful, more diffuse interference patterns at scattering angles of 3° and less, the so-called small angle X-ray scattering (SAXS) range (Glatter and Kratky, 1982, Guinier *et al.*, 1955). According to Bragg's

87

famous equation $\lambda = 2d \sin\theta$, the small scattering angles θ at X-ray wave lengths comparable to diffraction ($\lambda = 1$–2 Å) imply that the structural features (lattice spacings d) observed by SAXS are rather colloidal in size, typically between 1–100 nm, as opposed to atomic. It thus became possible to study larger lattice spacings, as found in certain minerals, polymers and biomolecules. Using longer X-ray wavelengths in order to obtain scattering patterns of colloidal objects at larger scattering angles is not possible, because X-rays of wave length above 2 Å are subject to absorption in matter.

A major pratical concern with SAXS is the spatial proximity of the scattered waves to the unscattered forward beam (Roe, 2000), which makes fine collimation of the primary and forward beam mandatory. Collimation implies loss of photon flux and thus, in principle, limits the signal-to-noise ratio. Modern experimental set-ups such as rotating anode and high brilliance synchrotron X-ray radiation haves alleviated such concerns.

SAXS has long been successfully applied in the past to structural analyses of soft polymeric matter (Brumberger, 1995, Glatter and Kratky, 1982, Guinier *et al.*, 1955, Stribeck, 2006), polymeric-inorganic hybrid materials (Glatter and Kratky, 1982, Guinier *et al.*, 1955, Stribeck, 2006), hard inorganic metal (Glatter and Kratky, 1982, Guinier *et al.*, 1955, Winter *et al.*, 2006) and metal oxide materials such as catalysts. The application of SAXS to study materials structure under electrochemical conditions, however, is a more recent development (Haubold *et al.*, 1997, Haubold *et al.*, 1999, Lee *et al.*, 2005, Smith *et al.*, 2008, Stevens and Dahn, 2000, Strasser *et al.*, 2010, Toney *et al.*, 2008, Yu *et al.*, 2008). In particular, the application of SAXS to membrane and catalyst components of single fuel cells is a new and exciting development (Koh *et al.*, 2006, Leisch *et al.*, 2006, Smith *et al.*, 2008, Toney *et al.*, 2008, Yu *et al.*, 2008).

SAXS is ideally suited for the non-invasive *in situ* analysis of structural changes of fuel cell components such as polymer membranes, porous supports and supported nanoparticle ensembles. This chapter first discusses some basic concepts of scattering and in particular SAXS, and then addresses special methods such as *in situ* SAXS and anomalous SAXS. The chapter then reviews selected important reports where SAXS was applied to analyze the structure of fuel cell components, in particular supported monometallic and bimetallic nanoparticle electrocatalysts. The chapter concludes with an outlook on the important future role of SAXS as an analytical tool in fuel cell research.

4.2 Principles and methods of small angle X-ray scattering (SAXS)

In this section some basic principles of X-ray scattering pertaining to SAXS will be reviewed. In doing so, we limit ourselves to the important equations and relations of elastic coherent X-ray scattering and skip detailed derivations for clarity.

4.2.1 X-ray scattering and interference

Structural information about condensed matter by X-ray analysis relies on the elastic scattering of incoming planar X-ray waves at electrons, followed by the interference of the resulting circular waves emenating from each scattering center. If J_0 and J denote the incoming planar wave flux (amount of energy per square area) and the scattered outgoing spherical wave flux (energy per angle), respectively, the scattered intensity I, also referred to as differential scattering cross-section $\left(\dfrac{d\sigma}{d\Omega}\right)$ becomes (Roe, 2000).

$$I = \frac{J}{J_0} = \frac{|A|^2}{|A_0|^2}, \tag{4.1}$$

where A denotes the amplitude of the respective wave. I has dimensions of area as the term 'cross-section' implies. The intensity is a function of the scattering angle 2θ, yet is more conveniently expressed as function of the scattering vector q rather than the scattering angle, because it is independent of the used wavelength. The magnitude of the scattering vector is related to the X-ray wavelength and the scattering angle according to

$$|\mathbf{q}| = \frac{4\pi}{\lambda}\sin\theta. \tag{4.2}$$

4.2.2 Scattering of electrons

The differential scattering cross-section (scattering intensity) of a single electron can be written as

$$\left(\frac{d\sigma}{d\Omega}\right)_e = b_e^2 = r_e^2\frac{1 + \cos^2\theta}{2} \tag{4.3}$$

where b_e denotes the θ-dependent scattering length of a single electron and r_e is the radius of the electron (2.818×10^{-15}m). The scattering intensity of a single electron is obviously highest in the forward direction ($\theta = 0$). The total scattering cross-section of a single electron, integrated over all angles, then becomes $\sigma_e = \dfrac{8}{3}\pi r_e^2$ and has a numerical value of 6.65×10^{-29} m^2.

The scattering and subsequent interference of X-rays at an ensemble of N electrons at fixed positions $\mathbf{r}_j (j = 1\dots N)$ creates a resulting X-ray wave with amplitude

$$A(\mathbf{q}) = A_0\, b_e \sum_{j=1}^{N} e^{-i\mathbf{qr}_j} \tag{4.4}$$

where A_0 is again the amplitude of the incident radiation. The scattered intensity of such an electronic ensemble is according to Eq. 4.5 (Roe, 2000)

$$I(\mathbf{q}) = \frac{J(\mathbf{q})}{J_0} = \frac{|A(\mathbf{q})|^2}{|A_0|^2} = b_e^2 \left| \sum_{j=1}^{N} e^{-i\,qr_j} \right|^2 \qquad [4.5]$$

For convenience, the intensity is often expressed in electron units b_e^2, which makes the intensity unitless and not dependent on any other experimental factor other than structure (Roe, 2000).

4.2.3   Scattering of atoms and particles

Within a given atom, the location of the electrons are typically expressed in terms of an electron density distribution, $n(\mathbf{r})$, rather than in terms of fixed postions. The vector $\mathbf{r}$ of spatial coordinates is here measured from the center of the atom. Accordingly, the amplitude of an outgoing X-ray wave scattered at an atom can be written as

$$A(\mathbf{q}) = A_0\, b_e \int_V n(\mathbf{r}) e^{-i q \mathbf{r}}\, d\mathbf{r} \qquad [4.6]$$

Where the integration is performed over the volume of the atom. The amplitude normalized to units of $A_0 b_e$ is referred to as f and can be written as

$$f(\mathbf{q}) = \int_V n(\mathbf{r}) e^{-i q \mathbf{r}}\, d\mathbf{r} \qquad [4.7]$$

We call this amplitude the atomic scattering factor. Formally, $f(\mathbf{q})$ in Eq. 4.7 is unitless; sometime units are given as (electrons/atom). If the electronic density distribution is spherically symmetrical, $f(\mathbf{q})$ is function of the scattering vector only. In the forward direction ($\mathbf{q} = 0$) all X-ray waves scattered from the atomic electron cloud of the atom are essentially in phase and hence add up, making $f(0)$ equal the atomic number Z. At finite angles, the scattered spherical waves from individual electrons start showing phase differences and the resulting intensity drops. $f(\mathbf{q})$ is hence a reflection of the shape of the electron density distribution within a given atom.

Let us now assume an ensemble of N identical atoms with Z electrons each (Roe, 2000). Figure 4.1 shows that, in a first step, we can express the position vector $\mathbf{r}_j$ of each electron as the sum of the position of its atomic center $\mathbf{r}_k (k = 1 \ldots N)$ and its position within the atom $\mathbf{r}_{k,m} (m = 1 \ldots Z)$. If the structure is then represented by an electron density distribution rather than fixed electron positions, the amplitude of an X-ray wave scattered from a particle can be formulated as (Roe, 2000).

$$A(\mathbf{q}) = A_0 b_e\, f(\mathbf{q}) = \int_V n_{atom}(\mathbf{r}) e^{-i q \mathbf{r}}\, d\mathbf{r} \qquad [4.8]$$

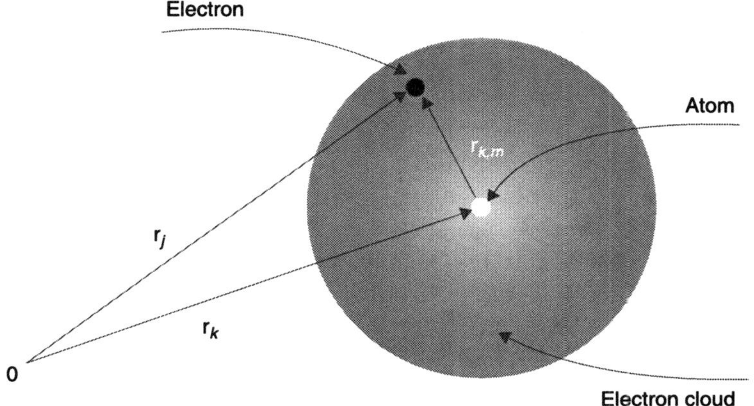

4.1 Vectors r_j and r_k denote the positions of an electron and the atomic center measured from the origin O, and $r_{k,m}$ denotes the position of the electron measured from the atomic center.

where $n_{atom}(\mathbf{r})$ is the density distribution of atomic centers. Equation 4.8 holds for one type of atoms. If there are multiple types of atoms, the amplitude is the sum of expressions like Eq. 4.8 for each type of atom.

It is common to consider a normalized amplitude $A_{norm}(\mathbf{q})$ in units of $A_0 b_e$ (electron units) and to introduce the scattering length density distribution $\rho(\mathbf{r})$ as

$$\rho(\mathbf{r}) = \Sigma_i f_i n_i(\mathbf{r}) \qquad [4.9]$$

where the subscript i refers to the various atomic species present. In Eq. 4.9, $\rho(\mathbf{r})$ has units of (m^{-3}); however, one should keep in mind that its name derives from the ratio of the effective scattering length of all atomic species present (units of m) and volume (m^3), such that units of (m^{-2}) result prior to normalizing with respect to the scattering length of a single electron b_e (units of m).

Using the normalized amplitude, the normalized scattered X-ray intensity reads

$$I_{norm}(\mathbf{q}) = \left| A_{norm}(\mathbf{q}) \right|^2 = \left| \int_V \rho(\mathbf{r}) e^{-i\mathbf{q}\mathbf{r}} d\mathbf{r} \right|^2 \qquad [4.10]$$

We will drop the subscript *norm* henceforth. The intensity has thus formally units of scattering intensity of the single electron. Oftentimes in literature, the observed scattered intensity is normalized also to the total sample volume such that the reported absolute intensity units are electron units (e.u.) divided by volume (Haubold *et al.*, 2003).

Equation 4.10 is essentially a complete description of the resulting scattering patterns of electromagnetic waves, in particular of X-rays. As a next step, we will transform Eq. 4.10 and recast it in terms of the so-called autocorrelation function of $\rho(\mathbf{r})$, $\Gamma_\rho(\mathbf{r}) = \int \rho(\mathbf{u}) \rho(\mathbf{u} + \mathbf{r}) d\mathbf{u}$, which is related to the average of the scattering length density distributions, $\rho(\mathbf{u})$ and $\rho(\mathbf{u} + \mathbf{r})$, at two points with distance $\mathbf{r}$, $\mathbf{u}$ and $\mathbf{u} + \mathbf{r}$, within the sample:

$$I(\mathbf{q}) = \int \Gamma_\rho(\mathbf{r}) e^{-i\mathbf{q}\mathbf{r}} d\mathbf{r} \qquad [4.11]$$

$\Gamma_\rho(\mathbf{r})$ specifies how the densities, $\rho(\mathbf{u})$ and $\rho(\mathbf{u} + \mathbf{r})$, separated by distance $\mathbf{r}$ are correlated on average (Roe, 2000). If $\rho(\mathbf{r})$ shows periodic maxima in space due to the presence of particles separated by $\mathbf{r}'$, for instance, $\Gamma_\rho(\mathbf{r})$ will show smeared-out centrosymmetric maxima at multiples of $\mathbf{r}'$, as well.

Following Eq. 4.11, the observed X-ray scattering intensity is the Fourier transform of the autocorrelation function of the scattering length density distribution carrying the structural information. More importantly, it can be shown that the intensity of the scattered X-ray wave may also be expressed as the autocorrelation function, $\Gamma_\eta(\mathbf{r})$, of the deviation of $\rho(\mathbf{r})$ from its mean, $\eta(\mathbf{r}) = \rho(\mathbf{r}) - \langle \rho \rangle$, within the sample according to (Roe, 2000)

$$I(\mathbf{q}) = \int \Gamma_\eta(\mathbf{r}) e^{-i\mathbf{q}\mathbf{r}} d\mathbf{r} \qquad [4.12]$$

This result is important, because it states that what is important in X-ray scattering is not the absolute value of the scattering length density distribution, but rather the fluctutations within the illuminated sample. If the intention is to increase the scattering intensity from the sample, the *contrast* in the scattering length density between the atoms or molecules or regions with the sample has to be increased (Roe, 2000). The concept of contrast is illustrated (Schnablegger and Singh, 2006) in Fig. 4.2a where the structural features of a face are shown against a white (left) and a black (right) background. Obviously, the black structure of interest against the black background does not offer any fluctuation in the scattering length density (here color) and hence provides zero contrast and consequently no structural information. Figure 4.2b presents a transmission electron microscopic (TEM) image of a dispersed carbon-supported Pt nanoparticle fuel cell catalyst material. The Pt particles appear clearly darker than the surrounding carbon matrix due to their larger electron scattering length density distribution. This electron density inhomogeneity of the catalyst is the basis for a successful SAXS analysis of its structural characteristics.

While the scattered intensity $I(\mathbf{q})$ can be computed from a given electronic scattering length density distribution, the reverse is not true. It is only the autocorrelation function of the electron density fluctuations, $\Gamma_\eta(\mathbf{r})$, which is accessible by an inverse Fourier transform of the scattered intensity.

This is why in X-ray scattering and, in particular, in SAXS, structural insight is sought by direct interpretation and comparison of experimental and model-based intensity patterns $I(\mathbf{q})$, while using additional information on the structure obtained from other experimental methods such as TEM. Most models used for structural SAXS analysis fall into one of three model classes, the most important for our purposes of fuel cells being (Roe, 2000):

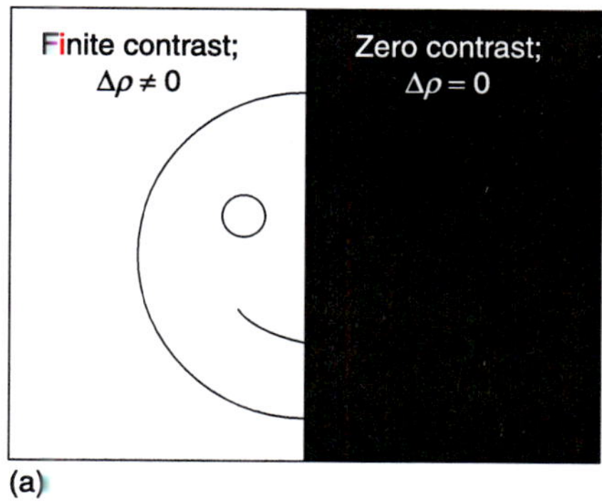

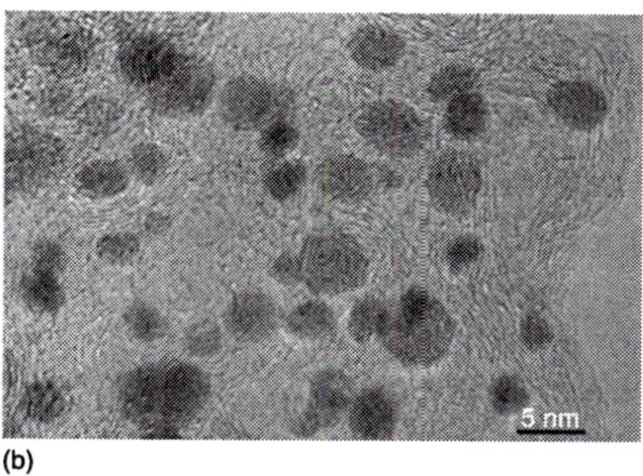

4.2 (a) Illustration of the concept of contrast. (Source: Schnablegger H and Singh Y, 2006) (b) Transmission electron micrograph (TEM) of a carbon-supported dispersed Pt nanoparticle fuel cell electrocatalyst. The electron density contrast between carbon support and Pt particles is directly observable.

- The dilute particle system model (catalytic nanoparticles dispersed on a support).
- The non-particular two phase system model.
- The soluble blend system model (hydrated fuel cell membrane).

4.2.4 Scattering of dilute, spherical, supported nanoparticles

The description of the scattering of X-rays from a sample of supported dilute nanoparticles, as found, for instance, in fuel cell electrodes, follows the dilute particulate system model. It is assumed that the particles are dispersed within a homogeneous supporting matrix devoid of any structure of its own. The important assumption is that the position of the particles are completely uncorrelated, such that the overall scattered intensity is the sum of the intensities of the individual particles.

The derivation of an equation describing the intensity $I_{sphere}(q)$ of an isotropic ensemble of supported spherical nanoparticles is particularly simple. Dispersed particles of radius R and of constant scattering length density distribution ρ are characterized by (Rayleigh, 1910, Roe, 2000)

$$I_{sphere}(q) = \rho^2 \left(\frac{4}{3}\pi R^3 \right)^2 \left[\frac{3(\sin qR - qR\cos qR)}{(qR)^3} \right]^2 \qquad [4.13]$$

The third term in Eq. 4.13 is the square of the particle form factor $P(q, R)$ of a sphere. For a monodisperse particle distribution, $P(q, R)$ and $I_{sphere}(q)$ show nearly periodic variations (Brumberger, 1995, Glatter and Kratky, 1982, Guinier *et al.*, 1955, Roe, 2000, Schnablegger and Singh, 2006) in q. For polydisperse particle size distributions, the periodic variations of $P(q, R)$ smooth out until a single 'hump' remains. The q value of this hump is roughly related to the mean particle diameter D via $q \approx \dfrac{2\pi}{D}$ (Yu *et al.*, 2008).

In case of non-negligible scattering contributions of a non-uniform support material or matrix, in which the particles are embedded, such as often found in irregularly intermixed, yet phase segregated polymers, two–phase model can be invoked, and the scattering intensity of N identical dilute particles within the matrix of scattering length density ρ_0 can be modeled using

$$I_{sphere}(q) = N(\rho - \rho_0)^2 \left(\frac{4}{3}\pi R^3 \right)^2 \left[\frac{3(\sin qR - qR\cos qR)}{(qR)^3} \right]^2$$

$$N\Delta\rho^2 8\pi^2 \left[\frac{R^2(1+\cos 1qR)}{q^4} - \frac{2R\sin 2qR}{q^5} + \frac{1-\cos 2qR}{q^6} \right]$$

[4.14]

Here, the term $(\rho - \rho_0)^2$ represents the contrast between the two phases.

For large scattering vectors q, Eq. 4.14 simplifies to

$$I_{sphere}(q) = \frac{2\pi\Delta\rho^2 S}{q^4}$$

[4.15]

where S represents the total surface area of the boundaries of the two phases, particles and carbon. Equation 4.15 is referred to as Porod's law, and is valid for any two-phase system with spherical boundaries, hence both for porous carbon supports with spherical pores as well as to spherical supported particles.

Figure 4.3 shows a typical experimental non background-corrected log-log intensity profile of a carbon-supported Pt nanoparticle fuel cell catalyst. A linear region at small and large values of q represents influence of the background and Porod's law (Eq. 4.15), respectively. The hump in the center of the profile represents the influence of the form factor of a polydisperse distribution of the Pt particles.

The fact that only fluctuations and inhomogeneities, that is, local differences in scattering length density distributions, carry structural information and are relevant in SAXS implies that an ensemble of pores in a two-phase system with given contrast will produce an identical scattering pattern like particles in a matrix with the same contrast (two-phase model concept). This is illustrated in Fig. 4.4, where the experimental scattering profiles of two porous materials with polydisperse distributions of pores of distinctly different mean pore size are shown. The q value of the hump indicates the relative mean size of the two distributions and the Porod law is seen at larger q values.

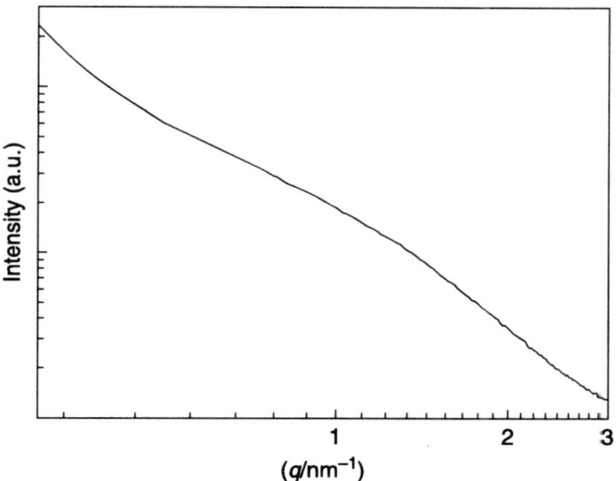

4.3 Experimental scattering profile (scattering intensity as function of scattering vector) of a carbon-supported Pt nanoparticle fuel cell catalyst material.

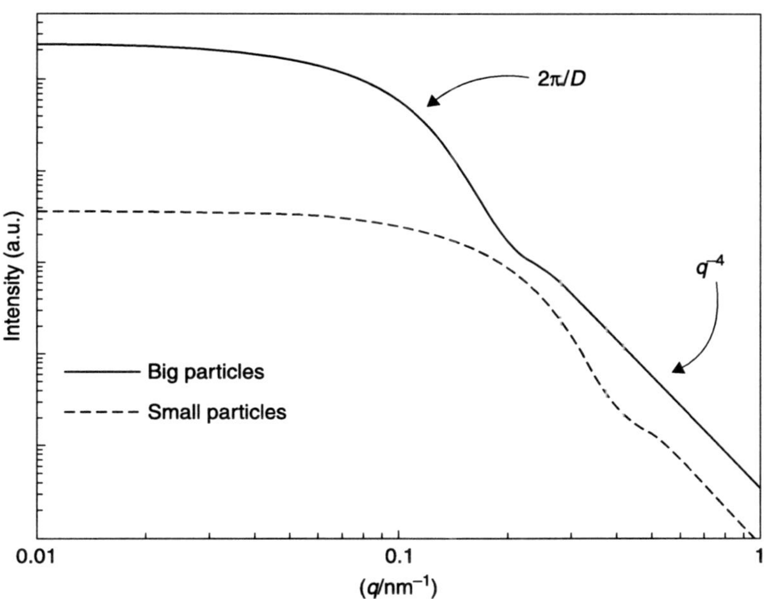

4.4 Scattering profiles of two polydisperse ensembles of small and large mean particle size.

4.2.5 Anomalous SAXS (ASAXS)

Small angle X-ray scattering does not inherently provide element-specific information. However, with the use of an anomalous scattering approach (Materlik *et al.*, 1994, Stuhrmann, 1985, Stuhrmann *et al.*, 1991, Winter *et al.*, 2006, Yu *et al.*, 2008), element specificity can be obtained. Consider a sample consisting of two different elements, for instance metal atoms 1 and metal atoms 2. Carrying out SAXS measurements at two wavelengths near an absorption edge of metal 1 is then equivalent to performing them on two identical systems having the same identical structure (volume fraction, interphase surface areas), but different electron densities for metal 1. Taking the difference of the two measurements now provides a scattering behavior of the metal 1 only, and its structural features can be determined independent of metal 2. So, more generally, in anomalous small angle X-ray scattering (ASAXS), the SAXS pattern is measured at a number (typically two or three) of X-ray energies near the X-ray absorption edge of the element of interest, e.g. Pt in case of a carbon-supported fuel cell electrocatalyst.

For Pt nanoparticles on a carbon support, for instance, we can write the intensity as (Toney *et al.*, 2008)

$$I(q) = \left| A_{support}(q) + A_{nanoparticle}(q) \right|^2 =$$
$$I_{support}(q) + I_{nanoparticle}(q) + 2\sqrt{I_{support}I_{nanoparticle}}\,\cos\omega \qquad [4.16]$$

Where A and I denote the (normalized) scattered wave amplitudes and (normalized) scattering intensities (scattering cross-sections) of the support and the nanoparticles. The last term in Eq. 4.16 is the interference between support and nanoparticles (ω is the phase factor); however, since the nanoparticles and the support are completely uncorrelated spatially, this term is so small that it can be ignored (phase factor averages to zero). Also, the Pt particles are much more electron-dense than the carbon support. Hence we can write

$$I(q) = I_{support}(q) + I_{nanoparticle}(q) =$$
$$\left| f_{support}(E) \right|^2 * S_{support}(q) + \left| f_{nanoparticle}(E) \right|^2 * S_{nanoparticle}(q) \qquad [4.17]$$

Here f denotes the atomic scattering factor (scattering strength) and S describes the reminder of the expression for the scattering cross-section of Eq. 4.8. The key to ASAXS is the energy dependence of the atomic scattering factor in Eq. 4.17. The atomic scattering factor of Pt, for instance, can be expressed as

$$f_{Pt}(E) = f_{Pt}^0 + f_{Pt}' + if_{Pt}'' \qquad [4.18]$$

where f_{Pt}^0 is the non-resonant (energy independent) scattering factor, f_{Pt}' is the (real) anomalous scattering factor, and f_{Pt}'' accounts for the absorption (imaginary part of the form factor). As Fig. 4.5 shows, the anomalous scattering factor of Pt

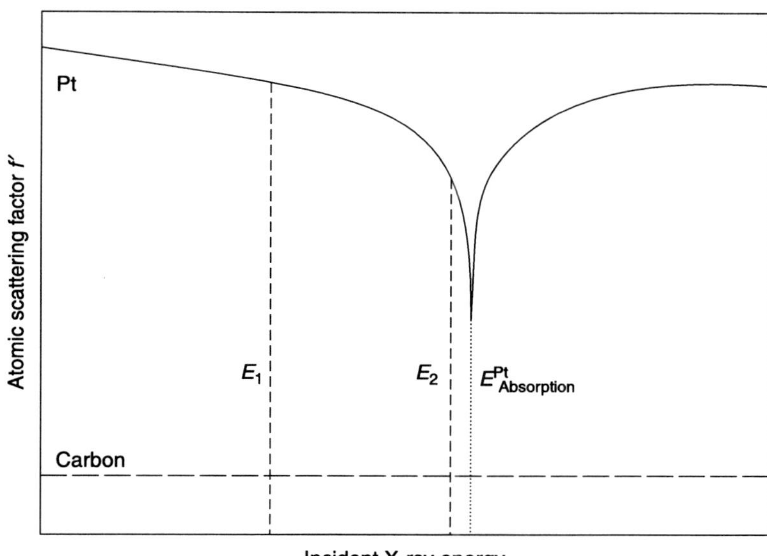

4.5 Change of anomalous atomic scattering factor f of Pt atoms near the energy of X-ray absorption of Pt (Pt absorption edge). SAXS measurements performed at E_1 and E_2 essentially differ in the scattering contribution of the Pt atoms.

drops by up to 20% as the absorption edge is approached (Haubold *et al.*, 1997). These results in a reduced scattering intensity at energy E_2 compared to incident energy E_1 in Fig. 4.5. The atomic scattering factor of carbon, however, is unchanged around the Pt absorption edge (Fig. 4.5). As a result of this and according to Eq. 4.17, the difference of the two intensities $I(E_1)-I(E_2)$ represents the scattering pattern of the Pt catalyst particles alone. Figure 4.6 illustrates the two measured ASAXS profiles of a carbon-supported electrocatalyst at the energies E_1 and E_2 and the difference profile (o) representing the contribution of the Pt nanoparticles alone.

4.2.6 SAXS data acquisition and analysis

A typical experimental point-collimated SAXS measurement set-up and data analysis work flow is illustrated in Fig. 4.7. The detailed individual steps of a SAXS model analysis may vary widely depending on the desired properties of the sample, the experimental conditions of the experiment, the nature of the sample (polymer blend or particulate system) under investigation and the detailed model employed.

The SAXS analysis workflow discussed below and represented in Fig. 4.8 is highly simplified and biased toward the analysis of supported metal fuel cell nanoparticles. Incident X-rays result in the scattering pattern on the two-dimensional

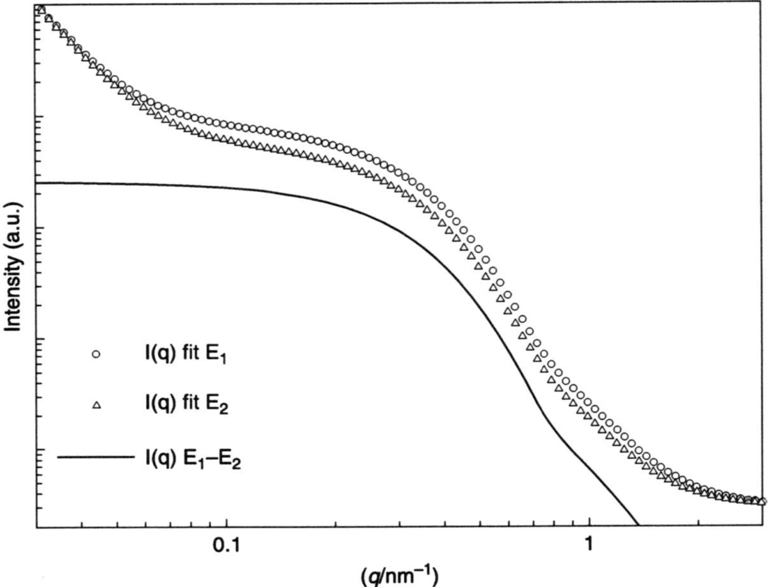

4.6 ASAXS intensity profiles at incident energies E_1 (o) and E_2 (Δ). The difference pattern represents the scattering of Pt particles only and can be fitted with a polydisperse particle size distribution model.

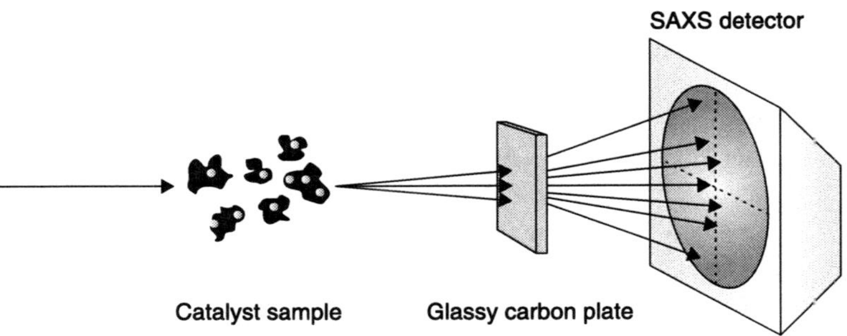

4.7 Transmission correction in laboratory-based SAXS systems without direct measurement of incident and outgoing primary beam. A glassy carbon (GC) plate is used and the integrated scattering intensity with and without GC plate is recorded and used to correct for transmission.

(2D) area detector, which is symmetric around the beam stop. The patterns are integrated to obtain one-dimensional raw I-q profiles and corrected for transmission, τ, and changes in the incident X-ray flux over the course of the entire experiment.

In case of synchrotron radiation, the X-ray intensity before and after the sample is usually measured with an ionization chamber, and one can directly read off the

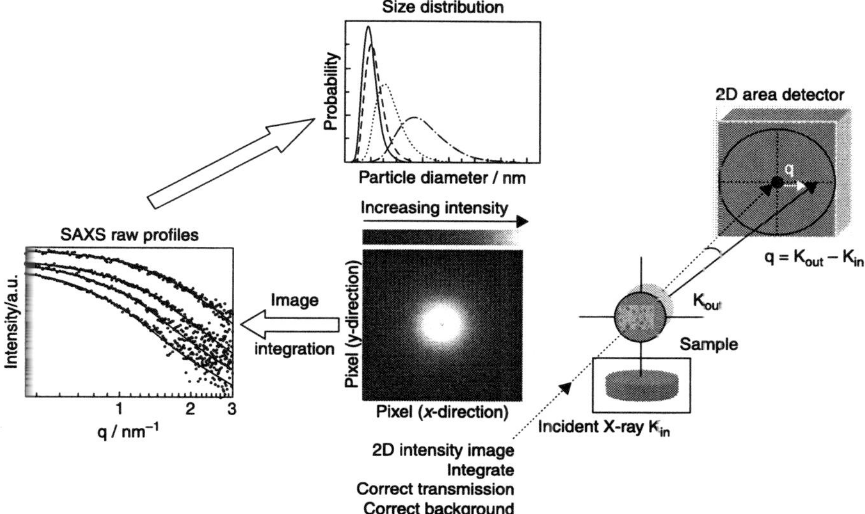

4.8 Point collimation SAXS measurement set-up and data analysis work flow. Measurement is performed in transmission geometry using an area detector (right). Raw scattering profiles are integrated to arrive at one-dimensional raw scattering (*I-q*) profiles (left); the raw profiles are corrected for transmission and background scattering. Finally, the profiles are fitted using structural models and particle size distributions are obtained.

incident X-ray intensity I_0 as well as the transmitted intensity I_1. The corrected intensity I_{corr} can then easily obtained by

$$I_{corr}(q) = \frac{I_{raw}(q)}{I_1 \tau} = \frac{I_{raw}(q)}{I_1 \frac{I_1}{I_0}} = \frac{I_{raw}(q)}{I_1}$$

[4.19]

The same correction is performed with the sample background profile, for instance, the pure carbon support without Pt nanoparticles. Thereafter, background-corrected scattering profiles of the Pt particles, I_{Pt} are obtained by subtraction:

$$I_{Pt}(q) = \frac{I_{Pt+Carbon}(q)}{I_1} - \frac{I_{Carbon}(q)}{I_1}$$

[4.20]

In case of synchrotron-based ASAXS analysis, the final element-specific scattering profile is obtained by subtracting profiles at two distinct energies as illustrated in Fig. 4.6. No additional background correction is necessary.

In the case of laboratory-based SAXS systems such as the Nanostar System (Bruker AXS) and others, there is neither an intensity measurement of the incident X-ray, nor of the transmitted primary beam, so that Eqs 4.19 and 4.20 cannot be applied. In this case, a trick is used in order to correct for varying transmissions of the sample and the background, as sketched in Fig. 4.8. In such lab-based SAXS systems, unlike synchrotrons, especially when a rotating anode is used, the intensity of the primary beam can be assumed to be essentially constant during the time scale of the measurements (several hours to days).

To obtain transmission-corrected laboratory-source SAXS data, the following version of Eq. 4.20 is used

$$I_{Pt}(q) = \frac{I_{Pt+carbon}(q)}{\tau_{Pt+carbon}} - \frac{I_{carbon}(q)}{\tau_{carbon}} \qquad [4.21]$$

To arrive at the transmission of Pt + carbon, $\tau_{Pt+carbon}$, and of carbon, τ_{carbon}, a glassy carbon (GC) plate is introduced in the beam path between sample and detector, and the integrated (q-independent) total scattering intensity (total counts for a given time) is recorded for each case. The transmission values can then be computed from the experimentally measured intensity values according to

$$\tau_{Pt+carbon} = \frac{I_{Pt+carbon+GC} - \tau_{GC}I_{Pt+carbon}}{I_{GC} - \tau_{GC}I_{air}} \qquad [4.22]$$

$$\tau_{carbon} = \frac{I_{carbon+GC} - \tau_{GC}I_{carbon}}{I_{GC} - \tau_{GC}I_{air}} \qquad [4.23]$$

Here, the subscripts specify the components that are in the X-ray beam path.

Substitution of Eq. 4.22 and Eq.4.23 into Eq. 4.21 results in an expression for the carbon-corrected scattering intensity of the Pt particles, I_{Pt}, multiplied with a constant c. The constant c comprises the scattering intensity value of the GC plate corrected for scattering in air.

Once the corrected scattering profile of the component of interest, for instance Pt nanoparticles, has been obtained, the profiles are fitted using structural models, and structural characteristics about the particulate material, such as radius of gyration (Aieta *et al.*, 2009, Brumberger, 1995, Glatter and Kratky, 1982, Guinier *et al.*, 1955, Schnablegger and Singh, 2006), size distributions (Goerigk *et al.*, 2003, Toney *et al.*, 2008, Walter *et al.*, 1985, Yu *et al.*, 2008), mean radius, dispersion, shapes (Damaschun *et al.*, 1969), volume fraction, spatial correlations (Ballauff and Jusufi, 2006, Jusufi and Ballauff, 2006, Stuhrmann, 1970), morphology (Aieta *et al.*, 2009), surface areas and others are obtained. We note that the detailed course of analysis depends strongly on the nature of the sample and the information needed. Here we focus on a particulate model of supported metal nanoparticles (Damaschun and Purschel, 1971, Moller *et al.*, 1995, Porod, 1951, Walter *et al.*, 1985). For systems where other models apply, such as polymer mixtures, the detailed course of data analysis may vary from the one indicated below. For instance, in the case of ASAXS data, information about the resonant part of the scattered intensity may be of primary interest (Ballauff and Jusufi, 2006, Jusufi and Ballauff, 2006, Lu *et al.*, 2006, Mei *et al.*, 2007).

Here, we use a (uncorrelated) particle model to illustrate how to fit experimental SAXS data such as that shown for pure Pt particles in Fig. 4.6. The expression for the intensity (cross-section) per sample volume of a polydisperse Pt particle model can be written as (Haubold *et al.*, 1997)

$$I(q) = n_{Pt}^2 f_{Pt}^2 \frac{N}{V_{sample}} \int_0^\infty D(R) V^2 P(R,q)^2 \, dR \qquad [4.24]$$

Where $I(q)$, V, f, R, and q have their usual meanings and n_{Pt}, N, V_{sample}, and P denote the Pt atmic volume density, the number of particles, the sample volume and the particle form factor, respectively. Unless there is strong evidence from other analytical techniques, such as TEM, that the particles are uniformly non-spherical (rods, disks etc.), a spherical particle form factor is employed (see Eq. 4.13). $D(R)$ is a particle size distribution function expressing the probability density of particles with a radius between R and $R + dR$. The shape of this function is not known *a priori* and may depend on the physical system under consideration.

Most reports in literature on particulate models, however, have assumed an asymmetric log-normal distribution function. This is supported by theoretical-computational studies (Ferreira *et al.*, 2005, Shao-Horn *et al.*, 2007) that showed that the microscopically accessible size distributions of metal nanoparticle over the course of electrochemical reactions are best described by log-normal type distributions with a sharp drop on the small diameter side and a longer tail on the larger diameter end. Furthermore, the rapid drop on the small diameter side of the curve is physically reasonable especially for particles with mean diameters of a few nm, because metal particles below 1nm in diameter touch the realm of individual atoms and as such would be highly unstable, experiencing rapid agglomeration toward larger sizes.

A frequently used expression for the log-normal particle size distribution function reads (Goerigk *et al.*, 2003, Haubold *et al.*, 1997, Haubold *et al.*, 1999, Haubold *et al.*, 2003, Haubold and Wang, 1995, Haubold *et al.*, 1996, Ingham *et al.*, 2009, Stevens and Dahn, 2000, Strasser *et al.*, 2010, Toney *et al.*, 2008, Vad *et al.*, 2002, Wen *et al.*, 2005, Yu *et al.*, 2008)

$$D(R) = \frac{1}{\sqrt{2\pi}\sigma R} e^{-\frac{\left[\ln\frac{R}{R_0}\right]^2}{2\sigma^2}} \qquad [4.25]$$

This function (Ingham *et al.*, 2009) has a maximum at $R_0 \exp(-\sigma^2)$, a mean of $R_0 \exp(+\sigma^2/2)$ and a variance of $R_0^2[\exp(2\sigma^2)-\exp(\sigma^2)]$. Expression of higher moments for the log-normal distribution can be found in a paper by Moller *et al.* (1995).

Figure 4.9 shows detailed pictures of the *in situ* SAXS cell used by Yu *et al.* (2008) to probe particle growth and coarsening and the formation of a nanostructured core-shell catalyst particles under electrochemical (half-cell) conditions with a liquid acid electrolyte. The catalyst film was deposited on a

4.9 Electrochemical SAXS cell for *in situ* and *ex situ* structural analysis of electrocatalysts in half-cell arrangements.

narrow strip of conductive carbon paper, which was contacted at the top with Cu tape interfacing to the potentiostat clamps. A sealed AgCl reference electrode and a Pt counter electrode served to complete the three-electrode set up.

4.3   Application of SAXS to fuel cell component characterization

This section will review selected SAXS-based structural studies of fuel cell components, in particular supported nanoparticle catalysts and polymer electrolytes such as Nafion. We will first focus on *in situ* and *ex situ* investigations of the growth dynamics of highly dispersed metal nanoparticle electrocatalysts in electrochemical environments, then address selected work on the structure of fuel cell-related polymers and finally close with a review of more general SAXS studies of metal nanoparticle catalysts under non-electrochemical and non-*in situ* conditions.

4.3.1   Growth and stability of nanoparticle fuel cell catalysts in electrochemical environments

Structural and compositional stability of a highly dispersed metal particle catalyst under the operating conditions of an industrial chemical process or a commercial device are a critical requirements for its practical later deployment. In the case of

PEMFCs, structural stability of the catalysts at anode and cathode translates into durability of both the pore structure of the porous carbon supports (Yu *et al.*, 2006) and the particle size distribution and dispersion of the catalytically active metal nanoparticles (He *et al.*, 2005, Sompalli *et al.*, 2007, Vielstich *et al.*, 2009). For alloy catalysts, structural stability also extends to the local composition and arrangements of dissimilar metal atoms at the surface and inside individual alloy nanoparticles (Alloyeau *et al.*, 2010). Especially at the cathode of PEMFCs, electrode potentials reach values where surface metal atoms of nanoparticles form surface oxides or soluble metal cations and subsequently leach into the electrolyte, resulting in structural and compositional changes of the nanoparticles during fuel cell operation (Borup *et al.*, 2006, Guilminot *et al.*, 2007, Iojoiu *et al.*, 2007, Virkar and Zhou, 2007). Oxidation and leaching of metal surface atoms typically result in an undesired net loss of catalytically active metal atoms. In accord with the Gibbs Thomson relation, redeposition of leached metal atoms can also result in an undesired change in the particle size distribution during fuel cell operation, typically favoring increasing mean particle sizes (particle coarsening) (Thiel *et al.*, 2009). Thus, the accurate measurement (Hasche *et al.*, 2010) and a thorough atomic-level mechanistic understanding of particle coarsening and growth processes (Finney and Finke, 2008, Shao-Horn *et al.*, 2007) is therefore a critical task in current fuel cell research in order to develop strategies to successfully mitigate catalyst degradation. Virtually all previous studies on fuel cell catalyst degradation involved macroscopic variables such as the electrochemical active surface area (ECSA) or the catalytic Pt-mass-based or the real Pt surface area-based activity combined with TEM-based analyses (Ferreira *et al.*, 2005, Ferreira and Shao-Horn, 2007, Shao-Horn *et al.*, 2007). Experimental *ex situ* TEM data of particle fuel cell catalysts before and after electrochemical treatment or testing were recently combined with kinetic rate modeling (Holby *et al.*, 2009, Rinaldo *et al.*, 2010, Shao-Horn *et al.*, 2007). The combined experimental and computational studies confirmed that Ostwald ripening (Granqvist and Buhrman, 1976, Levitan and Domany, 1998a, Levitan and Domany, 1998b, Rosenfeld *et al.*, 1998, Voorhees, 1985) could, to a large part, explain observed growth behaviors and suggested a critical mean particles size above which particle growth became very slow (Holby *et al.*, 2009). However, as detailed earlier, TEM analyses suffer from limited accuracy and only provide a pathological picture of the catalyst. SAXS-based analyses are an attractive alternative.

One of the first, perhaps the very first, report on the use of synchrotron-based ASAXS to study the behavior of platinum nanoparticle electrocatalysts under *in situ* electrochemical conditions dates back to 1996, and was authored by a group of scientists at the Forschungszentrum Juelich around H.-G. Haubold (Haubold *et al.*, 1996). In this landmark, although hardly cited publication, Haubold and his group pioneered the use of ASAXS to monitor changes in particle size distribution and mean particle radii of highly dispersed carbon-supported Pt nanoparticle catalysts during voltammetric cycling in an acidic electrolyte. Important findings included the increase of particle size upon cycling to more anodic electrode potentials. In combination with X-ray absorption results, the authors interpreted the observed size increases with the formation of a Pt oxide shell of about 1 nm thickness around a Pt core. At the end of this paper, the authors envisioned ASAXS as a future powerful element-specific analytical technique to study catalyst degradation, such as loss of catalysts, sintering, and compositional changes in alloys. In doing so, the authors laid the foundation for most fuel cell catalyst-related SAXS studies to the present day. Already a year earlier, in 1995, Haubold and his group demonstrated *ex situ* that the Pt atomic scattering amplitude of Pt nanoparticle catalysts can be varied up to 20% near the Pt absorption edge (Haubold and Wang, 1995). The same group extended their *in situ* ASAXS studies of carbon-supported Pt nanoparticle fuel cell catalysts to a number of catalysts with different Pt metal loadings, ranging from 5 wt% to 80 wt% Pt (Haubold *et al.*, 1997). For each catalyst, they determined the mean particle size and the volume of oxidized and reduced Pt particles. The authors close with an estimate of the oxide shell thickness similar to their previous study. In 1999, the Haubold group reported on combined ASAXS and XAS investigations of a 10 wt% Pt/C catalyst in the presence of iodide adsorption and methanol oxidation (Haubold *et al.*, 1999). The authors proposed SAXS/XAS analysis to be useful for obtaining information about adsorption of anions or reactive intermediates. Based on their presented data, the authors gave an estimate of the number of reaction steps of a surface reaction; a claim, which remained largely unsubstantiated from their presented data. Also in 1999, Benedetti *et al.* (Benedetti *et al.*, 1999) used *ex situ* ASAXS to investigate the particle size distributions of Au, Pd, and Au-Pd bimetallic alloy nanoparticles supported on an active carbon support. The report is an extension on an earlier study on Au nanoparticles (Benedetti *et al.*, 1997) and served to check the ASAXS findings including a large set of samples and comparing them to X-ray diffraction data. The study does not provide detailed insight in the particle size distribution, but rather stops short at providing mean diameters of the supported particles. No electrochemical environment or *in situ* size changes are considered.

In 2003, the Haubold group presented an interesting *in situ* SAXS and ASAXS study of the formation of Pt nanoparticles from metal organic precursors (Haubold *et al.*, 2003). Conventional SAXS revealed that the number of Pt particles with comparable size increased steadily during the synthesis process. To obtain more detailed insight into the absolute amount of Pt in the formed particles, ASAXS data were collected and analyzed. An exponential time dependence of the Pt particle mass was found. In another experiment, Pt particle networks linked by metal organic compounds were investigated, which resulted a mean interparticle distance. Also in 2003, Haubold's group in Juelich, together with his collaborators from the Max Planck Institute for Coal Research, the University of Bonn, the Center for Advanced Microstructure and the Technical University Clausthal published a micro review article emphasizing the powerful combination of SAXS with TEM, XAS, and other spectroscopic techniques to obtain atomic level insight in the formation and behavior of Pt nanoparticle ensembles under reactive conditions (Wen *et al.*, 2005).

In 2000, the group around Jeff Dahn reported on the use of conventional SAXS to monitor the electrochemical insertion of sodium into carbonaceous materials (Stevens and Dahn, 2000). Based on their results, the authors resolved an earlier controversy about the mechanism of such insertion reactions. Metal insertion at a chemical potential near that of the metals occurred in the pores of the carbons (pore filling mechanism), while at lower chemical potentials, the metal was inserted between graphene layers (intercalation mechanism). Later, the same group reported a conventional *ex situ* SAXS experiment on various carbon-supported Pt electrocatalysts with Pt weight loadings between 0–80 wt% and compared these results with WAXS and TEM studies of the same catalysts (Stevens *et al.*, 2003). SAXS proved to be more effective for determining reliable mean particle sizes at small particle diameters, where WAXS peaks tend to be become very broad.

The effect of pressure during hot pressing of a membrane electrode assembly (MEA) was investigated by Tsao and Chen (2004). The authors found evidence for collodial aggregation of the initial Pt particles at geometric Pt loadings of about 3 mg (Pt)/cm^2 leading to a 40% loss of surface-to-volume ratio.

In 2008, our group at the University of Houston and at the Technical University Berlin reported an ASAXS study of a nanostructured Pt-Cu bimetallic core-shell fuel cell cathode electrocatalysts (Yu *et al.*, 2008). Virtually all previous ASAXS and SAXS studies on fuel cell catalysts were carried out using pure Pt particles at the Pt absorption edge with the goal to elucidate particle size distribution changes of the Pt particle. In this paper, we reported, to our knowledge for the first time, an element-specific ASAXS investigation at both the Pt and the Cu absorption edge. The goal of this study was to test our hypothesis whether element-specific ASAXS would be able to probe the core-shell structure of dealloyed Pt-Cu bimetallic fuel cell electrocatalysts. If so, ASAXS would be an ideal complementary analytical tool to characterize core shell particle catalysts with the prospect of providing *in situ* information about shape and size changes with time.

Dealloyed Pt bimetallic nanoparticles for the 4-electron electroreduction of molecular oxygen at fuel cell cathodes represent a recently developed class of highly active polymer electrolyte membrane fuel cell cathode catalysts (Koh and Strasser, 2007, Srivastava *et al.*, 2007, Strasser, 2009a, 2009b, Strasser *et al.*, 2010). Starting from uniformly alloyed non-noble metal rich precursor particles, such as Pt$_{25}$Cu$_{75}$, Cu atoms are preferentially leached out of the nanoparticles by electrochemical potential cycling until a stable cyclic voltammogram and hence surface is established. During this 'dealloying' step, large quantitites of Cu are removed from the alloy while the particle surface is enriching in Pt (Strasser *et al.*, 2008). Microscopic and spectroscopic studies clearly suggested the formation of a Pt-enriched particle shell surrounding a Pt-Cu alloy core (Strasser *et al.*, 2010) (core-shell catalyst) as shown in Fig. 4.10.

Carbon-supported Pt$_{25}$Cu$_{75}$ alloy fuel cell catalysts were prepared at three different annealing temperatures, and the particle size distributions of the as-prepared catalysts were subsequently characterized by *ex situ* SAXS at the Pt

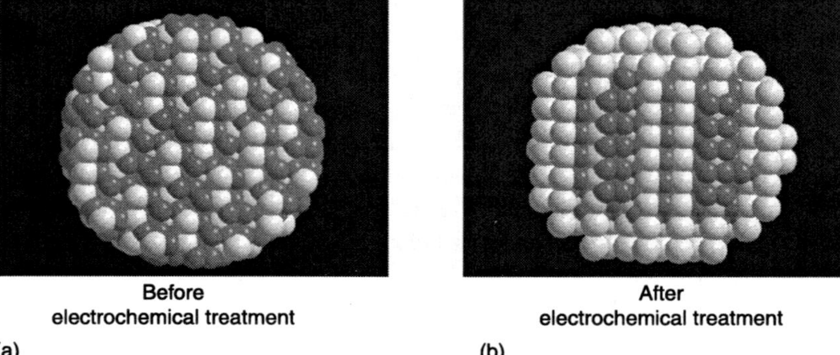

4.10 (a) Schematics of a uniformly alloyed Pt–Cu alloy nanoparticle giving rise to overlapping scattering profiles for the Cu and the Pt edge. (b) Dealloyed Pt–Cu with a Pt enriched shell surrounding a Pt–Cu core giving rise to a Cu-specific scattering profile shifted towards smaller mean diameter. (Source Yu *et al.* (2008).)

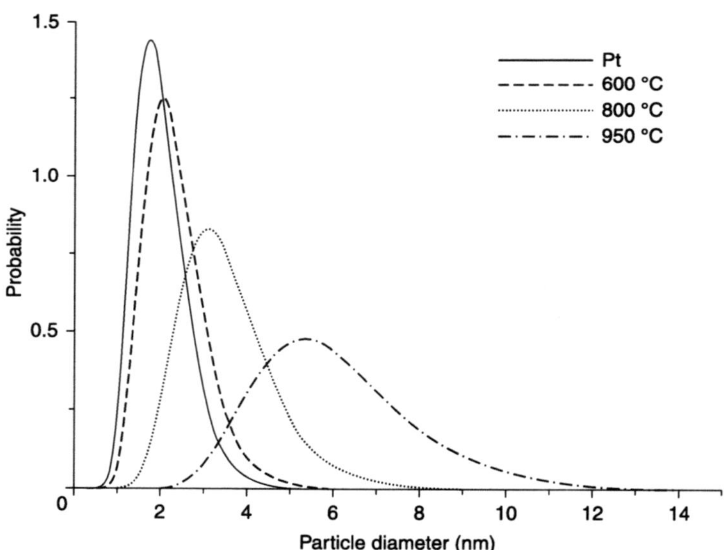

4.11 Particle size distributions (probability density) P(R) for three carbon-supported Pt-Cu nanoparticle alloy electrocatalyst precursors compared to the pure Pt nanoparticle catalyst used as a precursor during preparation of the three alloys. The temperatures indicate the annealing conditions of each catalyst. (Source Yu *et al.* (2008).)

edge. The results in Fig. 4.11 indicate how the mean particle size increases with increasing annealing temperature, consistent with expectations of increasing particle coarsening by coalescence and Ostwald ripening at elevated temperatures.

Figure 4.12 sketches the idea behind using ASAXS to probe core shell structures. A Pt-enriched near-surface region in the dealloyed core shell particle would give rise to a larger particle diameter when probed at the Pt absorption edge as compared to when probed at the Cu absorption edge, because the contiguous scattering domain for Pt extends across the entire particle, whereas that of Cu is confined to the particle core (Fig. 4.12a). When fitted with a log-normal size distribution function (Fig. 4.12b), the mean diameter of the Cu domain turned out to be significantly smaller than that of Pt, confirming a core shell structure and suggesting a Pt shell thickness of about 0.4 nm.

Also in 2008, the group of Deborah Myers at Argonne National Lab presented an *in situ* SAXS study of Pt particle growth during potential cycling (Smith *et al.*,

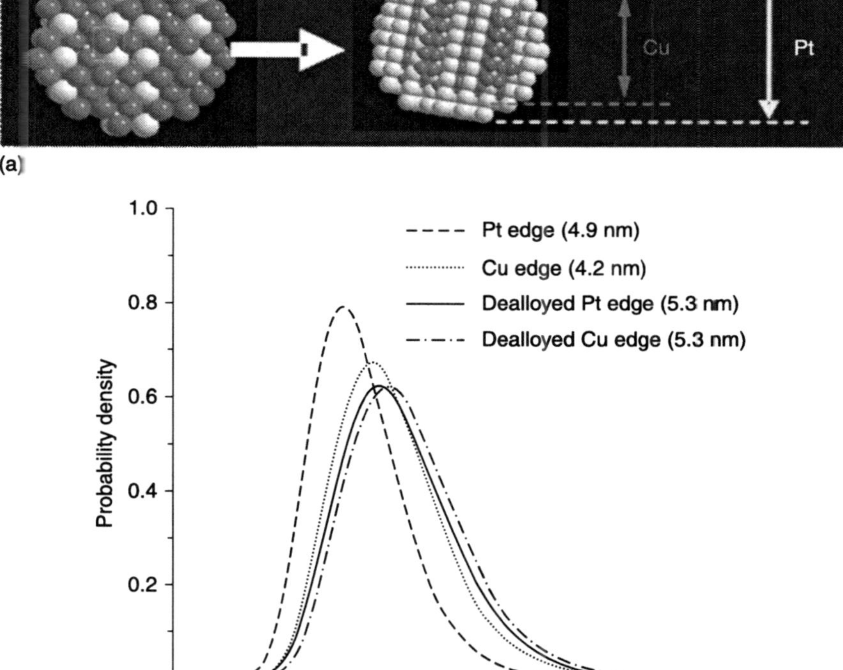

4.12 (a) A core-shell is characterized by a larger Pt-specific particle diameter compared to the Cu-specific particle diameter (Source Yu *et al.*, 2008). (b) Element specific log-normal particle size distributions of Pt and Cu before and after dealloying of the bimetallic precursor materials. The clear shift of the Cu edge to smaller diameters indicates the formation of core shell particles in the dealloyed catalyst.

2008). Here, growth trajectories of Pt particle ensembles were traced out over the course of a 16-hour cycling. A 20 wt% Pt on carbon catalyst showed a slowdown in particle growth beyond 3 nm, while a 40 wt% Pt/C fuel cell catalyst showed a steady growth throughout. In contrast to earlier results by Haubold, Smith *et al.* found a reversible particle size increase of only about 0.1 nm when the potential was swept from 0.4–1.4 V. The authors also observed a slight delay between the maximum of the potential cycles and the peak particle size, suggesting formation of oxide on the initial cathodic potential scan or a continued Pt deposition during after scan reversal at 1.4 V.

The research group around Fiechter recently presented an ASAXS-derived structural model of a selenium-modified ruthenium nanoparticle fuel cell cathode electrocatalyst (Haas *et al.*, 2010) (Fig. 4.13); the model was further supported by TEM, XRD, and extended X-ray adsorption fine-structure (EXAFS) analysis. The authors show that the surface of Se-modified Ru nanoparticles was not completely covered by Se, and instead forms patches on the Ru surface while the rest of the surface is covered by oxygen. This new model was in contrast to earlier core-shell models of this Pt-free fuel cell cathode catalyst for the electroreduction of oxygen.

Finally, Alexeyeva *et al.* recently reported on a combined electrochemical RDE, SAXS, grazing incident XRD study of a carbon nanotube-supported Au nanoparticle electrocatalyst for the oxygen reduction reaction (Alexeyeva *et al.*, 2010). Here, SAXS was used as an *ex situ* analytical tool to verify structural characteristics of the Au nanoparticles.

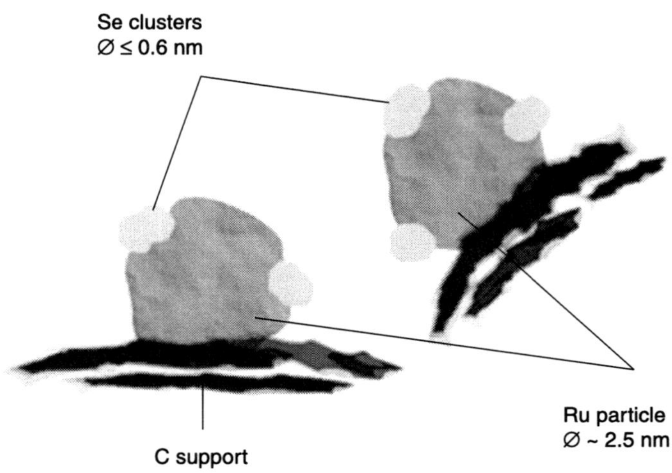

4.13 A structural model of selenium-modified ruthenium nanoparticles derived from anomalous small-angle X-ray scattering (ASAXS)8. (Source: Haas *et al.* (2010).)

4.3.2 Structural studies of fuel cell polymer electrolytes and fuel cell polymer membranes

Nafion is a polymer electrolyte membrane material giving polymer electrolyte membrane fuel cells (PEMFC)s, their name. It is undoubtly the most successful and widely-used polymeric proton conductor for low-temperature fuel cells (Mauritz and Moore, 2004). Developed at the DuPont Company, Nafion is a copolymer consisting of perfluorinated vinylether units and tetrafluoroethylene backbones. Depending on the equivalent weight of the polyelectrolyte, the ether side-chains carry sulfonic acid groups, which enable proton conduction. While early Nafion development efforts were geared toward a more efficient and selective membrane for the chlorine alkaline process, its applicability as a solid electrolyte for PEMFCs is driving current research to improve the chemical and mechanical characteristics of Nafion. It became clear that improving the electrochemical performance of a membrane electrode assembly (MEA) requires a detailed understanding of the chemical and physical processes inside the humidified Nafion membrane. In particular, the molecular mechanisms of proton conduction, the water management inside the membrane, its affinity to potential poisoning species and non-aqueous solvents, and its behavior at elevated temperature or low relative humidity, its mechanical and chemical resistance, and the electroosmotic water transport are important issues that researchers hoped to clarify and understand by studying the molecular polymeric structure of Nafion. Over the years, an incredible wealth of information was gathered about the morphology and molecular structure of Nafion in solubilized and cast form. Scattering, in particular neutron and X-ray scattering, has thereby played a pivotal role. As with all reciprocal space methods, scattering techniques always required the assumption of some sort of simple structural model, the validity of which had to be verified by alternative methods. This has always left room for ambiguities and opened the door for yet another structural model consistent with experimental scattering data.

During the 1970s and 1980s, general concepts and morphological and structural details of ionic aggregation emerged (Holliday, 1975), in large part based on SAXS studies of dry polyelectrolytes (Gierke *et al.*, 1981, Hsu and Gierke, 1982; MacKnight *et al.*, 1974). Noteworthy are early studies by Gierke and coworkers (Hsu and Gierke, 1982) who used a combination of SAXS and WAXS to clarify the structure of Nafion. They noticed a characteristic Bragg peak at about 1.6°, corresponding to a Bragg spacing of about 3–5 nm, characteristic of a system containing ionic clusters within a semicrystalline matrix. Based on characteristic shifts in intensity and angle with equivalent weight, Gierke *et al.* concluded that Nafion was best described by a model of ionic spherical clusters with an inverted micellar structure (Gierke *et al.*, 1981). The spherical clusters were further proposed to be interconnected by channels. This model has come to be known as the cluster-network model of Nafion and was the first seriously discussed model and has perhaps remained one of the most widely referenced models in the history of perfluorosulfonic ionomers (Mauritz and Moore, 2004). Other significant scattering models that have been proposed to date include a modified core-shell model proposed by Fujimura *et al.* (1982).

In conclusion, one can say that our understanding of Nafion polymer membranes is still evolving today, as more details continue to be uncovered on how ionic polymer chains aggregate as a function of the external conditions.

4.3.3   SAXS studies of (alloy) nanoparticle catalyst formation

Over the past decade, a large body of X-ray scattering – in particular SAXS – based structural studies have addressed the nucleation and growth kinetics of nanoparticles (Abecassis *et al.*, 2010, Cheong *et al.*, 2009, Grohn *et al.*, 2001, Harada and Katagiri, 2010, Kim *et al.*, 2007, Machado *et al.*, 2007, Pavlopoulou *et al.*, 2010, Sakamoto *et al.*, 2006, Sciortino *et al.*, 2011, Tristany *et al.*, 2008, Watt *et al.*, 2010). Nucleation is the process of generating small initial areas of a more stable phase (1) within a meta-stable starting phase (2); the initial fragments (nuclei) of phase 1 typically subsequently grow into spatially dominating portions of the available phase space (Akaiva and Meiron, 1995, Akaiwa and Meiron, 1996, Finney and Finke, 2008, Mer, 1952). The spontaneous nucleation of phase 1 within a mother phase 2 is driven by a net energy gain during the formation of the volume and the surface of the nuclei of phase 1. If the work gain by the formation of the volume exceeds that needed to form the nucleus surface, spontaneous nucleation occurs.

An early pseudo-*in situ* SAXS study of metal nanoparticle formation within a polymer matrix was reported by a Russian group. Polymer gels were combined with soluble metal precursor solutions and metal particle nucleation was induced by addition of reducing agents such as hydrazine or sodium borohydride (Svergun *et al.*, 2001). The authors studied the structure of the polymer gel concomitantly to the growth of Pt nanoparticles. The authors found that the polymer gel was able to stabilize the mean particle growth of the Pt nanoparticles and proposed SAXS as an effective tool to study such metal-polymer composite systems in the future.

Shortly after, Haubold and collaborators presented an anomalous SAXS study of the formation of a three-dimensional Pt nanoparticle network (Vad *et al.*, 2002). The authors prepared three-dimensional metal/organic hybrid nanostructures by cross-linking aluminium organic stabilized platinum nanoparticles with various organic spacer molecules. The authors demonstrated the advantage of ASAXS in separating the particle scattering from the scattering of the organic component, thus providing unbiased information about particle size distributions and inter-particle correlations.

Carrado and co-workers later presented an *in situ* SAXS study of the formation of Pt nanoparticles inside a polyethylene oxide film matrix. They monitored the formation of metallic Pt cyrstallites above 100 °C in a reducing atmosphere (Carrado *et al.*, 2005).

Other SAXS studies addressing the formation of noble metal nanoparticle formation inside micelles or polymers to yield organic-inorganic hybrid materials were presented in subsequent years (Batra *et al.*, 2007, Bumajdad *et al.*, 2007, Calandra *et al.*, 2006, Ceolin *et al.*, 2008, Grohn *et al.*, 2001, Kim *et al.*, 2007, Sakamoto *et al.*, 2006, Tristany *et al.*, 2008).

Recently, Polte and co-workers in the group of Kraehnert reported on an *in situ* SAXS study of the formation of Au nanoparticles during homogeneous mixing of continuous flows of metal precursors and reducing agents (Polte *et al.*, 2010). The authors combined a fluid micro mixer with an *in situ* SAXS cell for an online time-resolved characterization of the formation process of Au particles. From the combination of SAXS, XANES, SEM and ultraviolet-visible spectroscopy (UV-VIS) data, the authors deduced a two-step mechanism of gold nanoparticle formation. The first step is a rapid conversion of the gold precursor into Au metal nuclei, followed by particle growth via coalescence of smaller entities. In the same year, Harada and Katagiri reported a SAXS study of Ag nanoparticle formation in a poly(N-vinyl-2-pyrrolidone) (PVP) solution. The authors resolved the nucleation and agglomeration process and investigated the dependence of the rate of growth on external experimental parameters (Harada and Katagiri, 2010).

All the studies above addressed particle nucleation and growth in three-dimensional bulk media. In order to follow the deposition and growth of nanoparticles on an extended flat surface, Lee *et al.* (2005) used anomalous grazing incidence SAXS (AGISAXS). The authors showed that AGISAXS is an efficient method to subtract the background and to provide unbiased information about particles sized less than 1 nm. More recently, the group around Toney proposed GISAXS and related X-ray techniques to structurally probe interfaces and surfaces *in situ* (Fong *et al.*, 2010).

Finally, a number of *ex situ* SAXS studies of supported metal nanoparticles were published over the past decade, where the particle size distributions of carbon-supported or metal oxide-supported Pt or Pt alloy particles were investigated (Brumberger *et al.*, 2005a, Brumberger *et al.*, 2005b, Jeng *et al.*, 2007). Building methodologically on previous metal catalyst studies, these studies gave experimental information on the dispersion of the prepared metal nanoparticle catalysts.

Other ASAXS work presented by the group of Haubold and Schilling (Goerigk *et al.*, 1997) studied the solid-solid decomposition of bimetallic Cu-rich Co-Cu alloys. Those insights may prove relevant for a deeper understanding of dealloying processes (Schofield *et al.*, 2008) currently studied in Pt-Cu, Pt-Co, and Pt-Cu-Co ternary fuel cell alloys. The formation of Raney-type dealloyed Ni catalysts was investigated by the group around Goerigk using ASAXS (Bota *et al.*, 2002, Bota *et al.*, 2008). Following early work by Haubold, a large number of authors have combined various complementary analytical techniques with SAXS to arrive at a more complete description of supported nanoparticle catalysts. Common are combinations of SAXS, WAXS, and EXAFS (Beale *et al.*, 2006, Jeng *et al.*, 2007, Jokela *et al.*, 2002, Shinoda *et al.*, 2007) or SAXS, XRD, and TEM (Borchert *et al.*, 2005).

4.4   Future trends in SAXS-based fuel cell catalysis research

Future challenges and applications of SAXS in the context of fuel cell catalysis lie in an accurate *in situ* measurement and a thorough atomic-scale understanding of the formation of nanoparticle catalysts inside porous supports (Bota *et al.*, 2007, Huang E *et al.*, 2002, Huang QR *et al.*, 2002, Ingham *et al.*, 2011, Park and Toney, 2005) or inside soft templates (Grosso *et al.*, 2011, Pavlopoulou *et al.*, 2010, Sciortino *et al.*, 2011) and – perhaps more important for today's fuel cell research – the degradation of nanoparticle catalysts inside porous conductive supports (Koh *et al.*, 2006, Strasser, 2009c, Yu *et al.*, 2008). Many resources are presently dedicated to a more fundamental understanding of how a fuel cell electrode catalyst gradually loses its activity and surface area, and how these macroscopic observables are linked to microscopic properties such as dispersion, particle morphology, and particle composition. However, almost all studies on structural degradation of electrodes and particle electrocatalysts to date have emphasized voltammetric, electron microscopy techniques and/or EXAFS/XANES techniques, with SAXS studies of full single fuel cells being virtually absent. An emerging interest in SAXS-based degradation studies is therefore likely in the coming years.

SAXS-based nanoparticle catalyst degradation studies with high statistical accuracy will also serve as critically important input for the development of accurate computational descriptions of nanoparticle coarsening and growth. Simple kinetic rate models to account for particle coarsening under PEMFC conditions have been developed and fitted to voltammetric and TEM data (Holby *et al.*, 2009, Rinaldo *et al.*, 2010). Work in progress involves fitting simple coarsening model parameters to SAXS-based experimental growth trajectories. Future work on a detailed description of catalyst coarsening and degradation will also focus on the incorporation of first-principle molecular interaction energies into stochastic and deterministic simulations to better capture the various metal dissolution, particle coalescence, and metal ion redeposition mechanisms proposed to take place during the catalyst degradation process.

4.5   Sources of further information

A number of excellent theoretical and practical books on SAXS techniques are available for the interested reader. Excellent books on the fundamentals and applied aspects of SAXS are available by Glatter and Kratky (1982), Guinier *et al.* (1955) and Brumberger (1995). A practical guide to SAXS was issued by Anton

Paar GmbH by Schnablegger and Singh (2006). Reviews on how to extract scattering size parameters and relevant structural information are provided in a number of articles (Damaschun *et al.*, 1969, Damaschun and Purschel, 1971, Möller *et al.*, 1995, Porod, 1951, Stuhrmann, 1970, Walter *et al.*, 1985).

Further information on common SAXS data analysis software such as SAXSfit and SASfit can be found in publications by Ingham *et al.* (Ingham, 2008, Ingham *et al.*, 2009). Most synchrotron facilities around the world have at least one SAXS end station and additional information on the available technique and energy range is available on the relevant websites.

4.6   References

Abecassis B, Testard F, Kong Q Y, Francois B and Spalla O (2010), 'Influence of monomer feeding on a fast cold nanoparticles synthesis: time-resolved XANES and SAXS experiments', *Langmuir*, **26**, 13847–13854.

Aieta N V, Stanis R J, Horan J L, Yandrasits M A, Cooking D J *et al.* (2009), 'Clipped random wave morphologies and the analysis of the SAXS of an ionomer formed by copolymerization of tetraflourethylene and "CF2=CFO(CF2)4SO3H"', *Macromolecules*, **42**, 5774–5780.

Akaiva N and Meiron D I (1995), 'Numerical simulation of two-dimensional late-stage coarsening for nucleation and growth', *Phys. Rev. E*, **51**, 5408–5421.

Akaiwa N and Meiron D I (1996), 'Two-dimensional late-stage coarsening for nucleation and growth at high-area fractions', *Phys. Rev. E*, **54**, 13–16.

Alexeyeva N, Kozlova J, Sammelselg V, Ritslaid P, Mandar H *et al.* (2010), 'Electrochemical and surface characterisation of gold nanoparticle decorated multi-walled carbon nanotubes', *Applied Surface Science*, **256**, 3040–3046.

Alloyeau D, Prevot G, Bouar Y L, Oikawa T, Langlois C *et al.* (2010), 'Ostwald ripening in nanoalloys: when thermodynamics drives a size-dependent particle composition', *Phys. Rev. Lett.*, **105**, 255901.

Ballauff M and Jusufi A (2006), 'Anomalous small-angle X-ray scattering: analyzing correlations and fluctuations in polyelectrolytes', *Colloid Polym. Sci*, **284**, 1303–1311.

Batra D, Seifert S, Varela L M, Liu A C Y and Firestone M A (2007), 'Solvent-mediated plasmon tuning in a gold-nanoparticle-poly(ionic liquid) composite', *Advanced Functional Materials*, **17**, 1279–1287.

Beale A M, van der Eerden A M J, Jacques S D M, Leynaud O, O'Brien M G *et al.* (2006), 'A combined SAXS/WAXS/EXAFS setup capable of observing concurrent changes across the nano-to-micrometer size range in inorganic solid crystallization processes', *J. Am. Chem. Soc.*, **128**, 12386–12387.

Benedetti A, Polizzi S, Riello P, Pinna F and Goerigk G (1997), 'ASAXS investigation of a Au/C catalyst', *Journal of Catalysis*, **171**, 345–348.

Benedetti A, Bertoldo L, Canton P, Goerigk G, Pinna F *et al.* (1999), 'ASAXS study of Au Pd and Pd-Au catalysts supported on active carbon', *Catal. Today*, **49**, 485–489.

Borchert H, Shevehenko E V, Robert A, Mekis I, Kornowski A *et al.* (2005), 'Determination of nanocrystal sizes: A comparison of TEM, SAXS, and XRD studies of highly monodisperse CoPt3 particles', *Langmuir*, **21**, 1931–1936.

Borup R L, Davey J R, Garzon F H, Wood D L and Inbody M A (2006), 'PEM fuel cell electrocatalyst durability measurements', *J. Pow. Sourc.*, **163**, 76–81.

Bota A, Goerigk G, Drucker T, Haubold H G and Petro J (2002), 'Anomalous small-angle X-ray scattering on a new, nonpyrophoric Raney-type Ni catalyst', *J. Catal.*, **205**, 354–357.

Bota A, Varga Z and Goerigk G (2007), 'Biological systems as nanoreactors: Anomalous small-angle scattering study of the CdS nanoparticle formation in multilamellar vesicles', *Journal of Physical Chemistry B*, **111**, 1911–1915.

Bota A, Varga Z and Goerigk G (2008), 'Structural description of the nickel part of a Raney-type catalyst by using anomalous small-angle X-ray scattering', *Journal of Physical Chemistry C*, **112**, 4427–4429.

Brumberger H (1995), *Modern Aspects of Small Angle Scattering*, Dordrecht, Boston, London, Kluwer Academic Publishers.

Brumberger H, Hagrman D, Goodisman J and Findelstein K D (2005a), '*In situ* anomalous small-angle X-ray scattering from metal particles in supported-metal catalysts. I. Theory', *J. of Appl. Cryst.*, **38**, 147–151.

Brumberger H, Hagrman D, Goodisman J and Findelstein K D (2005b), '*In situ* anomalous small-angle X-ray scattering from metal particles in supported-metal catalysts. II. Results', *J. of Appl. Cryst.*, **38**, 324–332.

Bumajdad A, Eastoe J, Zaki M I, Heenan R K and Pasupulety L (2007), 'Generation of metal oxide nanoparticles in optimised microemulsions', *Journal of Colloid and Interface Science*, **312**, 68–75.

Calandra P, Giordano C, Longo A and Liveri V T (2006), 'Physicochemical investigation of surfactant-coated gold nanoparticles synthesized in the confined space of dry reversed micelles', *Materials Chemistry and Physics*, **98**, 494–499.

Carrado K A, Sandi G, Kizilel R, Seifert S and Castagnola N (2005), 'Platinum nanoclusters immobilized on polymer-clay nanocomposite films', *Applied Clay Science*, **30**, 94–102.

Ceolin M, Galvez N, Sanchez P, Fernandez B and Dominguez-Vera J M (2008), 'Structural aspects of the growth mechanism of copper nanoparticles inside apoferritin', *European Journal of Inorganic Chemistry*, **5**, 795–801.

Cheong S, Watt J, Ingham B, Toney M F and Tilley R D (2009), '*In situ* and *ex situ* studies of platinum nanocrystals: growth and evolution in solution', *Journal of the American Chemical Society*, **131**, 14590–14595.

Damaschun G, Muller J J, Purschel H V and Sommer G (1969), 'Computation of shape of colloid particles from low-angle X-ray scattering', *Monatshefte Fur Chemie*, **100**, 1701–1714.

Damaschun G and Purschel H V (1971), 'Rontgen-kleinwinkelstreuung von isotropen proben ohne fernordnung. I. Allgemeine theorie', *Acta Crystallographica Section A*, **27**, 193–197.

Ferreira P J, la O G J, Shao-Horn Y, Morgan D, Makharia R, Kocha S *et al.* (2005), 'Instability of Pt/C electrocatalysts in proton exchange membrane fuel cells – A mechanistic investigation', *J. Electrochem. Soc.*, **152**, A2256–A2271.

Ferreira P J and Shao-Horn Y (2007), 'Formation mechanism of Pt single-crystal nanoparticles in proton exchange membrane fuel cells', *Electrochem. Solid State Lett.*, **10**, B60–B63.

Finney E E and Finke R G (2008), 'Nanocluster nucleation and growth kinetic and mechanistic studies: A review emphasizing transition-metal nanoclusters', *J. Coll. Interface Sci.*, **317**, 351–374.

Fong D D, Lucas C A, Richard M I and Toney M F (2010), 'X-ray probes for *in situ* studies of interfaces', *Mrs Bulletin*, **35**, 504–513.

Friedrich W, Knipping P and Laue M V (June 8, 1912), *Interference Phenomena with Roengten Rays*, Munich, The Royal Bavarian Academy of Science – communicated by A Sommerfeld

Fujimura M, Hashimoto R and Kawai H (1982), *Macromolecules*, **15**, 136.

Gierke T D, Munn G E and Wilson F C (1981), *J. Polym. Sci., Polym. Phys.*, **19**, 1687.

Glatter O and Kratky O (1982), *Small Angle X-ray Scattering*, London, Academic Press.

Goerigk G, Haubold H-G and Schilling W (1997), 'Kinetics of decomposition in copper-cobalt: a time-resolved ASAXS study', *J. Appl. Cryst.*, **30**, 1041–1047.

Goerigk G, Haubold H-G, Lyonc O and Simon J-P (2003), 'Anomalous small-angle X-ray scattering in materials science', *J. Appl. Cryst.*, **36**, 425–429.

Granqvist C G and Buhrman R A (1976), 'Size distributions for supported metal-catalysts – coalescence growth versus Ostwald ripening', *J. Catal.*, **42**, 477–479.

Grohn F, Kim G, Bauer A J and Amis E J (2001), 'Nanoparticle formation within dendrimer-containing polymer networks: Route to new organic-inorganic hybrid materials', *Macromolecules*, **34**, 2179–2185.

Grosso D, Ribot F, Boissiere C and Sanchez C (2011), 'Molecular and supramolecular dynamics of hybrid organic–inorganic interfaces for the rational construction of advanced hybrid nanomaterials', *Chemical Society Reviews*, **40**, 829–848.

Guilminot E, Corcella A, Chatenet M, Maillard F, Charlot F *et al.* (2007), 'Membrane and active layer degradation upon pemfc steady-state operation – I. Platinum dissolution and redistribution within the MEA', *J. Electrochem. Soc.*, **154**, B1106–B1114.

Guinier A, Fournet G, Walker C B and Yudowitch K L (1955), *Small Angle X ray Scattering*, New York, Wiley.

Haas S, Zehl G, Dorbandt I, Manke I, Bogdanoff P *et al.* (2010), 'Directly accessing the nanostructure of carbon supported Ru-Se based catalysts by ASAXS', *J. Phys. Chem. C*, **114**, 22375–22384.

Harada M and Katagiri E (2010), 'Mechanism of silver particle formation during photoreduction using *in situ* time-resolved SAXS analysis', *Langmuir*, **26**, 17896–17905.

Hasche F, Oezaslan M and Strasser P (2010), 'Activity, stability and degradation of multi-walled carbon nanotube (MWCNT) supported Pt fuel cell electrocatalysts', *Phys. Chem. Chem. Phys*, **12**, 15251–15258.

Haubold H-G, Wang X H, Goerigk G and Schilling W (1997), '*In situ* anomalous small angle X ray scattering investigation of carbon supported electrocatalysts', *J. Appl. Cryst.*, **30**, 653–658.

Haubold H-G, Hiller P, Jungbluth H and Vad T (1999), 'Characterization of Electrocatalysts by *in situ* SAXS and XAS Investigations', *Jpn. J. Appl. Phys.*, **38**, 36–39.

Haubold H-G, Vad T, Waldoefner N and Boennemann H (2003), 'From Pt molecules to nanoparticles: *In situ* small angle X ray scattering studies', *J. Appl. Cryst.*, **36**, 617–620.

Haubold H G and Wang X H (1995), 'ASAXS studies of carbon supported electrcatalysts', *Nucl. Instr. and Meth. in Phys. Res. B*, **97**, 50–54.

Haubold H G, Wang X H, Jungbluth H, Goerigk G and Schilling W (1996), '*In situ* anomalous small-angle X-ray scattering and X-ray absorption near-edge structure investigation of catalyst structures and reactions', *J. Mol. Structure*, **383**, 283–289.

Haubold H G, Hiller P, Jungbluth H and Vad T (1999), 'Characterization of electrocatalysts by *in situ* SAXS and XAS investigations', *Japanese Journal of Applied Physics Part 1–Regular Papers Short Notes & Review Papers*, **38**, 36–39.

He C, Desai S, Brown G and Bollepalli S (2005), 'PEM fuel cell catalysts: cost, performance, and durability', *The Electrochemical Society Interface*, **14**, 41–44.

Holby E F, Sheng W C, Shao-Horn Y and Morgan D (2009), 'Pt nanoparticle stability in PEM fuel cells: influence of particle size distribution and crossover hydrogen', *Energy Environ. Sci*, **2**, 865–871.

Holliday L (1975), *Ionic Polymer*, London, Applied Science.

Hsu W Y and Gierke T D (1982), *Macromolecules*, **15**, 101.

Huang E, Toney M F, Volksen W, Mecerreyes D, Brock P *et al.* (2002), 'Pore size distributions in nanoporous methyl silsesquioxane films as determined by small angle x-ray scattering', *Applied Physics Letters*, **81**, 2232–2234.

Huang Q R, Volksen W, Huang E, Toney M, Frank C W *et al.* (2002), 'Structure and interaction of organic/inorganic hybrid nanocomposites for microelectronic applications. 1. MSSQ/P(MMA-co-DMAEMA) nanocomposites', *Chemistry of Materials*, **14**, 3676–3685.

Ingham B (2008), 'Small Angle X-ray Scattering – Manual for SAXSFit program (Version 2.3)', download at http://www.irl.cri.nz/SAXSfiles,

Ingham B, Li H, Allen E L and Toney M F (2009), 'SAXSFit: A program for fitting small-angle x-ray and neutron scattering data', arXiv:0901.4782v1, online.

Ingham B, Lim T, Dotzler C, Toney M F and Tilley R (2011), 'How nanoparticles coalesce – An *in situ* study of Au nanoparticle aggregation and grain growth', *ACS Nano*, in press.

Iojoiu C, Guilminot E, Maillard F, Chatenet M, Sanchez J-Y *et al.* (2007), 'Membrane and active layer degradation following PEMFC steady-state operation II. Influence of Pt(z+) on membrane properties', *J. Electrochem. Soc.*, **154**, B1115–B1120.

Jeng U-S, Lai Y-H, Sheu H-S, Lee J-F, Sun Y-S *et al.* (2007), 'Anomalous small- and wide-angle X-ray scattering and X-ray absorption spectroscopy for Pt and Pt–Ru nanoparticles', *J. Appl. Cryst.*, **40**, s418–s422.

Jokela K, Serimaa R, Torkkeli M, Etelaniemi V and Ekman K (2002), 'Structure of the grafted polyethylene-based palladium catalysts: WAXS and ASAXS study', *Chemistry of Materials*, **14**, 5069–5074.

Jusufi A and Ballauff M (2006), 'Correlations and fluctuations of charged colloids as determined by anomalous small-angle X-ray scattering', *Macromolecular Theory and Simulations*, **15**, 193–197.

Kim Y W, Lee D K, Lee K J, Min B R and Kim J H (2007), '*In situ* formation of silver nanoparticles within an amphiphilic graft copolymer film', *Journal of Polymer Science Part B–Polymer Physics*, **45**, 1283–1290.

Koh S, Leisch J, Toney M F and Strasser P (2006), Size and composition distribution dynamics of Pt alloy electrocatalysts probed using Small Angle X-ray scattering (ASAXS), The Electrochemical Society fall meeting, abstract #522, Cancun, Mexico.

Koh S and Strasser P (2007), 'Electrocatalysis on bimetallic surfaces: Modifying catalytic reactivity for oxygen reduction by voltammetric surface de-alloying', *J. Am. Chem. Soc.*, **129**, 12624–12625.

Lee B, Seifert S, Riley S J, Tikhonov G, Tomczyk N A *et al.* (2005), 'Anomalous grazing incidence small-angle x-ray scattering studies of platinum nanoparticles formed by cluster deposition', *J. Chem. Phys.*, **123**, 074701-1–074701-7.

Leisch J, Koh S, Toney M F and Strasser P (2006), *In situ* anomalous small X-ray scattering studies of high surface area electrocatalyst, in The Electrochemical Society, fall meeting, Cancun, Mexico.

Levitan B and Domany E (1998a), 'Ostwald ripening in two dimensions: treatment with pairwise interactions', *Journal of Statistical Physics*, **93**, 501–510.

Levitan B and Domany E (1998b), 'Ostwald ripening in two dimensions: Correlations and scaling beyond mean field', *Physical review E*, **57**, 1895–1911.

Lu Y, Mei Y, Drechsler M and Ballauff M (2006), 'Thermosensitive core–shell particles as carriers for Ag nanoparticles: modulating the catalytic activity by a phase transition in networks', *Angew. Chem. Int. Ed.*, **45**, 813 –816.

Machado G, Scholten J D, de Vargas T, Teixeira S R, Ronchi L H *et al.* (2007), 'Structural aspects of transition-metal nanoparticles in imidazolium ionic liquids', *International Journal of Nanotechnology*, **4**, 541–563.

MacKnight W J, Taggert W P and Stein R S (1974), 'A model for the structure of ionomers', *J. Polym. Sci.*, **45**, 113.

Materlik G, Sparks C J and Fischer K (1994), *Resonant Anomalous X-ray Scattering: Theory and Applications*, Amsterdam, North Holland.

Mauritz K A and Moore R B (2004), 'State of understanding Nafion', *Chem. Rev.*, **104**, 4535–4585.

Mei Y, Lu Y, Polzer F, Ballauff M and Drechsler M (2007), 'Catalytic activity of palladium nanoparticles encapsulated in spherical polyelectrolyte brushes and core-shell microgels', *Chem. Mater.*, **19**, 1062–1069.

Mer V L (1952), 'Nucleation in Phase Transitions', *Industry and Engineering Chemistsry*, **44**, 1270–1277.

Möller J, Kranold R, Schmelzer J and Lembke U (1995), 'Small-angle X-ray scattering size parameters and higher moments of the particle-size distribution function in the asymptotic stage of Ostwald ripening', *Journal of Applied Crystallography*, **28**, 553–560.

Park K W and Toney M F (2005), 'Electrochemical and electrochromic properties of nanoworm-shaped Ta2O5-Pt thin-films', *Electrochemistry Communications*, **7**, 151–155.

Pavlopoulou E, Portale G, Christodoulakis K E, Vamvakaki M, Bras W *et al.* (2010), 'Following the synthesis of metal nanoparticles within pH-responsive microgel particles by SAXS', *Macromolecules*, **43**, 9828–9836.

Polte J, Erler R, Thuenemann A F, Sokolov S, Ahner T *et al.* (2010), 'Nucleation and growth of gold nanoparticles studied via *in situ* small angle X-ray scattering at millisecond time resolution', *ACS Nano*, in press.

Porod G (1951), 'Die rontgenkleinwinkelstreuung von dichtgepackten kolloiden systemen .1', *Kolloid-Zeitschrift and Zeitschrift Fur Polymere*, **124**, 83–114.

Rayleigh L (1910), 'The incidence of light upon a transparent sphere of dimensions comparable with the wave-length', *Proceedings of the Royal Society of London. Series A, Containing Papers of a Mathematical and Physical Character*, **84**, 25–46.

Rinaldo S G, Stuemper J and Eikerling M (2010), 'Physical theory of platinum nanoparticle dissolution in polymer electrolyte fuel cells', *J. Phys. Chem. C*, **114**, 5773–5785.

Roe R-J (2000), *Methods of X ray and Neutron Scattering in Polymer Science*, New York, Oxford University Press.

Rosenfeld G, Morgenstern K, Beckmann I, Wulfhekel W, Laegsgaard E *et al.* (1998), 'Stability of two-dimensional clusters on crystal surfaces: From Ostwald ripening to single-cluster decay', *Surf. Sci.*, **404**, 401–408.

Sakamoto N, Harada M and Hashimoto T (2006), '*In situ* and time-resolved SAXS studies of Pd nanoparticle formation in a template of block copolymer microdomain structures', *Macromolecules*, **39**, 1116–1124.

Schnablegger H and Singh Y (2006), *A Practical Guide to SAXS*, Graz, Anton Paar GmbH.

Schofield E J, Ingham B, Turnbull A, Toney M F and Ryan M P (2008), 'Strain development in nanoporous metallic foils formed by dealloying', *Appl. Phys. Lett.*, **92**, 043118.

Sciortino L, Giannici F, Martorana A, Ruggirello A M, Liveri V T *et al.* (2011), 'Structural characterization of surfactant-coated bimetallic cobalt/nickel nanoclusters by XPS, EXAFS, WAXS, and SAXS', *Journal of Physical Chemistry C*, **115**, 6360–6366.

Shao-Horn Y, Sheng W C, Chen S, Ferreira P J, Holby E F *et al.* (2007), 'Instability of supported platinum nanoparticles in low-temperature fuel cells', *Top. Catal.*, **46**, 285–305.

Shinoda K, Sato K, Jeyadevan B, Tohji K and Suzuki S (2007), 'Local structural studies of directly synthesized L1(0) FePt nanoparticles by using XRD, XAS and ASAXS', *Journal of Magnetism and Magnetic Materials*, **310**, 2387–2389.

Smith M C, Gilbert J A, Mawdsley J R, Seifert S and Myers D J (2008), '*In situ* small-angle X-ray scattering observation of Pt catalyst particle growth during potential cycling', *J. Am. Chem. Soc.*, **130**, 8112–8113.

Sompalli B, Litteer B A, Gu W and Gasteiger H A (2007), 'Membrane degradation at catalyst layer edges in PEMFC MEAs', *J. Electrochem. Soc.*, **154**, B1349–B1357.

Srivastava R, Mani P, Hahn N and Strasser P (2007), 'Efficient oxygen reduction fuel cell electrocatalysis on voltammetrically de-alloyed Pt-Cu-Co nanoparticles', *Angew. Chem. Int. Ed*, **46**, 8988–8991.

Stevens D A and Dahn J R (2000), 'An *in situ* small angle X-ray scattering study of sodium insertion into a nanoporous carbon anode material within an electrochemical cell', *J. Electrochem. Soc.*, **147**, 4428–4431.

Stevens D A, Zhang S, Chen Z and Dahn J R (2003), 'On the determination of platinum particle size in carbon-supported platinum electrocatalysts for fuel cell applications', *Carbon*, **41**, 2769–2777.

Strasser P, Koh S and Greeley J (2008), 'Voltammetric surface dealloying of Pt bimetallic nanoparticles: An experimental and DFT computational analysis', *Phys. Chem Chem. Phys.*, **10**, 3670–3683.

Strasser P (2009a), 'Dealloyed core shell fuel cell electrocatalysts', *Rev. Chem. Eng.*, **25**, 255–295.

Strasser P (2009b), Dealloyed Pt bimetallic electrocatalysts for oxygen reduction, in *Handbook of Fuel Cells – Fundamentals, Technology and Applications*, vol 5, 6 eds W Vielstich, H A Gasteiger and H Yokokawa, Chichester, West Sussex, John Wiley & Sons Ltd.

Strasser P (2009c), Particle Size and Composition Dynamics of electrocatalysts probed by *in situ* SAXS, in San Francisco, North American Catalysis Society, abstract #KD03. http://www.nacatsoc.org/21nam/start.htm.

Strasser P, Koh S, Anniyev T, Greeley J, More K *et al.* (2010), 'Lattice-strain control of the activity in dealloyed core–shell fuel cell catalysts', *Nature Chem.*, **2**, 454–460.

Stribeck N (2006), *X–Ray Scattering of Soft Matter*, New York, Springer.

Stuhrmann H (1970), 'Interpretation of small-angle scattering functions of dilute solutions and gases. A representation of the structures related to a one-particle scattering function', *Acta Crystallographica Section A*, **26**, 297–306.

Stuhrmann H B (1985), *Adv. Polym. Sci.*, 67, 123.

Stuhrmann H B, Goerigk G and Munk B (1991), Anomalous X-ray Scattering, In: Ebashi, S., Koch, M., Rubenstein, E. (Eds.) *Handbook on Synchrotron Radiation*. Vol. 4, 17 Amsterdam: Elsevier, 555–580.

Svergun D I, Shtykova E V, Kozin M B, Volkov V V, Konarev P V *et al.* (2001), 'Small-angle X-ray scattering study of the structure of self-organized polymer matrices and formation of imbedded metal nanoparticles', *Crystallography Reports*, **46**, 586–595.

Thiel P A, Shen M, Liu D-J and Evans J W (2009), 'Coarsening of two-dimensional nanoclusters on metal surfaces', *J. Phys. Chem. C*, **113**, 5047–5067.

Toney M F, Koh S, Yu C and Strasser P (2008), 'Use of anomalous X-ray scattering for probing the Structure, Composition, and Size of Binary Alloy Nanoparticle Electrocatalysts', *ECS Transactions*, **16**, 1001.

Trstany M, Moreno-Manas M, Pleixats R, Chaudret B, Philippot K *et al.* (2008), 'Formation of nanocomposites of platinum nanoparticles embedded into heavily fluorinated aniline and displaying long range organization', *Journal of Materials Chemistry*, **18**, 660–666.

Tsao C S and Chen C Y (2004), 'Small-angle X-ray scattering of carbon-supported Pt nanoparticles for fuel cell', *Physica B-Condensed Matter*, **353**, 217–222.

Vad T, Haubold H-G, Waldoefner N and Boennemann H (2002), 'Three-dimensional Pt nanoparticle networks studied by anomalous small-angle X-ray scattering and X-ray absorption spectroscopy', *J. App. Cryst.*, **35**, 459–470.

Vielstich W, Gasteiger H A and Yokokawa H (2009), *Handbook of Fuel Cells: Advances in Electrocatalysis, Materials, Diagnostics and Durability*, Chichester, John Wiley & Sons Ltd.

Virkar A V and Zhou Y (2007), 'Mechanism of catalyst degradation in proton exchange membrane fuel cells', *J. Electrochem. Soc.*, **154**, B540–B547.

Voorhees P W (1985), 'The theory of Ostwald ripening', *J. of Stat. Phys.*, **38**, 231–252.

Walter G, Kranold R, Gerber T, Baldrian J and Steinhart M (1985), 'Particle size distribution from small-angle X-ray scattering data', *Journal of Applied Crystallography*, **18**, 205–213.

Watt J, Cheong S, Toney M F, Ingham B, Cookson J *et al.* (2010), 'Ultrafast growth of highly branched palladium nanostructures for catalysis', *Acs. Nano.*, **4**, 396–402.

Wen F, Waldöfner N, Schmidt W, Angermund K, Bönnemann H *et al.* (2005), 'Formation and characterization of Pt nanoparticle networks', *Eur. J. Inorg. Chem.*, **18**, 3625–3640.

Winter R, Messurier D L and Martin C M (2006), *Energy-dependent* in situ *small-angle X-ray scattering study of nano-ceramics*, Warrington, UK, Taylor & Francis.

Yu C, Koh S, Leisch J, Toney M T and Strasser P (2008), 'Size and composition distribution dynamics of alloy nanoparticle electrocatalysts probed by anomalous Small Angle X-ray Scattering (ASAXS)', *Faraday Discuss.*, **140**, 283–296.

Yu P T, Gu W, Makharia R, Wagner F T and Gasteiger H A (2006), 'The impact of carbon stability on PEM Fuel Cell start up and shutdown voltage degradation', *ECS Transactions*, **3**, 797.

5

X-ray absorption near edge structure ($\Delta\mu$ XANES) techniques for low temperature fuel cell characterization

D. E. RAMAKER, George Washington University, USA and
C. ROTH, Technische Universität Darmstadt, Germany
and Karlsruhe Institute of Technology, Germany

Abstract: In recent years, the more conventional extended x-ray absorption fine structure (EXAFS) data analysis technique has been complemented by the $\Delta\mu$ XANES technique, which uses a difference method to isolate the changes in the X-ray absorption near edge structure (XANES) due to adsorbates on a metal surface. With $\Delta\mu$ XANES it is possible to determine the adsorbate, the specific adsorption site and adsorbate coverage on metal catalysts. This chapter summarizes the fundamentals and then gives specific applications, including water activation and oxygen reduction at a Pt cathode, and methanol and ethanol oxidation at a Pt anode in an electrochemical cell or fuel cell.

Key words: X-ray absorption near edge structure, *in situ* adsorbate coverage, adsorbate binding site, methanol oxidation, ethanol oxidation.

5.1   Introduction

In current fuel cells, there is a critical need for new catalysts that not only enhance reaction rates but are more durable and less costly. However, the rational design of new catalysts meeting these requirements will require a more detailed understanding of the underlying reaction mechanisms occurring at the catalyst's surface. In the future, novel spectroscopic techniques will become more useful, as they provide significant insights into the behaviour of commercial electrocatalysts *in situ* under fuel cell operating conditions (i.e. *in operando* conditions). It would be ideal if these techniques provided some measure of the catalyst particle size and morphology as well as the adsorbate coverages of small molecules and atoms (as illustrated in Fig. 5.1) under operating conditions. However, there are very few techniques that lend themselves to these rigorous requirements. X-ray absorption spectroscopy (XAS) is one of very few spectroscopic techniques that meet these requirements.

XAS is extremely versatile because it is one of only a few probes that utilize photons exclusively (i.e. photons in and using transmission or fluorescence yield photons out) enabling its application under nearly all conditions. X-ray diffraction (XRD) is similar in this regard, but it requires long-range order and does not provide adsorbate coverage, although it can provide information on structure,

120

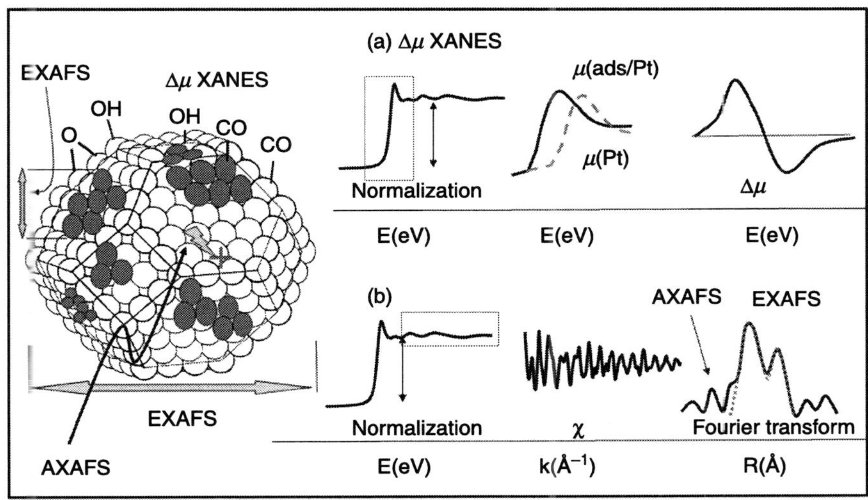

5.1 Illustration of the $\Delta\mu$ XANES (a) and AXAFS and EXAFS (b) analysis procedures and the information that can be obtained from each as depicted at the left. Background subtraction is critical in AXAFS/EXAFS, energy alignment is critical in $\Delta\mu$ XANES. (Source: Roth et al. (2005).)

particle size and phase identity. X-ray photoelectron spectroscopy (XPS), as well as any other electron spectroscopy such as Auger, provide electronic information but require ultra-high vacuum conditions.

XAS has been for the most part limited until recently to providing information on the metal island size and in the case of bimetallics some information on the morphology (Scott et al., 2007a, 2007b) via the conventional extended X-ray absorption fine structure (EXAFS) technique. However, recent advances have extended XAS to also provide electronic structure information via the atomic XAFS (AXAFS) technique and absorbate coverage and binding site information via a difference technique utilizing the X-ray absorption near edge structure (XANES), called here the $\Delta\mu$ XANES technique. All three of these techniques are illustrated in Fig. 5.1. The two new developments in XAS, along with the conventional EXAFS techniques, make XAS nearly ideal for the study of heterogeneous and electro-catalysts, both under realistic reaction conditions, as well as in vacuum.

The traditional EXAFS technique is summarized by Principi in Chapter 1 of this volume. AXAFS has been summarized recently in another review (Ramaker and Koningsberger, 2010a). This chapter will examine the second new technique, the '$\Delta\mu$' XANES analysis technique applied in the gas phase (Ramaker and Koningsberger, 2002; Ramaker et al., 1999, 2003; Oudenhuijzen et al., 2005; van Bokhoven et al., 2001), in an electrochemical cell (Teliska et al., 2004, 2005a, 2005b), and in an operating fuel cell (Scott et al., 2007b; Roth et al., 2005, 2007).

The $\Delta\mu$ XANES technique utilizes XANES data (denoted in this work as μ) for a 'reference' sample (when the metal, M, surface is free of adsorbate), which are subtracted from the data obtained after chemisorption of adsorbates (Ad): i.e., $\Delta\mu = \mu(Ad/M) - \mu(M)$ as described in Section 5.2.1 and illustrated in Fig. 5.1a. It will be shown in Section 5.2.2, with the help of full multiple scattering *ab initio* calculations, that this $\Delta\mu$ difference reflects the signature of the adsorbate binding site. Section 5.3 will discuss some applications in electrochemical or fuel cells. The $\Delta\mu$ XANES technique has been successfully applied to the adsorption of H, O, and OH on Pt (Teliska *et al.*, 2004, 2005a) and Pt-M (M = Cr, Fe, Co, and Ni) (Teliska *et al.*, 2005b) cathodes in an electrochemical cell, as well as CO, O and OH on bimetallic Pt-M (M = Ru, Sn, and Mo) electrodes in methanol (Roth *et al.*, 2005, Scott *et al.*, 2007a, 2010a,) and reformate, i.e. ppm level CO in H_2 (Scott *et al.*, 2007b). Even coverage levels of active and inactive H (overpotential deposited (opd) and underpotential deposited (upd) H, respectively) have been determined, and the role that co-adsorbed CO has on opd H has been understood (Scott *et al.*, 2007b, 2010b). It has been applied to non-Pt catalysts (Ziegelbauer *et al.*, 2006a, 2006b) and even to the competition for halide ion adsorption with O(H) adsorption on operating Pt/C cathodes (Arruda *et al.*, 2008). Finally Section 5.4 discusses some limitations to the method and future trends, and will give sources where more details and other applications can be found.

5.2   Basic principles, methods and theoretical calculations

It is a well-known phenomenon that the Pt L_3 or other metal core-level XANES are sensitive to the adsorption of various adsorbates, as is shown schematically in the analysis procedure in Fig. 5.1a. In the $\Delta\mu$ analysis procedure, we isolate the change in the metal XANES data due to the presence of the adsorbate. Figure 5.1a shows two different XANES spectra. Here $\mu(Ad/M)$ is the M (metal) edge spectrum in the presence of the adsorbate and $\mu(M)$ the L_3 edge spectrum at some different potential, current, or in vacuum, the latter serving as the reference. In the reference condition, the adsorbate coverage is either much smaller than, or at least very different from, the initial spectrum. The $\Delta\mu$ technique can therefore be considered as a subtractive technique with the difference, $\Delta\mu$, given as: $\Delta\mu = \mu(Ad/M) - \mu(M)$ to isolate the influence of the adsorbate. Essentially, careful normalization and subtraction of the XANES signals at applied potentials different from the 'ref' potential (i.e. in the double layer region) (Scott *et al.*, 2007a, 2007b; Teliska *et al.*, 2004, 2005a, 2005b; Roth *et al.*, 2005) or different from the reference temperature or pressure in a gas phase reactor (Koningsberger *et al.*, 2000; Guo *et al.*, 2010), will result in a spectrum that has completely eliminated the underlying chemically un-reactive bulk signal. Hence the resulting $\Delta\mu$ line shape reflects the chemistry, site symmetry, and magnitude of adsorbed species on the surface. It should be further noted that contributions to μ coming from surface sites not available for adsorption (e.g. those in contact with the support or with neighbouring particles, those covered by surface islands, or just not in the region of the required electrochemical three-phase boundary, or accessible to exposed gases) will cancel out when taking the difference, so the $\Delta\mu$ technique can even give an indication of the accessible surface Pt sites (i.e. the electrochemically active surface area, ECSA, in an electrochemical cell) as shown previously (Scott *et al.*, 2007a, 2007b, 2010).

A noteworthy point to be stressed is the incredible sensitivity of the $\Delta\mu$ technique. While direct analysis of the −50 to 50 eV XANES spectrum can reveal the presence of *strongly* interacting ligands, such as direct metal-oxygen covalent bonds, this type of interpretation is limited by the extent of the change in the white line region. Only qualitative information can be obtained from changes in the white-line intensity; for instance, it is not easily possible to distinguish between O and CO adsorption just from the white line (Roth *et al.*, 2007). However, identification of both the O and CO coverage is essential in many studies, such as in fuel cells when one wants to investigate the effect of CO poisoning and Pt surface oxidation. The $\Delta\mu$ analysis technique pushes the sensitivity to fractions of a percent, picks up different adsorbed species, and allows for direct spectroscopic observation of even weakly adsorbed species.

5.2.1   Methods for obtaining the experimental $\Delta\mu$

The $\Delta\mu$ XANES technique is applicable whenever the coverage of an adsorbate can be varied leaving the rest of the catalysts relatively unchanged. In the gas phase, the coverage is changed by varying the temperature and pressure of the adsorbate gas, such as H_2, O_2 or CO (Koningsberger *et al.*, 2000; Guo *et al.*, 2010). However, more often the adsorbate coverage is changed by simply varying the applied potential to an electrode in an electrochemical cell allowing the difference $\Delta\mu = \mu(V) - \mu(V_{ref})$ to be obtained (Teliska *et al.*, 2004, 2005a, 2005b; Roth *et al.*, 2005; Scott *et al.*, 2007a), or in an operating fuel cell (Roth *et al.*, 2005; Scott *et al.*, 2007b; Melke *et al.*, 2010a), when the current is varied allowing $\Delta\mu = \mu(I) - \mu(I_{ref})$. More recently (Melke *et al.*, 2010b; Ramaker and van Bokhoven, 2010) time resolved XANES data allowed the difference $\Delta\mu = \mu(T) - \mu(T_{ref})$ to be obtained, where the adsorbate coverage dependence with time could be observed.

Since the absorption coefficient can be expressed, $\mu = \mu_o (1 + \chi)$, the total change in the XANES due to the presence of an adsorbate (Ad) can be expressed as (Ramaker and Koningsberger, 2002; Ramaker *et al.*, 2003; Teliska *et al.*, 2004, 2005a):

$$\Delta\mu = \mu(\text{Ad/M}) - \mu(\text{M}) = \Delta\mu_o + \Delta(\mu_o\chi_{M-M}) + \mu_{o,\text{Ad/M}}\chi_{M-Ad} \qquad [5.1]$$

where the terms on the right include: $\Delta\mu_o$ equal to the change in the atomic XAFS due to the adsorbate , $\Delta(\mu_o\chi_{M-M})$ equal to the change in the M-M total scattering induced by chemisorption of the adsorbate, and $\mu_{o,\text{Ad/M}}\chi_{M-Ad}$ equal to the additional scattering by the adsorbate. These three individual contributions can be isolated using theory, and will be discussed further below.

To obtain the experimental $\Delta\mu$ spectrum, the pre-edge background has to be removed, followed by normalization over the 20–100 eV range (relative to the edge) typically for XANES analysis (as illustrated in Fig. 5.1a). The normalized data are energy calibrated using foil data of the same material studied (e.g Pt foil, Ru, foil, etc. for Pt or Ru edges, etc.). This energy calibration is critical for the success of the $\Delta\mu$ XANES technique, as full cancellation of the atomic contribution (i.e., except for the changing AXAFS, $\Delta\mu_o$, term in Eq. 5.1) in the XANES must be achieved when taking the difference to obtain $\Delta\mu$. This atomic contribution dominates μ so that $\Delta\mu$ is typically only about 3–5% of the total μ signal. However, the small $\Delta\mu$ is not generally a problem. The oscillations utilized in EXAFS are even smaller than those in the XANES, particularly at large k ranges. With modern synchrotrons it is possible to obtain signal-to-noise ratios at 15 eV above the absorption edge typically from 50/1 to 5/1. Data can also be smoothed to reduce noise levels to around 0.2%.

A problem that often enters when determining the experimental $\Delta\mu$ is incomplete cancellation of the background; i.e. the $\Delta\mu$ appears to have a non-zero slope or exist on top of a slowly varying contribution over a wide energy range such as illustrated in Fig. 5.2 for ethanol oxidation in a fuel cell using 'quick' EXAFS

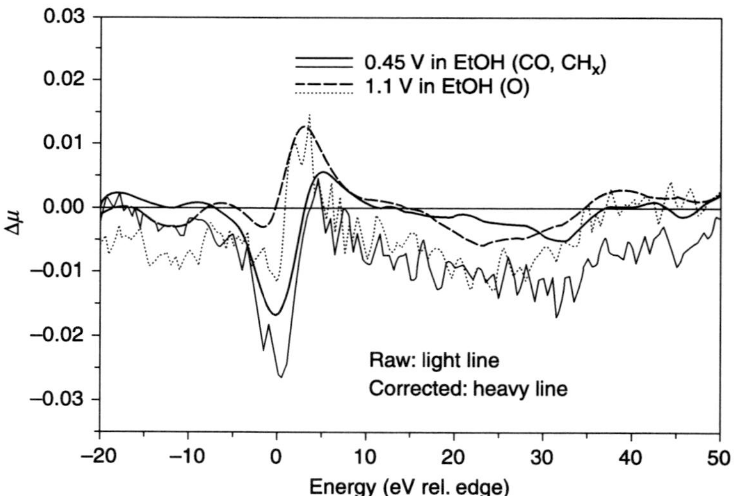

5.2 Plot of $\Delta\mu$ at the Pt L$_3$ edge for ethanol in HClO$_4$ at 0.45 V (reflecting (CO and CH$_x$)/Pt adsorbates) and at 1.1 V (reflecting O/Pt) as reported by Melke *et al.* (2010a). The corrected data (heavy lines) have had a background subtracted and then smoothed by 5% to remove random noise, where the background is obtained by simply smoothing the raw data by 40%; i.e. $\Delta\mu$(corrected) = smooth 5%($\Delta\mu$(raw) – $\Delta\mu$(smoothed 40%)).

(a case where the better time resolution sacrifices the signal-to-noise ratio) (Melke *et al.*, 2010b). This 'residual background' can easily be removed by performing a Savitzky–Golay (1964) smoothing of the $\Delta\mu$ data using 30–50% of the points. This is similar to the post-edge background removed in the normal EXAFS analysis; but now it can extend continuously through the edge region and well below the edge, since the step at the edge in μ has been removed by taking the $\Delta\mu$ difference. The same Savitzky–Golay smoothing – but now with say 2–3 % of the points – can be used to decrease the noise level if needed as shown in Fig. 5.2.

The results in Fig. 5.2 show that even when the signal-to-noise ratio is quite low, good quality $\Delta\mu$s can be obtained. Limitations of course exist. The $\Delta\mu$ technique isolates the contribution from adsorbates, so that the dispersion of the metal particles must be reasonably large ($\geq 25\%$) or particle diameter less than 4–5 nm. In a fuel cell often the electrochemically active surface area is only 10–20% of the total surface, yet we have routinely obtained reliable results (to be described below). On several occasions the $\Delta\mu$ magnitudes have been shown to be reproducible to within 10% even for similar but actually different samples on different runs at the synchrotron (e.g., Ramaker *et al.*, 2010a). We (all references attributed to Ramaker, Roth, Scott or Teliska) and others (Guo *et al.*, 2010) have reproduced the data for H and O on Pt dozens of times, and that for methanol on the anode several times using data from different research groups, on different synchrotrons, and for different samples. Likewise in the gas phase, H and O has been followed on Pt, Ir, Ru, and other metal catalysts, with $\Delta\mu$ signatures remarkably similar to those found in an electrochemical cell. We believe the technique to be quite robust and generally applicable to follow small molecule absorption *in situ* on the surface of metal and even metal-oxide catalysts. Some applications will be summarized below.

5.2.2 Theoretical calculations

The $\Delta\mu$ experimental difference spectra do not offer much information by themselves. Indeed, they require proper interpretation by comparing theoretical $\Delta\mu$ spectra to the experimentally derived curves. To generate the theoretical spectrum, a theoretical XANES spectrum is calculated for a three-dimensional model of the 'adsorbate-free' cluster (usually a Pt_6 cluster as described below) mimicking the catalyst of interest. Then a theoretical XANES spectrum is calculated for the same model cluster but now with an atom or molecule adsorbed onto a surface atom. Subtracting these two XANES spectra gives a theoretical $\Delta\mu$ spectrum analogous to the experimental $\Delta\mu$. The closer the model structure is to the experimental one, the more likely the theoretical 'signature' will match the experimental data. Comparison between experiment and theory then allows for interpretation of the experimental $\Delta\mu$ to provide adsorbate information such as the coverage and binding site geometry, since in the theoretical calculations one can move the adsorbate into different adsorption sites, and the theoretical calculations show a strong sensitivity to adsorption site (e.g. atop, bridged, fcc, etc). Further, in the theory one knows exactly the ratio of adsorbate to substrate atom, so the coverage of adsorbate in the experiment can be ascertained, at least semi-quantitatively, by comparing the relative magnitudes of the experimental and theoretical $\Delta\mu$s (Scott *et al.*, 2007a, 2007b, 2010a, 2010b).

The FEFF8 code (Ankudinov *et al.*, 1998) is normally used to carry out the theoretical calculations. FEFF8 performs real-space full multiple scattering calculations utilizing a muffin-tin potential calculated with the Hedin-Lundquist exchange correlation approximation, and implements self-consistent field potentials for the determination of the Fermi-level and the charge transfer. The absorption coefficient, μ, as a function of the photon energy is directly given by the code.

In our work a Pt_6 cluster (Janin *et al.*, 2000) has been used most often (Scott *et al.*, 2007a, 2007b, 2010a, 2010b; Teliska *et al.*, 2004, 2005a, 2005b) to model the experimental Pt clusters, as this highly asymmetrical cluster provides for all of the possible common binding sites (fcc, hcp, bridged, and atop) as illustrated in Fig. 5.3. This asymmetrical cluster does not introduce any 'surface resonances' (Ramaker and Koningsberger, 2002) that might arise from a more symmetric cluster, and the cluster is sufficiently large to account for the change in Pt-Pt scattering and new Pt-Ad scattering introduced by the adsorbate (Teliska *et al.*, 2004, 2005a). Figure 5.3 shows the $\Delta\mu$ L_3 edge FEFF8 results for the cluster models mimicking the adsorption of H on Pt with increasing Pt-ad coordination (1 for atop, 2 for bridged, and 3 for the three-fold fcc). The differences between $\Delta\mu$ for the *n*-fold coordinated cases (bridged, fcc, and hcp) are not significant enough to distinguish them from each other. Therefore, the signature or fingerprint for the atop site is unique, and only this site can be distinguished from the other sites for H on Pt. The atop signature for H/Pt has negligible negative contribution while the bridge/fcc signatures have a large negative contribution (Koningsberger *et al.*, 2003; Teliska *et al.*, 2004).

The FEFF8 calculations also reveal the source of the binding site sensitivity. The individual contributions in Eq. 5.1 can be identified with FEFF8, and the dominant contributions ($\Delta(\mu_o\chi_{Pt-Pt})$ and $\mu_{o,H/Pt}\chi_{Pt-H}$) are shown in Fig. 5.3 along with the total $\Delta\mu$ for H/Pt (Koningsberger *et al.*, 2003; Teliska *et al.*, 2004). The $\Delta\mu_o$ term is the smallest of the three terms and is greatly exaggerated by the calculations on such small clusters, so consistent with previous work (Koningsberger *et al.*, 2003) we ignore this term here. The $\mu_{o,H/Pt}\chi_{Pt-H}$ term changes in magnitude but not in spectral shape when the Pt-H coordination changes from 3 in the fcc case to 1 in the atop case. The $\Delta(\mu_o\chi_{Pt-Pt})$ term varies the most dramatically on changing the hydrogen adsorption site, and gives the important binding site sensitivity. When H atoms are adsorbed on the fcc site, hydrogen weakens the Pt-Pt bonds with the Pt atoms beneath the hydrogen adsorption site. This bond weakening has been called d-electron frustration by Feibelman (1997) and Pt-Pt destabilization by Papoian *et al.* (2000). Recent generalized

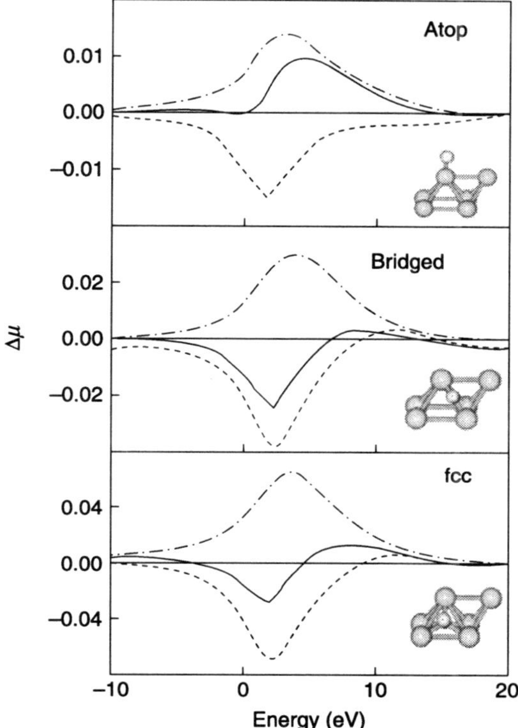

5.3 The theoretical $\Delta\mu = \mu(\mathrm{H/Pt_6}) - \mu(\mathrm{Pt_6})$ difference spectra, heavy solid line, and the two major components ($\mu_{o,H}\chi_{Pt\text{-}H}$, dashed-dotted line, and $\Delta(\mu_o\chi_{Pt\text{-}Pt})$, dotted line), obtained from FEFF8 calculations (Koningsberger *et al.*, 2000, 2003; Teliska *et al.*, 2004). The left-upper Pt atom on the $\mathrm{Pt_6}$ cluster (i.e., the one closest to the H atom adsorbate) was taken as the absorber atom in all cases. The hcp site is not shown here because the resultant $\Delta\mu$ signature is nearly identical to that for the fcc site.

valence-bond calculations by Kua and Goddard (1998) also revealed that Pt-H coordination in the fcc site results in a weakening of the Pt-Pt bonding beneath the H atom. Fig. 5.3 shows that the $\Delta(\mu_o\chi_{Pt\text{-}Pt})$ contribution is much smaller when the hydrogen is bonded in the atop position, which allows the Pt-H contribution to dominate. The weakening of the Pt-Pt bonding when H is bonded in the fcc site reduces the Pt DOS just above the Fermi level, making the $\Delta(\mu_o\chi_{Pt\text{-}Pt})$ more negative.

Figure 5.3 also helps us understand the differences between the $\Delta\mu$ for H and O or Pt. The Pt-Ad scattering is larger for O, because O has more electrons, so that the $\mu_{o,O/Pt}\chi_{Pt\text{-}O}$ contribution in Eq. 5.1 is much larger than the $\mu_{o,H/Pt}\chi_{Pt\text{-}H}$ contribution (Ramaker *et al.*, 2003). This makes $\Delta\mu$ for atop vs *n*-fold O/Pt more similar and positive in all the different sites. However, we can still distinguish the atop O from the *n*-fold O sites, since the position in energy is different as shown in Fig. 5.4.

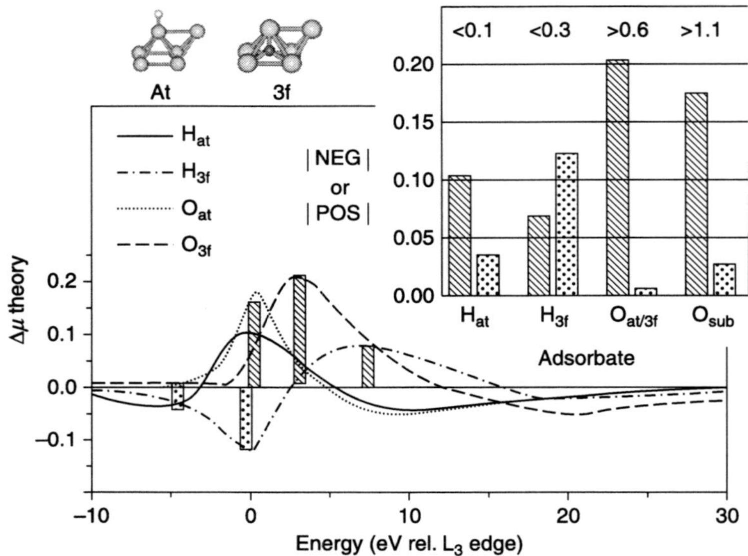

5.4 FEFF 8.0 calculated line shapes for the indicated adsorbates in the binding sites pictured (Ramaker *et al.*, 2010a). The |NEG| and |POS| magnitudes of all the signatures are highlighted with the shaded boxes. Subscripts 'at', '3f', and 'sub' indicate atop, fcc 3-fold, and subsurface sites, respectively. The bar chart insert illustrates the relative amplitudes of |POS| and |NEG| for the indicated adsorbates that dominate in the indicated potential (v) regions.

This difference turns out to be critical because it allows us to distinguish between atop OH from n-fold O atoms on Pt; both species critical in electrochemical applications to be discussed below. Calculations for atop OH (not shown) are nearly identical to atop O because H is a weak scatterer, but since O wants to be at least doubly bonded to something, OH (or adsorbed H_2O) generally exists in an atop site, and O in an n-fold site enabling them to be distinguished.

Agreement between the theoretical and experimental signatures, although generally not quantitative is sufficient to identify the binding site, as will be shown below with examples. Good agreement of the experimental $\Delta\mu$ signature with the theoretical $\Delta\mu$ signature confirms the validity of the entire technique. Since the experimental data is an average over a wide array of particles sizes and shapes in a catalyst, any further attempt to improve on the model cluster to find more quantitative agreement between the experimental and theoretical signatures is deemed to be fruitless.

Although the XAS process samples all Pt atoms in the beam's cross-sectional area, it is assumed above that the μ of Pt atoms not bonded to the adsorbate (i.e. the spectators) would not change so that $\Delta\mu$ would be zero for these atoms. Although this latter assumption, now referred to as the 'limited absorber' assumption (Stoupin, 2009) may not be rigorously correct for the small Pt_6 cluster, because the charge or electron distribution of nearby Pt atoms may be altered by the adsorbate (Smotkin *et al.*, 2008); in general the much larger (and therefore more metallic) clusters in the experimental catalysts are believed to make this effect small (Ramaker and Koningsberger, 2002). Indeed calculations reported by Lewis *et al.* (2009) have shown that when the Pt_6 cluster is embedded in much bigger clusters to allow for more metallic-like screening, the 'configurationally average model', which averages the FEFF8-calculated $\Delta\mu$ on all 6 Pt atoms, is quite similar to the limited absorber model. Hence the limited absorber model as utilized here is certainly adequate to mimic the experimental results and identify the adsorbate signatures.

5.3   Applications

Here the $\Delta\mu$ technique will be illustrated utilizing one of the most common systems in electrochemistry, a polycrystalline Pt electrode in acidic electrolyte as investigated in an electrochemical cell. Of course, this situation is also very much similar to a porous gas diffusion electrode, which is generally applied in a low-temperature fuel cell, and will also be discussed below. In these experiments, several adsorbates including H, O, OH, and CO could come down onto the Pt surface, depending on potential, fuel (H_2 or CH_3OH), and other working conditions, and compete for adsorption sites. A strength of the $\Delta\mu$ XANES technique is that the competition between these different adsorbates for free Pt sites can now be followed during operation.

5.3.1   Water activation: atop vs *n*-fold adsorbates (H and O).

The $\Delta\mu$ XANES technique was initially applied in heterogeneous catalysis to determine the Pt-H bond strength and binding site of chemisorbed hydrogen on supported Pt particles (Koningsberger *et al.*, 2000; Ramaker and Koningsberger, 2002, 2010a). This information has recently turned out to be crucial in understanding the mechanism and structure sensitivity of industrially important and very relevant metal catalysed hydrogenation reactions (Ramaker and Koningsberger, 2010a, 2010b). When, a bit later, the technique was to be applied in an electrochemical environment (Teliska *et al.*, 2004, 2005a, 2005b), many questions immediately needed to be answered: What does the electrolyte do (i.e., do we 'see' the water double layer or anions or if present the effect of the standard polymer electrolyte material such as Nafion®)? Would the changing potential in the cell change the edge energy? Do H and O have the same $\Delta\mu$ signature in an electrochemical cell as in the gas phase? The answers to these questions are given by examining simple water activation ($H_2O \rightarrow$ H/Pt at low potentials, and $H_2O \rightarrow$ O/Pt or OH/Pt at higher potentials) in an electrochemical cell.

The change in adsorbate binding site or identity as a function of potential can most conveniently be followed by plotting the absolute magnitude of the $\Delta\mu$

(which is negative for n-fold H) just below 0 eV. This magnitude is labelled |NEG|. In contrast atop H and O has large positive amplitude around 5 eV which we label |POS|. The |NEG| amd |POS| $\Delta\mu$ amplitudes obtained from FEFF8 for the various adsorbed species are highlighted in the bar chart (Fig. 5.4) to make this more explicit (Ramaker *et al.*, 2010a).

As a starting point, Fig. 5.5 shows a plot of both the |NEG| and |POS| amplitudes for Pt/VC (Pt nanoparticles supported on high surface area carbon such as Vulcan® carbon) obtained from experimental $\Delta\mu$ Pt L_3 data (Ramaker *et al.*, 2010a). This plot illustrates the detail possible as a function of potential. The |POS| amplitude here primarily reflects atop H adsorption below 0.3 V (point a), atop chemisorbed H_2O between 0.6–0.8 V, and O above 0.8 V (point c). The |NEG| amplitude primarily reflects three-fold H adsorption below 0.5 V (point b), H_2O adsorption (0.6–0.8 V) and subsurface O above 0.8 V (point d). Figure 5.5 therefore indicates that as one proceeds to lower potential, first three-fold fcc H_{upd} adsorbs, followed

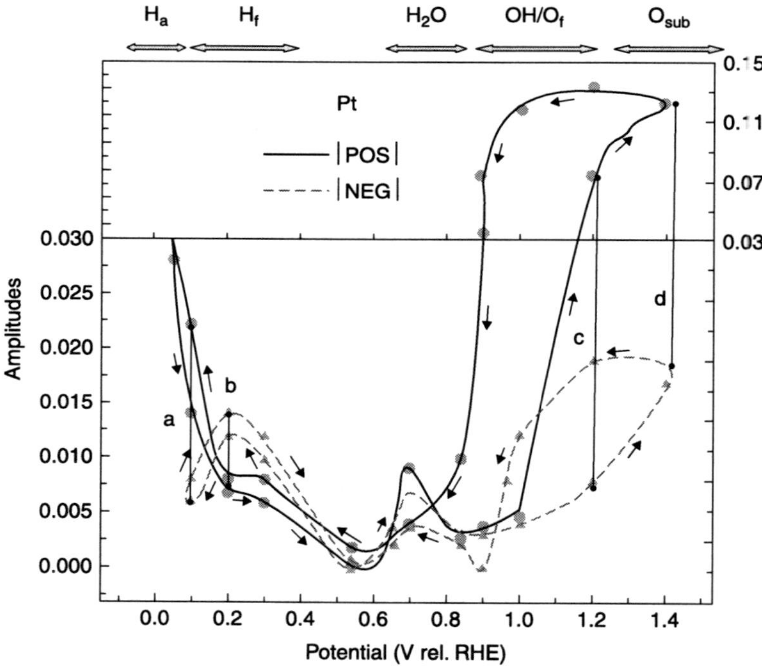

5.5 Plot of |POS| and |NEG| amplitudes from the experimental $\Delta\mu$ as a function of potential for Pt/VC (Ramaker *et al.*, 2010a). Note different scales for magnitudes below and above 0.03 to increase clarity. Arrows indicate direction of scan beginning and ending at the open circuit potential (OCP, around 0.95 V). The potential regions where specific adsorbates dominate are indicated at the top. Lines labelled a–d indicate paired |NEG| and |POS| points as discussed in the text.

by atop H_{opd} (upd and opd refer to under and over potential adsorbed H, respectively). In the O region, OH comes on above 0.8 V and some subsurface O appears only above 1.2 V, the latter occurring by place exchange with Pt atoms at the surface. Note the relative reversibility of the H adsorption compared to the O adsorption, consistent with that previously observed in many CV curves.

It is important to note the feature seen in both |POS| and |NEG| between 0.65 and 0.8V in Fig. 5.5. This feature is particularly striking since both |POS| and |NEG| indicate a rise and then a fall in magnitude before the adsorption of O(H) above 0.8 V. We suggest that this is due to atop O in adsorbed H_2O (having nearly identical $\Delta\mu$ signature as atop OH). In this region, it is surmised that the formation of a weakly bonded water layer develops and then it is disrupted as the adsorption of O(H) begins. Current-voltage data reported by Markovic *et al.* (1999) for Pt(111) in 0.1M $HClO_4$ report a feature at similar potentials and suggest it arises from an overlayer of H_2O with ClO_4^- anions, which we cannot rule out.

These results give us answers to the questions asked above. Water in the double layer is not observed, apparently because it is too disordered near the surface to contribute specific features in the scattering contributing to the XANES. The same goes for Nafion®. But just below 0.80 V, an ordered chemisorbed water layer (with or without ClO_4^-) is formed making it become visible as shown in Fig. 5.5. Likewise, we have shown that bisulfate anions, even though they are adsorbed on Pt/VC at nearly all potentials above 0.4 V in sulfuric acid electrolyte, are not observed until some OH is adsorbed above 0.8 V, forcing the bisulfate anions to adsorb in registry in unique sites between these OH/Pt making the bisulfate/Pt visible in the XANES (Teliska *et al.*, 2007). Further the $\Delta\mu$ signature for atop and *n*-fold H/Pt and O/Pt is essentially the same in the gas phase as in an electrochemical environment (Ramaker *et al.*, 2003).

5.3.2   The oxygen reduction reaction (ORR) on Pt

Another reaction highlighting the versatility of the $\Delta\mu$ XANES for *operando* studies in a fuel cell is the oxygen reduction reaction (ORR). The kinetic equations derived from the double-trap kinetic model of Wang *et al.* (2007a, 2007b, 2008) for the 4-e ORR in acidic media, include four elementary reactions:

$$1/2O_2 \rightarrow O_{ad} \qquad\qquad \text{Dissociative adsorption (DA)} \qquad [5.2]$$

$$1/2O_2 + H^+ + e^- \rightarrow OH_{ad} \qquad \text{Reductive adsorption (RA)} \qquad [5.3]$$

$$O_{ad} + H^+ + e^- \rightarrow OH_{ad} \qquad \text{Reductive transition (RT)} \qquad [5.4]$$

$$OH_{ad} + H^+ + e^- \rightarrow H_2O \qquad \text{Reductive desorption (RD)} \qquad [5.5]$$

These reactions can be summarized even more basically as follows:

$$\tfrac{1}{2} O_2 \rightarrow OH_{ad} \text{ (via RA or via DA + RT)} \rightarrow H_2O \text{ (via RD)} \qquad [5.6]$$

and as shown by Wang *et al.* (2007a, 2007b, 2008), the RA mechanism is dominant below 0.7 vs. reversible hydrogen electrode (RHE). Thus the adsorbed OH intermediate is the most important.

Figure 5.6 shows the typical Tafel plot of V vs. log *i*, and three slopes, ~60 mV/decade above 700 mV, ~125 mV/decade below, and then much larger slope again below 500 mV (Korovina *et al.*, 2010b). Wang *et al.* (2007b) has attributed the 60–120 mV slope change to the change from DA + RT to RA dominance in the production of OH. Figure 5.6 shows that right at this slope change the dominant coverage obtained from the $\Delta\mu$ XANES results switches from primarily O to OH on the surface consistent with this model. Figure 5.6 also shows that the sharp deviation from the 60 V/decade slope around 550 mV (i.e. the second slope change) occurs because additional oxygen-containing species, in particular peroxide, OOH, begin to appear on the surface at this point. The onset of H_2 adsorption crowds the surface and encourages the 2-e ORR, which causes the current to fall with potential at an even faster rate. Thus, the *operando* fuel cell measurements support in part the validity of the Wang *et al.* model, and further allow us to directly study by a spectroscopic method the coverages existing in the diffusion limited region and/or when other adsorbates such as H, OOH, and other anions are on the surface. Such studies are critical to understanding the complex electrode-electrolyte interface, which determines the surface diffusion, the concentration of other adsorbates on the electrode surface, and hence the performance of a fuel cell (Roth and Ramaker, 2010a).

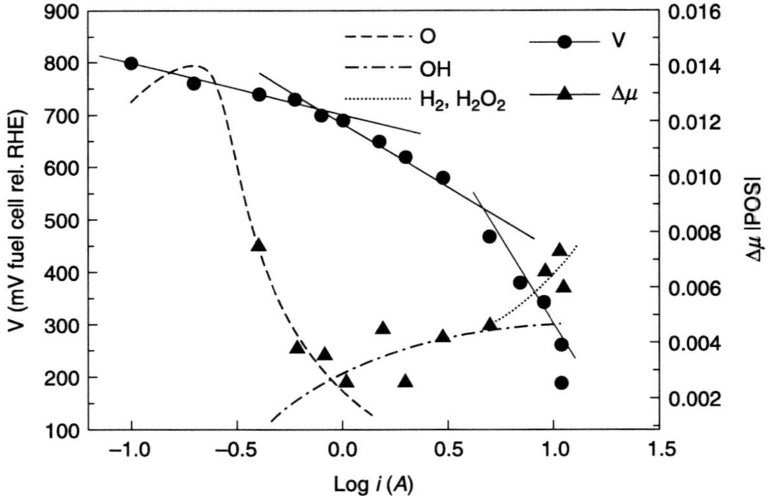

5.6 Tafel plot (V vs log *i*) for a Pt fuel cell in H_2 along with the $\Delta\mu$ magnitudes (right axis) reflecting dominant fcc O, atop OH, and OH + H_2/H_2O_2 coverage on the Pt cathode (Korovina *et al.*, 2010b). Note the change in slopes right where the dominant adsorbate changes character.

5.3.3 The direct methanol fuel cell (DMFC) on PtRu vs PtMo and effect of Ru island size

We also illustrate the $\Delta\mu$ XANES technique to follow the coverage of specific adsorbates such as CO, OH and O on the anode, on carbon supported Pt and different Pt/Ru bimetallic systems, to be described below, in direct methanol operation as a function of potential in an electrochemical cell (Scott *et al.*, 2007a, 2010a) or current in a fuel cell (Scott *et al.*, 2007b; Roth *et al.*, 2005). The primary adsorbate (poison) on the surface is CO resulting from incomplete oxidation of the methanol. It should be emphasized that the $\Delta\mu$ XANES technique does not 'see' temporal adsorbates, but only poisons or reaction intermediates existing in surplus on the surface. Figure 5.7 shows Pt L_3 $\Delta\mu$ data for a PtRu catalyst in 0.3 M methanol using the data obtained at 0.4 V in the cathodic sweep direction as the reference (Scott *et al.*, 2007a). At this potential the coverage of CO was found to be at a minimum although not necessarily zero. Figure 5.7 also shows theoretical signatures for CO, OH, and O in order to identify the changes seen in the experimental results with potential. Clearly CO is leaving as O or OH species build up on the surface; this is expected because the adsorbed OH reacts with the CO to create CO_2 thereby cleaning the surface.

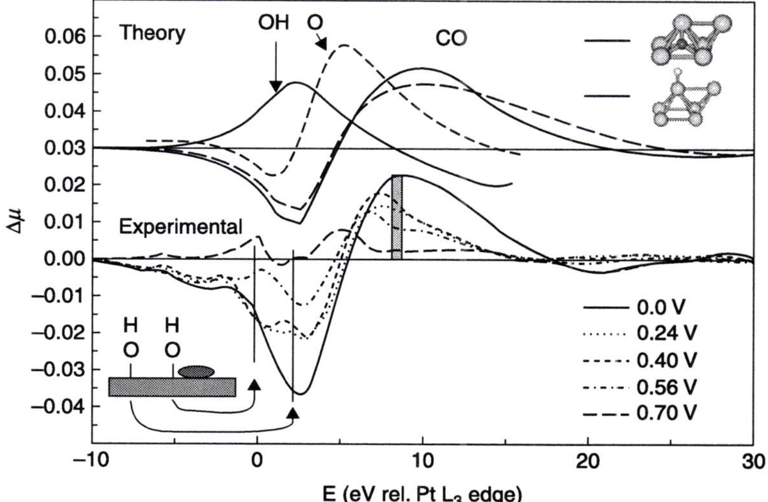

5.7 *Δμ* data at the Pt L$_3$-edge for a PtRu Watanabe sample using 0.40 V on the return cathodic scan as the reference (Scott *et al.*, 2007a). Theoretical signatures obtained from FEFF8.0 calculations are shown for CO adsorbed in atop and *n*-fold sites and for OH$_{at}$ and O$_{fcc}$. Shaded area indicates region utilized to determine relative CO *Δμ* magnitudes or coverage in Fig. 5.8. The peaks for OH$_{at}$/Pt near and far from Ru islands as described in text are indicated.

It is also possible to identify the neighbour atoms to the OH adsorption sites. It has been observed in previous work that the signature curves are significantly affected when OH is adsorbed on Pt near another 'foreign' metal atom such as Ni, Co or Ru (Teliska *et al.*, 2005b; Roth *et al.*, 2005). The fingerprint is found to be shifted downward in energy compared to OH adsorbed at a Pt site surrounded only by platinum atoms (see Fig. 5.7). This explains the three-peak structure between 0 and 6 eV; the peaks correspond respectively to atop OH/Pt near a Ru island, atop OH/Pt away from the islands, and bridged or fcc/hcp bonded O/Pt. The latter feature renders the $\Delta\mu$ XANES a suitable technique to unravel complex reaction mechanisms and synergistic actions that are often active in bimetallic catalysts.

As an example of this ability to unravel different reaction mechanisms, Fig. 5.8 compares the CO coverages obtained for three different PtRu catalysts in

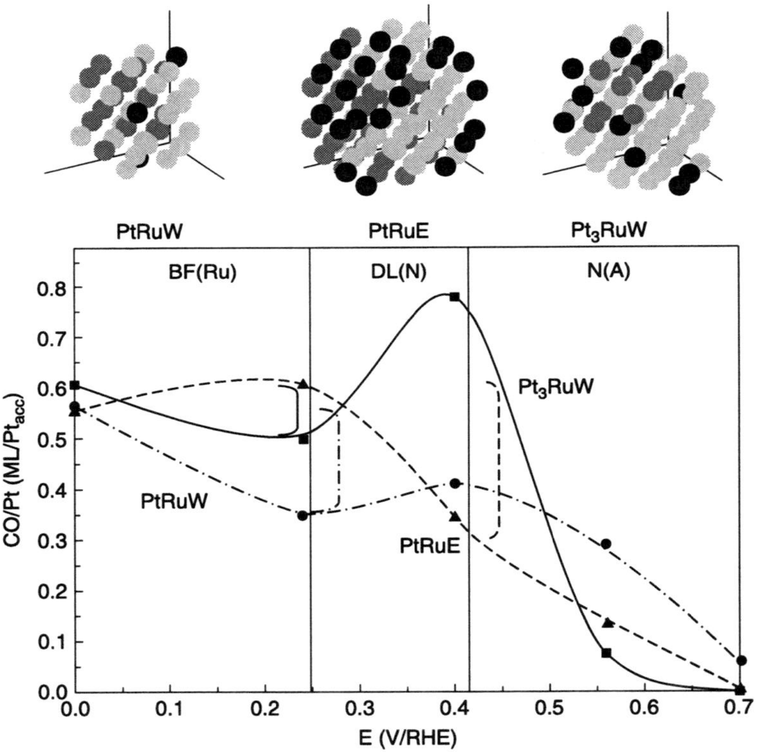

5.8 Estimated CO/Pt coverage in 0.3 M MeOH for the catalysts depicted (the morphologies determined from EXAFS analysis), and the indicated regions where each CO oxidation mechanism (BF(Ru) = bifunctional: OH/Ru, DL(N) = direct ligand: OH/Pt near Ru islands, and N(A) = normal: OH/Pt away from Ru islands) dominates (Scott *et al.*, 2007a). The brackets indicate the net amount of CO stripped via either the BF or DL mechanism due to mobile CO moving toward the Ru islands.

methanol, and the different regions where the CO poison resulting from incomplete oxidation of methanol is stripped from the Pt surface. These results were obtained from a similar $\Delta\mu$ XANES analysis as described above (Scott *et al.*, 2007a). The average PtRu particle morphology of the three different catalysts, as illustrated in Fig. 5.8, was determined by an EXAFS modeling procedure outlined previously (Scott *et al.*, 2007a). The mechanism for CO cleaning that dominates in each potential region is indicated by the source of the active OH (on the Ru, or on the Pt either Near or Away from surface Ru islands), which really differentiates the mechanisms

$$CO/Pt + OH/(Pt \text{ or } Ru) \rightarrow CO_2 + H^+ + e^- + 2^* \text{ (Pt or Ru)} \qquad [5.7]$$

and indeed is information provided by the $\Delta\mu$ XANES technique (* here indicates empty binding site). When the OH results from water activation on the Ru islands, this is generally referred to as the bi-functional mechanism; when it results from water activation near a RuO_x island at a lower potential than on normal Pt due to a ligand effect from the island, we refer to it as the 'direct ligand' mechanism (Scott *et al.*, 2007a). The more reactive Ru sites will activate water below 0.25 V, Pt sites near the Ru islands between 0.25–0.4 V because of the ligand effect from the nearby Ru atoms, and the 'normal' Pt sites away from Ru only well above 0.4 V. Since the CO coverage is depicted in Fig. 5.8, we are actually looking for a decrease in coverage, and defining the CO oxidation mechanism which dominates in producing that decrease in CO. Figure 5.8 shows that both Watanabe samples exhibit a significant bifunctional CO stripping component below 0.3 V, consistent with OH coming onto the Ru surface in this region. However, this fractional component is larger in the PtRuW case than in the Pt_3RuW case as indicated by the brackets in Fig. 5.8. This is consistent with the larger fraction of Ru in the former case, so that more CO/Pt is in the vicinity of the Ru islands. In contrast, the E-TEK sample exhibits a large direct ligand component and no bifunctional component.

These data strongly indicate that the bifunctional mechanism for CO oxidation is enhanced by the formation of smaller Ru islands (as exist in the PtRu W catalysts as illustrated in Fig. 5.8) with adsorption of CO in competition with OH adsorption at low potential, while the bigger RuO_x islands existing in the PtRu E catalysts are much more oxidized and adsorb no CO or OH. However, larger RuO_x islands are found to impose an increased ligand effect on the nearby Pt atoms enabling CO oxidation via OH directly adsorbed on the Pt. This interplay between Pt and Ru, apparently determined by the relative size of the Ru/RuO_x islands and Pt particles as suggested above, has been discussed previously but the data in Fig. 5.8 give the clearest evidence of these effects.

Figure 5.9 shows similar Pt L_3 $\Delta\mu$ data for 0.3 M methanol similar to Fig. 5.7, but for a PtMo catalyst (Scott *et al.*, 2010a). Note that now below 0.54 V, both H_{upd} and CO are observed. Estimates of the amount of each can be made from the magnitudes at the indicated regions in Fig. 5.9. The difference between Figs 5.7 and 5.9, showing the lack or presence of H_{upd}, results from the different behaviour

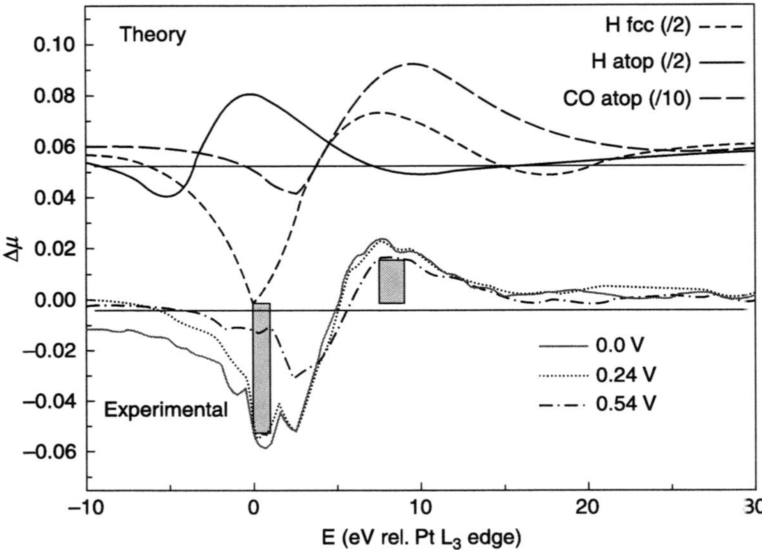

5.9 $\Delta\mu$ spectra at the Pt L_3 edge for a Pt_4Mo catalyst prepared as described by Scott *et al.* (2010a) using 0.4 V on the return cathodic scan as the reference for obtaining $\Delta\mu$. Experimental data are compared with theoretical signatures from FEFF 8.0 (scaled as indicated) for CO adsorbed in an atop site, and for atop and fcc H. Shaded areas indicate energy regions one could utilize to determine relative CO and fcc H_{upd} $\Delta\mu$ magnitudes or coverage. No atop H_{upd} is seen experimentally here.

of PtRu vs Pt_4Mo catalysts. The Pt-CO bond is apparently much weaker on the Pt_4Mo clusters enabling H to displace the CO at low potentials. In this example, one can clearly see the capability of the $\Delta\mu$ technique. It has been known for a long time that Pt-Ru and Pt-Mo show distinctly different activities in reformate and methanol operation (Mukerjee and Urian, 2002). The reasons for this behaviour, however, were largely unknown. With the help of the $\Delta\mu$ XANES analysis, a more weakened CO-bond on the Pt_3Mo catalyst is indicated compared to PtRu allowing H replacement at potentials below 0.3 V (Scott *et al.*, 2010a). This obviously aids CO removal in reformate, but much less so in methanol operation mode because the anode exists at higher potentials in methanol. Thus, by identifying basic mechanisms, the $\Delta\mu$ XANES technique can help in the design of tailor-made catalysts for specific requirements.

5.3.4   The direct ethanol fuel cell

Finally we give an example of an even more complex reaction, the ethanol oxidation reaction (EOR) in a fuel cell. Figure 5.10 compares the magnitudes of the $\Delta\mu$ signatures (i.e. |POS|) with potential (relative to DHE) in 1 M aqueous methanol and

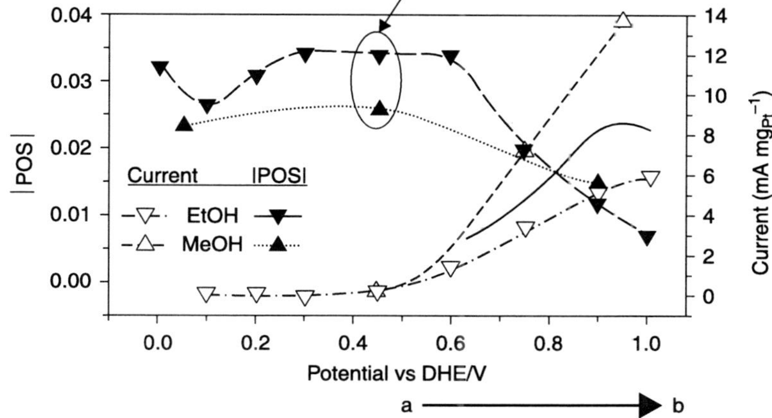

5.10 Steady-state coverage of CO,CH$_x$ and OH as indicated by $\Delta\mu$ amplitude |POS| along with the measured current in a fuel cell for either 1 M MeOH or EtOH solution vs dynamic hydrogen electrode (DHE) from Melke *et al.* (2010a). The arrow at the bottom indicates the sudden switch of the potential from point a at 0.45 V to b at 1.1 V used to obtain the results in Fig. 5.11. The circle with arrow highlights the larger |POS| for EtOH compared with MeOH as discussed in text.

ethanol solution as reported by Melke *et al.* (2010a). The |POS| $\Delta\mu$ amplitude for EOR initially decreases; we attribute this to H desorption from the surface between 0–0.2 V and the remarkable similarity between the $\Delta\mu$ signatures for H and CO makes it in this case difficult to separate these two. These were separable in Fig. 5.9 because the Pt-CO bond was weaker on PtMo as discussed above, and this altered the CO signature a bit making it more different from the H$_{upd}$ signature. From 0.1–0.3 V, |POS| increases, apparently due to increasing coverage of CH$_x$ and CO. Koper's group (Lai *et al.*, 2008) have found that CH$_x$ becomes a product during EOR at potentials as low as 0.1 V. We suggest the increase in |POS| here is due to the production of CH$_x$, and that together CH$_x$ plus CO makes $\Delta\mu$ larger even though the total coverage CO+CH$_x$ is comparable to that for CO in the methanol oxidation reaction (MOR) (Melke *et al.*, 2010a). The $\Delta\mu$ is larger for the EOR because CO is mostly in an atop singly coordinated site on the Pt, and the CH$_x$ is mostly doubly or even triply coordinated. FEFF8 calculations show that the $\Delta\mu$ signature (shape) for CH$_x$ and CO in an atop or *n*-fold coordination do not significantly differ, unlike that found for adsorbed H and O, but the magnitudes will differ if the coordination is different.

Above 0.6 V, the amplitude decreases again, now in parallel with the appearance of an O(H)-group signature (see Fig. 5.4). This decrease occurs at the same potential as the rise in current, indicating that the EOR is limited by the oxidation of CO or CH$_x$ to CO$_2$ just as the MOR is limited by CO oxidation. The conclusion regarding CO and/or CH$_x$ poisoning leads to an interesting question brought out by Fig. 5.10: why is the MOR current, $I = 13.7$ mA mg$_{Pt}^{-1}$, nearly three times larger at 0.9 V compared to that in EOR, $i = 5.5$ mA mg$_{Pt}^{-1}$, when both appear to be limited by carbon poisoning, and the $\Delta\mu$ carbon signature appears to be decreasing faster with potential for the EOR than for the MOR. This can be explained by several possible factors:

1 The Pt ensemble size, i.e., the number of sites needed for the reaction is probably larger in the EOR. Therefore the required larger Pt ensemble size could explain the slower increase in current with potential in the EOR.
2 As the C coverage decreases, C-C bond cleavage can also be a slower process than C-H cleavage, so the MOR may simply be faster than the EOR even when sufficient Pt sites are available.
3 Partial oxidation to other intermediates during the EOR may limit the current to considerably less than six electrons per initial adsorbate, as compared to the MOR.

The factor which dominates here may depend on the carbon coverage hence potential. The Pt ensemble size (1) will be more important when free Pt sites are minimal at lower potential, (2) when free Pt sites are plentiful at higher potential, and (3) possibly important at all potentials.

5.4   Advantages, limitations and future trends

XAS is an attractive *in situ* and *in operando* technique as shown by both past and present work, some of it reviewed above. In recent years, the more conventional EXAFS data analysis technique has been complemented by the newer AXAFS and $\Delta\mu$ XANES techniques. With AXAFS it is possible to follow the electronic effect that a support has on a metal particle as summarized elsewhere (Ramaker and Koningsberger, 2010a); with $\Delta\mu$ XANES it is now possible to determine the adsorbate, the specific adsorption sites and adsorbate coverage, which help a great deal to unravel the kinetic mechanisms operating in working reactors or fuel cell systems. Atop CO, opd and upd H, bridge-bonded or fcc O, and even OH, all possibly present on a metal surface in fuel cell-relevant operation, can now be monitored individually. However, the strength of XAS lies, in particular, in its ability to provide information on both nanoparticle morphology (EXAFS), electronic polarization (AXAFS), and adsorbate coverage ($\Delta\mu$ XANES), all *in situ* or under operating conditions with just one experimental spectroscopy. It also enables one to directly probe the structure-activity relationship even in bimetallic catalysts, e.g., a state-of-the-art PtRu fuel cell system.

There are of course significant limitations to the $\Delta\mu$ technique as suggested above. Theoretical FEFF8 results for several possible C-intermediates in the EOR have been performed (Melke *et al.*, 2010a). Unfortunately these calculations reveal that nearly all of the predicted $\Delta\mu$ signatures are essentially the same. This arises because the scattering is dominated by the adsorbate atom bonded to the photon absorbing Pt atom in the XAS process, and in this case the $\Delta\mu$ signatures for atop versus bridged sites are not significantly different, in contrast to that for H and O discussed above. Thus the $\Delta\mu$ technique cannot differentiate many complex carbon species on the surface during ethanol oxidation, although previously it was possible to differentiate between different molecular geometries (atop π-bonded, bridged di-σ bonded, and a dehydrogenated vertical ethylidyne species) as the temperature was increased, revealing the ability of the $\Delta\mu$ technique to differentiate between different bonding geometries (Bus *et al.*, 2007) on Pt(111) in the gas phase.

Other studies have shown that $\Delta\mu$ is applicable to other adsorbates on other metal surfaces. The combined $\Delta\mu$ XANES and EXAFS studies have recently been reported on many complex electrocatalysts, including RuS (Ziegelbauer *et al.*, 2006a, 2006b, 2009), Au/SnO_x at the Au and Sn edges (Gatewood *et al.*, 2008a), Pt/NbO_x at the Pt and Nb edges (Korovina *et al.*, 2010a), Pt/Polyvinyl Pyrrolidone (Shyam *et al.*, 2010) and Pt/TPPTP/C (TPPTP = triphenyl phosphine triphosphate) (Gatewood *et al.*, 2008b) catalysts, and more recently applied to $LiFePO_4$ (Love *et al.*, 2010) and Fe(VI) (Farmand *et al.*, 2010) cathodes in batteries to follow oxidation state and solid-electrolyte interface (SEI) layer changes. The $\Delta\mu$ XANES has also been used to study the effects of anion adsorption on Pt including bisulfate (Teliksa *et al.*, 2007) and Cl (Arruda *et al.*, 2008), and differentiate the various oxidation states of S during the electrochemical oxidation of S on Pt (Ramaker *et al.*, 2010a). It has also been used to follow the growth and morphology changes (i.e. aging) of PtRu particles with cycling in a fuel cell (Shyam *et al.*, 2009), as well as the deposition of Ru on a Pt cathode, and its effect on the oxygen reduction reaction (Arruda *et al.*, 2010). In all of these studies, unprecedented new details on the coverage and binding sites of adsorbates *in situ* or *in operando* have been revealed. Further, similar XAS studies in gas-phase catalytic reactors have been reported, including H on supported Pt clusters in zeolites and various flat supports (Ji *et al.*, 2007) as well as ethylene on Pt(111) as mentioned above, revealing the ability of the $\Delta\mu$ technique to differentiate between different bonding geometries (Bus *et al.*, 2007) on Pt, the ethylene-type intermediates in the hydro-epoxidation of propene over Au (Nijhuis *et al.*, 2009), and the H_2, H_2O, and CO intermediates in the water gas shift reaction (Guo *et al.*, 2010). As the complexity of the systems and reactions studied with the $\Delta\mu$ technique increase, biofuel cell investigations and biocatalysis experiments could also come into focus soon, with some experiments already underway (Arruda, 2009). We expect a much wider application of the combined $\Delta\mu$ XANES, AXAFS and EXAFS techniques in the future.

The ultimate goal is to study short-lived intermediates on the surface of a catalyst to obtain even more information on the kinetics occurring in a complex system such as, for example, a fuel cell with ethanol. This will require XAS data to be taken at a much faster rate than the normal data summarized above (about 30–50 minutes). Time-resolved data can be obtained either by quick EXAFS (QEXAFS) or by energy-dispersive XAS (EDXAS), both needing a dedicated beamline set-up with specific additional components. QEXAFS (taken in seconds up to minutes depending on scan rate and applied monochromator set-up) has been reported recently (Lutzenkirchen-Hecht *et al.*, 2009) with excellent results (Singh *et al.*, 2008; Melke *et al.*, 2010b; Ramaker and van Bokhoven, 2010). Time-dependent results for 1 M ethanol are summarized in Fig. 5.11, after the potential is suddenly switched from 0.45 V to 1.1 V as illustrated by the arrow in Fig. 5.10; i.e. when steady-state conditions show the electrode is covered mostly by CO,CH_x, to when it is covered by OH and O. How long does it take to convert from the initial coverage to the latter? Or how fast is the CO,CH_x oxidation process? The insert in Fig. 5.11 shows $\Delta\mu = \mu(\text{MeOH, t, 1.1 V}) - \mu(\text{H}_2\text{O, 0.45V})$ where the reference to evaluate $\Delta\mu$ is the same electrode in H_2O at 0.45 V when the Pt surface is relatively free of adsorbates. Note the expected change in the $\Delta\mu$ signature (as indicated by the FEFF8 calculations in Fig. 5.7) from that initially reflecting CO to that reflecting OH or O at later times. Data are recorded every 2

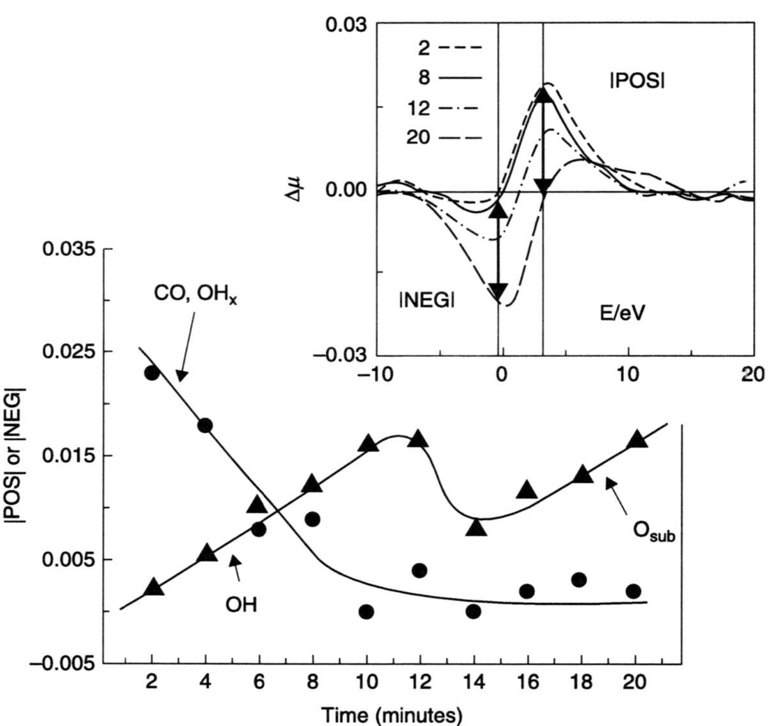

5.11 Plot of |POS| ∝ CO,CH_x and |NEG| ∝ OH,O as function of time after the potential was suddenly switched from 0.54 V, when the steady state coverage is mostly CO,CH_x in 1M EtOH, to 1.1 V when the steady state coverage is mostly OH,O as shown in Fig. 5.10 (Melke *et al.*, 2010b). The inset shows the $\Delta\mu = \mu(t) - \mu(\text{H2O}, 0.54\text{V})$ and the points at 0 and 3 eV where the amplitudes |NEG| and |POS| were measured.

minutes, and Fig. 5.11 shows it takes about 10 minutes to convert from mostly CO, CH$_x$ to OH,O under the existing fuel cell conditions. Beyond 10 minutes (i.e. when the OH adsorption is in excess), the adsorbed OH quickly begins to place exchange with the Pt to produce subsurface O. The bar chart in Fig. 5.4 shows that O$_{sub}$ has a smaller |POS| than OH, thus the sudden decrease in the |POS| magnitude shown in Fig. 5.11. We suspect similar time dependent results during the MOR would show a different change-over from CO to OH, O, assuming the CO oxidation rate is different than the CH$_x$, but unfortunately these results were not reported by Melke *et al.* (2010b).

The results in Fig. 5.11 were recorded every 2 minutes. Even higher speeds are possible on the Super-XAS beamline at the Swiss Light Source, as reported by Frahm *et al.* (2009) (about 5 s) and Singh *et al.* (2008), but this might still be too slow for some short-lived intermediates. Tada *et al.* (2007) have reported *in operando* time resolved results on a Pt/C cathode in a fuel cell using a 'time gated' QEXAFS (TG-QEXAFS) procedure giving a time resolution of 1 s. Still shorter time intervals are allowed using the energy-dispersive EXAFS (ED-EXAFS) technique with time resolution of the order 0.1–10 s. However, to date the signal-to-noise ratio with ED-EXAFS has been too small for good EXAFS or $\Delta\mu$ analysis, in particular when considering more inhomogeneous catalyst samples and less well-defined *in situ* studies (Allen *et al.*, 1995; Ressler *et al.*, 2000; Ramaker *et al.*, 2007). However, surely this will improve in the future.

We also expect spatially resolved data to be utilized in the future, since degradation processes in fuel cells and reactor beds in heterogeneous catalysis would especially benefit from regionally dependent information. Particularly exposed regions, e.g. at the fuel inlet in a direct methanol fuel cell, could be identified and probed. With sufficient resolution and using a µm-sized beam, spatially resolved $\Delta\mu$ XANES results should be possible. Studies on the variations in the electrode-electrolyte water content and its effect on the catalysts' long-term performance in a fuel cell are already underway (Roth *et al.*, 2009; Ramaker *et al.*, 2011). Finally, in addition to the original references given in this work, the $\Delta\mu$ technique has been summarized in several previous reviews with different examples given (Ramaker and Koningsberger, 2010a; Roth and Ramaker, 2010a, 2010b).

We are convinced that the future will hold many more interesting developments, since ever more advanced catalysts and system designs will force the characterization methods to keep pace.

5.5   References

Allen PG, Conradson SD, Wilson MS, Gottesfeld S, Raistrick ID *et al.* (1995), 'Direct observation of surface oxide formation and reduction on platinum clusters by time-resolved X-ray absorption spectroscopy', *J. Electroanal. Chem.*, **384**, 99–103.

Ankudinov AL, Ravel B, Rehr JJ and Conradson SD (1998), 'Real-space multiple-scattering calculation and interpretation of x-ray-absorption near-edge structure', *Phys. Rev. B*, **58**, 7565–7576.

Arruda T (2009), 'Characterization of Laccase of *Trametes Versicolor*: An enzyme for the electrochemical reduction of dioxygen for biological fuel cells', **5**, 185–244, PhD Thesis, Northeastern University.

Arruda T, Shyam B, Ziegelbauer J, Mukerjee S and Ramaker DE (2008), 'Investigation into the competitive and site-specific nature of anion adsorption on Pt using *in situ* x-ray absorption spectroscopy', *J. Phys. Chem. C*, 112, 18087–18097.

Arruda TM, Shyam B, Lawton JS, Ramaswamy N, Budil DE *et al.* (2010), 'Fundamental aspects of spontaneous cathodic deposition of Ru onto Pt/C electrocatalysts and membranes under direct methanol fuel cell operating conditions: An *in situ* X-ray absorption spectroscopy and electron spin resonance study', *J. Phys. Chem. C*, **114**, 1028–1040.

Bus E, Ramaker DE and van Bokhoven JA (2007), 'Structure of ethene adsorption sites on supported metal catalysts from *in situ* XANES analysis', *J. Am. Chem. Soc.*, **129**, 8094–8102.

Farmand M, Jiang D, Wang B, Ghosh S, Ramaker DE *et al.* (2011), 'Super-iron nanoparticles with facile cathodic charge transfer' *Electrochem. Commun.*, **13**, 909–912 (2011).

Feibelman PJ (1997), '*d*-electron frustration and the large fcc versus hcp binding preference in O adsorption on Pt(111)' *Phys. Rev. B*, **56**, 10532–10537.

Frahm R (2009), personal communication to Roth CR, Grenoble (2009).

Gatewood D, Ramaker DE, Sasaki K and Swider-Lyons KE (2008a), 'Support effects on water activation and oxygen reduction over Au–SnO$_x$ electrocatalysts observed with X-ray absorption spectroscopy', *J. Electrochem. Soc.* **155**, B834–B842.

Gatewood DS, Schull TL, Baturina O, Pietron JJ, Garsany Y *et al.* (2008b), 'Characterization of ligand effects on water activation in triarylphosphine-stabilized Pt nanoparticle catalysts by X-ray absorption spectroscopy', *J. Phys. Chem. C*, **112**, 4961–4970.

Guo N, Fingland BR, Williams WD, Kispersky VF, Jelic J *et al.* (2010), 'Determination of CO, H$_2$O and H$_2$ coverage by XANES and EXAFS on Pt and Au during water gas shift reaction', *Phys. Chem. Chem. Phys.*, **12**, 5678–5693.

Janin E, von Schenck H, Gothelid M, Karlsson UO and Svensson M (2000), 'Bridge-bonded atomic oxygen on Pt(110)', *Phys. Rev. B*, **61**, 13144–13149.

Ji Y, Koot V, van der Eerden AMJ, Weckhuysen BM, Koningsberger, DC *et al.* (2007), 'A three-site Langmuir adsorption model to elucidate the temperature, pressure, and support dependence of the hydrogen coverage on supported Pt particles', *J. Catal.*, **245**, 415–427.

Koningsberger DC, Mojet BL, van Dorssen GE and Ramaker DE (2000), 'XAFS spectroscopy; fundamental principles and data analysis', *Top. Catal.*, **10**, 143–155.

Koningsberger DC, Oudenhuijzen MK, van Bokhoven JA and Ramaker DE (2003), '*In situ* X-ray absorption spectroscopy as a unique tool for obtaining information on hydrogen binding sites and electronic structure of supported Pt catalysts: towards an understanding of the compensation relation in alkane hydrogenolysis', *J. Catal.*, **216**, 178–191.

Korovina A, Garsany Y, Epshteyn A, Swider Lyons KE and Ramaker DE (2010a), 'Effect of Niobium Oxide and Oxyphosphate Supports on the Reactivity of Pt Nanoparticles as Revealed by Electrochemical and *in situ* XAS Data', to be submitted.

Korovina A, Croze V, Giebeler L, Melke J, Mazurek M *et al.* (2010b), 'Correlation of fuel cell current with O, OH, and H$_2$O$_2$ adsorbate coverage on the Pt cathode using $\Delta\mu$ XANES', to be submitted.

Kua J and Goddard WA (1998), 'Chemisorption of organics on platinum. 1. The interstitial electron model', *J. Phys. Chem. B.*, **102**, 9481–9491.

Lai SC, Kleyn SEF, Rosca V and Koper MTM (2008), 'Mechanism of the Dissociation and Electrooxidation of Ethanol and Acetaldehyde on Platinum As Studied by SERS', *J. Phys. Chem. C*, **112**, 19080–19087.

Lewis EA, Segre CU and Smotkin ES (2009), 'Embedded cluster-XANES modeling of adsorption processes on Pt', *Electrochimica Acta* **54**, 7181–7185.

Love CT, Korovina A, Ramaker DE and Swider-Lyons KE (2010), '*In situ* EXAFS Study of SEI Formation on $LiCoO_2$ and $LiFePO_4$ Cathodes', Proc. Fall 2010 ECS meeting, Las Vegas.

Lützenkirchen-Hecht D, Wagner R, Haake U, Watenphul A and Frahm R (2009), 'The materials science X-ray beamline BL8 at the DELTA storage ring', *J. Synchr. Rad.*, **16**, 264–272.

Markovic NN, Schmidt TJ, Grgur BN, Gastieger HA, Ross PN *et al.* (1999), 'Effect of temperature on surface processes at the Pt(111)-liquid interface: hydrogen adsorption, oxide formation, and CO oxidation', *J. Phys. Chem. B*, **103**, 8568–8577.

Melke J, Schökel A, Dixon D, Cremers C, Ramaker DE *et al.* (2010a), 'Ethanol oxidation on carbon-supported Pt, PtRu, and PtSn catalysts studied by operando X-ray absorption spectroscopy', *J. Phys. Chem. C*, **114**, 5914–5925.

Melke J, Croze V, Schoekel A, Dixon D, Cremers C *et al.* (2010b), 'Time-resolved *in operando* X-ray absorption spectroscopy investigations and modelling of the ethanol oxidation reaction on Pt', *J. Phys. Chem. C*, submitted.

Mukerjee, S and Urian, RC (2002), 'Bifunctionality in Pt alloy nanocluster electrocatalysts for enhanced methanol oxidation and CO tolerance in PEM fuel cells: electrochemical and *in situ* synchrotron spectroscopy', *Electrochimica. Acta*, **47**, 3219–3231.

Nijhuis TA, Sacaliucb E, Beale AM, van der Eerden AMJ, Schouten JC *et al.* (2009), 'Spectroscopic evidence for the adsorption of propene on gold nanoparticles during the hydro-epoxidation of propene', *J. Catal.*, **258**, 256–264.

Oudenhuijzen MK, van Bokhoven JA, Miller JT, Ramaker DE and Koningsberger DC (2005), 'Three-site model for hydrogen adsorption on supported platinum particles: Influence of support ionicity and particle size on the hydrogen coverage', *J. Am. Chem. Soc.*, **127**, 1530–1540.

Papoian G, Norskov JK and Hoffman R (2000), 'A comparative theoretical study of the hydrogen, methyl, and ethyl chemisorption on the Pt(111) surface', *J. Am. Chem. Soc.*, **122**, 4129–4144.

Ramaker DE, Mojet BL, Garriga Oostenbrink MT, Miller JT and Koningsberger DC (1999), 'Contribution of shape resonance and Pt–H EXAFS in the Pt $L_{2,3}$ X-ray absorption edges of supported Pt particles: Application and consequences for catalyst characterization', *Phys. Chem. Chem. Phys.*, **1**, 2293–2302.

Ramaker DE, Teliska M, Zhang Y, Stakheev AYU and Koningsberger DC (2003), 'Understanding the influence of support alkalinity on the hydrogen and oxygen chemisorption properties of Pt particles: Comparison of X-ray absorption near edge data from gas phase and electrochemical systems', *Phys. Chem. Chem. Phys.*, **5**, 4492–4501.

Ramaker DE, Gatewood D, Beale AM and Weckhuysen BM (2007), 'ED-XAS data reveal *in situ* time-resolved adsorbate coverage on supported molybdenum oxide catalysts during propane dehydrogenation', *AIP Conf. Proc.*, **882**, 619–621.

Ramaker DE, Gatewood D, Korovina A, Garsany Y and Swider-Lyons KE (2010a), 'Resolving sulfur oxidation and removal from Pt and Pt_3Co electrocatalysts using *in situ* X-ray absorption spectroscopy'. *J. Phys. Chem. C.*, **114** (27), 11886–11897.

Ramaker DE and van Bokhoven JA (2010), 'A time-resolved $\Delta\mu$ XANES and mass spectroscopy study of CO oxidation over Pt/Al_2O_3', unpublished.

Ramaker DE and Koningsberger DC (2002), 'Comment on effect of hydrogen adsorption on the X-ray absorption spectra of small Pt clusters', *Phys. Rev. Lett.*, **89**, 139701.

Ramaker DE and Koningsberger DC (2010a) 'The atomic AXAFS and $\Delta\mu$ XANES techniques as applied to heterogeneous catalysis and electrocatalysis', *Phys. Chem. Chem. Phys.* **12**, 5514–5534.

Ramaker DE and Koningsberger DC (2010b), 'Mechanism of aromatic saturation on Pt, dual role of adsorbed hydrogen', *J. Catal.*, submitted.

Ramaker DE, Habereder A, Dixon D, Farmand M, Roth C (2011), 'Space-resolved, *in operando* X-ray absorption spectroscopy investigations on both the anode and cathode in a DMFC', ECS Meeting, Boston, MA.

Ressler T, Timpe O, Neisius T, Find J, Mestl G *et al.* (2000), 'Time-resolved XAS investigation of the reduction/oxidation of MoO_{3-x}', *J. Catal.*, **191**, 75–85.

Roth C, Benker N, Buhrmester T, Mazurek M, Loster M *et al.* (2005) 'Determination of O[H] and CO Coverage and Adsorption Sites on PtRu Electrodes in an Operating PEM Fuel Cell', *J. Am. Chem. Soc.*, **127**, 14607–14615.

Roth C, Benker N, Mazurek M, Scheiba F and Fuess H (2007), 'Pt-Ru fuel cell catalysts subjected to H_2, CO, N_2 and air atmosphere: An X-ray absorption study', *Appl. Catal. A: General*, **319**, 81–90.

Roth C, Dixon D, Ettingshausen F, Schökel A and Scheiba F (2009), 'Analysis of DMFC Catalysts', MRS Meeting, Boston, MA, 1–5 December.

Roth C and Ramaker DE (2010a), 'Operando XAS Techniques: Past, Present, and Future' in Wieckowski A and Nørskov JK, *Fuel Cell Science: Theory, Fundamentals, and Biocatalysis,* John Wiley & Sons, **17**, 511–544.

Roth C and Ramaker DE (2010b), 'XAS investigations of PEM fuel cells' in Vayenas C, *Modern Aspects of Electrochemistry,* Springer, **49**, 159–201.

Savitzky A and Golay MJE (1964), 'Smoothing and differentiation of data by simplified least squares procedures', *Anal. Chem.*, **36**, 1627–639.

Scott FJ, Mukerjee S and Ramaker DE (2007a), 'CO coverage/oxidation correlated with PtRu electrocatalyst particle morphology in 0.3 M methanol by *in situ* XAS', *J. Electrochem. Soc*, **154**, A396–A406.

Scott FJ, Roth C and Ramaker DE (2007b), 'Kinetics of CO poisoning in simulated reformate and effect of Ru island morphology on PtRu fuel cell catalysts as determined by operando X-ray absorption near edge spectroscopy', *J. Phys. Chem. C*, **111**, 11403–11413.

Scott FJ, Mukerjee S and Ramaker DE (2010a), 'Contrast in metal-ligand effects on Pt_nM electrocatalysts with M equal Ru vs Mo and Sn as exhibited by *in situ* XANES and EXAFS measurements in methanol', *J. Phys. Chem. C*, **114**, 442–453.

Scott FJ, Roth C and Ramaker DE (2010b), 'The effect of particle size, crystal face, and Ru ligand effect on the hydrogen oxidation reaction over Pt and PtRu', *J. Phys. Chem. C*, to be published.

Shyam B, Arruda TM, Mukerjee S and Ramaker DE (2009), 'Effect of $R_uO_xH_y$ island size on PtRu particle aging in methanol', *J. Phys. Chem. C*, **113**, 19713–19721.

Shyam B, Susit C, Tong YY and Ramaker DE (2010), 'Probing the influence of PVP on Pt/C nanoparticles in 0.1M $HClO_4$ using *in situ* X-ray absorption spectroscopy', *J. Phys. Chem. C*, (to be submitted).

Singh J, Alayon EMC, Tromp M, Safonova OV, Glatzel P *et al.* (2008), 'Generating highly active partially oxidized platinum during oxidation of carbon monoxide over Pt/Al_2O_3: *In situ*, time-resolved, and high-energy-resolution X-ray absorption spectroscopy', *Angew. Chem. Int. Ed.*, **47**, 9260–9264.

Smotkin E, Stoupin S, Lewis EA, Rivera H, Grice CR *et al.* (2008), 'Subtractively normalized XANES of Pt/C in operating fuel cells: theory-experiment', *Proceedings 214th ECS*, Honolulu, HA.

Stoupin S (2009), 'Influence of adsorbate-free atoms on $\Delta\mu$ XANES signatures', *J. Chem. Theory Comput.* **5**, 1337–1342.

Tada M, Murata S, Asaoka T, Hiroshima K, Okumura K *et al.* (2007), '*In situ* time-resolved dynamic surface events on the Pt/C cathode in a fuel cell under operando conditions', *Angew. Chem. Int. Ed.*, **46**, 4310–4315.

Teliska M, O'Grady WE and Ramaker DE (2004), 'Determination of H adsorption sites on Pt/C electrodes in $HClO_4$ from Pt L_{23} X-ray absorption spectroscopy', *J. Phys. Chem. B*, **108**, 2333–2344.

Teliska M, O'Grady WE and Ramaker DE (2005a), 'Determination of O and OH adsorption sites and coverage *in situ* on Pt electrodes from Pt L_{23} X-ray absorption spectroscopy', *J. Phys. Chem. B*, **109**, 8076–8084.

Teliska M, Murthi VS, Mukerjee S and Ramaker DE (2005b), 'Correlation of water activation, surface properties, and oxygen reduction reactivity of supported Pt–M/C bimetallic electrocatalysts using XAS', *J. Electrochem. Soc.* **152**, A2159–A2169.

Teliska M, Murthi VS, Mukerjee S, Ramaker DE (2007), 'Site-specific vs specific adsorption of anions on Pt and Pt-based alloys', *J. Phys. Chem. C*, **111**, 9267–9274.

van Bokhoven JA, Koningsberger DC and Ramaker DE (2001), 'The interaction of aromatic molecules with the acid site in zeolite H-Beta as observed through a newly discovered Fano resonance at the Al K edge', *J. Phys. Cond. Matter*, **13**, 10383–10393.

Wang JX, Springer TE, Liu P, Shao M and Adzic RR (2007a), 'Hydrogen oxidation reaction on Pt in acidic media: Adsorption isotherm and activation free energies', *J. Phys. Chem. C*, **111**, 12425–12433.

Wang JX, Zhang J and Adzic RR (2007b), 'Double-trap kinetic equation for the oxygen reduction reaction on Pt(111) in acidic media', *J. Phys. Chem. A*, **111**, 12702.

Wang JX, Uribe FA, Springer TE, Zhang J and Adzic RR (2008), 'Intrinsic kinetic equation for oxygen reduction reaction in acidic media: the double Tafel slope and fuel cell applications', *Faraday Discussions* **140**, 347–362.

Ziegelbauer JM, Gatewood D, Gullá AF, Ramaker DE and Mukerjee S (2006a), 'X-Ray absorption spectroscopy studies of water activation on an Rh_xS_y electrocatalyst for oxygen reduction reaction applications', *Electrochem. Solid-State Lett.*, **9**, A430–A434.

Ziegelbauer JM, Gatewood D, Ramaker DE and Mukerjee S (2006b), '*In situ* X-ray absorption spectroscopy studies of water activation on novel electrocatalysts for oxygen reduction reaction in acid electrolyte', *ECS Trans.*, **1**, 119–128.

Ziegelbauer JM, Gatewood D, Gullá AF, Ramaker DE, Mukerjee S *et al.* (2009), 'Fundamental investigation of oxygen reduction reaction on rhodium sulfide-based chalcogenides', *J. Phys. Chem. C*, **113**, 6955–6968.

Part II
Characterization of water and fuel management in polymer electrolyte membrane and direct methanol fuel cells

6

Characterization and modeling of interfaces in polymer electrolyte membrane fuel cells

T. SWAMY, The Pennsylvania State University, USA and
E. C. KUMBUR, Drexel University, USA

Abstract: Research on polymer electrolyte fuel cells (PEFCs) has shown the technology to be a viable clean energy solution for commercial and automotive applications. Further elucidation is required on the role of interfaces in performance and durability noting that interfacial regions are of critical importance. The chapter summarizes current research and understanding of the role of interfaces in PEFCs. Experimental and modeling efforts are discussed, emphasizing how these interfaces affect cell performance and durability. State-of-the-art interface characterization and imaging techniques are discussed. The contribution of PEFC interfaces to ohmic resistance and water transport and the techniques used are detailed. Future research to enable proper treatment of interfaces in PEFC models is discussed.

Key words: durability, interface; polymer electrolyte fuel cells; water management.

6.1 Introduction

Polymer electrolyte fuel cells (PEFCs) have tremendous potential as a power source for various applications. However, certain performance limitations have yet to be fully explained and optimized (Mench, 2008). In particular, ohmic and mass transport losses can originate at the various interfaces that exist between the fuel cell components, including the micro-porous layer and the catalyst layer (MPL|CL) interface, the micro-porous layer and the diffusion media (MPL|DM) interface and lastly, the diffusion media and the bipolar plate (DM|BP) interface. However, little is known or understood regarding interfacial structure and related effects on the performance and durability of PEFCs.

Due to the rough nature of the various layers in a PEFC, the contact between these layers is imperfect under compression. This imperfect contact may result in the formation of interfacial gaps between the contacting surfaces, and consequently a reduction in contact area (Fig. 6.1). The reduced contact area may interfere with the flow of electrons across the various interfaces and cause an increase in the electronic contact resistance. In locations such as the MPL|CL interface, the absence of electrochemical reactions and heat generation in the interfacial gaps may result in thermal disruptions, which will impact multi-phase flow as well. In addition, the interfacial gaps at the various interfaces may also act as water accumulation sites (pooling locations), which could have adverse affects on the performance of the

149

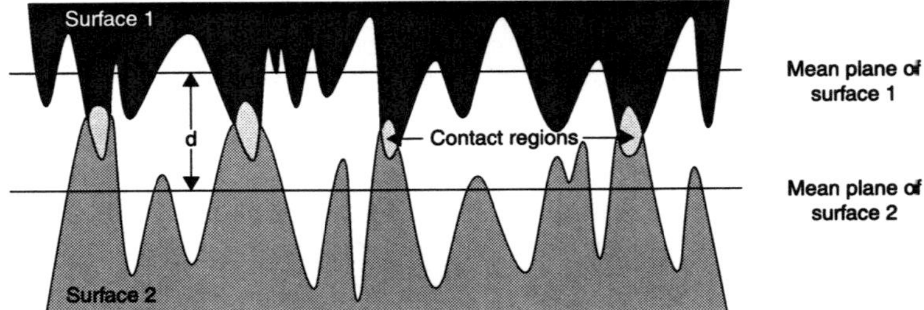

6.1 Two-dimensional (2D) schematic of the interfacial contact. (Source: adapted from Swamy *et al.*, 2009.)

PEFC. In order to better understand the aforementioned phenomena, emphasis is placed on evaluating the role of the interfacial morphology, compression, and material properties of these interfacial regions on PEFC performance.

6.2 Characterization of interfacial morphology in polymer electrolyte fuel cells (PEFCs)

There exists a gap in the literature regarding how the morphology of the interfaces affects the various fuel cell polarization losses. This can be attributed to the experimental limitations involved in the three-dimensional (3D) characterization of the highly irregular surfaces of the contacting layers. In the reported studies, optical microscopy, scanning electron microscopy (SEM) and atomic force microscopy (AFM) have been commonly employed to characterize the surface topography of the various layers in a PEFC (Kleemann *et al.*, 2009, Wang *et al.*, 2006, Zhang *et al.*, 2006). The reader is advised to refer to the aforementioned articles for a detailed description of characterization techniques.

Among the techniques used to characterize surface morphology, optical profilometry is one viable solution as it can provide three-dimensional and quantitative measurements of surfaces with a vertical resolution on the order of several nanometers, while scanning areas as large as several hundred thousand square micrometers in less than a minute without having direct contact with the surface (Podczeck, 1998, Lamb and Zecchino, 1999). However, a major disadvantage of optical profilometry is its inability to scan light-dispersive or poorly reflective surfaces. To achieve improved data acquisition, it is common practice to sputter poorly reflective surfaces, such as the CL, MPL and the DM, with a thin layer of gold (Au) for enhanced reflectivity (Podczeck, 1998, Chopra *et al.*, 2002). The gold layer thickness needs to be optimized to avoid introducing an artificial roughness to the surface by sputtering which can change surface structural properties (Podczeck, 1998, Chopra *et al.*, 2002, Faith and Horsfield, 2006). The method employed to determine the optimum sputtering time is detailed in the next section.

6.2.1 Determination of optimum sputtering time for optical profilometry measurements

Special sputtering protocols were developed for the non-reflective layers, i.e., the CL, MPL and DM, to determine the optimal sputtering time for improved profilometry measurements. For the catalyst layer, six different CL samples were sputtered with Au for 40 s, 1, 2, 3, 4 and 6 minutes, respectively. Arbitrary areas were scanned on each sample, and two-dimensional (2D) optical profilometry images of the scanned areas were captured for further analysis (Fig. 6.2). The number of black points on the images (missing data points due to low reflectivity or physical pores in the surface) was observed to dramatically decrease with increasing sputtering time up to 4 minutes. For sputtering times longer than 4 minutes, a significant improvement in optical profilometry images was not observed, indicating that 4 minutes of sputtering is sufficient to acquire the necessary set of surface scan data for the tested CL surface.

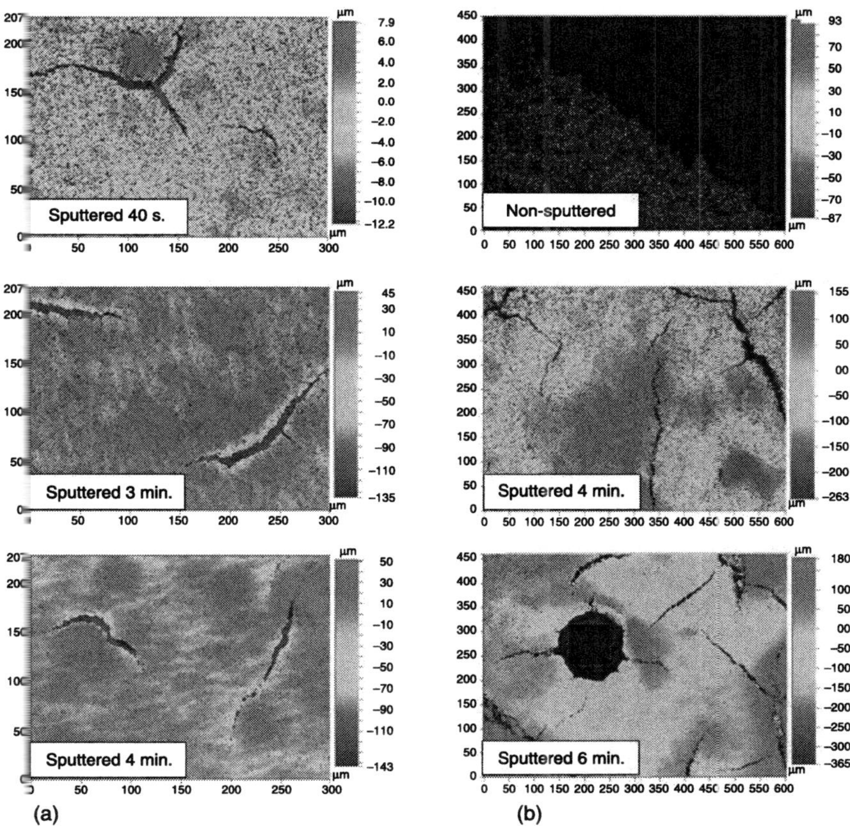

6.2 Two-dimensional optical profilometry images. (a) Naturally cracked CL. (b) MPL side of SGL 10BB. (Source: adapted from Hizir *et al.*, 2010.)

It should be noted that after 4 minutes of sputtering, the vast majority of black data points represents the surface pores of the tested CL samples.

Once the critical sputtering time was determined, the effects of increasing gold coating thickness on the surface morphology of the CL were investigated. This was accomplished by post-mortem SEM image analysis. Initially, the SEM images of a specific region on a virgin CL specimen were captured without any sputtering using a Hitachi S-3000H scanning electron microscope. Then, the SEM images of the same region were captured after sputtering of the same sample for 2, 3 and 4 minutes via the application of a series of sequential sputtering processes. The captured SEM images indicated a relatively small growth in surface particle size (on the order of 0.01 µm) and a small decrease in surface pore size (on the order of 0.01 µm) for 4 minutes sputtering (Fig. 6.3). This suggests that 4 minutes

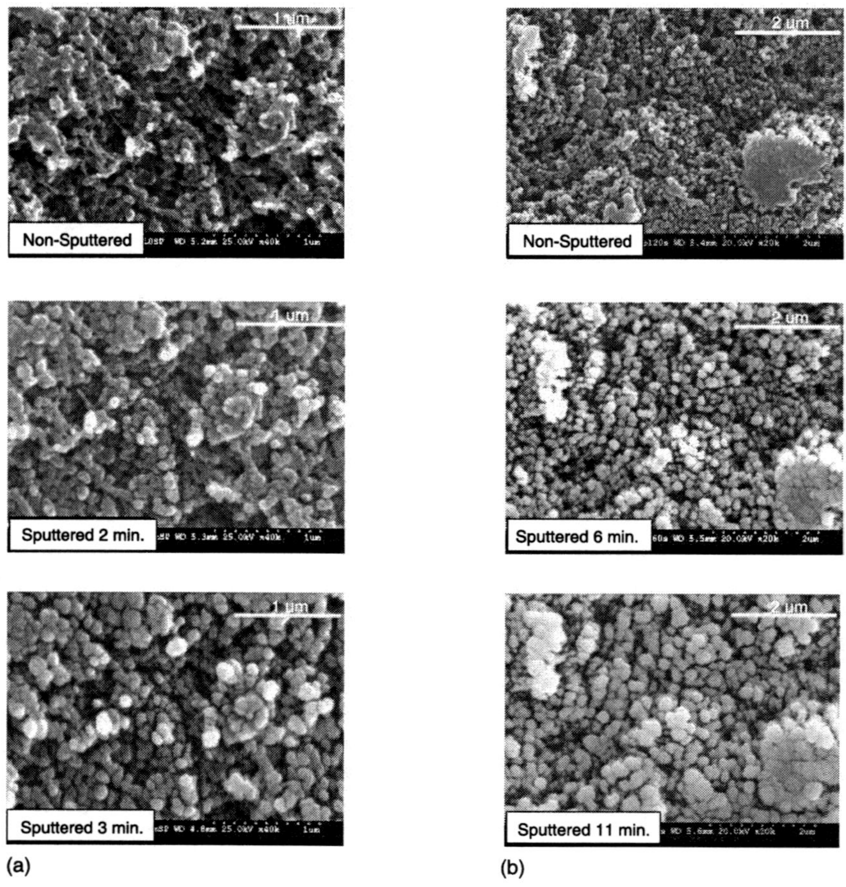

6.3 SEM images. (a) naturally cracked CL. (b) MPL side of SGL 10BB. (Source: adapted from Hizir *et al.*, 2010.)

of sputtering does not significantly change the surface morphology of CL samples. Therefore, for the material tested, it was determined that optical profilometry measurements can be conveniently performed on a 4 minute sputtered CL sample. Further information on the technique can be found in Hizir *et al.* (2010).

The steps mentioned above were similarly employed to determine the optimum sputtering time for the MPL and DM layers, which were found to be 6 minutes for either surface. Optical profilometry and SEM images of the MPL can be seen in Fig. 6.2b and Fig. 6.3b, respectively. Figures 6.4 and 6.5 depict the BP and post-sputtered DM images. The reader is encouraged to refer to Hizir *et al.* (2010) and Swamy *et al.* (2011) for further information on the images presented.

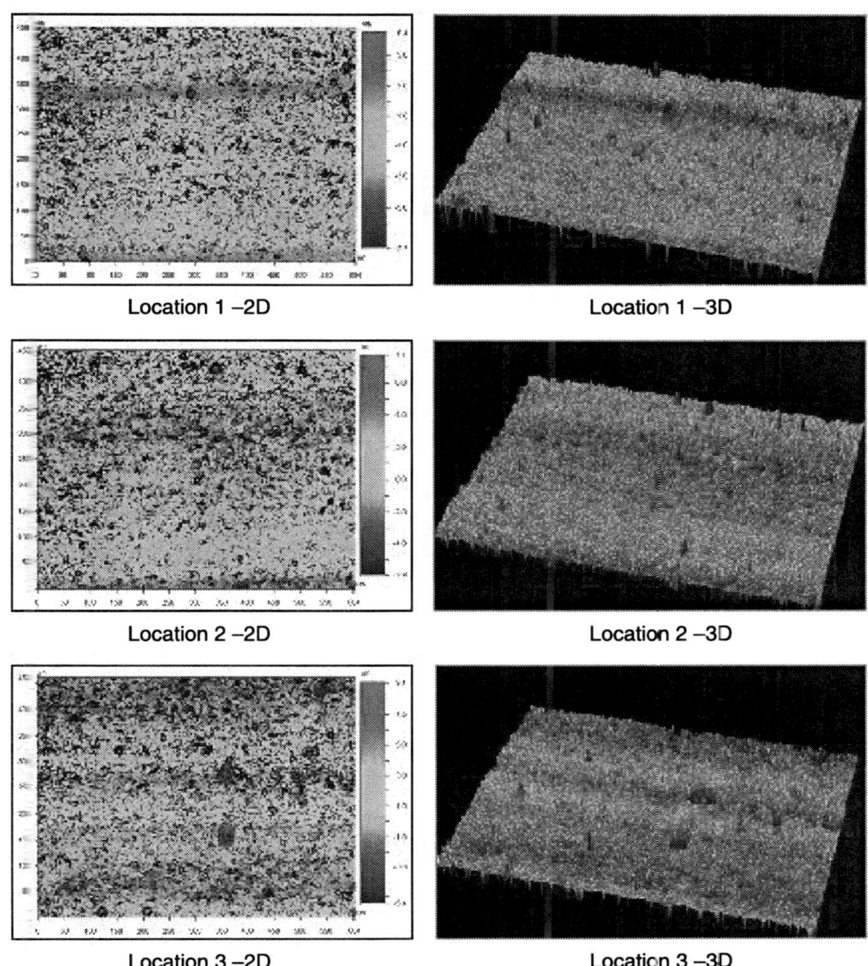

6.4 2D and 3D optical profilometry images of graphite BP surface at various locations.

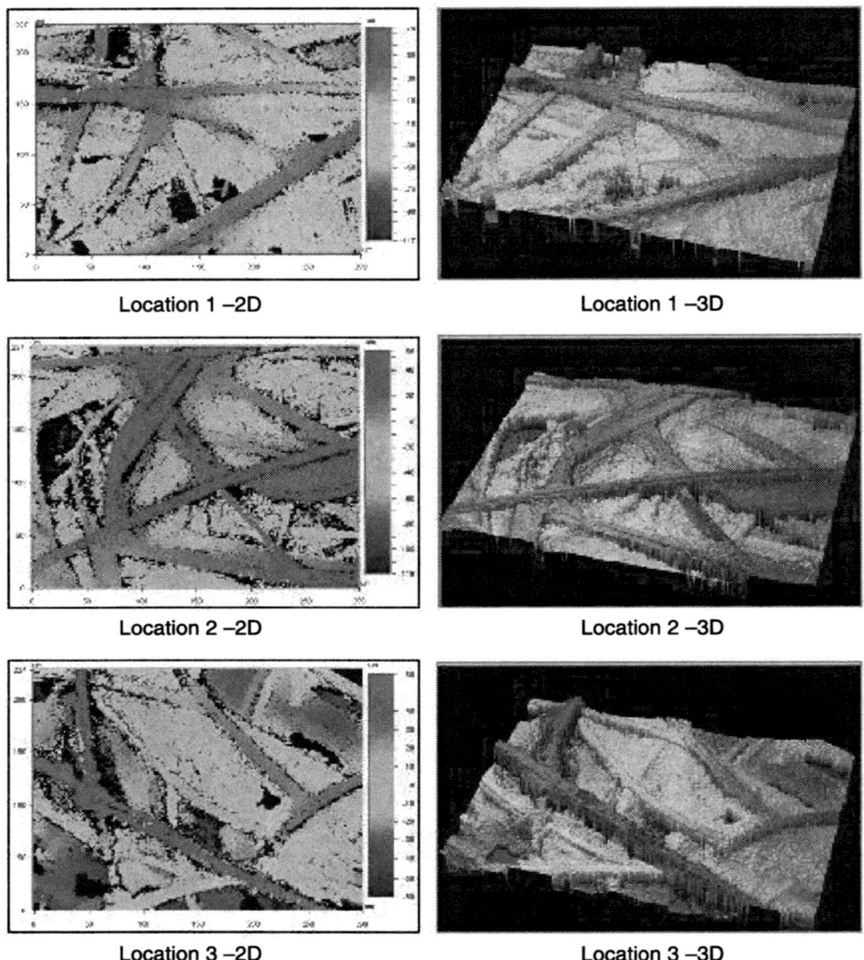

Location 1 –2D   Location 1 –3D

Location 2 –2D   Location 2 –3D

Location 3 –2D   Location 3 –3D

6.5 2D and 3D optical profilometry images of SGL 10BB DM surface at various locations.

6.2.2   Results and discussion

Surface morphological characteristics

Optical profilometry results indicate that the BP, DM, MPL and CL surfaces exhibit a relatively high degree of roughness, having perturbations in the form of high hills, large valleys and deep cracks on the surface. Representative optical profilometry images of the MPL and CL are shown in Fig. 6.6 and 6.7. 3D surface

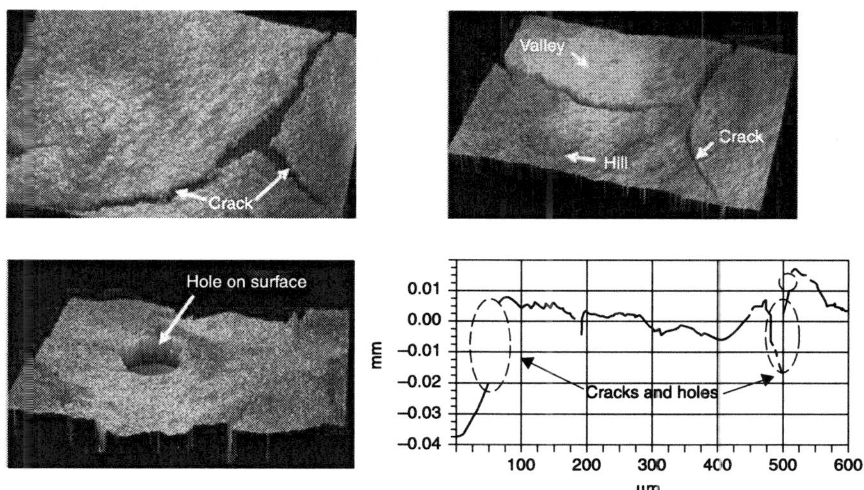

6.6 Representative three-dimensional images and surface profile of MPL of SGL 10BB obtained using optical profilometry. (Source: adapted from Hizir *et al.*, 2010.)

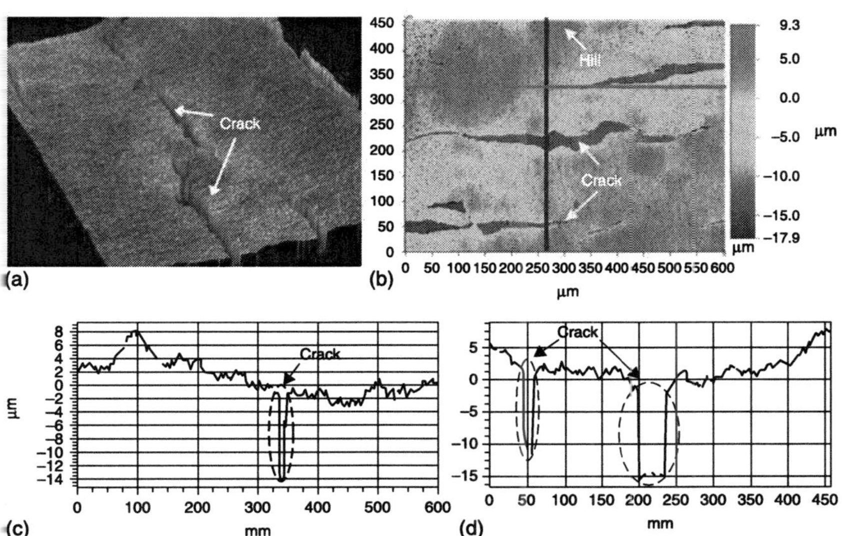

6.7 Representative three-dimensional image (a), two-dimensional image (b), surface profile in *x*-direction (c), and surface profile in *y*-direction (d) of cracked catalyst layer obtained using optical profilometry. (Source: adapted from Hizir *et al.*, 2010.)

morphology of the MPL can be seen in Plate IV(a) in the colour section between pages 252 and 253. Since the use of a single amplitude parameter can lead to dubious conclusions (Lamb and Zecchino, 1999), a set of six amplitude parameters (average roughness (R_a), root mean square roughness (R_q), average maximum height of the profile (R_z), maximum profile peak height (R_p), maximum profile valley depth (R_v), maximum height of the surface (R_t)) were used to compare surface roughness of MPL and CL surfaces. The reader is referred to British Standard 1134 (1972) for a detailed description of the amplitude parameters. The amplitude parameters, which essentially provide a measure of the surface roughness, were observed to be the highest for the DM, followed by the MPL, CL and BP, listed in a decreasing order. The three principal average roughness metrics for the various PEFC components are provided in Table 6.1.

SEM images show that there are significant differences between the cracks on the MPL and CL surfaces in terms of their orientation, size, shape, depth and density (Figure 6.8). Cracks on the tested CL surface are in the shape of thin ribbons that are aligned with respect to each other. The SEM images and the height values in the profilometry data indicate that cracks can extend through the entire CL thickness, having an average crack density of 3.4% ± 0.2%. On the contrary, cracks on the MPL surface are randomly oriented with larger and

Table 6.1 Measured surface roughness data of PEFC components (arithmetic mean values ± standard deviation of measurements)

Sample	Average roughness, λ_a (μm)	RMS roughness, λ_q (μm)	Crest-trough roughness, λ_t (μm)
MPL	5.35 ± 1.96	7.39 ± 3.20	68.32 ± 19.95
Cracked CL	2.19 ± 0.20	3.63 ± 0.21	26.56 ± 3.19
BP	1.38 ± 1.12	1.64 ± 0.84	14.36 ± 1.43
DM	19.65 ± 2.15	25.21 ± 1.43	161.56 ± 23.54

Source: adapted from Hizir *et al.*, 2010 and Swamy *et al.*, 2011.

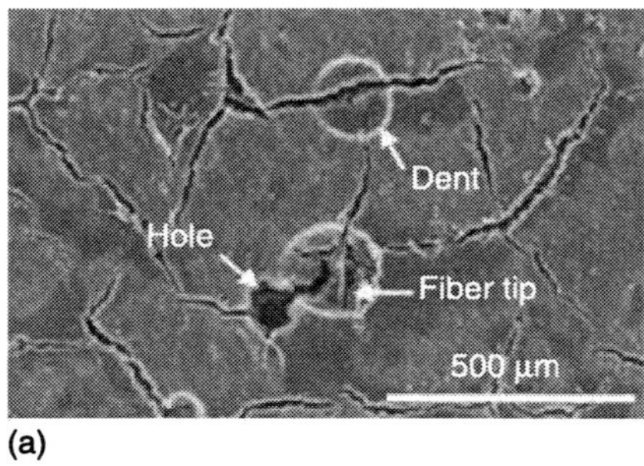

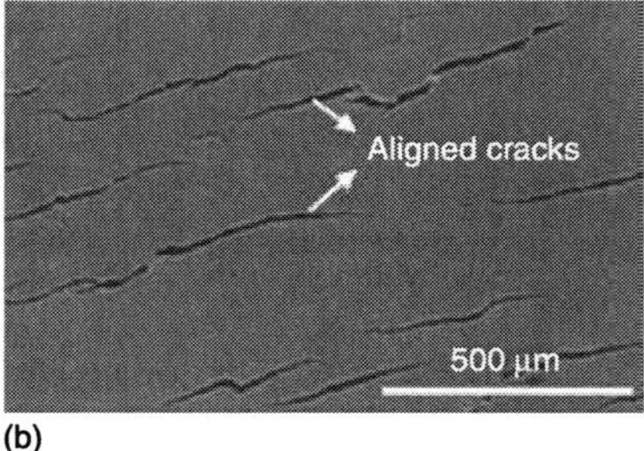

(a)   (b)

6.8 Representative SEM images. (a) MPL side of SGL 10BB. (b) Cracked CL. (Source: adapted from Hizir *et al.*, 2010.)

variable width, which can be as large as 60 μm. Holes and dents with diameters on the order of 100 μm are also present on the MPL surface. These frequent and relatively large cracks could conceivably play an important role in the multiphase flow along and through the interfaces, a topic that is in great need of study. Although optical profilometry data regarding crack depth is missing, the data obtained from few points on the crack base indicates that the crack depth can be as large as the MPL thickness. This can be observed in the captured SEM images, where the fibers belonging to the DM are visible through some of the cracks on the MPL surface. These observations form strong evidence that a certain portion of the cracks on the MPL surface can extend through the entire MPL thickness, reaching to the DM|MPL interface.

Optical profilometry images of the BP and DM show that there are significant differences between the surface morphological features of the BP and DM in terms of their orientation, size, shape and depth, as shown in Fig. 6.4 and 6.5. Figure 6.4 clearly highlights the constituent fiber matrix in the DM materials (which has randomly oriented fibers), which is responsible for the observed contours on the surface of the DM. 3D surface morphology of the DM can be seen in Plate IV(b). It is very likely that the existence of the randomly oriented fiber matrix on the DM surface results in large void regions and irregular surface perturbations (Fig. 6.4), yielding relatively high degrees of surface roughness patterns. The presence of large void regions indicates the existence of possible water accumulation sites at the BP|DM interfaces. While the DM exhibits considerable surface roughness, Fig. 6.5 indicates that the graphite BP possesses a relatively smoother surface. The reader is encouraged to refer to Swamy *et al.* (2011) for further information.

Implications of the results

As the peak height and crack depth of the MPL and CL are on the same order of magnitude as their thickness, the volume of the interfacial gaps at the MPL|CL interface is expected to be comparable to the total pore volume of the MPL and CL. This suggests that the interfacial gaps at the MPL|CL interface could have a large water storage capacity and consequently a strong impact on the mass transport losses. Moreover, in sub-freezing environments, water in the interfacial gaps may result in ice lens formation and growth between the contacting layers. With repeated freeze/thaw cycles, this may lead to interfacial delamination between the contacting layers, permanent deformation of the DM and fracturing of its carbon fibers resulting in reduced longevity and operational stability.

Although the MPL is typically considered to act as a capillary barrier that directs liquid flow from cathode to anode (Lin and Nguyen, 2006, Weber and Newman, 2005), the macro-cracks and holes in the MPL structure may act as water transport channels that carry the liquid water from cathode CL to the cathode DM|MPL interface (Fig. 6.9). Gostick *et al.* (2009) argued that cracks in the MPL

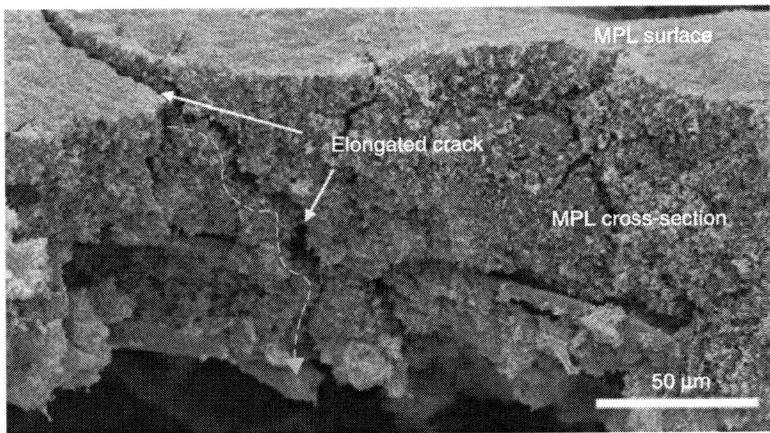

6.9 Cross-sectional SEM image of MPL of SGL 10BB. Arrow indicates the flow path formed by the cracks. (Source: adapted from Hizir *et al.*, 2010.)

could be beneficial, as they have the potential to reduce the DM saturation by limiting the number of water-inaccessible clusters formed in the DM. They suggested that water entry into the DM takes place mainly through the cracks on the MPL surface. Hence, the pores on the remaining portions of the DM face (where there is no overlap with the cracks on the MPL face) remain inaccessible to water and form paths for gas-phase reactant transport.

Furthermore, cracks in the MPL could alter the saturation and water mass distribution in the MPL substantially by holding a significant amount of water. Approximating the crack depth to be as large as the MPL thickness for all the cracks in the MPL and the MPL thickness to be 80 µm, the MPL cracks can hold 0.22–0.71 mg/cm^2 of water when fully filled for the calculated area crack density of the MPL. In his recent neutron imaging work, Turhan *et al.* (2010) observed a reversed saturation profile at the DM|MPL interface with the MPL having a higher level of saturation than the DM, in contrast to the modeling studies that predict a higher level of DM saturation for similar conditions. In addition to phase-change-induced flow (Kim and Mench, 2009), this unexpected increase in the MPL water saturation can be explained by the liquid water held by the MPL cracks. Cross-sectional X-ray radiography results (Hartnig *et al.*, 2008) indicating the existence of significant water accumulation around the MPL|CL and DM|MPL interfaces may also be attributed to the wide MPL cracks that are filled with liquid water. Similarly, Nam *et al.* (2009) claimed that the size of the pores on the MPL surface controls the interfacial saturation and size of the droplets formed at the MPL|CL interface, supporting the theory that MPL cracks could promote droplet growth, coalescence and accumulation. Hizir *et al.* (2010) provides further information on the morphology related phenomena at the interfaces.

6.3   Experimental investigation of interfaces in PEFCs

As mentioned earlier, the impact of the interfaces in a PEFC, and in particular the mating surface morphology, on water storage and motion are not yet well understood. Recent visualization studies have hinted that interfacial effects may be responsible for altering the saturation profile and water storage near the MPL|CL and DM|MPL interfaces (Turhan *et al.*, 2006, Hartnig *et al.*, 2008, Swamy *et al.*, 2009, Hizir *et al.*, 2010, Bajpai *et al.*, 2010, Turhan *et al.*, 2010). Discussed herein are *in situ* experiments conducted on unaltered and laser-perforated DM to further understand the effect of modified pore characteristics and irregularities (*i.e.*, tailored perforations) within the connected pore network on performance and transport characteristics of PEFCs. The effect on water pooling at the various component interfaces is also highlighted.

6.3.1   Experimental methods and setup

Laser perforation

A ytterbium fiber laser was used to introduce evenly-spaced through holes, each with a diameter of 300 μm, to the cathode-side DM and MPL of a PEFC, covering approximately 15% of the total geometric surface area (Fig. 6.10a, b). The laser technique has the advantage of physically burning carbon/binder material away, instead of simply translating material to the surrounding areas as may happen with micromachining techniques (Gerteisen *et al.*, 2008). Using the laser, however, causes a 'heat-affected zone', seen in Fig. 6.10a as the dark region surrounding the perforation. The diameter of the heat-affected zone is dependent on the key laser parameters previously mentioned. The heat-affected zone was analyzed using energy dispersive X-ray spectroscopy (EDS) and an environmental scanning electron microscope (ESEM). The EDS showed only a dominant carbon peak in the heat-affected zone (Fig. 6.10c), indicating the absence of PTFE, and the unaffected regions (beyond the heat-affected zone) displayed carbon and fluorine peaks (Fig. 6.10d), showing the presence of PTFE among the carbon. The ESEM showed a marked decrease in contact angle of water droplets in the heat-affected zone, indicating that the heat-affected zone was more hydrophilic than the non-affected portion of the MPL surface. The employed experimental testing techniques are discussed below.

In situ polarization testing

The materials used for testing were SGL 10 BB series DM with MPL (SIGRACET Gas Diffusion Layer™) and commercially available membrane electrode assemblies to investigate the effect of laser-perforated DM (herein, 'perforated DM') on PEFC performance. The anode-side DM in each experiment remained unaltered (herein, 'virgin DM'). *In situ* diagnosis was performed on an in-house-designed, 5 cm^2 active area, double serpentine fuel cell using a Scriber Associates

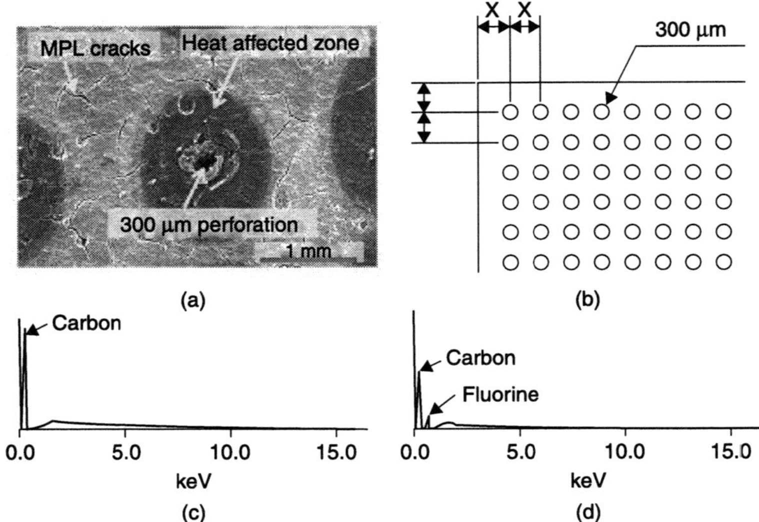

6.10 (a) SEM image of the MPL surface with laser perforations. A heat-affected zone and 300 µm perforations are highlighted as key modifications due to the laser treatment, whereas the MPL cracks occur on both laser-treated and virgin samples. (b) Schematic of laser perforations, where x = 0.67 mm for polarization testing and 0.97 mm for NR testing. (c) An EDS spectrum of MPL surface within the heat affected zone. There is no fluorine peak, indicating the absence of PTFE. (d) An EDS spectrum of the virgin MPL surface, showing dominant carbon and fluorine peaks. (Source: adapted from Manahan *et al.*, 2011.)

850C fuel cell test system. Tests were conducted with over-humidified (120%/120% anode/cathode) and low-humidity (50%/50% anode/cathode) inlet reactant gases, as well as at high (75 °C) and low (50 °C) temperatures with both nitrox and heliox cathode gases (N_2 and He as the inert cathode gas mixed with 21% O_2 in a dry state, respectively) at constant flow rates of 139 and 332 sccm anode and cathode, respectively, with atmospheric exit pressure. This technique is described in Manahan *et al.* (2009) and summarized in Manahan *et al.* (2011). In summary, a variety of conditions were chosen to elucidate the gas- and liquid-phase transport phenomena associated with the introduction of laser-perforated DM.

Neutron radiography

Neutron radiography (NR) testing was conducted in order to observe and quantify alterations in water management characteristics of both virgin and perforated DM. The high-resolution imaging system had a resolution of *ca.* 13 µm, and images had a nominal pixel pitch of 10 µm. The fuel cell was imaged in the through-plane direction, and a schematic of the cell and perforated DM is shown in Fig. 6.11.

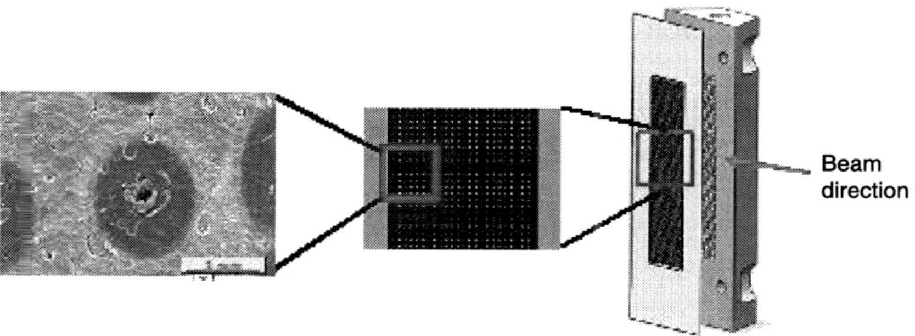

6.11 Rendering of the neutron radiography test cell with expanded view of perforated DM schematic and SEM image. (Source: adapted from Manahan *et al.*, 2011.)

After appropriate break-in procedures, each cell (virgin and perforated DM) was operated at a constant current for 30 minutes at 65 °C at an inlet relative humidity of 100%/100% and 50%/50% (anode/cathode) at a stoichiometry of 2/2 hydrogen/ nitrox (air). In high humidity cases, the current density selected was 0.2 A cm^{-2} and 1.2 A cm^{-2}; in the low humidity case the current densities of 0.2 A cm^{-2} and 1.7 A cm^{-2} were chosen. During this 30-minute interval, neutron images and performance data were collected. This spectrum of low and high current density under low and high humidity conditions characterized the performance and water content differences between the virgin and perforated DM cases.

6.3.2   Results and discussion

Effect of perforations on performance and water accumulation

The following sections describe the results of low and high humidity polarization testing of perforated and unaltered DM cases.

Low humidity conditions – polarization testing

In these tests, cell fixtures with a perforated cathode-side DM and with a virgin (unaltered) cathode-side DM were exposed to low humidity conditions (50% relative humidity) with nitrox (21% O_2 and 79% N_2), which was used as an inlet cathode gas. Figure 6.12 shows the IR-compensated polarization data of these two cases obtained at low humidity conditions.

While the perforated DM shows an increase in performance in the *lower* current regions, the significant performance drop in the *higher* current regions (i.e. greater than 1.4 A cm^{-2}) indicates the existence of increased mass transport losses, possibly due to the perforations in the DM (Fig. 6.12). While comparable error bars between the two cells are seen at 1.6 A cm^{-2}, the virgin cell is observed to quickly recover

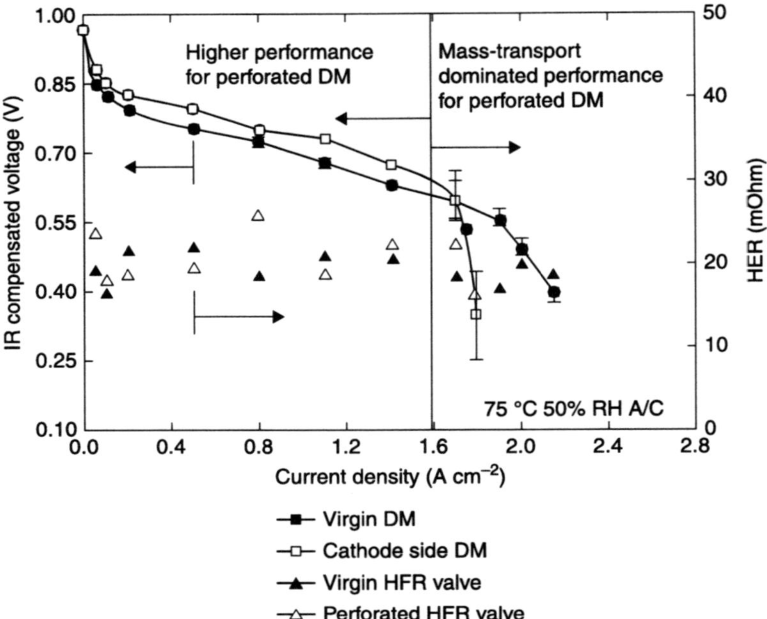

6.12 Performance data for 75 °C, 50% relative humidity, nitrox conditions with virgin DM, 15% 300 μm perforated cathode-side DM, and corresponding HFR values for the virgin and perforated cases. Perforated DM shows on average 38 mV higher potential in current densities less than 1.4 A cm^{-2} but fail to achieve steady state at current densities greater than 1.4 A cm^{-2}. (Source: adapted from Manahan *et al.*, 2011.)

the mass transport losses and regain stability as it effectively removes liquid water build-ups from reaction sites. The observed difference in performance characteristics of these two cases at low humidity conditions can be attributed to the fact that while the unmodified pore structure of the virgin DM can handle more effectively liquid water removal from the electrochemical active regions, the cell with perforated DM can store significant water most probably in the perforated regions, promoting severe local flooding at high current densities. A further analysis of the observed phenomena can be found in Manahan *et al.* (2011).

Low humidity conditions – neutron radiography testing

To further analyze the potential impact of perforations on the cell performance, neutron imaging tests were performed under low humidity conditions. For low current densities, neutron images indicate little qualitative distinction between the water amounts of the virgin and perforated DM, as shown in Fig. 6.13a. At the higher current, however, the perforated DM is observed to have significantly higher liquid water mass in all locations, most likely due to the increased water pooling in the perforation sites.

In Fig. 6.13, $x = 0$ mm corresponds to a row of pixels normal to a centrally-located land|channel interface, and $x = 7$ mm corresponds to the length of four lands and three channels (each 1 mm wide). The secondary x-axis is included to clarify the locations from which water mass values were extracted. In both low and high current density conditions, the periodic peaks in the perforated DM are observed at locations corresponding to $x = 0.56$, 1.53, 2.48, 3.37, 4.22, 5.16, 6.07, 6.99 mm (Fig. 6.13). These periodic peaks, which are not observed in the virgin DM case, indicate the possible pooling locations induced by the perforations, which had a center-to-center distance of *ca.* 0.97 mm. In a recent study by Gerteisen *et al.* (2008), it has been suggested that perforations may create paths of least resistance for water to flow through the DM towards the channels. Therefore, it can be hypothesized that under low-humidity operating conditions, the perforated DM exhibits the potential benefit of dictating the preferred path of liquid water in the water management process.

When the data in Fig. 6.13b are integrated across the x-axis, the results show that the perforated DM contains 46% more water mass than the virgin DM at 1.7 A cm^{-2}. From Fig. 6.14, it is clear that this additional 46% water mass is a result of the accumulated water in the perforated regions at the cathode. The water accumulation into perforated regions can be explained by analyzing the water transport modes inside the cell. It was shown by Kim and Mench (2009) that the phase change induced (PCI) flows tend to increase with hydrophobic content in the DM regardless of the temperature gradient. Since the laser pyrolyzes the PTFE coatings at the vicinity of the perforated regions (i.e. making these regions more hydrophilic), it is very likely that the water removal by PCI flow would become less favorable in the perforated DM. In addition, it is very likely that the capillary flow can be inhibited by the perforations, such that introducing large perforations (300 μm, as in this study) significantly decreases the capillary pressure due to the increased pore size, impeding the removal of liquid water via capillary-driven transport within the cell. As a result, water accumulation is more likely to dominate the performance of the perforated DM cell at higher current operations. A further analysis of the described phenomena can be found in Manahan *et al.* (2011).

High humidity conditions – polarization testing

Figure 6.15 shows the high humidity condition (120% inlet relative humidity anode and cathode) performance data for the same setup previously described. As seen in Fig. 6.15, the perforated DM case exhibits extremely poor performance under high humidity condition, showing a 183 mV lower potential than the virgin DM at a current density of 0.1 A cm^{-2}. Furthermore, it was not possible to reach stable operation higher than 0.1 A cm^{-2} for the perforated DM. The virgin DM,

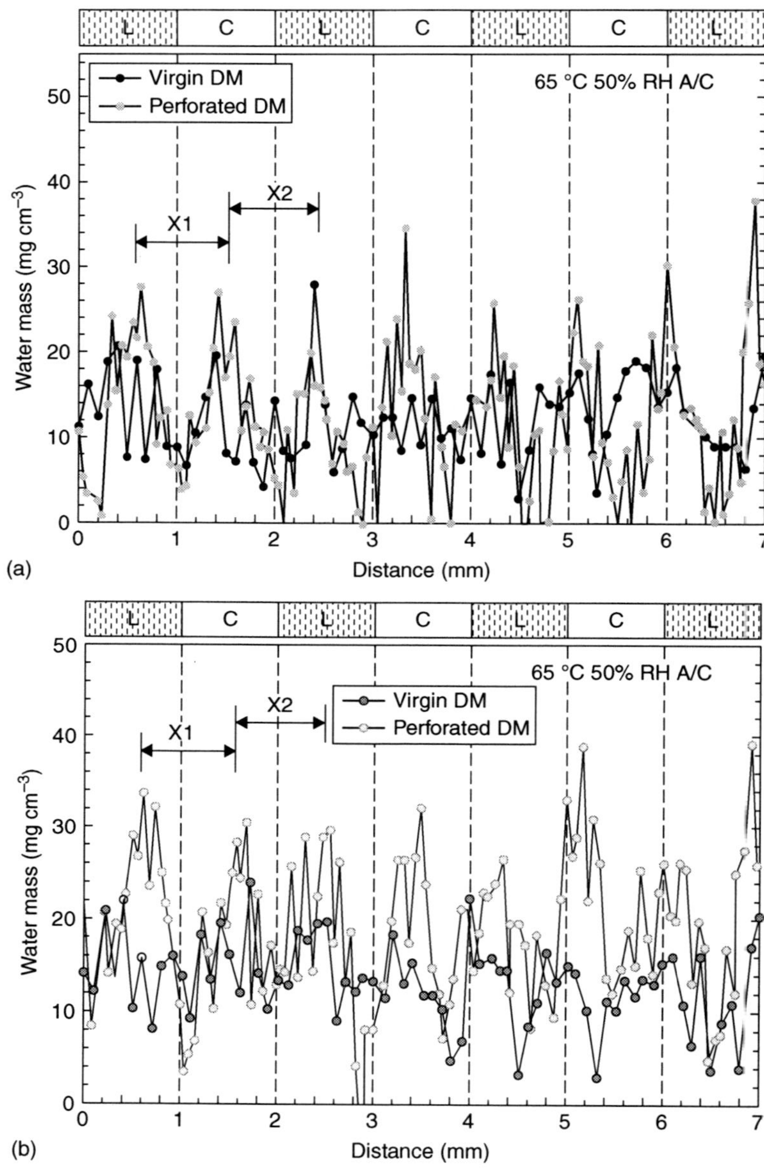

6.13 Neutron data for water mass per volume distributions in the in-plane direction at low inlet relative humidity condition. L represents land, and C represents channel. (a) Water distributions for virgin and perforated DM under low current density (0.2 A cm^{-2}) testing operation. (b) Water distributions for virgin and perforated DM under high current density (1.7 A cm^{-2}) testing operation. A reduced number of data points are shown to improve clarity. (Source: adapted from Manahan *et al.*, 2011.)

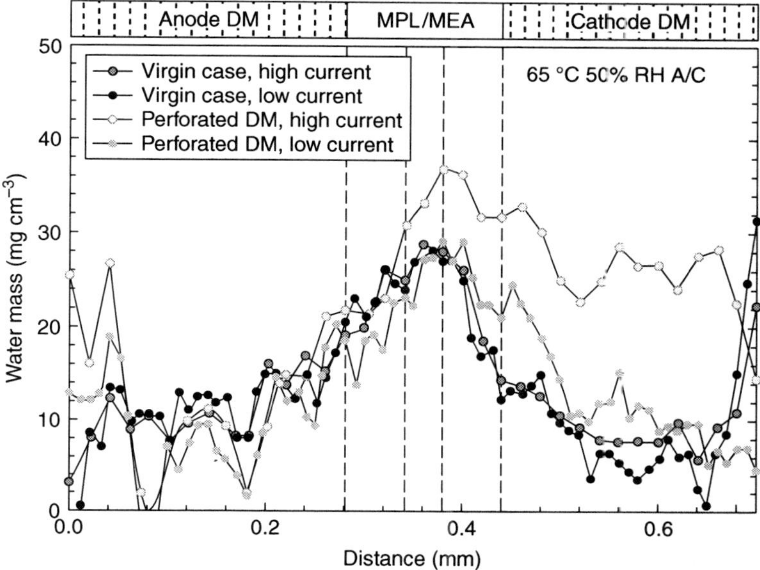

6.14 Neutron data for water mass per volume distributions in the through-plane direction at low inlet relative humidity condition. A reduced number of data points are shown to improve clarity. (Source: adapted from Manahan *et al.*, 2011.)

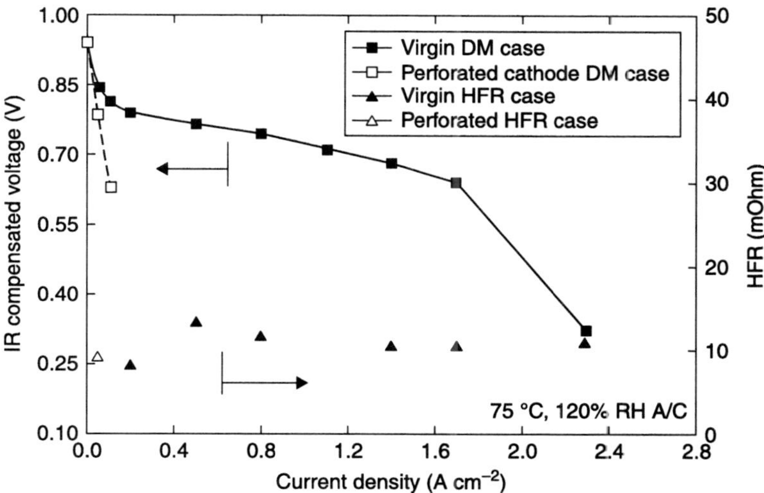

6.15 Performance data of 75 °C, 120% relative humidity, nitrox conditions for virgin DM case, 15% 300 µm perforated cathode DM case, and corresponding HFR values for the virgin and perforated cases. The DM perforations cause drastic performance losses (55 and 183 mV at 0.05 and 0.1 A cm^{-2}, respectively), indicating poor water management in high humidity conditions. (Source: adapted from Manahan *et al.*, 2011.)

however, was capable of achieving limiting current densities of *ca.* 2.3 A cm^{-2}. This drastic performance difference can be directly attributed to the over humidification of the cell facilities and the condensation of water generated during operation. The increase in condensed liquid water enhances water accumulation and promotes the formation of liquid water buildups, which prevent reactants accessing the reaction sites. As a result, the mass transport losses dominate the polarization of the cell, even at low current densities. While the cell with perforated DM cannot handle the excess liquid water, the PCI flow and capillary-driven transport appear to properly manage the water distribution within the cell for the virgin DM case.

High humidity conditions – neutron radiography testing

Figure 6.16 displays neutron images obtained under high humidity (100% inlet relative humidity) conditions. Figures 6.16a and 6.16c show neutron data for the virgin DM at high and low current, respectively, whereas Fig. 6.16b and 6.16d are the perforated DM case at high and low current operation, respectively. Qualitatively, the perforated DM images show significantly more water accumulation than the virgin DM case. Most notable observation is the significant water accumulation (both under lands and channels) for the perforated DM case, which clearly indicates that the perforations enhance the liquid water storage in the cell regardless of humidity condition.

Figures 6.17 and 6.18 show the quantitative values of in-plane and through-plane water content of these cases obtained from the neutron images shown in Fig. 6.16. For all the tested conditions, the total water mass per volume of the DM was observed to be higher in the perforated case than the virgin DM. As previously discussed, the EDS and ESEM analysis showed that the area surrounding the laser

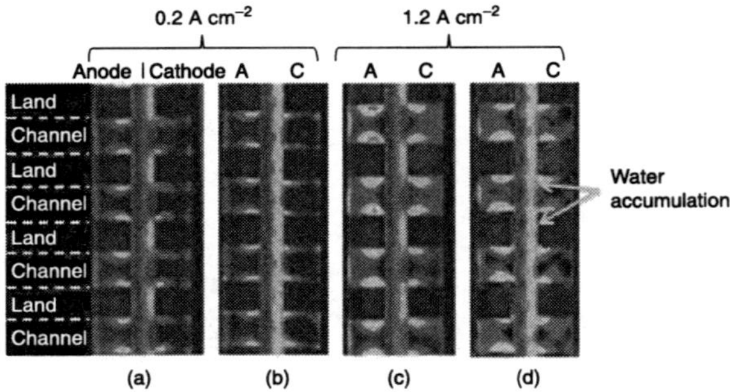

6.16 Neutron images from high humidity (100% inlet relative humidity anode and cathode) testing. (a) High current (1.2 A cm^{-2}) for virgin DM case. (b) High current for perforated DM case. (c) Low current (0.2 A cm^{-2}) for virgin DM case. (d) Low current for perforated DM case. In each image, the right-hand-side represents the cathode. (Source: adapted from Manahan *et al.*, 2011.)

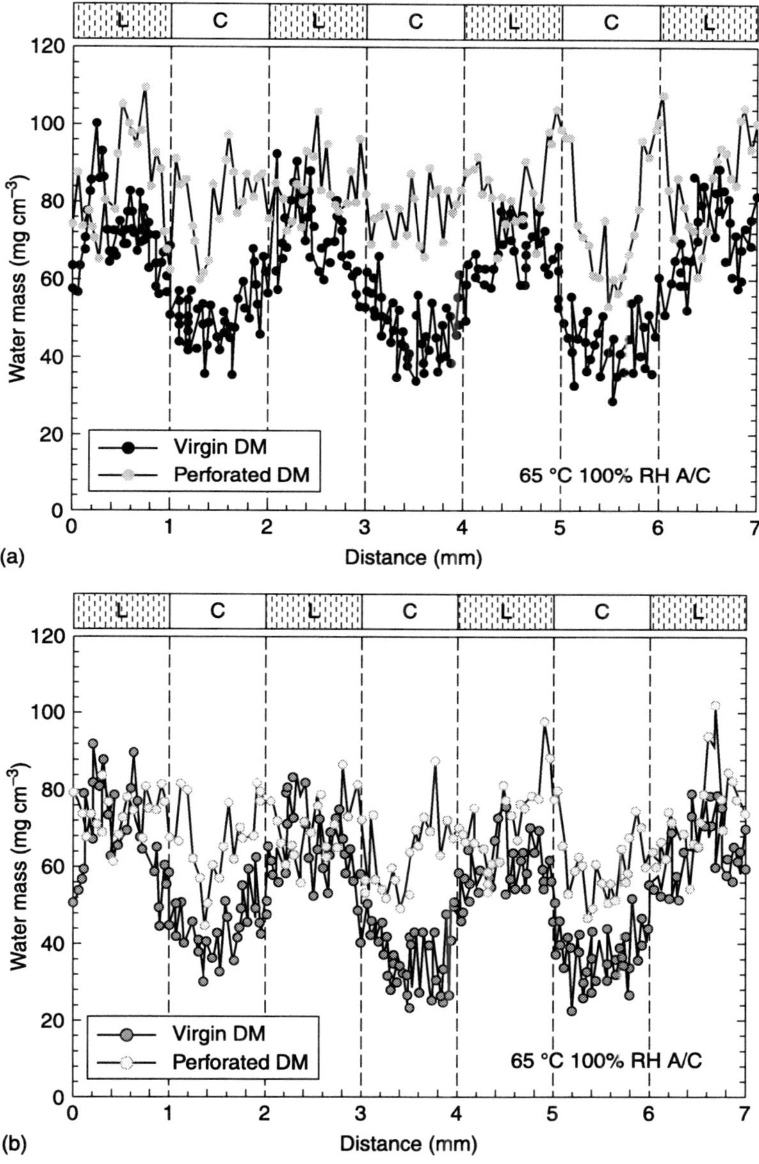

6.17 Neutron data for water mass per volume distributions in the in-plane direction at high inlet relative humidity condition. L represents land, and C represents channel. (a) Water distributions for virgin and perforated DM under low current density (0.2 A cm^{-2}) testing operation. (b) Water distributions for virgin and perforated DM under high current density (1.7 A cm^{-2}) testing operation. A reduced number of data points are shown to improve clarity. (Source: adapted from Manahan *et al.*, 2011.)

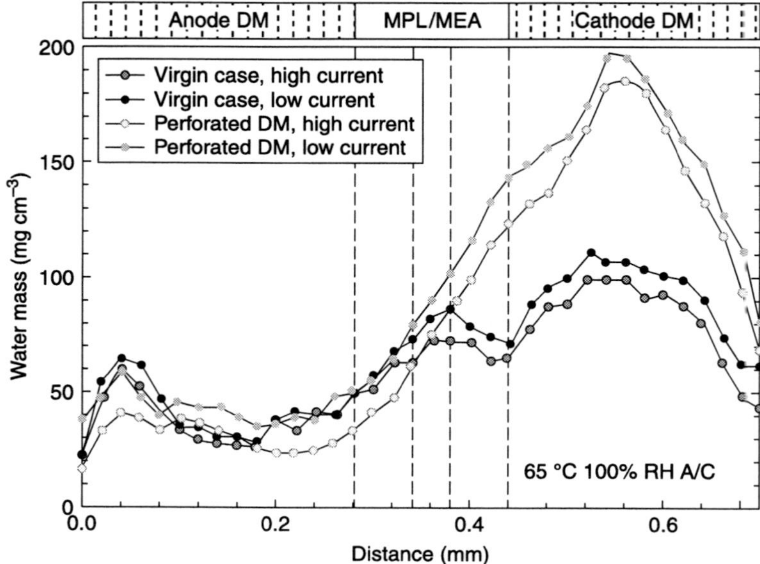

6.18 Neutron data for water mass per volume distributions in the through-plane direction at high inlet relative humidity condition. A reduced number of data points are shown to improve clarity. (Source: adapted from Manahan *et al.*, 2011.)

perforations was more hydrophilic due to the removal of the PTFE during the laser-perforation process. It is very likely that these highly hydrophilic regions may facilitate the condensation of water vapor in these regions, especially at high relative humidity condition. The reader is encouraged to refer to Manahan *et al.* (2011) for additional information.

Effects of perforations on reactant transport

The effect of the perforations on the diffusivity of the reactants was studied by variation of the inlet cathode streams. The ratio of the limiting currents measured using two different cathode-stream gases, namely nitrox and heliox (21% O_2 and 79% He), can be compared to the ratio of the molecular diffusivity values of oxygen in nitrogen and helium. Comparing these ratios can yield information regarding the importance/dominance of Fickian diffusion as compared to other forms of diffusion. Detailed information regarding this analysis can be found in Manahan *et al.* (2009).

Figure 6.19 highlights the 17% increase in limiting current observed with virgin DM when the inert gas is changed from nitrogen to helium at cathode. This was also observed by Manahan *et al.* (2009). Figure 6.20 shows the same analysis for the perforated DM, where the nitrox case is observed to unexpectedly achieve a higher limiting current density (0.8 A cm^{-2}) than the heliox case (0.5 A cm^{-2}). It is anticipated that due to the increased water content in the perforated DM, the reactant flow rates become a key factor in convective removal of accumulated water, especially in the perforated regions. In these tests, the heliox flow rate was set to be 25% lower than the nitrox flow rate due to the adjustments required for the mass flow controllers to maintain the equivalent oxygen concentration. This resulted in reduced water removal by convection and shear/drag effects in the flow channel. Therefore, it can be hypothesized that the performance data for the perforated DM is mostly dominated by the mass transport losses, indicating the fact that the channel flow rate conditions play a key role in establishing effective water management strategy for the perforated DM case.

6.4   Modeling of interfaces in PEFCs

Although interfacial modeling has been scarcely dealt with in PEFCs (Springer *et al.*, 1993, Mishra *et al.*, 2004, Makharia *et al.*, 2005, Ramasamy *et al.*, 2008, Nitta *et al.*, 2008, Avasarala and Halder, 2009), much is yet to be explained regarding the morphological aspects of interfaces. Therefore, microscopic and macroscopic models were developed to bring forth qualitative and quantitative descriptions of electronic and mass transport mechanisms resulting from interfacial morphological characteristics, and to investigate their effects on PEFC performance. These models are detailed in the following sections.

6.4.1   Microscopic modeling

Microscopic contact models have been extensively developed and studied in the fields of tribology and electrical contact mechanics (Holm, 1958, Greenwood and Williamson, 1966, Nayak, 1971, Bush *et al.*, 1975, O'Callaghan and Cameron, 1976, McCool, 1986, Majumdar and Tien, 1990, Majumdar and Bhushan, 1991, Majumdar and Tien, 1991). To date, the most common approach to simulate interfacial contact is based on the Hertzian theory of contact and probabilistic modeling techniques, which can account for the stochastic nature of the mating surface profiles (Greenwood and Williamson, 1966, Nayak, 1971, Bush *et al.*, 1975, O'Callaghan and Cameron, 1976, McCool, 1986). The concept behind the model was developed by Greenwood and Williamson (1966), and is therefore referred to as the GW model. Since the model parameters are dependent on the resolution of the roughness measurements, these approaches can be employed in cases where the root mean square (RMS) roughness of the contacting surfaces is proportional to the sampling interval of the measuring instrument (Swamy *et al.*, 2009). Therefore, the GW modeling framework was employed to simulate the MPL|CL interface, the results of which are discussed below. Details of the model formulation can be found in Swamy *et al.* (2009). However, for cases where the roughness metrics are sensitive to the adopted sampling interval of the measuring instrument (as is the case of the DM, where the average RMS roughness of the DM (~25.2 μm) is much greater than the sampling interval of the surface measurements), a fractal geometry-based approach is needed to describe the fractal nature of the rough surfaces that are in contact (Majumdar and Tien, 1990, Majumdar and Bhushan, 1991, Majumdar and Tien, 1991). Therefore, in this study, the fractal modeling approach was used to simulate the BP|DM interface. Details of the model formulation can be found in Swamy *et al.* (2011).

This section discusses the results of an analytical model developed to predict the contact resistance and the potential water accumulation capacity of the MPL|CL and BP|DM interface under compression. The model is also capable of digitally reconstructing the contact morphology of the two interfaces via control volume allocation, which can be further incorporated into a macroscopic fuel cell model to enable a more accurate quantification of the governing losses and the PEFC performance.

Results and discussion: micro-porous layer (MPL)|catalyst layer (CL) interface

Figure 6.19 shows the predicted MPL|CL interfacial structure under homogeneous compression of 1.5 MPa (e.g. under a typical land). The cracks in the MPL and CL are clearly visible, and it was observed that some of the cracks in the CL penetrate into the entire layer (Hizir *et al.*, 2010). The separation between mean planes, *d*, was determined to lie in the range of 5–10 μm. The maximum gap width was found to be close to 20 μm. It is evident from Fig. 6.19 that the MPL|CL interfacial

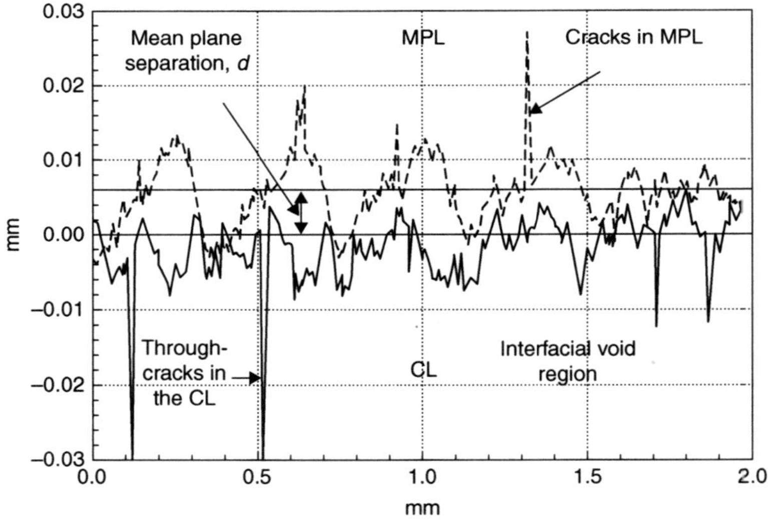

6.19 A typical MPL|CL interface cross-section. (Source: adapted from Swamy *et al.*, 2009.)

gaps are relatively larger in dimension compared to the average pore size in the two layers, as reported in Mench (2008). This observation suggests that the MPL|CL interface could act as a locally low capillary pressure region to store liquid water in its voids, which could result in significant mass transport losses across the interface.

Figure 6.20 shows the variation of the predicted MPL|CL contact resistance as a function of the applied compression pressure for different degrees of uncompressed roughness of the mating surface profiles. Details of the roughness parameter alteration process can be found in Swamy *et al.* (2009). It can be seen from Fig. 6.20 that lowering the roughness of the MPL and CL surfaces by 50% results in nearly a 40% drop in the MPL|CL interfacial resistance. This significant drop in the contact resistance can be attributed to the fact that as the roughness of the mating surfaces is decreased (i.e. smoother contact), the number of contact points increases, which in turn, facilitates the electron flow across the MPL|CL interface. It should also be noted that, since the MPL surface exhibits a higher degree of roughness when compared to the CL tested (Hizir *et al.*, 2010), the surface characteristics of the MPL is expected to dominate the interfacial region, and consequently the MPL|CL contact resistance.

Figure 6.21 shows the impact of the variation of Young's modulus of the backing DM layer on the MPL|CL contact resistance for a homogeneous compression pressure of 1.5 MPa, which is a typical compression value measured under the land. The reason for the choice of the DM Young's modulus over that of

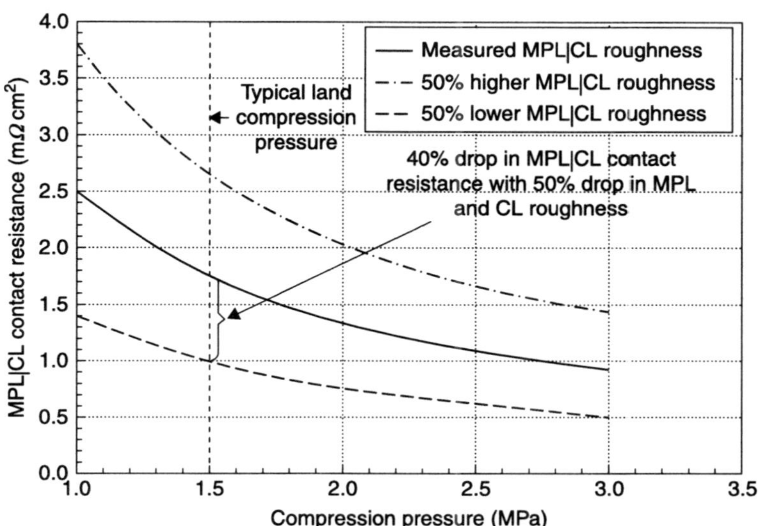

6.20 Predicted MPL|CL contact resistance versus compression pressure for different degrees of MPL and CL roughness. (Source: adapted from Swamy *et al.*, 2009.)

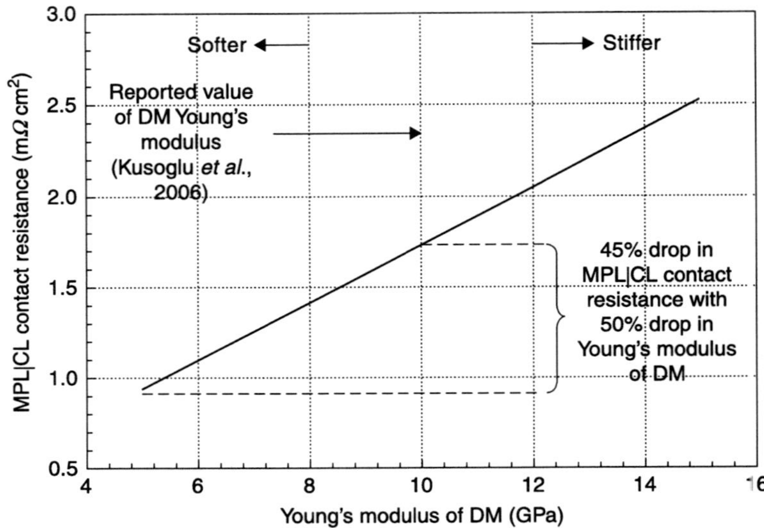

6.21 Effect of the Young's modulus of the backing DM layer on the predicted MPL|CL contact resistance for homogeneous compression pressure of 1.5 MPa, under a typical land. (Source: adapted from Swamy *et al.*, 2009.)

the MPL and CL is explained in Swamy *et al.* (2009). Young's modulus appears to have a significant impact on the MPL|CL interfacial resistance, since it is observed that a 50% drop in the Young's modulus of the DM yields approximately a 45% drop in contact resistance. This strong dependence of the contact resistance on the material property of the backing DM layer can be attributed to the fact that a drop in the Young's modulus of the DM results in the softening of the material, which allows the thin MPL and CL surfaces to exhibit a higher degree of conformation under compression. However, it should be noted that considering the channel, a stiffer material might be preferred, since it may sustain the land compression under the channel, and avoid severe loss of contact under the channel.

Figure 6.22 shows the variation of the predicted MPL|CL contact resistance under one set of mid-land to mid-channel configuration. The model predicts that there is approximately a 2100% increase in the MPL|CL contact resistance at mid-channel location as compared to contact resistance under the land. This drastic increase can be attributed to the non-homogeneous compression pressure distribution. For instance, while moving into the channel and away from the land, the compression pressure drops quite drastically and is minimal at mid-channel position. To suppress this drastic increase in the MPL|CL contact resistance, using a stiff DM may be a better option, since it could counteract the drop in the compression pressure moving into the channel, maintaining better interfacial

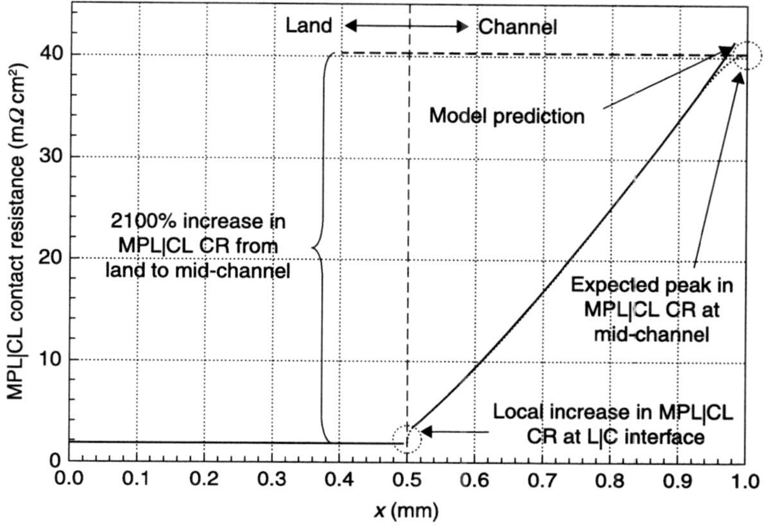

6.22 Predicted MPL|CL contact resistance (CR) as a function of the distance (x) under one set of land (L) and channel (C) affected due to inhomogeneous compression. (Source: adapted from Swamy *et al.*, 2009.)

contact. It was mentioned earlier that a softer DM may be preferred under the land; however considering the existence of competing Young's modulus effects under the land and the channel, the Young's modulus of the DM layer can be optimized to minimize the losses. Further information is detailed in Swamy *et al.* (2009).

Utilizing the mean plane separation, d (obtained as a model output), and the MPL|CL surface profile information, the capacity of potential water accumulation in the MPL|CL interfacial gaps can be predicted. Figure 6.23 shows the variation of the predicted maximum water content as a function of the applied compression pressure in a PEFC. Figure 6.23 indicates that the MPL|CL interface can hold approximately 0.9–3.1 mg/cm^2 of liquid water when fully filled. Neutron imaging results, reported in (Pekula *et al.*, 2005, Turhan *et al.*, 2006, Turhan *et al.*, 2008), suggest that nearly 5–15 mg/cm^2 of water can be stored in a PEFC under normal operating conditions. Therefore, considering the neutron imaging results and current model predictions, it is estimated that the MPL|CL interface can store nearly 6–18% of the total water content in the PEFC, which is significant considering the overall water balance in PEFCs. Cross-sectional X-ray radiography results in Hartnig *et al.* (2008) also indicate that the MPL|CL interface can store a considerable amount of liquid water (600 μm water thickness at 600 mA/cm^2) (Fig. 6.24), which is in good agreement with current predictions.

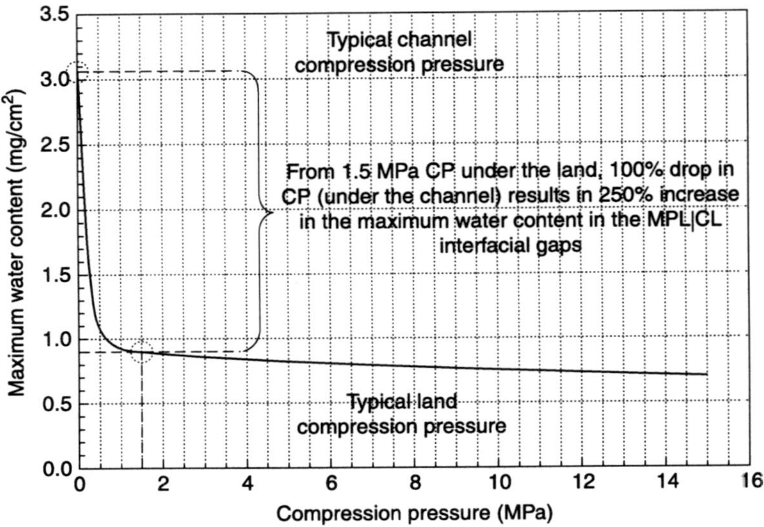

6.23 Effect of the compression pressure (CP) on the maximum water content in the MPL|CL interfacial gaps. (Source: adapted from Swamy *et al.*, 2009.)

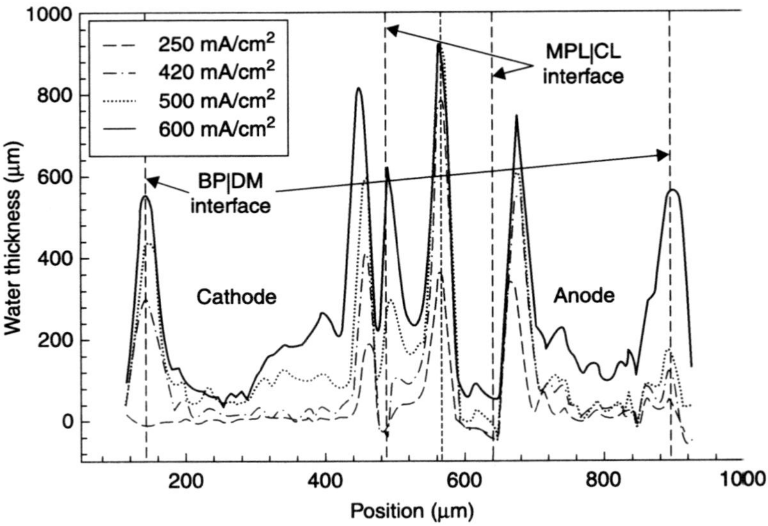

6.24 Quantification and location of liquid water in a PEFC. (Source: adapted from Hartnig *et al.*, 2008.)

Results and discussion: bi-polar layer (BP)\diffusion media (DM) interface

Figure 6.25 shows the virtually-cloned BP|DM interfacial structure under homogeneous compression of 1.5 MPa (e.g. under a land). The virtual interface structure shown in Fig. 6.25 clearly depicts the morphological features of the two contacting surfaces, demonstrating the relatively smoother nature of the BP surface as compared to the DM surface. The virtual interface also shows the existence of large interfacial gaps and some deep cracks, which most likely originate from the highly rough nature of DM surface and heterogeneous fiber matrix constituent within the DM material. The separation between mean planes, d, was determined to lie in the range of 15–20 μm. The maximum gap width was found to be close to 30 μm. It is evident from Fig. 6.25 that the BP|DM interfacial gaps are relatively larger in dimension as compared to the average pore size in the DM layer, which supports the results reported in Mench (2008). This observation suggests that the BP|DM interface can act as a locally low capillary pressure region, which can store a significant amount of liquid water (e.g. especially under landings, as observed in neutron imaging studies).

Model results show that under a landing, lowering the roughness of the DM surface by 50% results in nearly a 40% drop in the BP|DM interfacial resistance. This significant drop in the contact resistance can be attributed to the fact that as the roughness of the mating surfaces is reduced (i.e. smoother contact), the number of contact points at the interface increases which, in turn, facilitates the electron flow across the BP|DM interface. A similar result is also expected for thermal contact resistance within this interfacial region. Details of the roughness parameter alteration process can be found in Swamy *et al.* (2011). It should be

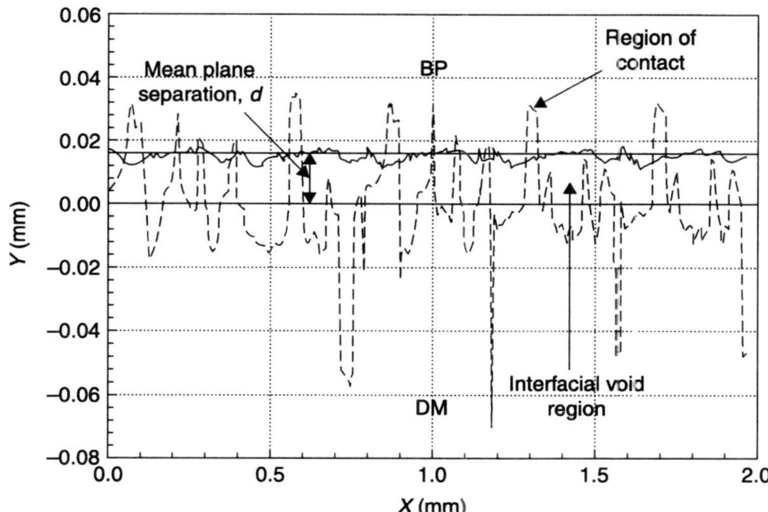

6.25 Virtually created BP|DM interface structure using the fractal geometry approach. (Source: adapted from Swamy *et al.*, 2011.)

noted that since the DM surface exhibits a higher degree of roughness as compared to the BP surface, the surface characteristics of the DM is expected to dominate the interfacial region, and consequently the BP|DM contact resistance.

For a homogeneous compression pressure of 1.5 MPa (a typical compression under the land), it is observed that a 50% drop in the Young's modulus of the DM and BP yields approximately a 38% drop in BP|DM contact resistance. The strong dependence of the contact resistance on the Young's modulus of the interfacing materials can be attributed to the fact that a drop in the Young's modulus of the DM and BP surfaces results in the softening of these materials, enabling the surfaces to exhibit a higher degree of conformation under compression. The increased conformation of these layers enhances the local interfacing percentage. As a result, a higher number of asperities come into contact, and the BP|DM contact resistance is lowered, as predicted by the present model.

Utilizing the mean plane separation, d (obtained as a model output), and the BP|DM surface profile information, the capacity of potential water accumulation (i.e. interfacial water storage capacity) in the BP|DM interfacial gaps can be predicted. Model results indicate that the BP|DM interface can hold approximately 0.85–3.5 mg/cm^2 of liquid water when fully filled. It is estimated that the BP|DM interface can store nearly 5–16% of the total water content in the PEFC, which is significant considering the overall water balance in PEFCs. The result is supported by findings reported in Hartnig *et al.* (2008), depicted in Fig. 6.24. As it has been shown in neutron imaging studies conducted by Turhan *et al.* (2008) that water tends to accumulate under lands, minimizing the interfacial gaps via more smooth contact at the BP|DM interface would be one potential solution to reduce excess liquid water storage in PEFCs.

6.4.2 Macroscopic modeling

This section discusses the results generated by a two-dimensional, single-phase, non-isothermal, anisotropic numerical model, which was developed to investigate the impact of interfacial morphology on PEFC performance. The reader is encouraged to study the detailed model formulation given in Bajpai *et al.* (2010). The MPL|CL interfacial morphology was discretized using the previously mentioned microscopic model and integrated into the computational framework of the macroscopic model as a unique interface layer. The objective of the macroscopic model described herein is to obtain a better understanding of the impact of the rough MPL|CL interface on the local ohmic, thermal and gas-phase mass transport losses.

To analyze the role of the interface layer on the PEFC performance, three cases were defined. In Case 1, the interface layer is not accounted for, and a perfect contact is assumed. This is similar in nature to most existing performance models. In Cases 2 and 3, the interface layer consisting of the MPL, CL and interfacial voids is integrated into the computational domain. Appropriate properties of each component are specified in each grid, depending on the appropriate region and can be found in Bajpai *et al.* (2010). The void region is assumed to be filled with water vapor for Case 2 and liquid water for Case 3. Table 6.2 provides a list of property values used for the two cases. These three cases were simulated at different voltage boundary conditions (ranging from 0.4–1 V) to investigate their impacts on the mass, charge and energy transport.

Results and discussion

The single-phase model with the interface layers was compared with measured experimental data given in Manahan *et al.* (2009). The performance curve from the model results and the experimental results are shown in Fig. 6.26. The model

Table 6.2 Interface layer properties for the discussed cases

Properties	Water vapor vapor-filled interface layer (case 2)	Liquid water water-filled interface layer (case 3)
Thermal conductivity (W/mK)	0.028	0.62
Electronic conductivity (S/m)	–	–
Oxygen diffusivity (m^2/s)	3.23×10^{-5}	–
Hydrogen diffusivity (m^2/s)	1.1×10^{-4}	–
Water vapor diffusivity (m^2/s)	7.35×10^{-5} (anode)	7.35×10^{-5} (anode)
	1.1×10^{-4} (cathode)	1.1×10^{-4} (cathode)

Source: adapted from Bajpai *et al.*, 2010.

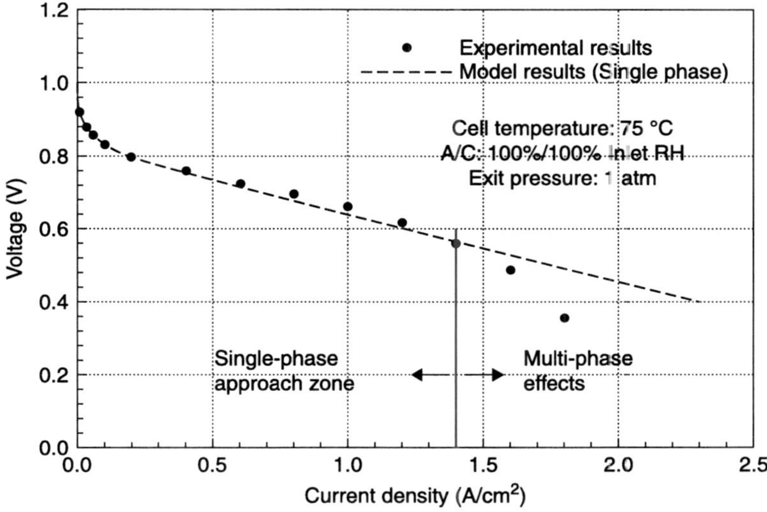

6.26 Comparison of model results with experimental data found in Manahan *et al.*, 2009 (Gore 5710 series MEA, SGL 10BB DM). (Source: adapted from Bajpai *et al.*, 2010.)

results show a reasonable agreement with the experimental data (~3% error) in the low current density region (less than 1.4 A/cm^2), considering the negligible effects of two-phase flow. For higher current densities (greater than 1.4 A/cm^2), the two-phase effects play an important role, causing an increase in divergence between model results and experimental results.

Impact of the water vapor-filled interface layer

Figure 6.27a shows the comparison between polarization curves for Case 1 and Case 2. Case 1 simulates perfect contact, and Case 2 has real interfacial layers with voids filled with water vapor. The polarization curve for both the cases were obtained by varying voltage from 1–0.4 V with steps of 0.1 V. It can be seen that

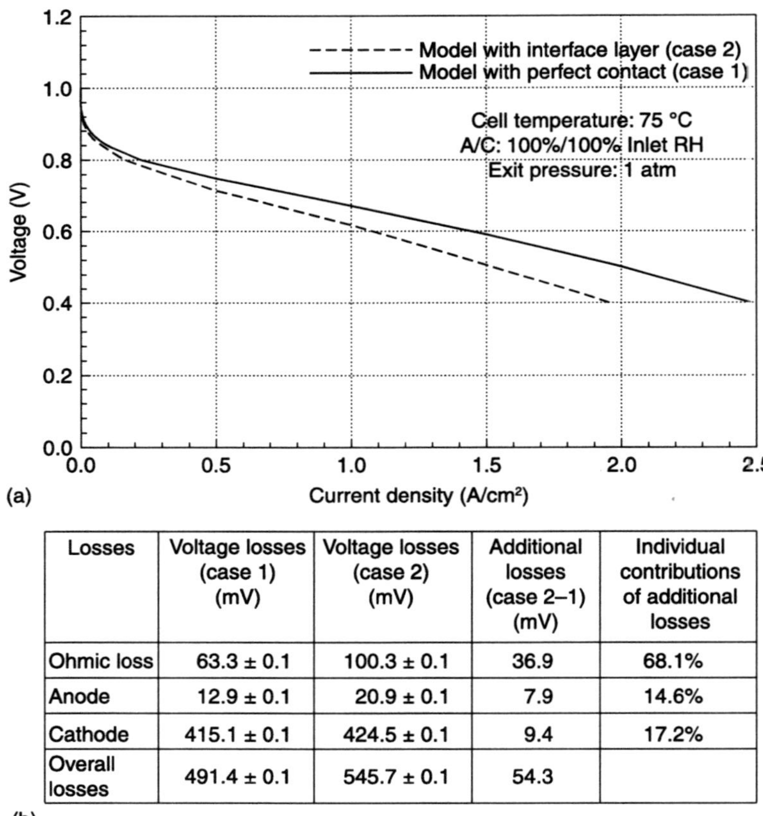

(a)

(b)

Losses	Voltage losses (case 1) (mV)	Voltage losses (case 2) (mV)	Additional losses (case 2–1) (mV)	Individual contributions of additional losses
Ohmic loss	63.3 ± 0.1	100.3 ± 0.1	36.9	68.1%
Anode	12.9 ± 0.1	20.9 ± 0.1	7.9	14.6%
Cathode	415.1 ± 0.1	424.5 ± 0.1	9.4	17.2%
Overall losses	491.4 ± 0.1	545.7 ± 0.1	54.3	

6.27 (a) Polarization curve comparison of model with interface layer and model with perfect contact. (b) Loss contributions at 1 A/cm^2 with inclusion of water vapor-filled interface layer. (Source: adapted from Bajpai *et al.*, 2010.)

there is a decrease in performance due to the addition of interface layers into the computational framework. In the activation region, the polarization curve for Case 2 is lower than Case 1, indicating more kinetic losses after the inclusion of interfaces in the model. This can be attributed to the decrease in the CL volume due to voids present at the interface layer. The gap between PEFC performance curves increase with an increase in current density, indicating a potential increase in the ohmic losses for Case 2. Since two-phase phenomena are not taken into account, the model is valid only for the lower current density region (< 1.4 A/cm^2). At 1 A/cm^2, additional voltage drop of ~54 mV is estimated for the Case 2 in comparison to Case 1. The individual anode over-potential, cathode over-potential and ohmic losses are shown for Case 2 in Fig. 6.27b. Also, the additional individual losses and individual contributions are given. As the void region is filled with water vapor for Case 2, the major contribution to the losses comes from ohmic over-potential. The model predicts 68% ohmic contribution in overall additional voltage drop at 1 A/cm^2, which is ~37 mV. The ohmic loss increase for Case 2 can be attributed to an increased effective mean electron current flow path at the MPL|CL interface. The small impact on the anode over-potential and cathode over-potential is also seen in Fig. 6.27b. This can be attributed to the decrease in CL volume because of the voids present in the catalyst layer surface, which will affect the anode and cathode kinetics in PEFC. The model predicts a 14.6% contribution of anode over-potential, and 17.2% contribution of cathode over-potential in the overall additional voltage drop at 1 A/cm^2. The ohmic contribution in additional voltage losses for Case 2 is expected to increase for current densities greater than 1 A/cm^2, and decrease for current densities less than 1 A/cm^2. Bajpai *et al.*, (2010) provides a further discussion on the above.

Impact of liquid water pooling at interfacial voids

Figure 6.28a shows the comparison between polarization curves for Case 1 and Case 3. Case 1 has no interface layer and Case 3 has an incorporated interface layer with voids filled with liquid water. As described, the polarization curves for both cases are obtained by changing the voltage from 1 to 0.4 V. Similar to Fig. 6.27, a decrease in performance is observed after the inclusion of a liquid water-filled interface layer into the model. For low current densities, the polarization curve for Case 3 is overlaps with polarization curve of Case 2. This result implies that kinetic losses are similar for Case 2 and Case 3, yet higher than Case 1, due to the voids present in the catalyst layer. But at high current densities where the mass transport losses become more important, the polarization curve for Case 3 is predicted to be lower than the polarization curve of Case 2. This behavior is attributed to the blockage of species transport because of the liquid accumulation at the interface layer. At 1 A/cm^2, an additional voltage drop of ~65 mV is estimated for Case 3 in comparison to Case 1. Figure 6.28b shows the individual anode over-potential, cathode over-potential and ohmic losses for

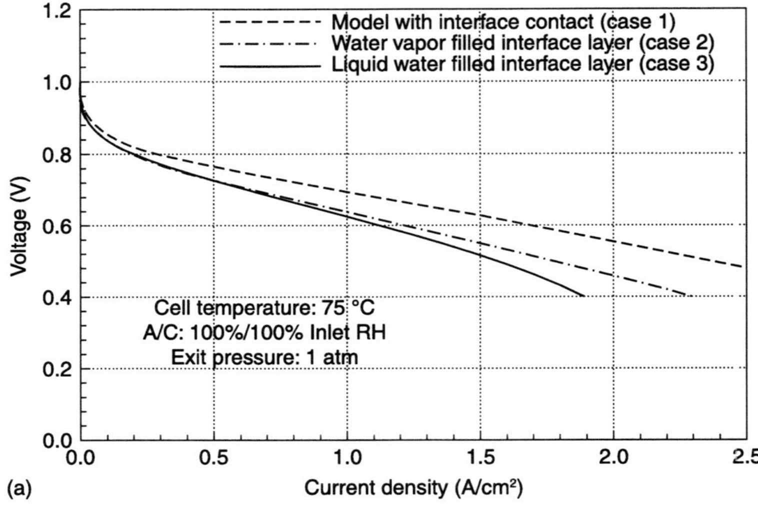

(b)

Losses	Voltage losses (case 3) (mV)	Additional losses (case 3–1) (mV)	Individual contributions of additional losses	Percentage change in additional losses (case 3 and 2)
Ohmic loss	97.1 ± 0.1	33.9	50.9%	8.4% decrease
Anode	20.9 ± 0.1	7.9	11.9%	No change
Cathode	439.7 ± 0.1	24.6	37.0%	162.8% increase
Overall losses	557.8 ± 0.1	54.3		

6.28 (a) Performance comparisons with liquid water-filled interfacial layer. (b) Contributions at 1 A/cm² because of inclusion of interface layer (liquid water-filled) and comparison of individual contribution with water vapor-filled and liquid water-filled interface layer at 1 A/cm². (Source: adapted from Bajpai *et al.*, 2010.)

Case 3 at 1 A/cm². The model predicts ~34 mV of increase in ohmic loss, ~8 mV of increase in anode over-potential and ~25 mV of increase cathode over-potential. The highest contribution comes from ohmic losses (51%) in additional voltage loss of Case 3 in comparison to Case 1. This is followed by cathode over-potential (37%) and anode over-potential (12%). Figure 6.28b also shows the comparison between the additional individual voltage losses at 1 A/cm² for Case 2 and 3 in comparison to Case 1. It can be seen that Case 3 has more additional overall losses in comparison to Case 2. This can be attributed to additional mass transport losses in Case 3. Therefore, an increase in the current results in an increase in the voltage loss between Case 2 and 3. Because of liquid water accumulation at the interface, reaction rates adjacent to liquid water-filled gaps are slower in comparison to other region in the catalyst layer. This leads to decrease in in-plane motion of electron current for Case 3 which causes an 8.5% decrease in additional ohmic loss in comparison to Case 2. As hydrogen diffusivity and mole fraction on the anode side are high in comparison to oxygen, negligible increase (0.2%) is seen on the anode over-potential. Alternatively, significant increase (162.8%) in cathode over-potential is observed for Case 3 in comparison to Case 2. This is attributed to the blockage of oxygen transport on the cathode side of the PEFC.

As discussed previously, a negligible increase in anode over-potential and a significant increase in cathode over-potential is predicted by the model for Case 3 in comparison to Case 2. This is due to the mass transport losses caused by the liquid water-filled interface layer that exists on both, the anode and the cathode side. Figure 6.29 shows the disruptions in species transport caused by the liquid

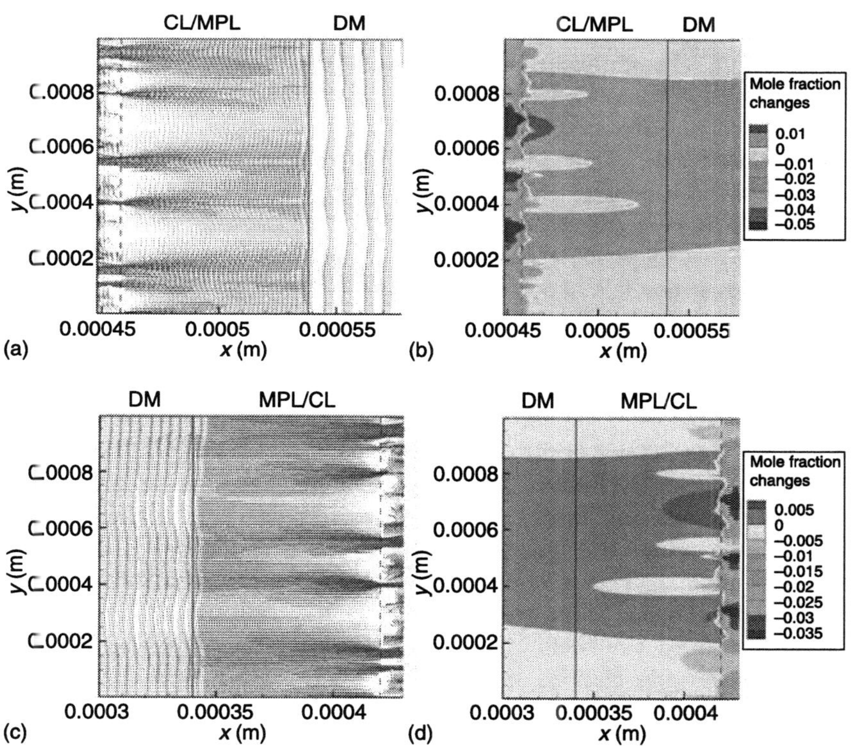

6.29 (a) Anode side hydrogen diffusion vector distortions caused by liquid water-filled interface layer. (b) Mole fraction change of hydrogen between Case 3 and Case 1. (c) Cathode-side oxygen diffusion vector distortions caused by liquid water-filled interface layer. (d) Mole fraction change of oxygen between Case 3 and Case 1. (Source: adapted from Bajpai *et al.*, 2010.)

water-filled interface layer on the anode and cathode side of the model. Liquid water blocks the diffusion of species transport causing a decrease in mole fraction of species in reaction sites. The change in mole fraction of species for Case 3 in comparison to Case 1 is shown in the contour plot in Fig. 6.29. A maximum decrease of 0.035 mole fraction of hydrogen is estimated adjacent to liquid filled gaps on the catalyst layer with slight increase (0.005 mole fraction) on the diffusion media and micro-porous layer. Similar behavior is observed for oxygen transport on the cathode side for which a maximum decrease of 0.05 mole fraction is estimated on the catalyst layer with slight increase (0.01 mole fraction) in the diffusion media and micro-porous layer. Higher impact on oxygen mole fraction can be attributed to the low diffusivity of oxygen in comparison to hydrogen.

Effects of interfacial gap width and location

Separate cases are simulated to evaluate the impact of interfacial gap width and location on PEFC performance. At 0.5 A/cm^2, Fig. 6.30 shows the variation of additional voltage losses with the increase in gap width at the MPL|CL interface. An interfacial gap is placed in between MPL and CL at the center of channel on the cathode side of the PEFC. Additional voltage loss is plotted with a non-dimensional parameter defined as gap width, W/δ (W is the interfacial gap width and δ is the catalyst layer thickness). The parameter W/δ is defined because the catalyst layer resistance is a key parameter for overall MPL|CL interfacial resistance (Chacko *et al.*, 2008). For Fig. 6.30a, the gap region is assumed to be filled with water vapor. It is modeled to possess infinite resistivity for electron and proton transport. Gas channel diffusivity is used for species transport and water vapor thermal conductivity is used for thermal transport. For this case, additional ohmic losses are found to be dominant. The model results predict that as the width of gap increases, additional ohmic losses increase in a non-linear fashion. The additional ohmic loss increase is attributed to an electrical insulation barrier due to the interfacial gap, which causes voltage and current distortion. It is also observed that there is no increase in additional ohmic loss up till $W/\delta = 10$, which is referred to as the critical gap width, after which additional ohmic losses start to play a role. Alternatively, the gap region is assumed to be filled with liquid water in Fig. 6.30b. The region is modeled to have infinite resistivity for electron and proton transport. Almost zero diffusivity is used for species transport, and liquid water thermal conductivity is used. For this case, significant effects are seen on the additional ohmic losses and additional cathode overpotentials. Figure 6.30b shows the variation of additional overall losses, additional cathode over-potential and additional ohmic losses with the increase in gap width. Bajpai *et al.* (2010) provides a detailed analysis on the aforementioned.

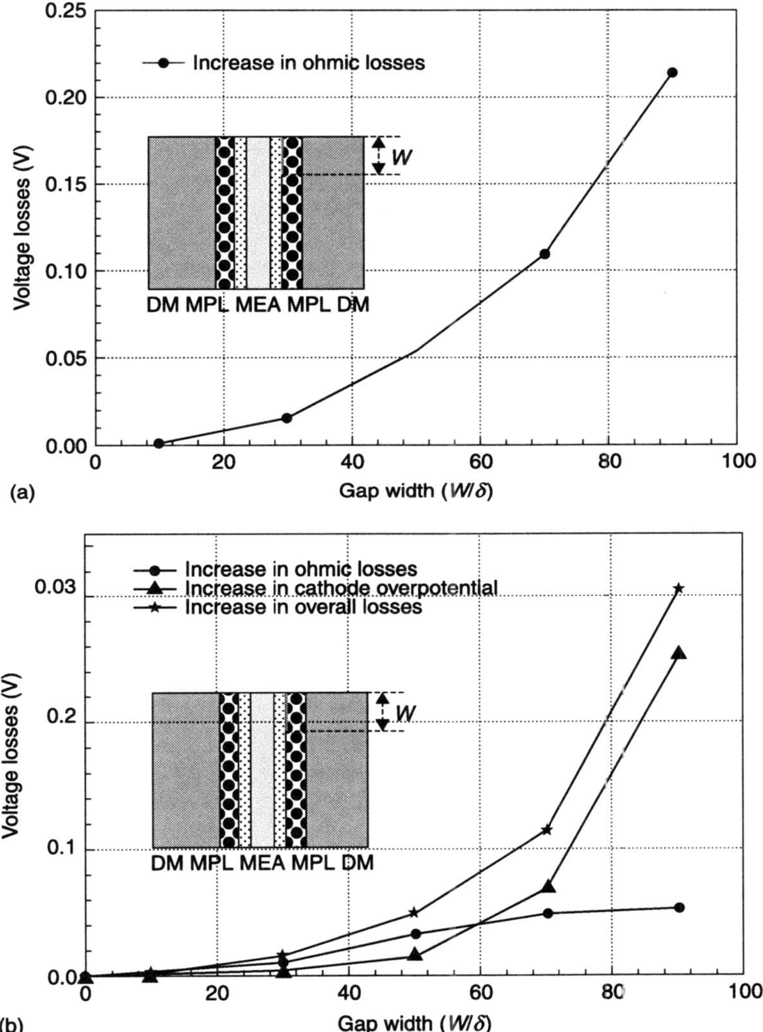

6.30 (a) Increase in ohmic losses due to increase in interfacial gap width. Gap assumed to be filled with water vapor. (b) Gap assumed to be filled with liquid water at 1 A/cm^2. (Source: adapted from Bajpai *et al.*, 2010.)

Temperature distribution

Figure 6.31 shows the temperature distribution in a PEFC for Case 2 at 1 A/cm^2. It can be seen that the temperature at the cathode side is higher than at the anode side, due to the reversible and irreversible entropy production. Also it can be seen

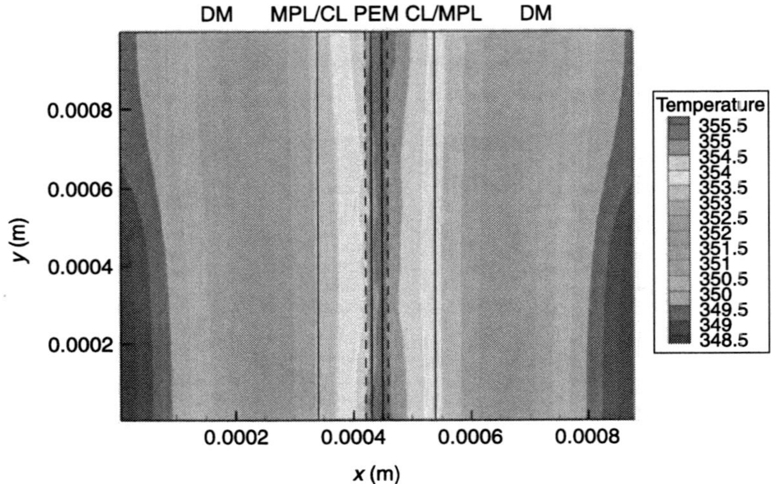

6.31 Temperature distribution inside a PEFC with an included water vapor-filled interface layer. (Source: adapted from Bajpai *et al.*, 2010.)

for Case 2 in Fig. 6.31 that there are some hot spots formed close to the contact points between the MPL|CL interface on the cathode side. A maximum temperature difference of 1 °C is observed on the cathode CL. Reaction regions in cathode CL close to the contact points at MPL|CL interface are expected to be more active than the reaction regions just behind the voids present at the interface. This may cause more heat generation due to the reaction close to the contact points. There is also a high amount of charge flow at the contact points at the MPL|CL interface. Temperature variation in cathode CL may also be attributed to the heat generation due to joule heating of a large amount of charge flowing at the contact points. Further information on temperature effects can be found in Bajpai *et al.* (2010).

6.5   Future work

Although recent studies shed light into the morphological features and related performance effects of interfaces in PEFCs; there are still many unknowns yet to be resolved. The following studies are needed to further the current understanding of interfacial effects on PEFC performance:

- Characterization of actual compressed morphology will allow for better estimation of metrics associated with interfacial morphology.
- *In situ* characterization of PEFC polarization using components tailored to a variety of different morphological specifications will allow for quantitative isolation of performance barriers.

- Incorporation of component porosity and anisotropicity into the current microscopic models will allow for improved prediction of interfacial phenomena.
- Further experimental and mathematical analysis of surface defects such as cracks is required in order to fully understand gas and mass transport mechanisms.
- Development of multi-phase macroscopic PEFC models, coupled with microscopic interfacial models will allow for better estimation of polarization losses at high current densities, as well as provide a better understanding of the phenomena occurring therein.
- Characterization of the MPL|DM interface and its incorporation into the macroscopic modeling framework is expected to shed additional light on the role of interfaces in PEFCs.

6.6   References

Avasarala B. and Haldar P. (2009), Effect of surface roughness of composite bipolar plates on the contact resistance of a proton exchange membrane fuel cell, *J. Power Sources*, **188**, 225–229.

Bajpai H., Khandelwal M., Kumbur E. C., and Mench M. M. (2010), A computational model for assessing impact of interfacial morphology on polymer electrolyte fuel cell performance, *J. Power Sources*, **195**, 4196–4205.

British Standard 1134 (1972), *Method for the Assessment of Surface Texture*, British Standard Institution, London.

Bush A. W., Gibson R. D., and Thomas T. R. (1975), The elastic contact of a rough surface, *Wear*, **35**, 87–111.

Chacko C., Ramasamy R., Kim S., Khandelwal M., and Mench M. M. (2008), Characteristic behavior of polymer electrolyte fuel cell resistance during cold start, *J. Electrochem. Soc.*, **155**, B1145–B1154.

Chopra R., Podczeck F., Newton J. M., and Alderborn G. (2002), The influence of film coating on the surface roughness and specific surface area of pellets, *Part. Syst. Char.*, **19**, 277–283.

Faith D. and Horsfield C. J. (2006), Characterization of pore size of trimethylol propane triacrylate (TMPTA) polymer foam by pulsed sputter coating and SEM analysis, *J. Mater. Sci.*, **41**, 3973–3977.

Gerteisen D., Heilmann T., and Ziegler C. (2008), Enhancing liquid water transport by laser perforation of a GDL in a PEM fuel cell, *J. Power Sources*, **177**, 348–354.

Gostick J. T., Ioannidis M. A., Fowler M. W., and Pritzker M. D. (2009), On the role of the microporous layer in PEMFC operation, *Electrochem. Commun.*, **11**, 576–579.

Greenwood J. A. and Williamson J. W. P. (1966), Contact of nominally flat surfaces, *Proc. R. Soc. London Ser. A*, **295**, 300–319.

Hartnig C., Manke I., Kuhn R., Kardjilov N., and Banhart J. (2008), W. Lehnert, Cross-sectional insight in the water evolution and transport in polymer electrolyte fuel cells, *Appl. Phys. Lett.*, **92**.

Hizir F. E., Ural S. O., Kumbur E. C., and Mench M. M. (2010), Characterization of interfacial morphology in polymer electrolyte fuel cells: Micro-porous layer and catalyst layer surfaces, *J. Power Sources*, **195**, 3463–3471.

Holm R. (1958), *Electrical Contacts Handbook*, Springer-Verlag, Berlin, Göttingen, Heidelberg.

Kim S. and Mench M. M. (2009), Investigation of temperature-driven water transport in polymer electrolyte fuel cell: Phase-change-induced flow, *J. Electrochem. Soc.*, **156**, B353–B362.

Kleemann J., Finsterwalder F., and Tillmetz W. (2009), Characterisation of mechanical behaviour and coupled electrical properties of polymer electrolyte membrane fuel cell gas diffusion layers, *J. Power Sources*, **190**, 92–102.

Kusoglu A., Karlsson A. M., Santare M. H., Cleghorn S., and Johnson W. B. (2006), Mechanical response of fuel cell membranes subjected to a hygro-thermal cycle, *J. Power Sources*, **161**, 987–996.

Lamb C. and Zecchino M. (1999), *Wyko Surface Profilers Technical Reference Manual*, Veeco Metrology Group.

Lin G. and Nguyen T. V. (2006), A two-dimensional two-phase model of a PEM fuel cell, *J. Electrochem. Soc.*, **153**, A372–A382.

Majumdar A. and Bhushan B. (1991), Fractal model of elastic-plastic contact between rough surfaces, *J. Tribology*, **113**, 1–11.

Majumdar A. and Tien C. L. (1990), Fractal characterization and simulation of rough surfaces, *Wear*, **136**, 313–327.

Majumdar A. and Tien C. L. (1991), Fractal network model for contact conductance, *J. Heat Transfer Trans. ASME*, **113**, 516–525.

Makharia R., Mathias M. F., and Baker D. R. (2005), Measurement of catalyst layer electrolyte resistance in PEFCs using electrochemical impedance spectroscopy, *J. Electrochem. Soc.*, **152**, A970–A977.

Manahan M. P., Hatzell M. C., Kumbur E. C., and Mench M. M. (2011), Laser perforated fuel cell diffusion media Part I: Related changes in performance and water content, *J. Power Sources*, submitted.

Manahan M. P., Kim S., Kumbur E. C., and Mench M. M. (2009), Effects of surface irregularities and interfacial cracks on polymer electrolyte fuel cell performance, *ECS Transactions*, **25**, 1745–1754.

McCool J. I. (1986), Comparison of models for the contact of rough surfaces, *Wear*, **107**, 37–60.

Mench M. M. (2008), *Fuel Cell Engines*, New York, John Wiley & Sons.

Mishra V., Yang F., and Pitchumani R. (2004), Measurement and prediction of electrical contact resistance between gas diffusion layers and bipolar plate for applications to PEM fuel cells. *ASME J. Fuel Cell Sci. Technol.*, **1**, 2–9.

Nam J. H., Lee K. J., Hwang G. S., Kim C. J., and Kaviany M. (2009), Microporous layer for water morphology control in PEMFC, *Int. J. Heat Mass Transfer*, **52**, 2779–2791.

Nayak P. R. (1971), Random process model of rough surfaces, *J. Lubr. Tech.*, **93**, 398–407.

Nitta I., Himanen O., and Mikkola M. (2008), Contact resistance between gas diffusion layer and catalyst layer of PEM fuel cell, *Electrochem. Commun.*, **10**, 47–51.

O'Callaghan M. and Cameron M. A. (1976), Static contact under load between nominally flat surfaces in which deformation is purely elastic, *Wear*, **36**, 79–97.

Pekula N., Heller K., Chuang P. A., Turhan A., Mench M. M. *et al.* (2005), Study of water distribution and transport in a polymer electrolyte fuel cell using neutron imaging, *Nucl. Instr. and Meth. A*, **542**, 134–141.

Podczeck F. (1998), *Particle-Particle Adhesion in Pharmaceutical Power Handling*, Imperial College Press, London.

Ramasamy R. P., Kumbur E. C., Mench M. M., Liu W., Moore D. *et al.* (2008), Investigation of macro- and micro-porous layer interaction in polymer electrolyte fuel cells. *Int. J. Hydrogen Energy*, **33**, 3351–3367.

Springer T. E., Wilson M. S. and Gottesfeld S. (1993), Modeling and experimental diagnostics in polymer electrolyte fuel cells, *J. Electrochem. Soc.*, **140**, 3513.

Swamy T., Kumbur E. C., and Mench M. M. (2009), Characterization of the interfacial structure in PEFCs: Water storage and contact resistance model, *J. Electrochem. Soc.*, **157**, B77–B85.

Swamy T., Kumbur E. C., and Mench M. M. (2011), Investigation of Bipolar Plate and Diffusion Media Interfacial Structure in PEFCs: A Fractal Geometry Approach, *Electrochim. Acta*, under review.

Turhan A., Heller K., Brenizer J. S., and Mench M. M. (2006), Quantification of liquid water accumulation and distribution in a polymer electrolyte fuel cell using neutron imaging, *J. Power Sources*, **160**, 1195–1203.

Turhan A., Heller K., Brenizer J. S., and Mench M. M. (2008), Passive control of liquid water storage and distribution in a PEFC through flow field design, *J. Power Sources*, **180**, 773–783.

Turhan A., Kim S., Hatzell M., and Mench M. M. (2010), Impact of channel wall hydrophobicity on through plane water distribution and flooding behaviour in a polymer electrolyte fuel cell, *Electrochim. Acta*, **55**, 2734–2745.

Wang X. L., Zhang H. M., Zhang J. L., Xu H. F., Tian Z. Q. *et al.* (2006), Micro–porous layer with composite carbon black for PEM fuel cells, *Electrochim. Acta*, **51**, 4909–4915.

Weber A. Z. and Newman J. (2005), Effects of microporous layers in polymer electrolyte fuel cells, *J. Electrochem. Soc.*, **152**, A677–A688.

Zhang J., Yin G., Wang Z., and Shao Y. (2006), Effects of MEA preparation on the performance of a direct methanol fuel cell, *J. Power Sources*, **160**, 1035–1040.

Neutron radiography for high-resolution studies in low temperature fuel cells

D. S. HUSSEY and D. L. JACOBSON, National Institute of
Standards and Technology, USA

Abstract: Due to the nature of the interaction between neutrons and
material, neutron radiography has played a critical role in the elucidation
of water management problems in proton exchange membrane fuel cells
(PEMFCs). Recent advances in neutron imaging detector spatial resolution
have enabled the study of through-plane water content in PEMFCs. With
the ability to discriminate between water in the anode and cathode gas
diffusion layers from that in the membrane electrode assembly, fundamental
studies have included visualizing the role of thermo-osmosis in water
transport, identifying where within the fuel cell sandwich ice forms during
operation below 0 °C, and measuring mass transfer resistance of the
membrane. This chapter discusses types of high-resolution detector systems,
typical sources and sizes of measurement uncertainties, and reviews recent
experiments.

Key words: neutron imaging and radiography, non-destructive *in situ*
evaluation, porous media, proton exchange membrane fuel cell (PEMFC),
through-plane water transport.

7.1   Introduction

Water management in proton exchange membrane fuel cells (PEMFCs) is a
critical engineering problem: present-day sulfonic acid membranes require
adequate hydration for the cell to deliver high power required in automotive
applications; yet water retention in the gas flow fields and gas diffusion layer
(GDL) both impedes reactant flow and results in material degradation on
shutdown. Since the typical length scales of the PEMFC components that impact
water management range from 1 μm to 10 cm, it is essential to have imaging
systems that can span this range. This is a difficult task for anyone imaging
modality to achieve both the high spatial and fast temporal resolution to capture
all facets of water transport phenomena. In particular, the metallic construction
and conductive nature of PEMFCs means that *in situ* measurement of the water
content in PEMFCs is challenging with common methods such as X-ray imaging,
optical microscopy or magnetic resonance imaging (MRI). The advantage of
these methods is the high spatial resolution that is achievable. Standard laboratory
X-ray tomography instruments have spatial resolution on the order of 1 μm, while
synchrotron X-ray sources can produce images with spatial resolution of the order

of 10 nm. By tuning the X-ray spectrum, it is possible to enhance the sensitivity to water compared to carbon, which is useful for studying the water distribution in GDLs. The high intensity of synchrotron sources enables visualizing the water content at high spatial resolution at high frame rates (Hartnig *et al.* 2009). However, the high-intensity X-ray source produces radiation damage in the polymer membrane, limiting the overall exposure time (Albertini *et al.* 2009, Buechi *et al.* 2010). With confocal microscopy, it is possible to obtain three-dimensional images of the water ejection process with about a 100 μm depth into the GDL, depending on the porosity and structure of the GDL (Gao *et al.* 2009). MRI measurements of the water content of an operating PEMFC requires the use of non-metallic flow fields in order to measure the nuclear magnetic resonance (NMR) signal from the water. Present MRI technology enables spatial resolution less than 10 μm along the through-plane direction of the fuel cell, with coarser spatial resolution in the in-plane directions in order to have sufficient NMR signal. Since hydrogen and deuterium have different nuclear spins, MRI has isotopic sensitivity enabling labeling experiments (Tsushima and Hirai 2009, Dunbar and Masel 2008).

While much progress has been made in applying these methods, only neutron radiography has been able to study full-scale, commercially viable PEMFC test sections. This is due to the fact that neutrons primarily interact with the nuclei of material rather than electrons, which gives rise to different image contrast than that available in X-ray imaging or MRI. In particular, the sensitivity to water has been used to study a variety of water transport questions in PEMFCs, and other chapters in this volume focus on the use of neutron imaging detectors with spatial resolution sufficient to resolve changes in the in-plane water content (that is, discriminating water content changes between the channels and lands), which is in the order of 100–200 μm (Manke and Kardjilov 2012, Schneider *et al.* 2011, Wippermann and Schröder 2012).

The primary focus of this chapter is the study of the through-plane water content of the fuel cell (that is the water distribution from the anode to cathode through the fuel cell sandwich) with high-resolution neutron radiography. This is a recent development, as new detector technology has become available enabling spatial resolutions approaching 10 μm (Siegmund *et al.* 2007, Boillat *et al.* 2010), and only experiments conducted at two facilities have produced publications in this field, the ICON facility at the Paul Scherrer Institut in Switzerland, and the BT2 neutron imaging facility at the NIST. Further, a 10 μm spatial resolution represents a near 30-fold improvement in the spatial resolution within a four year period and this improvement is demonstrated in Fig. 7.1. With this resolution, it is possible to directly study water transport phenomena in the anode and cathode gas diffusion layers (with typical through-plane thickness of 250 μm) and in thick membranes, such as Nafion 117,[†] or fundamental studies.

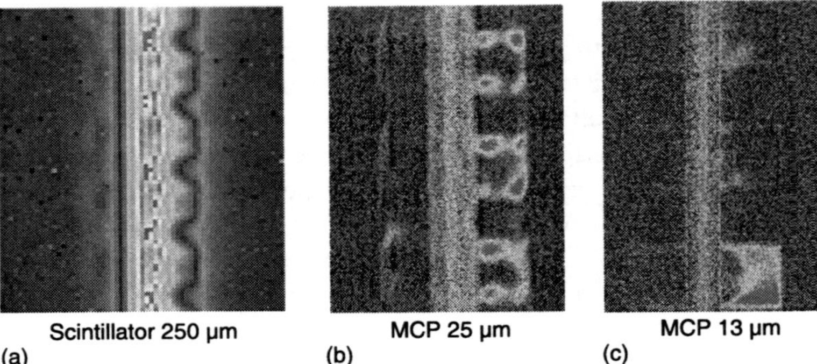

7.1 The through-plane water content of a test section at nominally the same operating conditions with spatial resolution: (a) 250 μm, (b) 25 μm, (c) 13 μm.

7.2 Experimental layout of a high-resolution neutron imaging beamline

In order to obtain high resolution neutron images of PEMFCs, one must ensure proper experimental conditions exist with regard to the neutron beamline, the dector, and the PEMFC test section.

These conditions are discussed in detail below.

7.2.1 Required beam characteristics

There are two primary sources of neutrons, fission reactors and spallation sources. In both cases, the high-energy neutrons that result from the reaction are reduced in energy through moderation, a process that involves inelastic scattering typically from hydrogenous materials. After multiple scattering, the neutrons come to thermal equilibrium with the moderator material, and the moderated neutron source has an energy spectrum described by a Maxwell-Boltzmann distribution (Byrne 1994). There are two typical temperature regimes: thermal neutrons have a characteristic energy of 25 meV and cold neutrons have a characteristic energy of less than 4 meV. The attenuation coefficient of thermal neutrons is lower than that of cold neutrons; thus thermal neutron beams penetrate more deeply through materials than do cold neutrons, which is an advantage for studying the through-plane water content of PEMFCs. However, the lower attenuation applies as well to the detector, which means that the detection efficiency of thermal neutrons is lower than that of cold neutrons, usually by about a factor of two.

Since the neutron optical index of refraction differs from unity by about one part in 10^{-5} to one part in 10^{-6}, it is not practical to focus a neutron imaging beam with conventional refractive optics. Although other optical focusing methods like Fresnel zone plates that are used at X-ray synchrotron sources in principle work with neutrons too, the low efficiency of these devices and the relatively low neutron

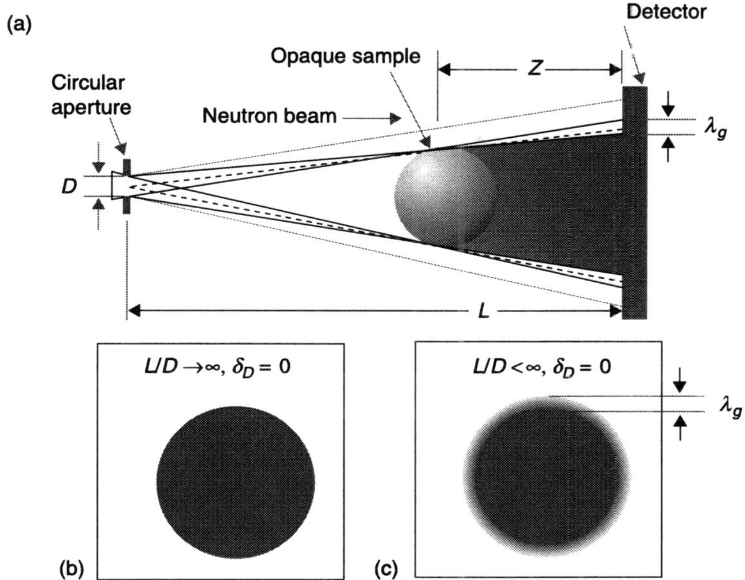

7.2 Schematic layout of a neutron radiography experiment (a) demonstrating the difference between no geometric blur (b) and finite geometric blur (c).

intensity available from the brightest neutron sources, limits their practical application in neutron imaging. Instead, neutron imaging beam lines utilize pinhole optics to define the neutron beam. Neutrons radiate from the source in a ballistic manner, and the trajectories can be modeled by ray-tracing. To define the beam, an aperture with opening D is placed downstream from the source, which produces a pinhole of the neutron source at the detector position, a distance L from the aperture. A sample is placed close to the detector, with separation z. As shown in Fig. 7.2, the geometry of the beamline limits the spatial resolution, and the geometric blur λ_g described as a full-width-half-maximum is found from similar triangles to be:

$$\lambda_g = z\frac{D}{L-z} \approx z\frac{D}{L}. \qquad [7.1]$$

Since the neutron fluence rate (intensity) at the detector position scales as:

$$\phi = \frac{D^2}{L^2}, \qquad [7.2]$$

one must optimize the imaging conditions to obtain sufficient spatial resolution while maintaining reasonable temporal resolution. This is the primary limitation of high-resolution neutron radiography, as low-noise images require long exposure times, in the order of several minutes. An effective optimization for neutron imaging of PEMFC is to use slit apertures, where the small dimension is parallel

Table 7.1 Apertures and beam properties at the NIST neutron imaging facility. Apertures 3, 4, and 5 are circular, and the dimension is the diameter; apertures 1 and 2 are slits

Aperture	Aperture dimension (mm)	$\approx$ L/D (x,y)	Fluence rate (cm^{-2} s^{-1})
5	15	450	1.38×10^7
4	10	600	4.97×10^6
3	3	2000	5.23×10^5
2	10×1	6000, 600	6.54×10^5
1	1×10	600, 6000	7.17×10^5

to the PEMFC through-plane direction (Boillat *et al.* 2010, Hussey and Jacobson 2010); this yields a small λ_g along the through-plane direction, while maintaining a sufficient fluence rate to minimize image exposure time. An example of the effect of aperture size on fluence rate is given in Table 7.1.

7.2.2 High-resolution neutron imaging detector systems

Since the neutron is a neutral particle, neutron detection occurs through multiple steps: absorption of a neutron; the release of charged particles as a byproduct of the neutron absorption; interaction of the charged particle with a second medium which produces either light or a charge avalanche; and detection of the light or charge. There are a few isotopes that have large neutron absorption cross-sections; those most often employed in neutron imaging detectors are ^{6}Li, ^{10}B, and ^{nat}Gd. In the case of neutron capture by ^{6}Li, the emitted charged particles are ^{3}H and He with a total energy of about 4.8 MeV. In the case of Gd, a simple description of the neutron capture is that the decay of the excited nucleus results in the emission of energetic gamma rays up to 7 MeV in energy and electrons (which produces additional conversion electrons through the Auger process) with a total of about 50 keV of energy. The technique of high-resolution neutron radiography has greatly benefited from recent developments in detector technology in charge-coupled-devices (CCDs) which view a scintillator via a lens and in high-resolution microchannel plates (MCPs), and each detector scheme is shown in Fig. 7.3.

The most common neutron imaging detector is a CCD coupled to a scintillator via a standard camera lens. Because CCDs are easily damaged by radiation from the main beam, it is necessary to view the scintillator via a 45° mirror, as shown in Fig. 7.3a. By changing the focal distance of the lens, one changes the field of view and resolution of the detector system, so that the detector system can be optimized for the object of study. The field of view of the CCD depends on the reproduction ratio of the lens and the size pixel array of the CCD; with a lens reproduction ratio of 1:1, a pixel pitch of 13.5 μm, and a pixel array size of 2048 × 2048, the field of view is about 2.76 cm × 2.76 cm. Modern CCD cameras have excellent cooling to below −100 °C which eliminates the dark current and enables exposure times of 10 s or longer with no real decrease in the signal-to-noise ratio. Another improvement

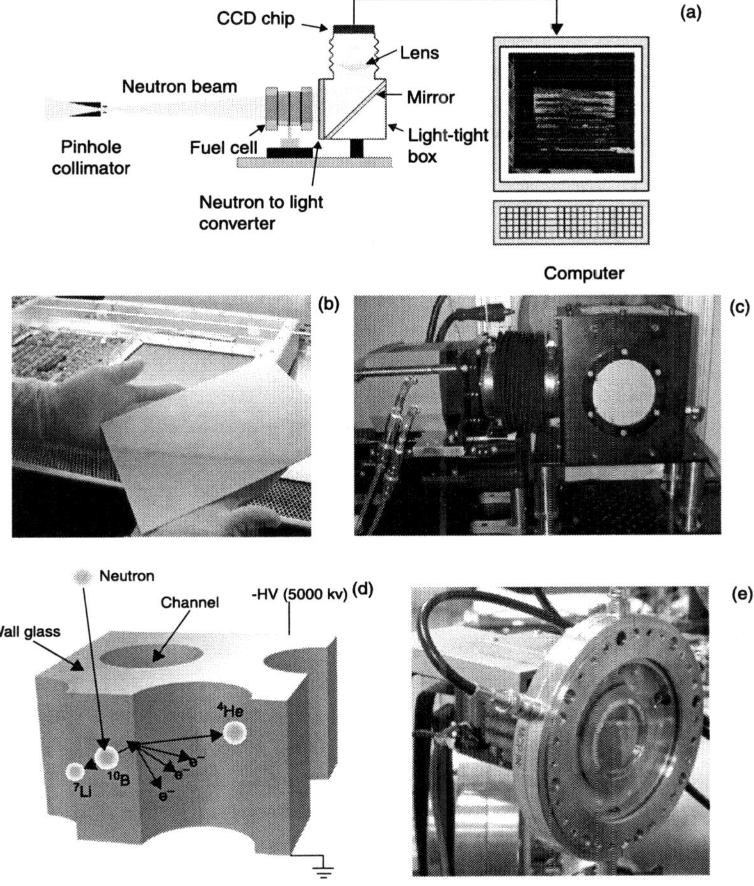

7.3 Neutron detection schemes. (a) Sketch of a neutron imaging setup with a CCD and scintillator. (b) Photo of a Li:ZnS scintillator and amorphous silicon detector. (c) Photo of a CCD and Gadox detector system. (d) Schematic of neutron detection with an MCP detector. (e) Photo of a 40 mm MCP detector with cross-strip readout that achieves a spatial resolution of 13.5 μm.

in CCDs has been the array read-out time, which for standard CCDs is of the order 1 MHz, resulting in an image read-out time of about 4 s for a 4 megapixel camera. The primary scintillator materials used in high-resolution neutron imaging are ^{6}LiF doped in ZnS (Li:ZnS) and Gadolinium oxysulfide (Gadox). There are three important properties of a neutron scintillator for imaging: the spatial resolution, the neutron capture efficiency, and the light output per neutron capture. The light output from a scintillator depends on the amount of absorbed energy. The massive charged particles from ^{6}Li have a short range (<10 μm) in ZnS, and all of the energy is absorbed by the scintillator resulting in about 10^5 photons per absorbed neutron. In the case of Gadox scintillators, only the electron deposits appreciable energy, resulting in about 100 times less light output than ^{6}Li doped into ZnS. Since the scintillation photons are emitted in all directions, the spatial resolution is proportional to the thickness of the scintillator. However, due to self attenuation of the light the overall neutron detection efficiency asymptotically approaches a maximum value as the thickness is increased. Since Gadox is composed of Gd, it has a very high capture efficiency for thin scintillator thickness, in the order of 30% for a thickness of about 10 µm. With such a scintillator, and a lens reproduction ratio of 1:1 the spatial resolution is limited by the light sensor pixel pitch per the Nyquist sampling theorem. In the case of Li:ZnS, the ^{6}Li content is only a few percent by volume, and in order to have an appreciable thermal neutron capture efficiency (15–20%), scintillators in the order of 100 µm are required, though thinner scintillators can be used for cold neutron beams. In order to overcome the Nyquist sampling theorem limit on the spatial resolution, the group at PSI has developed a tilted scintillator system (Boillat *et al.* 2010). By tilting the scintillator so that the normal to the scintillator plane is at an 87° angle to the neutron beam axis, an image spatial resolution of about 9 µm was reported, with a field of view of less than 1.5 mm along the through-plane fuel cell direction. Larger fields of view can be obtained with this system at the sacrifice of spatial resolution.

MCP detectors have only recently been developed for neutron imaging and are used solely for high-resolution imaging, with spatial resolutions of 25 µm to 13 µm having been reported (Hussey *et al.* 2007, Hussey and Jacobson 2010). The neutron detection mechanism of an MCP is shown schematically in Fig. 7.3d. The MCP glass contains Gd or ^{10}B, which on neutron capture release energetic, charged particles that have a range of approximately 5 µm in the MCP glass. When the charged particle enters a pore, electrons are stripped from the surface. A high electric field (3–5 kV) accelerates the electrons down the channel; subsequent wall collisions result in more electrons being stripped, and thus with a sufficiently long channel the initial charge is amplified by a factor of 10^6–10^7. The centroid of the charge cloud is determined with a two-dimensional (2D) resistive anode. Since the detection is event-based, there is a limit to the global count rate before events are lost; this is known as deadtime, which is a percentage of events lost at a given input rate; 10% deadtime loss is typically the maximum acceptable loss. For the NIST 40 mm cross-strip MCP detector, the 10% deadtime occurs for a global rate of about 1 MHz; thus for a given fluence rate there is a maximum field of view, which is an area about 3 cm^2 at the NIST neutron imaging facility.

7.2.3   Test section design considerations

While neutron imaging permits the use of common PEMFC materials construction, attention must be paid to the design of a test section for use in high-resolution neutron imaging studies to minimize geometric blur, provide a stable mounting, yield high neutron transmission, and provide flat interfaces for alignment (see

Section 7.3.2) and component identification. The primary design consideration is the effect of the geometric blur, see Eq. 7.1. If the test section is far from the detector, or has a long extent parallel to the neutron beam path, the geometric blur will limit the spatial resolution of the image. To prevent this, it is necessary to use a test section with a narrow active width (parallel to the neutron beam axis). To place the test section as close to the detector as possible, it is necessary to place gas inlet and outlets, the current and voltage leads, the thermocouples, and the heaters away from the face nearest the detector. Examples of cell designs can be found in Mukundan *et al.* (2007) and Preston *et al.* (2011). The test section must be rigidly mounted in a manner that is thermally stable and electrically insulating. If the test section will be imaged at multiple operating temperatures, the mounting block should be made of Invar, which has a negligible coefficient of thermal expansion. This ensures that thermal expansion of the cell and mounting hardware will not result in poor image registration between the dry and wet images. While neutrons penetrate deeply into most materials, in high-resolution imaging, the neutron beam passes through a thick section; thus choosing optimal materials increases the neutron transmission and improves the counting statistics. Good material choices are aluminum flow fields that are coated with gold and utilize polytetrafluoroethylene (PTFE) gaskets, since both Al and PTFE have low neutron absorption cross-sections. Gold provides corrosion protection for the Al flow fields and although it has a significant absorption cross-section, the amount of gold is very small and does not adversely affect the images. Some research has been conducted on graphite flow fields. Although carbon has a factor of four larger scattering cross-section than aluminum, graphite is more stable than the gold-coated aluminum, especially under high humidity conditions. However, unsealed graphite plates are very porous and will absorb large amounts of water making quantitative image analysis impossible with neutrons. Unsealed plates also leak gases making it difficult to hold back pressure on the cell. Therefore it is critical to seal graphite flow fields with a sealant that will not adversely affect the electrical or thermal conductivity of the graphite or the thermal neutron transmission properties (Hickner *et al.* 2006). Other gasket materials, such as Furon or silicone should be avoided due to the relatively high hydrogen concentration, which obscures water content in the diffusion media. Some gasket materials also have fiber reinforcement, which gives rise to small pores that can wick water during operation through capillary forces and make it difficult to quantitatively analyze the data.

7.3   Image acquisition and analysis

In principle, neutron radiography is a simple technique to measure the water in a PEMFC. In this section, image analysis and a discussion of measurement uncertainties are presented.

7.3.1 Obtaining water content

The overall interaction of a neutron with a material can be described with a collision cross-sectional area referred to as the total scattering cross-section σ_T;

$$\sigma_T = \sigma_a + \sigma_s = \sigma_a + \sigma_c + \sigma_i \qquad [7.3]$$

where σ_a is the absorption cross-section, σ_s is the scattering cross-section, which is the sum of the coherent cross-section σ_c and the incoherent scattering cross-section σ_i. In radiography, one measures the local transmission $T(x,y)$ of the neutron beam. The transmission is given by the Lambert-Beer law of attenuation as:

$$T(x,y) = I(x,y)/I_0(x,y) = \exp[-N\sigma_T t(x,y)], \qquad [7.4]$$

where $I(x,y)$ is the background corrected intensity in the object state of interest, $I_0(x,y)$ is the background corrected intensity of the reference state, N is the material number density, t is the material thickness through which the neutron beam passes, $\mu = N\sigma_T$ is referred to as the attenuation coefficient, and $N\sigma_T t(x, y)$ is the local optical density. In the case of a composite material the attenuation coefficient μ is given by:

$$\mu = \sum_j N_j \sigma_T, j, \qquad [7.5]$$

where the sum is over all species j that compose the material. In general, σ_T can depend on neutron energy and in some molecular species, such as water, σ_T depends on the rotational and vibrational excitation modes; thus μ is not always given by Eq. 7.5, rather a more complicated model is required (Edura and Morishima 2004) or calibration measurements using well-characterized stepped wedges is performed (Hussey *et al.* 2010). The contrast in a neutron image is determined by the μ of the materials present; for the thermal neutron beam at the NIST imaging facility the values for water, aluminum and carbon are approximately: $\mu_{water} = 0.37$ mm^{-1}; $\mu_{Al} = 0.009$ mm^{-1}; $\mu_C = 0.055$ mm^{-1}.

The neutron transmission through water is in general slightly more complicated than Eq. 7.3, as the total scattering cross-section is neutron energy-dependent; σ_T increases with decreasing neutron energy. This means that the neutron beam that emerges from a section of water has a more energetic spectrum than the incident beam and is more penetrating. The result is that the optical density ($N\sigma_T t_w$) is non-linear in t_w, an effect known as beam hardening which is frequently encountered in X-ray imaging (Barrett and Swindell 1981). As shown in Fig. 7.4, beam hardening for water is well modeled by a quadratic function where the optical density as a function of water thickness is $OD(t_w) = \mu\, t_w + \beta\, t_w^2$. The water thickness is then obtained from the transmission image via

$$t_w = -\sqrt{-\ln(T)/\beta + \mu^2/4\beta^2} - \mu/2\beta \qquad [7.6]$$

While the water content is accurately obtained, beam hardening limits the maximum thickness of water that can be measured, $t_w^{max} = -\mu/2\beta$; for the NIST

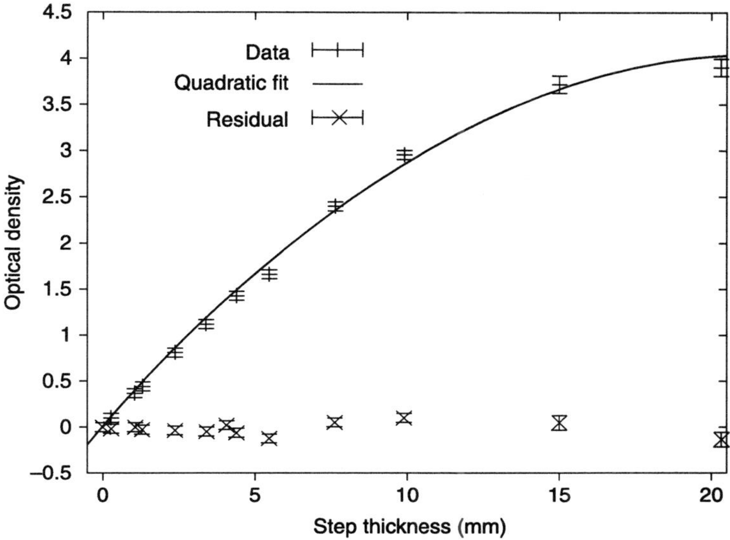

7.4 Calibration measurement of the neutron attenuation by water for the NIST 13 μm MCP detector. As the residual shows, the quadratic fit $OD(t_w) = \mu t_w + \beta t_w^2$ accounts for beam hardening, and the fit parameters are $\mu = (0.374 \pm 0.005)$ mm^{-1}, $\beta = (-0.0086 \pm 0.0004)$ mm^{-2}, and the reduced $\chi^2 = 1.58$. The uncertainties in the optical density are due to counting statistics and the fit parameters were determined from a non-linear least squares fit.

high-resolution MCP detector $t_w^{max} = 21.7$ mm, which is thicker than typically encountered in PEMFC imaging.

There are a few implicit assumptions in Eqs 7.3 and 7.5 including that there is perfect registration between the dry and wet images, that the incident intensity is the same for both image sets, and that there is no water in the dry image of the fuel cell. This variation is accounted for by normalizing the dry and wet images by the average intensity of a region in the PEMFC that does not change in time, i.e. the aluminum compression plates, away from the gas channels. The issue of residual water in the dry image is discussed in more depth in Section 7.3.2. Image registration means that the static components of the fuel cell (flow fields, interface between DM and lands) are in the same position in the dry and wet images. There are a few mechanisms that can result in poor registration. Thermal expansion of the mounting hardware, or using a dry reference image acquired at a different temperature than the operating image results in poor image registration. With the high spatial resolution, even small temperature changes result in noticeable thermal expansion. Another source of poor image registration is due to the operation of a nuclear reactor (and not a spallation source such as at the Paul Scherrer Institut) wherein the control mechanism results in a small shift in the neutron flux distribution in the core of the reactor (Hussey *et al.* 2011). This changes the axis of the neutron beam by a small angle (on the order of microradians) and changes the position of the image on the detector by a small amount. In both cases, the effect is usually manifest at the interface between a weak and strong absorber, for instance the channel edges as shown in Fig. 7.5 or the interface of the aluminum flow fields and the diffusion media. Since the image shift is often a fraction of a pixel, and the detector field response is not generally uniform, one cannot simply shift one raw image with respect to another to correct the registration. Rather, to correct poor image registration, the dry and wet images are normalized by the flat field to remove the detector structure and either the images or the through-plane transmission profiles can be shifted using standard interpolation methods. Thus it is possible to correct for these sources of image artifacts.

In order to discriminate anode from cathode, it is necessary to align the test section. This can be done by mounting the test section on a rotation or tilt stage (depending on the desired orientation) and acquiring images at several rotation (or tilt) angles. When the test section approaches optimal alignment, the measured width varies approximately quadratically with the tilt angle, as shown in Fig. 7.6. The minimum of this quadratic is the aligned position. The required precision of the alignment is determined by the active width of the test section, w, and the image spatial resolution δ_i by $\Delta\theta = \tan^{-1}(2\delta_i/w)$, or $\Delta\theta = 0.15°$ for a 1 cm-wide test section and 13 µm spatial resolution.

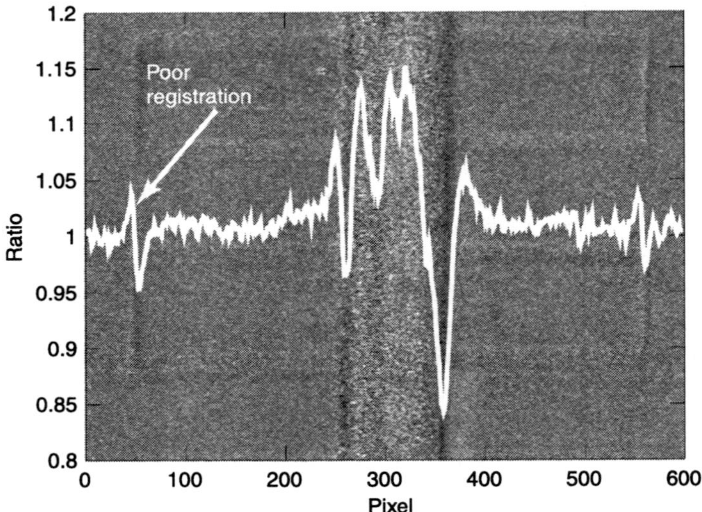

7.5 Comparison of two images, taken at different times that show evidence of poor image registration; the ratio of the two images has a sawtooth pattern at the channel edges (near pixel 40) showing an increase followed by decrease, and the sense is the same one the other electrode at pixel 570.

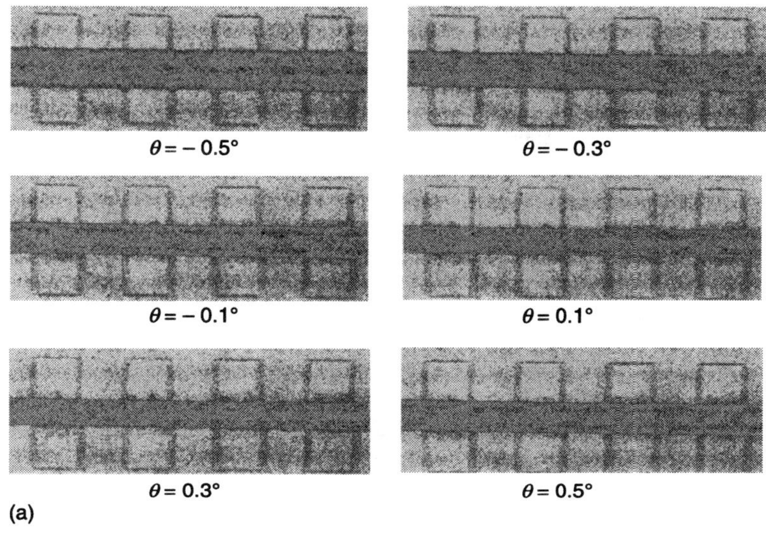

(a)

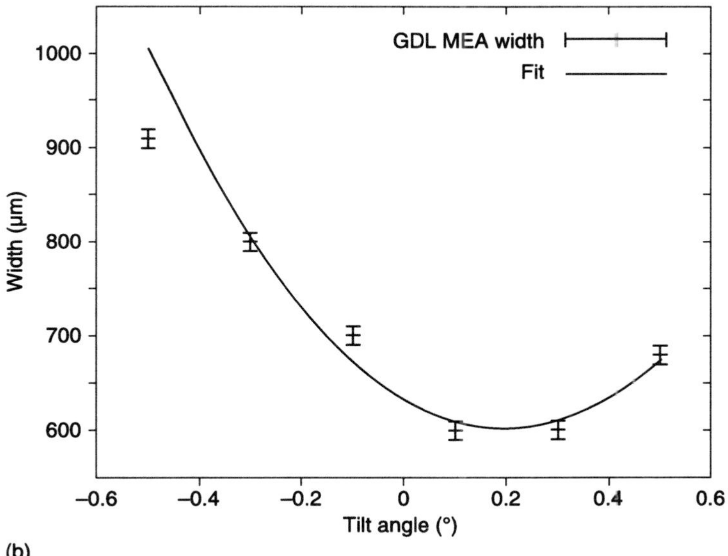

(b)

7.6 Images illustrating test section alignment (a). Close to the aligned position, the width of the through-plane direction changes quadratically with tilt angle (b).

7.3.2 Measurement uncertainty and comparison with model prediction

There are three primary sources of liquid water measurement uncertainty in high-resolution neutron imaging: counting statistics, spatial resolution and residual membrane water content. These effects are understood and can be well-modeled, thus the neutron image data provide an accurate, precise measurement of the PEMFC through-plane water content.

Counting statistics, bias, backgrounds

The primary source and limit of uncertainty in the measurement of liquid water in an operating PEMFC is due to Poisson counting statistics (shot noise). The uncertainty σ_{tw} in liquid water thickness has been quantified for an older MCP detector at NIST (Hussey *et al.* 2010), and in general can be estimated from:

$$\sigma_{t_w} = \frac{1}{\mu_w}\sqrt{\frac{1}{I_w}} = \frac{1}{\mu_w}\sqrt{\frac{\exp\left(\mu_w t_w + \beta t_w\right)}{\phi A \tau \eta}} \qquad [7.7]$$

where ϕ is the neutron fluence rate, A is the area of analysis, τ is the exposure time, and η is the detector efficiency (typically $\eta = 20\%$ for the NIST MCP detectors). The uncertainty improves with both measurement time and area over which one averages. Since fuel cells have some degree of uniformity along the in-plane direction, one can maintain reasonable exposure times by averaging along the in-plane direction. Thus, two-dimensional (2-D) images are often converted to one-dimensional plots of the water content along the through-plane direction.

In high-resolution neutron imaging, the number of neutrons in a pixel per image tends to be of the order 10 or less. Since the distribution of detected neutrons is given by a Poisson distribution, there is a reasonable probability that there will be many empty pixels or zeros in a given image. This gives rise to a somewhat subtle effect; a bias is introduced into the analysis if the average water content were obtained from a 'density image'

$$V_{t,l} = -\frac{A_p}{\mu_w}\sum_{j=1}^{N} \ln\left(\frac{I_{w,j}}{I_{d,j}}\right), \qquad [7.8]$$

rather than taking the log transform of the averaged intensities in the wet and dry images:

$$V_{t,a} = -\frac{NA_p}{\mu_w}\left(\ln\sum_{j}^{N} I_{w,j} - \ln\sum_{j}^{N} I_{d,j}\right). \qquad [7.9]$$

In Eqs 7.8 and 7.9 j indexes pixels in the averaging region. One can understand the source of the bias simply by considering what happens when a pixel contains a zero. If the zero value is in the numerator, the log transform would indicate an infinite amount of water; if the zero value is in the denominator, an undefined value is returned. The relative size of the bias decreases to about 10^{-4} once the mean number of neutrons detected is greater than 50. This can be achieved by increasing the image exposure time, or by averaging along the in-plane direction of the fuel cell.

Spatial resolution

The primary source of systematic uncertainty stems from the finite image spatial resolution. The effect of the spatial resolution of the system is a blurring of the true image $f(x,y)$. This blurring is described by the convolution of the system point spread function (PSF) with the true image. Since convolution in the frequency domain is a multiplication rather an integral, the Fourier transform of the PSF, known as the modulation transfer function (MTF), is often encountered. The PSF of neutron imaging detectors is well-modeled by a Gaussian so that

$$PSF(\vec{\mathbf{x}}) = \exp(-\vec{\mathbf{x}}/2\sigma^2) \qquad [7.10]$$

$$MTF(\vec{\mathbf{q}}) = \exp(-2\pi^2\vec{\mathbf{q}}^2\sigma^2) \qquad [7.11]$$

where $q = x^{-1}$. The spatial resolution δ_i is determined from the frequency at which $MTF(q) = 0.1$ giving

$$\delta_i = \frac{\pi\sigma}{\sqrt{2\ln(10)}}. \qquad [7.12]$$

The measured image $g(x,y)$ is then

$$g(x,y) = PSF(\vec{\mathbf{x}}) ** f(\vec{\mathbf{x}}) = IFT\{MTF(\vec{\mathbf{q}}) * FT\{f(\vec{\mathbf{x}})\}\} \qquad [7.13]$$

where FT (IFT) stands for the (inverse) Fourier Transform. Thus the effect of the MTF is to suppress the high-frequency components of the image. In the case of through-plane imaging of PEMFCs, the water is concentrated in the membrane. Shown in Fig. 7.7 is the impact on the measured neutron transmission, assuming a spatial resolution of 25 µm. For a region that is thick (perpendicular to the beam direction), the water depth along the beam is fully resolved in the center; for a thin region the neutron transmission never reaches the true value. Since the water content is obtained by taking the natural log transform of the transmission, this reduction for a thin slab represents a potentially large uncertainty in the water thickness measurement.

While it is possible to deconvolve the images by application of the inverse of the MTF, one must consider that high-resolution neutron images contain a

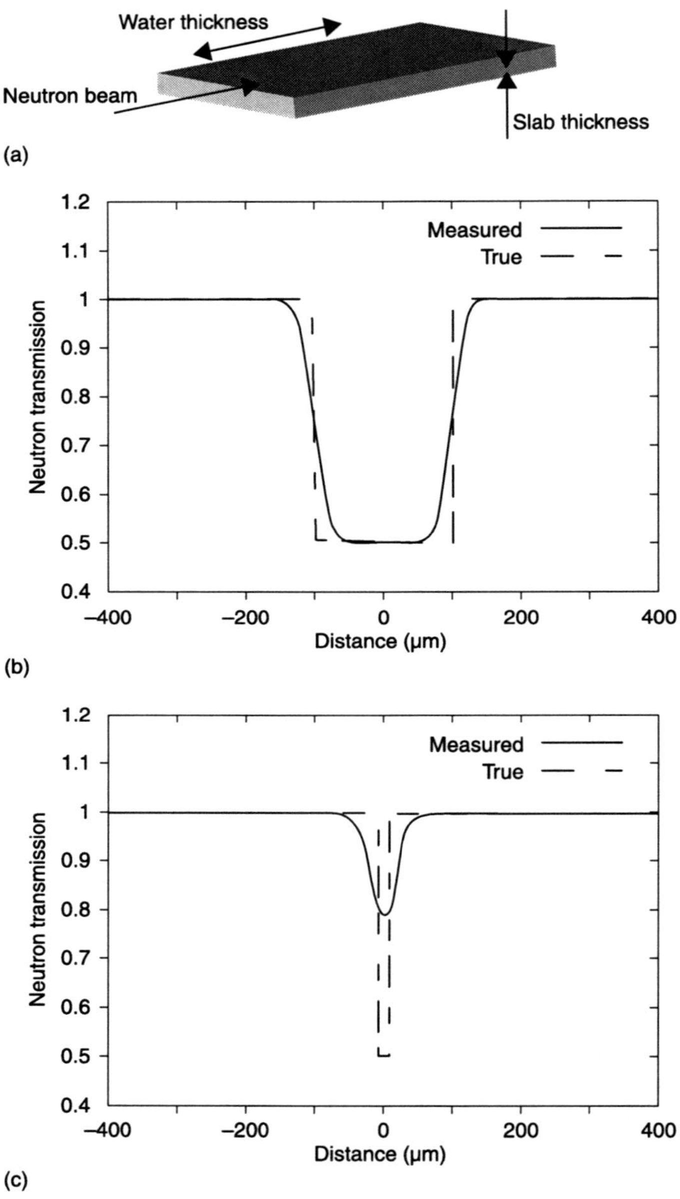

7.7 Illustration of the effect of resolution on the measured water content. (a) Schematic of the setup, where the slab thickness effects the measurement of the water content. Assuming a spatial resolution of 25 µm, the effect on the measured water content for a slab thickness of (b) 200 µm and (c) 25 µm.

significant amount of random shot noise, which has a uniform distribution in the frequency domain; the inverse of the MTF would then amplify this noise, resulting in unrecognizable images. Other filtering techniques such as Weiner filtering apply a frequency cut-off, which is typically lower than the spatial frequency of the imaging system and further degrades the spatial resolution. Thus, one cannot correct for this systematic effect, rather the effect of spatial resolution must be incorporated into any analysis that compares model output data of the water content to the image data.

In order to compare high-resolution imaging data to model predictions of saturation or membrane water content one must simulate the image data from the model predictions. The simulation requires that the imaging system be well-characterized in terms of pixel pitch, spatial resolution, geometry of the setup, and measurement of the $OD(t_w)$. The spatial resolution of the model can be as fine as desired to capture the details of the water transport, but the image data will only provide useful information if there are discernable effects on the simulated profiles. Since the image data is only sensitive to the thickness of water along the through-plane direction (as opposed to saturation, etc.) the output of the model should be in water thickness. Further, since the images are blurred, the water content is smoothed, especially at interfaces, so that expected sharp discontinuities in the water content, e.g. at the micro-porous layer (MPL)/DM interface, will not be realized.

Beam hardening

Another source of systematic uncertainty stems from beam hardening and a partially hydrate membrane in the dry image. Since Nafion is hygroscopic, the typical procedure to fully dehydrate the membrane includes placing the membrane under vacuum at 105 °C (Zawodzinski *et al.* 1991). However, due to time constraints at the imaging beamline and potential damage to the membrane, it is not practical to use this method to dry out the test section and membrane. Rather, the typical method is to flow dry nitrogen at 80 °C for 1–2 h, which does not fully dehydrate the membrane. This residual water thickness t_{res} changes the neutron spectrum, as discussed above in Section 7.3.1, so that the conversion from neutron attenuation to water thickness in the membrane is slightly altered from Eq. 7.5 in that the linear term is reduced by t_{res}:

$$t_w = -\sqrt{-\frac{\ln(T)}{\beta} + \frac{(\mu + 2\beta t_{res})^2}{4\beta^2}} - \frac{\mu + 2\beta t_{res}}{2\beta}. \qquad [7.14]$$

One can account for t_{res} either by direct measurement with a water wedge, or rely on *ex situ* measurement of the water content from gravimetric measurements. The effect was measured by performing the water calibration measurement with several membrane thicknesses at various states of hydration. The measured relative change in the slope $2\beta t_{res}/\mu$ versus the estimated water content is shown in Fig. 7.8. The impact of neglecting to account for the residual water in the image analysis is shown in Fig. 7.8b where the data are compared to previously measured water sorption at 30 °C (Springer *et al.* 1991) and 80 °C (Hinatsu *et al.* 1994).

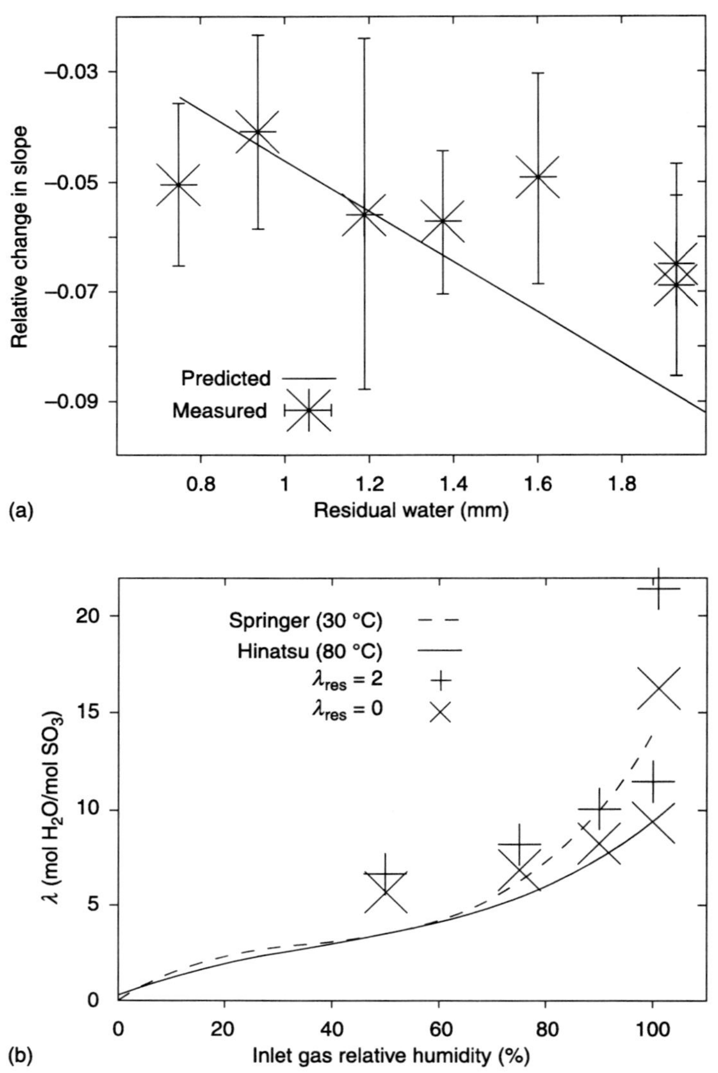

7.8 (a) The effect of residual water membrane water content in the dry image is to affect the slope of the calibration measurement.
(b) Incorporating this correction into the image analysis is critical for high water contents, especially for liquid equilibrated membranes. Size of the measured points indicate the uncertainty due to counting statistics.

7.4   Review of recent experiments

Since high-resolution neutron detectors have only recently been available, there are only a few published examples of the application of the method to the study of the through-plane water content in fuel cells. These initial experiments have explored detector usability (Hussey *et al.* 2007, Boillat *et al.* 2008a, Boillat *et al.* 2010) (Section 7.4.1), water or hydrogen uptake in membranes (Boillat *et al.* 2008b, Spernjak *et al.* 2010) (Section 7.4.2), the impact of diffusion media and operating conditions on water content distribution (Preston *et al.* 2011, Hickner *et al.* 2008, Weber and Hickner 2008, Mukundan *et al.* 2007, Spendelow *et al.* 2008, Turhan *et al.* 2010, Seyfang *et al.* 2010, Aaron *et al.* 2011) (Section 7.4.3), effects of thermal gradients (Kim and Mench 2009, Fu *et al.* 2011) (7.4.4), and freeze studies (Mukundan *et al.* 2008) (Section 7.4.5).

7.4.1   Detector demonstration experiments

Three experiments were performed to demonstrate that the high-resolution detector technology could resolve features of the through-plane water content. Using an MCP-based detector with an overall image spatial resolution of better than 30 μm, Hussey *et al.* (2007) showed that under low current density operation (less than 0.2 A cm^{-2}) the water content was seen to increase with increasing current. Boillat *et al.* (2008a, 2010) showed that with a tilted scintillator and slit apertures, and using thinner scintillators, a spatial resolution of ~10 μm spatial resolution could be achieved along the through-plane direction and still maintain an appreciable neutron fluence rate by sacrificing spatial resolution along the in-plane direction. These experiments showed the potential of the new detector technologies for exploring the through-plane water content.

7.4.2   Mass transport in thick membranes

Since the best achievable spatial resolution is about 10 μm, it is not possible to resolve features within commercially competitive membranes, whose thickness range is 10 μm to 20 μm. This requires using thicker membranes, such as N117 (Nafion, 1100 equivalent weight, 175 μm (0.007") thick) in order to resolve through-plane features. While this is not directly applicable to automotive materials, it is possible to explore fundamental heat and mass transport phenomena. In this spirit, there have been two published experiments to look at transport in the membrane. In the first experiment, Boillat *et al.* (2008b) exploited the factor of 10 difference in scattering between hydrogen and deuterium to measure the exchange current density in the hydrogen oxidation reaction. The experiment consisted of operating a test section at two current densities, 0.02 A cm^{-2} and 0.8 A cm^{-2}, and switching from hydrogen to deuterium gas (but not changing the water in the humidifiers from light water to heavy water). The primary observation was that the hydrogen in the membrane was rapidly replaced by deuterium, and this rate was independent of the current density; thus suggesting the exchange current density is driven by diffusion from the surface to the bulk, is large and is limited only by the transport of hydrogen from the gas channels to the anode. In the second experiment, Spernjak *et al.* (2010) performed *in situ* water sorption measurements in order to verify the validity of the gravimetrically determined correlation between water activity and water uptake in Nafion membranes. The most important result from this work was that one must account for the beam hardening due to the residual water in the membrane during the dry image. A secondary correction is to account for the in-plane water diffusion during long exposure times. By including these effects in the image analysis, a number of observations were made: there is reasonable agreement with the historic *ex situ* measurements; there is evidence of Schröder's paradox; membrane compression does not reduce the water uptake for pressures up to 300 kPa except under liquid saturation. These data provide further confirmation that high-resolution neutron radiography provides accurate, repeatable, *in situ* measurements of the through-plane water content and all neutron-related systematic measurement uncertainties are understood and well-quantified.

7.4.3 Diffusion media through-plane water distribution during operation

There have been numerous studies analyzing the DM water content. An enzymatic fuel cell was investigated in Aaron *et al.* (2011) and the observed performance decay was well correlated with the dehydration of the cathode compartment. The effect of the surface treatment of the channel lands was studied by Turhan *et al.* (2010). It was observed that a hydrophilic treatment resulted in a laminar sheet that continuously wicked water to the channels, while a hydrophobic treatment resulted in unsteady water removal and hence a varying DM water content. Preston *et al.* (2011), demonstrated that the rate of property change at the interface between the MPL and the substrate can have a significant effect on the through-plane water distribution; a sharp interface results in a higher water content in the diffusion media, whereas a broad interfacial region results in a lower water content. The lower water content is in better agreement with the neutron radiography data, and SEM images of the GDL/MPL demonstrate that there is in fact a broad interfacial region.

In a follow-on experiment to investigate the effects of local heating, Hickner *et al.* (2008) measured the water content over a wide current range and at two different stoichiometric flow rates. These data were further analyzed by Weber and Hickner (2008) in an effort to qualitatively understand the water distribution in the diffusion media. The primary results of these data and analysis is that there is a non-trivial thermal gradient between the membrane and flow fields, up to 6 °C, which results in a vapor transport of the product water away from the membrane. This vapor transport can present a mass transport resistance to the reactants and thus contribute to flooding. The vapor transport is discussed in more depth below in Section 7.4.4. Another feature of the data is evidence for a non-uniform porosity in the diffusion media, where the interior had a higher porosity than the surfaces, indicating the need for better material characterization in order to properly describe the observed water content.

The research group from Los Alamos has explored the effects of flow configuration and the surface treatment of the diffusion media (Mukundan *et al.* 2007, Spendelow *et al.* 2008). For the conditions tested, a lower PTFE concentration in the cathode MPL was beneficial to performance as a higher PTFE content resulted in greater water accumulation in the cathode GDL. In co-flow, there was also more water accumulation in the regions of the outlets, the cathode GDL, the anode flow channel, and the MEA/GDL above the lands when compared to the inlets, anode GDL, cathode flow channel, and MEA/GDL above the channel respectively. As well, significant water was observed in the anode GDL, indicating that back diffusion was significant enough to keep the membrane hydrated even under relatively dry inlet (50% RH or less) conditions. In counter-flow, it was observed that the water content in the membrane was more uniform along the channel than in co-flow. The effect of gravity was also investigated, and it was seen that when the anode was above the cathode, significantly (factor of 2–3) less water accumulated in the anode diffusion media, while the membrane hydration was unaffected.

A micro PEMFC was studied (Seyfang *et al.* 2010) to evaluate the possibility of operating with no diffusion media. The PEMFC consisted of a catalyst-coated membrane compressed between flow fields that consisted of trapezoidal channels in the order of 100 μm in dimension. It was observed that at a repeatable limiting density, a water sheet was formed at the cathode catalyst/channel interface. Further, the operation with no diffusion media resulted in membrane creep into the channels, which both damaged the catalyst layer as well as reduced the open diameter of the channel. The authors mention that there are still difficulties in quantifying the water content using the PSI system and only present relative measures of water thickness.

7.4.4   Thermal gradients

In two sets of experiments, Kim and Mench (2009) and Hatzell *et al.* (2011) investigated the nature of temperature-driven flow across diffusion media. Kim and Mench coupled an extensive set of *ex situ* balance measurements with qualitative neutron imaging to determine that, in diffusion media, the phase change induced flow transports water in the vapor phase from the hot side to the cold side, while in membranes the direction of the flow is reversed. It was further observed that the presence of a hydrophilic region strongly influenced the flow, and that the arrangement in the through-plane sandwich was also a factor. By measuring the water flux as a function of time, Hatzell *et al.* were able to show that there is a balance between phase change-induced transport and capillary action. Initially, there is insufficient water for there to be a connected pore network, and thus the water transport is due to the phase change induced flow. Once a minimum saturation of about 20% is reached, it is seen that a connected pore network exists, and the much more effective capillary action dominates the water transport, resulting in a more homogeneous distribution of water.

The study of thermal gradients was extended in Fu *et al.* (2011) by assessing a model fit of the effects of a temperature gradient through a thick (245 μm) Nafion membrane with no catalyst layer and Toray gas diffusion media using data derived from water balance measurements and neutron radiography, to evaluate the effect of temperature gradients to drive the flow of water in a fuel cell MEA operated under differential flow conditions. By not using a catalyst layer and therefore not applying a load, the complication of electro-osmotic drag on the analysis was avoided. The anode was supplied with 100% RH and the cathode RH was stepped from 40% RH, 64% RH and 95% RH at two different temperature gradients (anode/cathode: case A (62 °C/70 °C) and case B (50 °C/70 °C)). The dominant mechanism of transport identified in this study was diffusion-driven transport due to a water concentration gradient. This mode was identified by the authors as being a transport mode that contributes an order of magnitude higher flux of water than thermo-osmosis. The model quantitatively predicted that the water flux was positive (from the cold side to the hot side) for cathode RHs below 50% and negative (from the hot side to the cold side) for RHs above 50%. Measurements of the through-plane water distribution with neutron radiography showed good qualitative agreement with the overall quantity of water in the membrane, but did not predict the measured shape. Multiple sources of uncertainty were identified including neutron radiography systematic uncertainties such as beam hardening, and fuel cell systematic effects such as cell uniformity and the quality of fuel cell materials characterizations as the source of the discrepancy between the neutron data and the model.

7.4.5   Ice formation after isothermal operation at temperatures below 0 °C.

One of the remaining challenges in PEMFCs is operation and durability at temperatures below 0 °C. One of the key metrics defined by the Department of Energy performance targets is to start a fuel cell stack within 30 s using less than 5 MJ of energy at an ambient temperature of −20 °C (DOE 2007). In order to achieve this, the product water must be stored within the GDM or MEA until the stack temperature is above 0 °C. While the membrane and to some extent the catalyst layer can store water and still be an efficient proton conductor, it is anticipated the use of thinner membranes and catalysts will be too small a volume. Thus, exploring where ice is formed within the sandwich provides insight as to how to accommodate

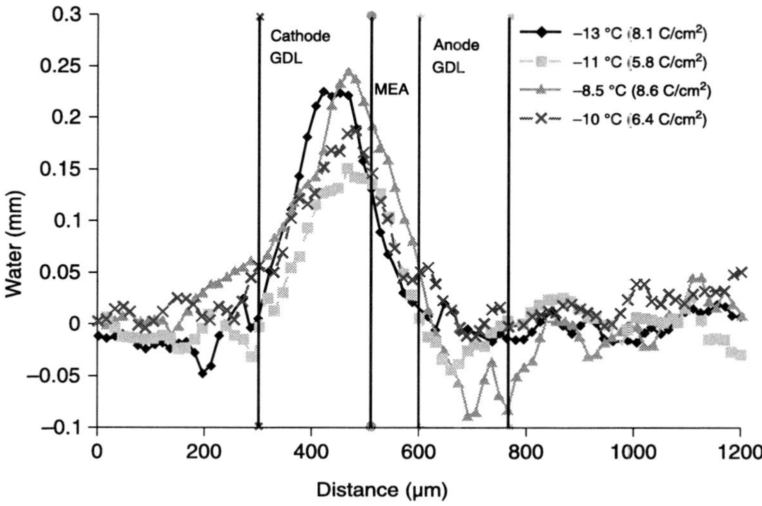

7.9 Location of ice formation after cell failure.

or prevent ice formation. In the first experiment to explore this topic with neutron radiography (Mukundan *et al.* 2008), an MCP detector and small area fuel cell were placed inside an environmental chamber to provide a dry and ambient temperature about −5 °C; the test section was further cooled using a water/ethylene glycol mixture. The experiment procedure was as follows: the test section was operated at 80 °C, purged with dry gas, cooled below 0 °C, imaged for 15 minutes to obtain a reference image, operated isothermally in current control mode until the cell failed to produce voltage, then imaged a second time to determine the location of the product water. As shown in Fig. 7.9, at temperatures near −10 °C, the product water escapes into the GDL and then freezes. At lower temperatures (−19 °C), the ice was seen to form in the MEA and as a result significantly less charge (factor of 2–5) could be obtained from the cell before failure – the more charge extracted, the longer the cell has to heat itself to above 0 °C.

7.5   Outlook and conclusions

High-resolution neutron imaging has provided valuable data about the through-plane water content, indicating the important role thermal gradients play in the water transport from the catalyst layer to the channel, as well as the need to have accurate characterization data of the diffusion media. The advantage of imaging the diffusion media is that the state-of-the-art in detector resolution can resolve features within this region. As well, by using slit apertures, the neutron intensity is reduced by a factor of a few, enabling image acquisition in a reasonable time scale (1–30 min). In order to apply the technique of neutron imaging to catalyst layers or membrane with ~1 μm resolution will require significant development in the spatial resolution and possibly in test section design. In the case of imaging detectors, the range of the charged particle is several micrometres, meaning that there is a limit to improved spatial resolution by thinning a scintillator or reducing the pore size in a MCP; in fact the current MCP technology is at its ultimate limit of about 10 μm. With higher detector resolution, it will be necessary to use smaller beam-defining apertures in order to reduce the blurring due to the beam to be less than or comparable to the blurring due to the detector resolution; thus reducing the neutron fluence rate at the detector. This means that in order to obtain sufficient measurement uncertainty, longer image exposure times will be needed, and the test section will need to operate at steady state for several hours. Alternatively, it might be possible to reduce the in-plane dimension so as to place the test section even closer to the detector, thereby reducing λg, as well as increasing the transmission through the test section. Efforts to improve detector spatial resolution are on-going at several neutron imaging facilities (Downing 2008, Boillat *et al.* 2010) and in the near future it may be possible to image the water distribution in the catalyst layer.

7.6   Note

† Certain trade names and company products are mentioned in the text or identified in an illustration in order to adequately specify the experimental procedure and equipment used. In no case does such identification imply recommendation or endorsement by the National Institute of Standards and Technology, nor does it imply that the products are necessarily the best available for the purpose.

7.7   References

Aaron, D. S., A. P. Borole, D. S. Hussey, D. L. Jacobson, S. Yiacoumi *et al.* (2011) 'Quantifying the water content in the cathode of enzyme fuel cells via neutron imaging'. *Journal of Power Sources*, **196**, 1769–1775.

Albertini, V. R., B. Paci, F. Nobili, R. Marassi and M. Di Michiel (2009) 'Time/space-resolved studies of the Nafion membrane hydration profile in a running fuel cell'. *Advanced Materials*, **21**, 578–583.

Barrett, H. H. and W. Swindell (1981) 'Radiological imaging. The theory of image formation, detection and processing'. In *Radiological Imaging. The Theory of Image Formation, Detection and Processing*, 2 vol. xliv+693.

Boillat, P., G. Frei, E. H. Lehmann, G. G. Scherer and A. Wokaun (2010) 'Neutron imaging resolution improvements optimized for fuel cell applications'. *Electrochemical and Solid State Letters*, **13**, B25–B27.

Boillat, P., D. Kramer, B. C. Seyfang, G. Frei, E. Lehmann *et al.* (2008a) '*In situ* observation of the water distribution across a PEFC using high resolution neutron radiography'. *Electrochemistry Communications*, **10**, 546–550.

Boillat, P., G. G. Scherer, A. Wokaun, G. Frei and E. H. Lehmann (2008b) 'Transient observation of H-2 labeled species in an operating PEFC using neutron radiography'. *Electrochemistry Communications*, **10**, 1311–1314.

Buechi, F. N., J. Eller, F. Marone and M. Stampanoni. (2010) 'Determination of local GDL saturation on the pore scale by in-situ X-ray tomographic microscopy'. In *218th Electrochemical Society Meeting*. Las Vegas, NV, USA.

Byrne, J. (1994) *Neutrons, Nuclei and Matter: An Exploration of the Physics of Slow Neutrons*. Philadelphia, PA: Institute of Physics Pub.

DOE, U. (2007) *Hydrogen, Fuel Cells and Infrastructure Technologies Program Multi-Year RD&D Plan*.

Downing, R. G. (2008) 'High resolution position sensitive neutron detector (HRPSND)'. *Neutron Radiography*, 616–621.

Dunbar, Z. W. and R. I. Masel (2008) 'Magnetic resonance imaging investigation of water accumulation and transport in graphite flow fields in a polymer electrolyte membrane fuel cell: Do defects control transport?' *Journal of Power Sources*, **182**, 76–82.

Edura, Y. and N. Morishima (2004) 'Cold and thermal neutron scattering in liquid water: cross-section model and dynamics of water molecules'. *Nuclear Instruments & Methods in Physics Research Section a-Accelerators Spectrometers Detectors and Associated Equipment*, **534**, 531–543.

Fu, R. S., J. S. Preston, U. Pasaogullari, T. Shiomi, S. Miyazaki *et al.* (2011) 'Water transport across a polymer electrolyte membrane under thermal gradients'. *Journal of the Electrochemical Society*, **158**(3) B303–B312.

Gao, B., T. S. Steenhuis, Y. Zevi, J. Y. Parlange, R. N. Carter *et al.* (2009) 'Visualization of unstable water flow in a fuel cell gas diffusion layer'. *Journal of Power Sources*, **190**, 493–498.

Hartnig, C., I. Manke, J. Schloesser, P. Kruger, R. Kuhn *et al.* (2009) 'High resolution synchrotron X-ray investigation of carbon dioxide evolution in operating direct methanol fuel cells'. *Electrochemistry Communications*, **11**, 1559–1562.

Hazell, M. C. T., A., S. Kim, D. S. Hussey, D. L. Jacobson and M. M. Mench. (2011) 'Quantification of temperature driven flow in a polymer electrolyte fuel cell using high-resolution neutron radiography'. *Journal of the Electrochemical Society*, **158**(6) B717–B726. DOI: 10.1149/1.3577597.

Hickner, M. A., N. P. Siegel, K. S. Chen, D. S. Hussey, D. L. Jacobson *et al.* (2008) '*In situ* high-resolution neutron radiography of cross-sectional liquid water profiles in proton exchange membrane fuel cells'. *Journal of the Electrochemical Society*, **155**, B427–B434.

Hickner, M. A., N. P. Siegel, K. S. Chen, D. N. McBrayer, D. S. Hussey *et al.*(2006) 'Real-time imaging of liquid water in an operating proton exchange membrane fuel cell'. *Journal of the Electrochemical Society*, **153**, A902–A908.

Hinatsu, J. T., M. Mizuhata and H. Takenaka (1994) 'Water-uptake of perfluorosulfonic acid membranes from liquid water and water-vapor'. *Journal of the Electrochemical Society*, **141**, 1493–1498.

Hussey, D. S., E. Baltic and D. L. Jacobson. (2011) 'Changes in the optical axis of a neutron imaging beam'. *Nuclear Instruments and Methods, Section A*, **651**, 73–76.

Hussey, D. S. and D. L. Jacobson. (2010) 'High resolution neutron radiography analysis of proton exchange membrane fuel cells'. In *Modeling and Diagnostics of Polymer Electrolyte Fuel Cells*, eds. U. Pasaogullari and C.-Y. Wang, 175–199. New York, NY: Springer.

Hussey, D. S., D. L. Jacobson, M. Arif, K. J. Coakley and D. F. Vecchia (2010) '*In situ* fuel cell water metrology at the NIST neutron imaging facility'. *Journal of Fuel Cell Science and Technology*, **7**.

Hussey, D. S., D. L. Jacobson, M. Arif, J. P. Owejan, J. J. Gagliardo *et al.* (2007) 'Neutron images of the through-plane water distribution of an operating PEM fuel cell'. *Journal of Power Sources*, **172**, 225–228.

Kim, S. and M. M. Mench (2009) 'Investigation of temperature-driven water transport in polymer electrolyte fuel cell: Phase-change-induced flow'. *Journal of the Electrochemical Society*, **156**, B353–B362.

Manke, I. and N. Kardjilov. (2012) 'Neutron tomography for polymer electrolyte membrane fuel cell characterisation'. In *Polymer Electrolyte Membrane and Direct Methanol Fuel Cell Technology. Volume 2:* In Situ *Characterisation Techniques for Low Temperature Fuel Cells*, eds C. Hartnig and C. Roth, Woodhead Publishing Ltd, Cambridge.

Mukundan, R., J. R. Davey, R. W. Lujan, J. S. Spendelow, Y. S. Kim *et al.* (2008) 'Performance and durability of PEM fuel cells operated at sub-freezing temperatures'. *Proton Exchange Membrane Fuel Cells 8, Pts 1 and 2*, **16**, 1939–1950.

Mukundan, R., J. R. Davey, T. Rockward, J. S. Spendelow, B. Pivovar *et al.* (2007) 'Imaging of water profiles in PEM fuel cells using neutron radiography: Effect of operating conditions and GDL composition'. *ECS Transactions*, **11**, 411–422.

Preston, J. S., R. S. Fu, U. Pasaogullari, D. S. Hussey and D. L. Jacobson. (2011) 'Consideration of the role of micro-porous layer on liquid water distribution in polymer electrolyte fuel cells'. *Journal of the Electrochemical Society*, **158**(2) B239–B246. DOI: 10.1149/1.3525626.

Schneider, I. A., M. H. Bayer, S. von Dahlen (2012), 'Submillimeter resolved transient techniques for polymer electrolyte membrane fuel cell characterisation: local in situ diagnostics for channel and land areas'. In *Polymer Electrolyte Membrane and Direct Methanol Fuel Cell Technology. Volume 2*: In Situ *Characterisation Techniques for Low Temperature Fuel Cells*, eds C. Hartnig and C. Roth, Woodhead Publishing Ltd, Cambridge.

Seyfang, B. C., P. Boillat, F. Simmen, S. Hartmann, G. Frei *et al.* (2010) 'Identification of liquid water constraints in micro polymer electrolyte fuel cells without gas diffusion layers'. *Electrochimica Acta*, **55**, 2932–2938.

Siegmund, O. H. W., J. V. Vallerga, A. Martin, B. Feller, M. Arif *et al.* (2007) 'A high spatial resolution event counting neutron detector using microchannel plates and cross delay line readout'. *Nuclear Instruments & Methods in Physics Research Section a-Accelerators Spectrometers Detectors and Associated Equipment*, **579**, 188–191.

Spendelow, J. S., R. Mukundan, J. R. Davey, T. Rockward, D. S. Hussey *et al.* (2008) 'High resolution neutron radiography imaging of operating PEM fuel cells: Effect of flow configuration and gravity on water distribution'. *Proton Exchange Membrane Fuel Cells 8, Pts 1 and 2*, **16**, 1345–1355.

Spernjak, D., P. P. Mukherjee, R. Mukundan, J. Davey, D. S. Hussey *et al.* (2010) 'Measurement of water content in polymer electrolyte membranes using high resolution neutron imaging'. *ECS Transactions*, **33**, 1451–1456.

Springer, T. E., T. A. Zawodzinski and S. Gottesfeld (1991) 'Polymer electrolyte fuel-cell model'. *Journal of the Electrochemical Society*, **138**, 2334–2342.

Tsushima, S. and S. Hirai (2009) 'Magnetic resonance imaging of water in operating polymer electrolyte membrane fuel cells'. *Fuel Cells*, **9**, 506–517.

Turhan, A., S. Kim, M. Hatzell and M. M. Mench (2010) 'Impact of channel wall hydrophobicity on through-plane water distribution and flooding behavior in a polymer electrolyte fuel cell'. *Electrochimica Acta*, **55**, 2734–2745.

Weber, A. Z. and M. A. Hickner (2008) 'Modeling and high-resolution-imaging studies of water-content profiles in a polymer-electrolyte-fuel-cell membrane-electrode assembly'. *Electrochimica Acta*, **53**, 7668–7674.

Wippermann, K. and Schröder, A. (2012). 'Neutron radiography for the investigation of reaction patterns in direct methanol fuel cells'. In *Polymer Electrolyte Membrane and Direct Methanol Fuel Cell Technology. Volume 2:* In Situ *Characterisation Techniques for Low Temperature Fuel Cells*, eds C. Hartnig and C. Roth, Woodhead Publishing Ltd, Cambridge.

Zawodzinski, T. A., M. Neeman, L. O. Sillerud and S. Gottesfeld (1991) 'Determination of water diffusion-coefficients in perfluorosulfonate ionomeric membranes'. *Journal of Physical Chemistry*, **95**, 6040–6044.

8

Neutron radiography for the investigation of reaction patterns in direct methanol fuel cells

K. WIPPERMANN and A. SCHRÖDER, Forschungszentrum
Jülich GmbH, Germany

Abstract: Neutron radiography allows the investigation of local fluid
distribution in direct methanol fuel cells (DMFCs) under operating conditions.
Spatial resolutions in the order of some tens of micrometers over the full test
cell area are achieved. This offers the possibility of studying large stack cells as
well as small test cells. Measurements can be performed in both through-plane
and in-plane mode. In the through-plane mode, an overview of local water and
gas distribution in the flow field channels is obtained. Combined studies of high
resolution neutron radiography and segmented cell measurements enable a
correlation of local fluid distribution and local performance. The knowledge of
this interdependency is essential to achieve high performance and durability of
DMFCs.

Key words: high-resolution neutron radiography, direct methanol fuel cells,
current distribution, fluid distribution, wettability, hydrophobic properties, gas
diffusion layer.

8.1 Introduction

One of the main problems concerning the development of direct methanol fuel
cells (DMFCs) is the uneven distribution of fluids. While CO_2 bubbles may inhibit
the methanol supply on the anode side, water droplets on the cathode side block
the oxygen supply. In combination with the concentration decrease of the reactants
over the active area along the flow field channels, these effects lead to a pronounced
inhomogeneous current distribution as well as to a power loss of the fuel cell. An
uneven current distribution may cause high local current densities associated with
a high local degradation rate. For that reason, it is obvious, that uneven current
distribution does not only reduce the performance, but also the durability of
DMFCs. With the knowledge of the local distribution and transport phenomena of
CO_2 bubbles and water droplets at different operation modes, investigations on
the improvement of cell components and cell design are possible. Additionally,
detrimental operating conditions can be identified and avoided during DMFC
operation. The final goal is to reach a media and current distribution as
homogeneous as possible.

On the cathode side of a liquid-fuelled DMFC, a partial flooding of either the
cathode gas diffusion layer (GDL) and/or the cathode flow field channels is
responsible for a blocking of the oxygen transport, causing a decrease of local

214

current. The cathode flooding effect in a liquid-fuelled DMFC is even more pronounced than in a PEFC, since the water is not only produced on the cathode side but permeates through the membrane additionally, driven by a concentration gradient and electro-osmotic drag. On the anode side, the blocking effect of CO_2 bubbles may lead to a disturbance of the methanol supply. This effect is detrimental especially to the anode catalyst, since local depletion of methanol causes a local increase of anode over-potential, leading at worst to a corrosion of the platinum/ruthenium catalyst and/or the carbon support at these areas.

The gas diffusion layer as part of both DMFC electrodes plays an important role with regard to blocking effects of either gas or liquid, because it is the mediator between the nanostructured catalyst layers and the flow field with structures in the mm range. Besides providing passages for methanol on the anode side and oxygen on the cathode side, the GDL is responsible for the product removal from the electrochemically active area to the flow field channels. These two-phase flow effects do not only play a crucial role inside the GDL but also in the flow field channels. Due to the interaction between GDL and flow field, different GDLs result in altered fluid distributions inside the cell and therefore in different operating behaviours.[1]

In order to investigate the effects leading to inhomogeneous current distributions, a measurement technique that allows the *in situ* observation of the current, CO_2 and water distribution simultaneously is mandatory. A method which allows the accurate measurement of both current and impedance distribution in fuel cells is based on compensated sensor resistors and printed circuit boards (PCBs). The features and advantages of this technology, which has been developed by Sauer *et al.*, is described[2, 3] (see also Section 8.3.2).

Several methods have been reported in the literature to observe the CO_2 and water distribution *in situ* under operating conditions of PEFCs and DMFCs. Some authors used cells with a transparent cover to observe the carbon dioxide evolution and the two-phase flow behaviour visually.[4–11] Transparent anode covers allow for the observation of different CO_2 flow patterns in DMFCs as a function of flow field design, current density, temperature, flow rate, pressure drop, and orientation.[4–7] A special set-up using a porphyrin dye compound reveals the distribution of oxygen in the cathode channels of a DMFC.[8] Observations of water transport phenomena in the cathode channels of PEFCs are also possible.[9–11] Applying transparent covers is a straightforward method, but requires substitution of flow field materials such as graphite or steel by acrylic glass. As the material properties such as wettability, electrical conductivity and thermal conductivity are different, the behaviour of the test cells, especially the flow patterns of CO_2 and liquid water, might be influenced.

A further method is synchrotron X-ray radiography.[12–15] This method allows the *in situ* observation of the fluid distribution in polymer membrane fuel cells on the micrometre scale, with a resolution down to 3 μm. In the field of hydrogen-fed PEFC, the formation, growth and transport of water droplets have been studied.[12,13,15] Through-plane measurements by Manke *et al.*[12] proved the existence of different transport mechanisms: The first was via a continuous GDL pore filling, indicating a continuous 'capillary-tree-like' transport, and the second was via an eruptive transport process. In-plane measurements by Hartnig *et al.*[13,14] enable a cross-sectional insight into the evolution and transport of water and thus a distinction between the different layers of the fuel cell. Both through-plane and in-plane techniques were applied to DMFCs under operating conditions.[15] The authors studied carbon dioxide evolution and bubble formation at the DMFC anode. They found that the dynamics of bubble formation and detachment of the bubbles from the position of formation is strongly correlated with the current density.[15]

However, the high absorption of X-rays by flow field and end plate materials, usually graphite and steel, requires the development of special cell designs including windows for the X-ray beam. It is hardly possible to examine DMFCs (and PEFCs) by synchrotron radiography with unmodified, common stack cells. In contrast, neutron radiography is a completely non-invasive method for the observation of CO_2 and water and has proven its applicability to a variety of questions.[16–26] This method is based on the high attenuation coefficient of hydrogen compared to the attenuation coefficient of most metals and carbon. It means that the neutron beam can penetrate the metallic end plates and graphitic flow fields almost unattenuated whereas liquid water leads to a strong attenuation of the beam. Thus, the distribution of hydrogen-rich species such as water and methanol can be observed during operation. On the other hand, areas where no liquid can be detected must be interpreted in terms of a gas phase (air, carbon dioxide).

In 1999, Bellows *et al.* used neutron radiography to study water transport profiles across Nafion in operating PEFCs.[16] Six years later, Kramer *et al.* investigated the two-phase flow inside the anodic compartment of an operating direct methanol fuel cell.[17] In the same year, Pecula *et al.* reported a procedure to utilise neutron imaging for the visualisation of two-phase flow within an operating PEFC.[18] Manke *et al.*[19] presented a quasi-*in situ* neutron tomography on PEFC cell stacks including a cell-by-cell detection of liquid-water agglomerates. A study of combined neutron radiography and locally resolved current density measurements of operating PEFCs was reported by Hartnig *et al.*[20] They correlated the water distribution with the local activity of the respective area. In 2008, Manke *et al.*[21] investigated the liquid-water exchange in two-phase flows within hydrophobic porous gas diffusion materials of PEFCs by spatially resolved hydrogen–deuterium contrast neutron radiography. Based on these results, they derived a new model for water transport based on an eruptive mechanism. Hickner *et al.*[22] and Boillat *et al.*[23] carried out high-resolution neutron radiography using the in-plane imaging mode. They obtained detailed information on cross-sectional water distribution in the MEA components and the gas flow channels. Schröder *et al.*[24] identified the combination of *in situ* high-resolution neutron radiography and segmented current distribution measurement as a suitable tool to correlate current and fluid distribution in DMFCs. It was found that strongly inhomogeneous current distributions during cathodic flooding processes results in a performance loss of up to 30% of the initial value.[24] Turhan *et al.* studied the impact of channel wall hydrophobicity on through-plane water distribution and flooding behaviour in a PEFC.[25] Schröder *et al.* investigated the influence of GDL wettability on the performance of a DMFC by combined local current distribution and high-resolution neutron radiography.[26] They found the hydrophobicity of cathode carbon cloth to be important for a fast removal of water droplets.[26]

8.2   Principle of neutron radiography imaging

Figure 8.1 shows a schematic representation of the principle of neutron radiography imaging. The incident neutron beam passing the DMFC is attenuated preferentially by liquid water/methanol in the flow field channels and, to a lesser extent, by water/methanol in the gas diffusion electrodes (GDEs). Despite the thickness of flow fields and end plates in the range of mm–cm, liquid in the flow field channels can clearly be detected. This is due to the following reasons:

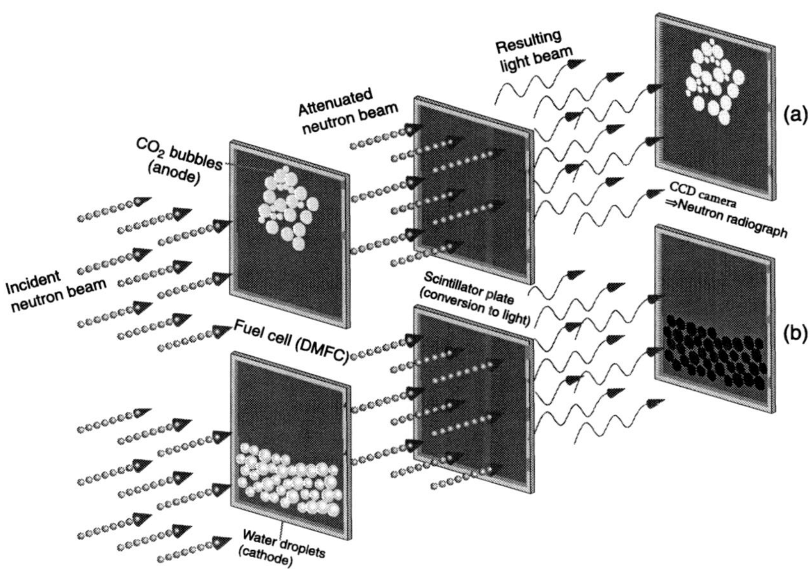

8.1 Schematic representation of the principle of neutron radiography imaging. (a) Stationary carbon dioxide bubbles in the anode flow field channels. (b) Stationary water droplets in the cathode flow field channels.

- The attenuation coefficient of ^1H is a factor of 100 higher than that of carbon (graphite flow fields, printed circuit boards) and about three orders of magnitude higher compared to that of aluminium (end plates).
- The attenuation of test cell materials can be eliminated by normalising each neutron radiograph by means of a reference radiograph (see Section 8.3.1).

Under normal DMFC operating conditions, CO_2 bubbles are generated at the anode and water droplets are produced at the cathode across the entire cell area. However, there is a tendency of increasing amounts of CO_2 from the bottom to the top of the anode compartment. On the other hand, water tends to accumulate in the middle and bottom part of the cathode compartment (see Section 8.4.3). Because high-resolution neutron radiography imaging requires an exposure time of several seconds per image to achieve a spatial resolution in the order of about 10 μm, only stationary CO_2 bubbles can be detected. This means that a distinction between CO_2 and methanol solution in the anode flow field channels is not possible if CO_2 bubbles are moving through the channels. In this case, only an average signal of gas and liquid phases is measurable. Depending on the flow field geometry and operating conditions, average flow velocities between less than 10 mm/s and more than 100 mm/s are possible. The same holds for stationary and moving water droplets in the cathode flow field channels.

Figure 8.1 illustrates the attenuation of the neutron beam in the case of stationary carbon dioxide bubbles in the anode flow field channels (Fig. 8.1a) and stationary water droplets in the cathode flow field channels (Fig. 8.1b). For the sake of simplicity it is presumed that there is a uniform fluid distribution in the rest of the cell. If the neutron beam passes spots of non-moving carbon dioxide bubbles (Fig. 8.1a), the neutron beam is less attenuated compared to the water-rich environment. Consequently, more light is generated in the scintillator screen at the spots behind the carbon dioxide bubbles, the CCD camera records a higher light intensity, and the final neutron radiograph shows light spots. Stationary water droplets in the cathode flow field channels cause a higher attenuation of the neutron beam compared to their environment (Fig. 8.1b). At these spots, a lower light intensity is generated by the scintillator and detected by the CCD camera, resulting in dark spots on the neutron radiograph.

More details concerning the principles of neutron radiography imaging are to be found in Chapters 6, 7, 9 and 12 of this book.

8.3   Development of combined high-resolution neutron radiography and local current distribution measurements

The combination of high-resolution neutron radiography and segmented current distribution measurement under operating fuel cell conditions allows the correlation of local water, gas and current distribution in direct methanol fuel cells. It therefore provides valuable information about local performance in dependence on operating parameters, cell design and material properties. In the following sections, the basic features of this technique, including high-resolution neutron radiography, segmented cell technology as well as design, choice and preparation of test cell components are described in detail.

8.3.1   High-resolution neutron radiography

The high-resolution neutron radiography technique used for combined neutron radiography and local current distribution measurements on DMFCs[24, 26] was developed at the neutron tomography instrument CONRAD/V7 at Helmholtz Centre Berlin (formerly Hahn-Meitner Institute) in Germany (Fig. 8.2). The instrument is located at the end of a curved neutron guide which provides a cut-off for neutrons with wavelengths larger than 2 Å. In this way only the cold neutrons (spectral maximum at 3.5 Å) – which provide a much higher contrast than thermal neutrons – are transmitted through the neutron guide. High-energetic neutrons and gamma radiation are almost completely eliminated. It is based on a pinhole geometry where apertures with diameters D from 1–3 cm can be used. The distance L between the aperture and the sample is 5 m. The larger the L/D ratio, the better is the collimation of the neutron beam and thus the spatial resolution. A high collimation is ensured by the large value of L. For the performed experiments,[24, 26] an aperture with a diameter of 3 cm was chosen, because a higher diameter of the aperture means higher neutron flux density, lower exposure times and thus higher temporal resolution. The main part of the detector system is

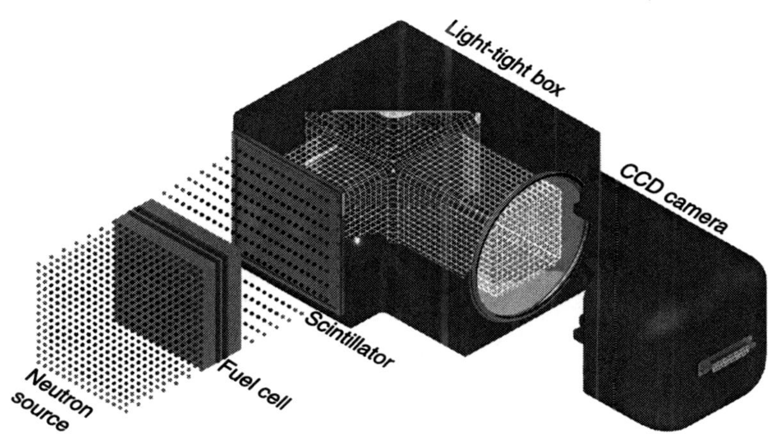

8.2 Scheme of the imaging set-up at CONRAD/V7 (Helmholtz Centre Berlin) used for combined neutron radiography and local current distribution measurements on DMFCs.[24, 26]

a 16-bit low-noise CCD camera (Andor DW436N with 2048 × 2048 pixel^2).[27] The camera is focused by a lens system on a neutron sensitive scintillator screen mounted close to the fuel cells (distance of 1 cm) to ensure high spatial resolutions down to 60 μm.

Suitable materials for the neutron sensitive scintillator screen are Gadox (Gd_2O_2S:Tb and ^{6}LiF-ZnS). Gadox is deposited on a quartz glass plate and lithium fluoride is prepared on an aluminium plate. Gadox has the advantage of a higher spatial resolution, whereas lithium fluoride provides a better light yield, which improves the signal-to-noise ratio. Furthermore, lithium fluoride is less sensitive to background gamma irradiation. Gamma irradiation is generated by interaction of neutrons with test cell materials causing white spots on the neutron radiographs. This effect is demonstrated in Fig. 8.3. When using Gadox a large number of small white spots appear on the picture (Fig. 8.3a), whereas this undesired effect is strongly reduced when using lithium fluoride (Fig. 8.3b).

The field of view was either 60 mm × 60 mm or 200 mm × 200 mm, corresponding to active cell areas of 18 cm^2 and 315 cm^2, respectively. The dynamic processes in the fuel cells were investigated by a time resolution of 15 s per image (10 s exposure time, 5 s readout) in the case of the small field of view and 65 s per image (60 s exposure time, 5 s readout) in the case of the large field of view. The higher exposure time is necessary because of the lower neutron flux density.

For a better interpretation of the neutron radiographs, normalised radiographs are calculated. This procedure is demonstrated in Fig. 8.4 with the example of a test cell with grid flow fields. A reference radiograph is taken with anode channels completely filled with methanol solution and cathode channels completely filled with air (see Fig. 8.4a). The grey values of the radiographs taken during the measurements (see Fig. 8.4b) are divided by the grey values of the reference

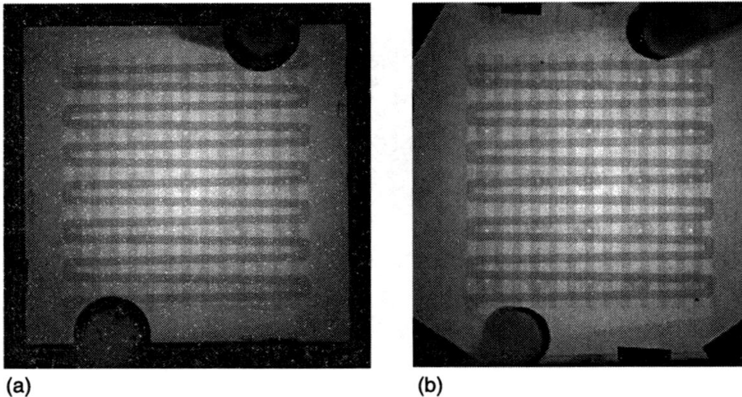

(a)   (b)

8.3 Effect of background gamma irradiation using different scintillator materials. (a) Gadox (Gd_2O_2S:Tb). (b) ^{6}LiF-ZnS.

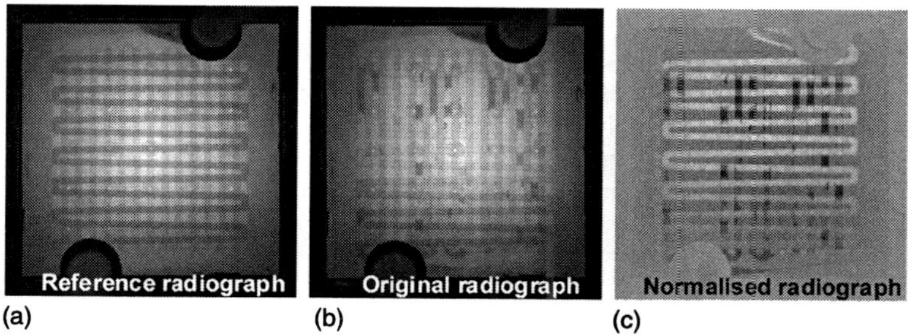

8.4 Example of the procedure for calculating normalised neutron radiographs. (a) Reference radiograph, anode channels completely filled with methanol solution and cathode channels completely filled with air. (b) Original radiograph. (c) Normalised radiograph obtained by division of Fig. 8.4b by Fig. 8.4a. If a meander geometry is used for the anode flow field, single CO_2 bubbles cannot be identified because of their fast movement through the anode channel.

radiographs (Fig. 8.4a). Only the changes in a radiograph compared to the reference radiograph emerge. The formation of CO_2 in the anode channels results in a light-grey value and the formation of liquid water in the cathode channels results in a dark-grey value in the normalised radiograph shown in Fig. 8.4c.

8.3.2   Segmented cell technology

The electrochemical set-up, including the electronics, of the segmented cell technology test cell was developed and constructed by the Institute of Power Electronics and Electrical Drives (ISEA) of RWTH Aachen University (Germany). The current and temperature distribution was measured with a special measurement system, consisting of a printed circuit board (PCB) that was inserted between the aluminium end plate and the graphite flow field on the anode side and special electronics outside the cell. A diagram of this set-up is shown in Fig. 8.5. The segmented contact area was connected with separate measurement and excitation wires to external electronics measuring the segment currents and controlling the potential of each segment so that all segments together form an equipotential surface in order to eliminate a smoothing of the current distribution caused by the measurement system, which would occur if shunt resistors alone were used for measuring the segment currents. A detailed description of the measurement system can be found in Sauer *et al.*[2]

This measurement system was also used to control the cell voltage and current by adjusting the voltage between the segmented contact area on the PCB and the cathode end plate. The whole measurement system was controlled with a central computer-based sequential control program that also controlled the mass-flow controllers for the air flow and the peristaltic pumps for the methanol flow.

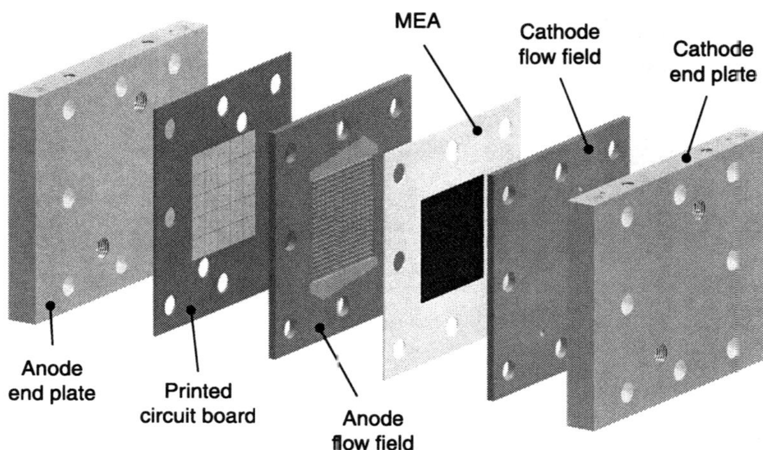

8.5 Exploded assembly drawing of the test cell including the adapter printed circuit board (PCB) used.[24, 26]

A separate program was used to record the temperature reading from the temperature sensors on the PCB inside the cell and to display the temperature distribution.

8.3.3 Test cell components

Preparation of membrane electrode assemblies (MEAs)

The functional layers of the MEAs, with an active area of 42 mm × 42 mm ($A_{electrode}$ = 17.64 cm^2), were prepared onto carbon cloth (Ballard). The carbon cloth was mostly hydrophobised by impregnation with a stable mixture of PTFE in a liquid. Only for studies of the influence of GDL wettability on DMFC performance (see Section 8.4.3.) carbon cloth was used as received. For functional layers, a hydrophobised microporous layer and then either the anode or cathode catalyst layer were prepared on the carbon cloth substrates by a knife-over-roll technique to obtain the complete electrodes. The microporous layer consisted of 60 wt% carbon (Cabot, 2.1 mg cm^{-2}) and 40 wt% PTFE (1.4 mg cm^{-2}). The anode catalyst consisted of 75 wt% Pt/Ru and 25 wt% carbon (Johnson Matthey). The Pt/Ru loading of the anodes was about 2 mg cm^{-2}. The cathode catalyst had a composition of 57 wt% Pt and 43 wt% carbon (Johnson Matthey) with a Pt loading of about 2 mg cm^{-2}. Finally, the electrodes were hot-pressed on both sides of a Nafion 115 membrane. The hot-pressing temperature was 130 °C and the pressure 0.5 kN cm^{-2}.

For studies of the influence of GDL wettability (see Section 8.4.3.), either the anode or cathode GDL was vertically split into an untreated left part and a hydrophobised right part (see Fig. 8.6). To avoid an interaction between anode

8.6 Scheme of GDL partitioning by vertically splitting the carbon cloth into an untreated and a hydrophobised part.[26] (a) partitioning of anode GDL (b) partitoning of cathode GDL. The neutron radiographs always represent the view from the cathode side. The splitting of anode and cathode was done in such a way, that the untreated carbon cloth (CC) always appears on the left hand side, and the hydrophobised carbon cloth is indicated in the right hand side of the neutron radiographs.

and cathode effects, the corresponding counter-electrode was always undivided and hydrophobised. The vertical splitting of the electrodes of the MEAs was realised by placing two GDE parts (each 21 mm × 42 mm), one with hydrophobised carbon cloth and the other one with untreated carbon cloth, side by side on a Nafion 115 membrane to obtain a complete electrode (42 mm × 42 mm). On the other side of the membrane, the undivided counter-electrode was placed. These three GDEs were subsequently hot-pressed on the membrane (conditions as described above).

Flow field designs and flow conditions

For the experiments described in this chapter, graphite flow fields with a thickness of 3 mm were used. Each flow field covers an electrode area of 17.64 cm^2 (42 mm × 42 mm). The geometries of the four flow field designs used here are presented in Table 8.1 and Fig. 8.7a–d. A special feature of grid flow fields (Fig. 8.7a) is the two-dimensionality of their geometry, which allows media flow to circumvent stationary, blocking water droplets or CO_2 bubbles. It means that – unlike channel or meander geometries – single liquid droplets or gas bubbles do not block complete channels. Therefore, performance of cells with grid flow fields is less affected if local flooding or accumulation of gas bubbles occurs. For example, the grid structure allows observation of stationary CO_2 bubbles on the anode side at low current densities. The grid and two-fold meander (Fig. 8.7b) flow fields were used for symmetric test cells with overlapping anode and cathode channels. In contrast, the single meander (Fig. 8.7c) and channel (Fig. 8.7d) flow

Table 8.1 Geometries of flow field designs

Flow field design[‡]	Used for anode	Used for cathode	Column cross-section (mm)	Column height (mm)	Channel width (mm)	Channel depth (mm)	Channel length (mm)	$A_{channel}/A_{land}$
(a) Grid	×	×	1 × 1	1				2.65
(b) Two-fold meander	×	×			1.0	1.0	300[†]	0.52
(c) Single meander	×				1.5	1.5	546	0.86
(d) Channel		×			1.5	1.5	41	0.95

[†]Per channel. [‡]See Fig. 8.7
Source: Schröder *et al.*[24, 26]

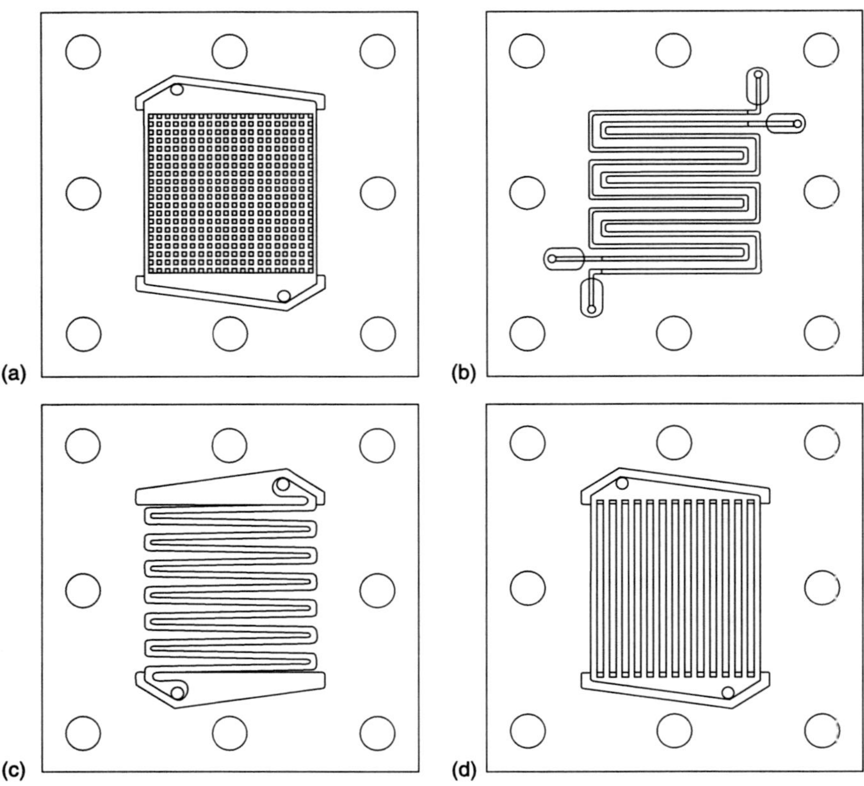

8.7 Flow field designs used:[24, 26] (a) Grid. (b) Two-fold meander. (c) Single meander. (d) Channel.

fields were applied for a more practice-oriented, asymmetric cell design (see Fig. 8.3 and 8.11).

All the experiments were run under counter-flow conditions, i.e. air was fed in at the top and methanol at the bottom of the cell. The measurements were always carried out at a temperature of 70 °C and ambient pressure. The anode was constantly fed by a methanol solution with a concentration of 1 mol l^{-1} and the cathode was supplied with air. The flow rates and stoichiometric factors respectively are indicated in the captions of Fig. 8.9–8.16.

Printed circuit boards

The adapter PCBs inside the cell assembly were equipped with a gold-plated segmented contact area including 25 or 28 segments, depending on the flow-field design, on the side towards the graphite flow field and one digital temperature sensor for each segment on the other side. Figure 8.8 shows the two PCBs with either 5 × 5 = 25 (Fig. 8.8a) or 4 × 7 = 28 (Fig. 8.8b) segmented gold contacts, the surrounding PCB tracks, the holes for screws and pipes and the current plug connector on the right side of the PCBs. The PCB with 25 segments was used for all measurements except the experiments with two-fold meander flow fields. In the latter case, only the 4 × 7 geometry allows each segment to cover both channels of the meander flow field.

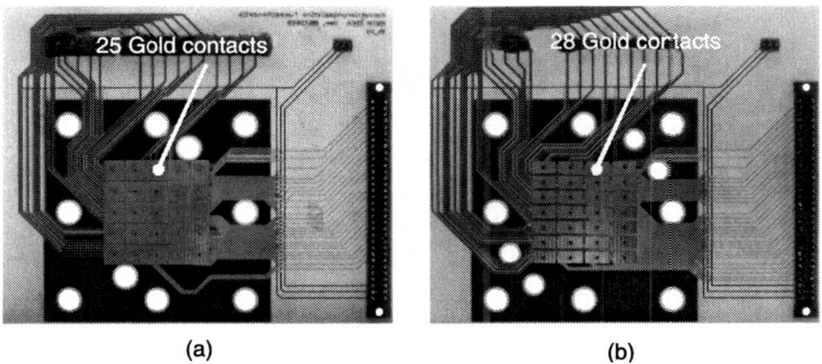

(a)   (b)

8.8 Adapter PCBs used:[24, 26] (a) PCB with 5 × 5 = 25 gold contacts. (b) PCB with 4 × 7 = 28 gold contacts.

8.4 Combined neutron radiography and local current distribution measurements

The first combined high-resolution neutron radiography and local current distribution measurements of operating DMFCs were published.[24,26] These experiments comprise the observation of water droplets and CO_2 bubbles including flooding effects, media and current distribution during bi-functional operation and the influence of GDL wettability on DMFC performance.

8.4.1 Observation of water droplets and CO_2 gas bubbles

Test cells with grid-flow field geometries on both electrodes were investigated with three different average current densities applying constant flow rates. The two-dimensional structure of the grid enables the formation and sticking of CO_2 bubbles. The gas evolution rate at 50 mA cm^{-2} is small enough that most of the CO_2 bubbles hardly move during exposure time (Fig. 8.9a). On the cathode side, there is no visible water content because channel water is removed in the gaseous state and liquid water in the cathode GDE is hardly perceptible. At 150 mA cm^{-2}, according to the current density, the CO_2 and water evolution rate is higher (Fig. 8.9c). Single CO_2 bubbles cannot be observed as they move faster and the temporal resolution is too low. The formation of liquid water occurs in the lower part of the flow field. At 300 mA cm^{-2}, the CO_2 evolution is further increased and the formation of liquid water expands to about two-thirds of the flow-field area (Fig. 8.9e).

If two-fold meander flow fields are used instead of grid-flow fields, water and CO_2 formation grow with increasing current density as well (Fig. 8.9b, d, f).

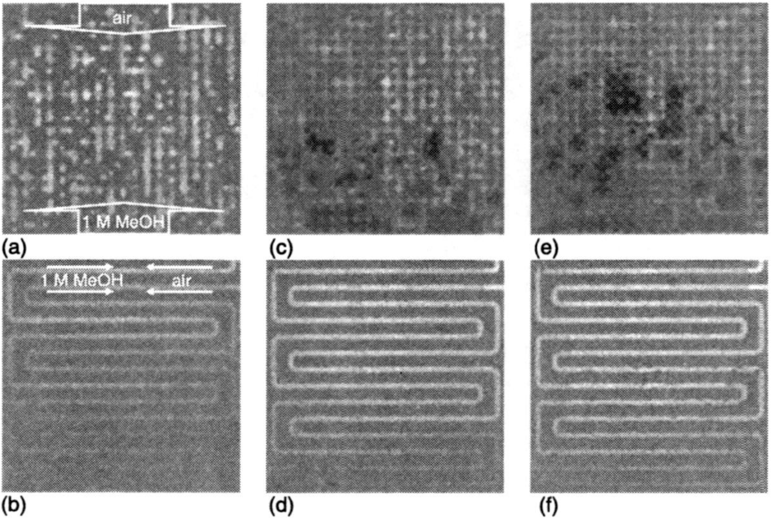

8.9 Comparison of water and gas distribution as function of current density in a grid structure and two-fold meander flow field.[24] (a, b) 50 mA cm^{-2}. (c, d) 150 mA cm^{-2}. (e, f) 300 mA cm^{-2}. Methanol flow rate: 2.19 ml min^{-1}, air flow rate: 378 ml min^{-1}.

However, there is an obvious difference. As the average velocity of the fluids in the channels is much higher than in the case of a grid-flow field, even at a low current density of 50 mA cm^{-2}, stationary CO_2 bubbles are not observable. Instead of single CO_2 bubbles, a light grey-scale value of the channel reflecting the average neutron beam attenuation of the moving liquid and gas phase is observed. With increasing current, the volume fraction of CO_2 grows and the anode channel appears brighter. At medium and high current densities, only small water droplets can be observed as a result of the high air velocity.[28]

Under the chosen conditions, a temporary flooding effect on the cathode side within the grid flow field channels has been observed. Figure 8.10 shows the neutron radiographs (Fig. 8.10a, c, e, g) and the corresponding current distributions (Fig. 8.10b, d, f, h). The higher the local current, the lighter is the grey-scale value. Before starting the measurement, the anode side was filled with methanol solution and the cathode side was completely filled with air. Initially, the current load was set to 300 mA cm^{-2} and water droplets began to form homogeneously in the bottom part of the flow field. During the 10 s exposure time of the first radiograph, (Fig. 8.10a), the droplets were still small, allowing a sufficient oxygen supply, resulting in a homogeneous current distribution ($I_{max}/I_{min} = 1.7$, Fig. 8.10b) and a high power density ($p = 113$ mW cm^{-2}). Growing bigger, the droplets blocked the oxygen supply significantly (Fig. 8.10c). The current distribution was highly inhomogeneous ($I_{max}/I_{min} = 6.3$, Fig. 8.10d) and the power density attained its minimum ($p = 78$ mW cm^{-2}). The spontaneous discharge of water partially reversed the flooding effect only 12 seconds later ($I_{max}/I_{min} = 2.3$, $p = 100$ mW cm^{-2}, Fig. 8.10e, f). Leaving the channels irregularly at different areas, the water droplets led to a reduced water content of the cathode flow field and then to more

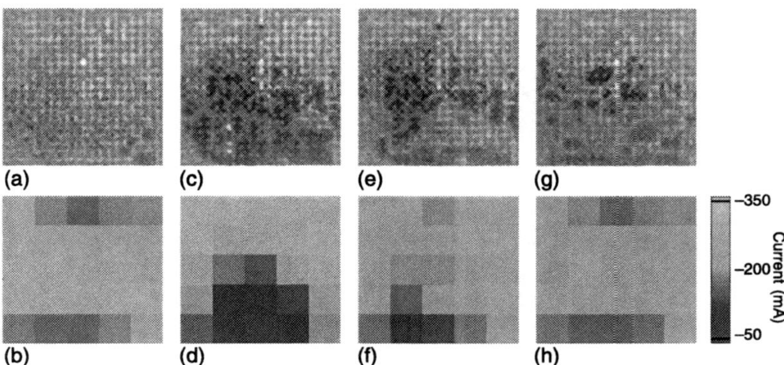

8.10 Neutron radiographs and corresponding current distributions during flooding effect at 300 mA cm^{-2} average current density.[24] Liquid water can be identified as dark spots along the bright channel area. (a, b) $t = 0$ s. (c, d) $t = 270$ s. (e, f) $t = 282$ s. (g, h) $t = 471$ s. Methanol flow rate: 2.19 ml min^{-1}, air flow rate: 378 ml min^{-1}.

homogeneous conditions. At 471 seconds after the start of the measurement the current distribution and power density nearly achieved their initial values ($I_{max}/I_{min} = 1.8$, $p = 110\,\mathrm{mW\,cm^{-2}}$, Fig. 8.10g, h). At the same operating conditions, the average velocity of air inside the channels of the two-fold meander flow field was significantly higher so that no large water amounts could build up and therefore no flooding effect occurred.

Another example of a flooding occurrence is shown in Fig. 8.11. In this case, a more application-oriented geometry of the flow fields was chosen including the meander structure on the anode side and parallel channels (41 mm × 1.5 mm × 1.5 mm) on the cathode side. This example illustrates what happens if channel geometries have improper dimensions, in this case an over-sized channel depth of 1.5 mm. At a current density of 50 mA cm^{-2}, $\lambda_{air} = 4$ and the chosen channel geometry, the pressure drop is too low to remove water droplets. As seen from Fig. 8.11, two cathode channels in the right part of the cell are completely blocked by water droplets, causing a pronounced decrease of current in this part and a significant decrease of cell performance. Moreover, under the hatched segments, negative currents indicate electrolytic domain, corresponding to an oxygen concentration of zero (bi-functional regime, see Section 8.4.2). It should be emphasised that air velocity increases with channel length. Thus, a channel depth of 1.5 mm may be a proper value for large stack cells with a higher channel length. This means that the same flow conditions in small single cells and stack cells are not achievable by choosing the same channel depths. It suggests performing combined high-resolution neutron radiography and local current distribution measurements with large stack cells also (see Section 8.5, Fig. 8.16).

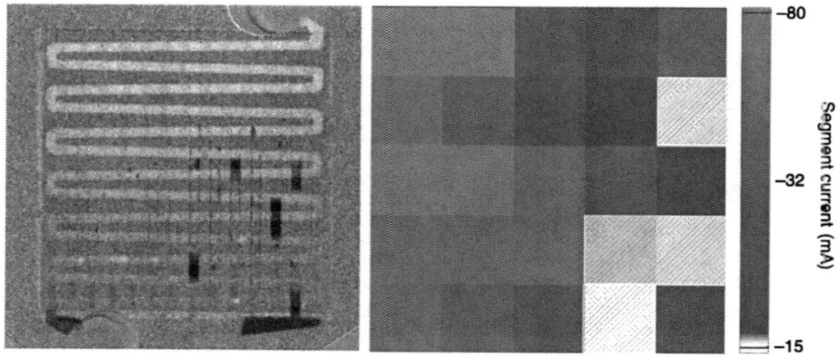

8.11 Neutron radiographs and corresponding current distributions during a blocking of air flow in cathode flow field channels by water droplets in the middle part of the cell. Hatched area indicates electrolytic domain; for flow field geometry, see Table 8.1; Operating conditions: $T = 70\,°C$, $j = 50$ mA cm^{-2}, $\lambda_{air} = \lambda_{MeOH} = 4$.

8.4.2   Water and current distribution during bi-functional operation

Several studies of direct methanol fuel cells[2, 29, 30–35] have shown that at low air flow rates and current densities, the active cell area splits up into two domains: The domain close to the inlet of the oxygen channel operates in normal fuel cell mode (galvanic domain). The domain close to the outlet of the oxygen channel works in electrolysis mode: it consumes current produced in the galvanic domain to convert methanol to hydrogen (electrolytic domain). This bi-functional regime of cell operation arises due to crossover of methanol from the anode to the cathode. In the galvanic domain, methanol permeated through the membrane is consumed on the cathode side in a direct catalytic combustion with oxygen. In the electrolytic domain, far from the inlet of the oxygen channel, no oxygen is left and here methanol is split on the cathode side into protons, electrons and CO_2. Protons move through the membrane to the anode side where they recombine with electrons to produce hydrogen. This part of the cell thus operates as an electrolytic cell. Thus, in the electrolytic domain the anodic reaction occurs on the cathode side, whereas the anode works as a hydrogen-evolving electrode.

Electrolytic domain, cathode:

$$CH_3OH + H_2O \rightarrow CO_2 + 6H^+ + 6e^-$$
[8.1]

Electrolytic domain, anode:

$$6H^+ 6e^- \rightarrow 3H_2$$
[8.2]

The cell is activated in the electrolytic domain.[29]

The visualisation of water and gas distribution during bi-functional DMFC operation was reported by Schröder *et al.*[24] In the cathode flow fields of the square-shaped test cells used in this work, the air flows from the top to the bottom of the cell. In conditions of bi-functional operation, a galvanic domain is expected at the top of the cell and an electrolytic domain at the bottom of the cell. This is confirmed by synchronous current distribution measurements and neutron radiography for the different cathode flow-field geometries, i.e. grid-flow field (Fig. 8.12a, b) and two-fold meander flow field (Fig. 8.12c, d). Figure 8.12a and c depict the fluid distribution while Fig. 8.12b and d show the grey-scale map of the respective current distribution. Independent of the flow field geometry, a negative (electrolytic) current results in the bottom part of the cells (hatched area), while a positive (galvanic) one is observed in the upper part of the cells. Since formation of water only takes place in the galvanic regime, water droplets in the cathode flow field channels emerge only in the top part of both cells (dark spots). The channels in the bottom part of the cells appear light because no water droplets are formed in the cathode channels and hydrogen gas is evolved in the anode channels.

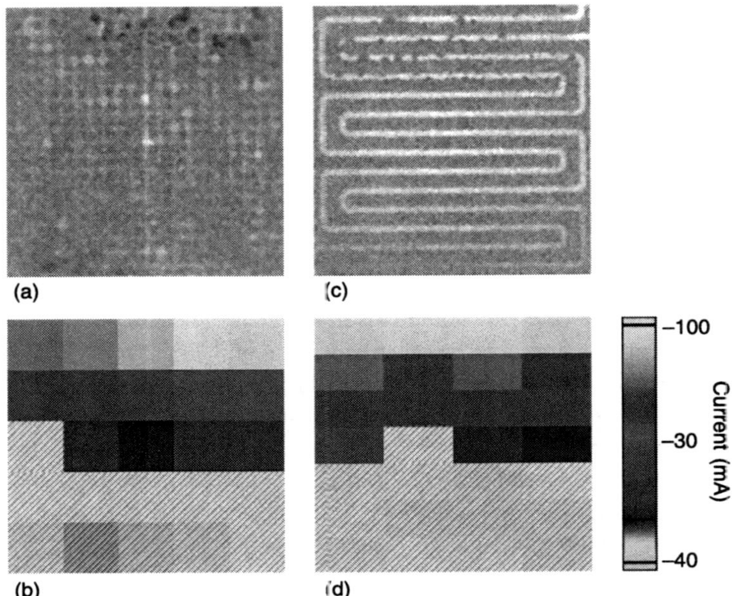

8.12 Neutron radiographs and corresponding current distributions during bi-functional operation;[24] dark spots indicating liquid water can only be observed in the upper (galvanic) region of the electrochemically active area. Hatched area indicates electrolysis domain. (a, b) Grid flow field. (c, d) Two-fold meander flow field. Average current density: 10 mA cm^{-2}, methanol flow rate: 2.56 ml min^{-1}, air flow rate: 20 ml min^{-1}.

8.4.3   Influence of gas diffusion layer (GDL) wettability on direct methanol fuel cell (DMFC) performance

For the above-mentioned reason, the wettability of the gas diffusion layer (GDL) plays an important role regarding the fluid transport in DMFCs. An option to visualise the difference between untreated and hydrophobised GDLs was presented in Schröder et al.[26] Either the anode or cathode GDL is vertically split into a less and a more hydrophobic part by using stripes of untreated and PTFE impregnated carbon cloth (see Section 8.3.3).

The common current distribution on the left and the right side of cells with undivided membrane electrode assemblies (MEAs) is approximately mirror-inverted, suggesting similar fluid distributions on the left and the right side of the cells. This condition is a precedent for the procedure of vertically splitting a GDL into two parts each with a different hydrophobicity, because unequal current or media distributions on the left and the right parts of the cell can be attributed to dissimilar hydrophobic properties of the two GDL parts. In contrast, a horizontal splitting of the GDLs would not be useful, since the media and current distribution of the upper and lower parts of the cell is considerably different, caused by an accumulation of CO_2 in the top part of the cell and flooding in the middle and bottom parts of the cell.

The results of MEAs with vertically split anode carbon cloths are shown in Fig. 8.13. The normalised radiographs (Fig. 8.13a, b, c) and the corresponding

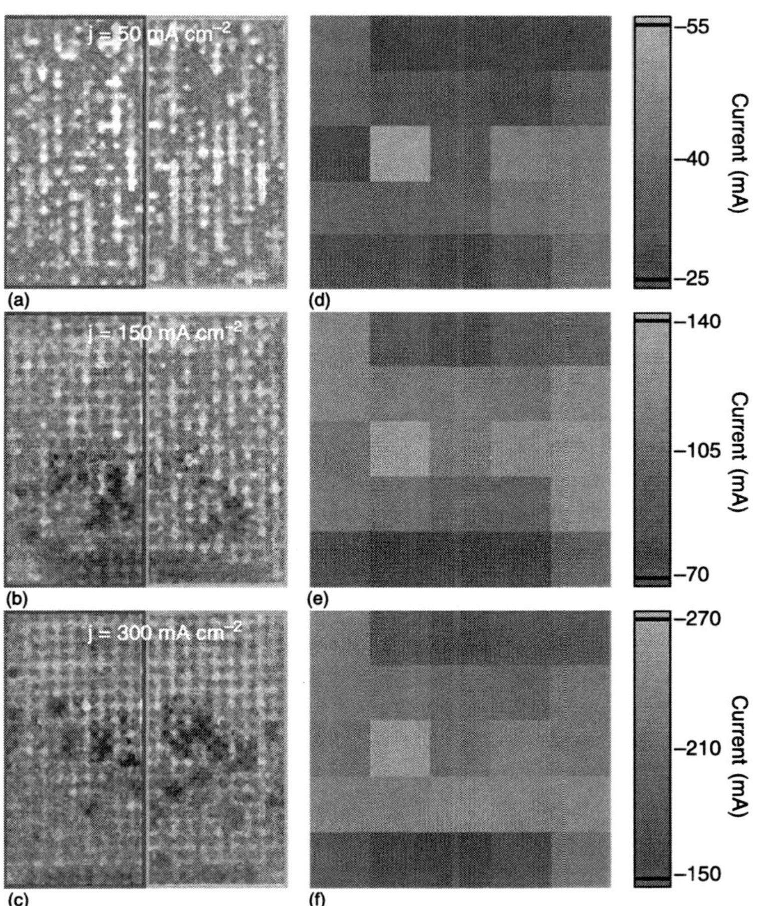

8.13 Normalised radiographs (a, b, c) and the corresponding current distributions (d, e, f) of a MEA with a vertically split anode carbon cloth at different current densities.[26] (a) 50, (b) 150 , (c) 300 mA cm⁻². Constant flow rates of 1 molar methanol solution (2.19 ml min⁻¹) and air (378 ml min⁻¹), corresponding to methanol and oxygen stoichiometry factors of (a) 24, (b) 8 and (c) 4. All the neutron radiographs and current distributions were recorded under steady-state conditions.

current distributions (Fig. 8.13d, e, f) are shown for three different average current densities and constant flow rates. All the neutron radiographs and current distributions shown in Fig. 8.13 were recorded under steady-state conditions, i.e. about 1 hour after starting the galvanostatic experiment. At the lowest average current density of 50 mA cm^{-2} and highest methanol and air stoichiometry of 24 (see Fig. 8.13a, d), only clusters of CO_2 bubbles appear in the anode flow field (see light spots). However, no flooding of cathode channels is observed under these conditions. The distribution of CO_2 bubbles across the anode is quite homogeneous: there is no visible difference between the left, untreated part (see dark grey frame) and the right, hydrophobised part (see light grey frame). The same is true for the current distribution, which appears to be more or less symmetrical. Thus, similar values of power generation are achieved for the left side (51%) and the right side (49%).

If the current density is increased to 150 mA cm^{-2} (see Fig. 8.13b), water droplets appear in the bottom part of the cathode channels (see dark spots), but there is virtually no difference between the untreated and the hydrophobised side. Again, similar currents and power densities are obtained for both parts of the MEA. The formation of water droplets in the bottom part of the channels may be explained by an increasing uptake of water vapour by the air stream from the top to the bottom of the cell. In the case of a current density as low as 50 mA cm^{-2}, the small amount of water is solely removed as a gas: The dew point is not exceeded even in the bottom part of the cell, where the water concentration is highest (see Fig. 8.13a). If the current increases from 50 mA cm^{-2} to 150 mA cm^{-2}, a three-fold higher amount of water has to be removed by the same air volume. Under these experimental conditions, the dew point is exceeded at the so-called 'water boundary'. Above this boundary, water is removed as vapour. Below the upper boundary of the water droplets, oversaturation occurs and water droplets are generated in the cathode flow field channels (see Fig. 8.13b). The removal of water droplets is governed by gravitation (droplet size), capillary forces (GDL wettability properties and pore size distribution) and the interaction of water droplets with the wall of the cathode flow field channels (channel wettability properties and geometry).

At the highest current density of 300 mA cm^{-2} and the lowest air stoichiometry of 4, there is even more water in the cathode channels, preferentially in the middle part of the cell (see Fig. 8.13c). A tentative explanation for the latter observation is the increasing size of water droplets and their faster removal because of gravity, when they flow down and merge. Still, there is no significant difference concerning the liquid and current distribution between the untreated and hydrophobised part of the MEA. From the results of a vertically split anode GDL, it can be concluded that the wettability of anode carbon cloth only has a minor influence on the fluid transport in the DMFC anode. Therefore, MEAs with vertically split anode carbon cloth have the same performance (110 mW cm^{-2}, 300 mA cm^{-2}) as undivided MEAs with hydrophobised anode carbon cloth.

The same experiments were performed by vertically splitting the cathode carbon cloth. Again, the neutron radiographs and the corresponding current distributions are shown for three different average current densities (see Fig. 8.14). As above, only the data obtained under steady-state conditions are shown. At the lowest current density of 50 mA cm^{-2}, only the evolution of CO_2 bubbles, but no formation of water droplets in the cathode gas channels is observable (see neutron radiograph in Fig. 8.14a). Nevertheless, there is a dominant power generation of about 60% in the right, hydrophobised part of the cell (see current distribution in Fig. 8.14f). Hence, there is no correlation between the fluid distribution shown in the neutron radiograph and the current distribution under

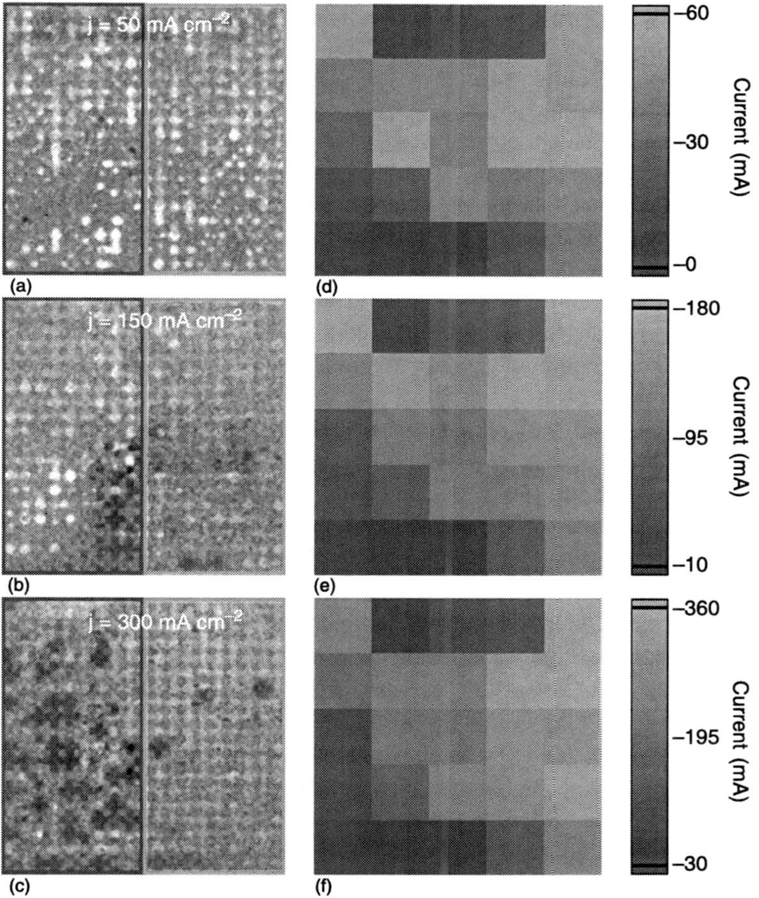

8.14 Normalised radiographs (a, b, c) and the corresponding current distributions (d, e, f) of a MEA with a vertically split cathode carbon cloth,[26] operating conditions as described for Fig. 8.13.

these conditions. This result indicates that it is not only the fluid distribution in the flow field channels that has to be considered, but also the media distribution in the underlying porous gas diffusion electrodes. The media distribution in the GDE cannot be resolved in these experiments and is also supposed to have a large influence on current distribution. Most probably, a partial flooding of the untreated carbon cloth in the left side of the MEA and subsequent blocking of the oxygen transport is responsible for the low performance of this part. If the average current density is increased to 150 mA cm^{-2}, water droplets appear preferentially in the left, untreated part of the cathode channels. Again, the left part of the cell contributes to only 40% of the overall power generation. The effects of water accumulation in the left part of the cathode flow field channels and the lower power generation in this part of the cell are even more pronounced at the highest average current density of 300 mA cm^{-2} (see Fig. 8.14f). Water droplets appear in the cathode channels over the entire left side of the cell, causing an additional power loss: the untreated part of the cell contributes to only 38% of the total power generation.

The time-dependent current density, power density and cell voltage during the experiment are presented in Fig. 8.15. Concerning current and power density, the total values as well as the data for the untreated (left) part and hydrophobised (right) part are shown. As the total current is constant during the experiment, inverse curves of the currents in the untreated and the hydrophobised part of the cell are indicated in Fig. 8.15. Additionally, neutron radiographs are shown for different, time-dependent states of media distribution in the flow field channels. They can be compared with a neutron radiograph taken at steady-state conditions in Fig. 8.14c.

Immediately after switching on the current, only very few, small water droplets are visible in the neutron radiograph (see Fig. 8.15a, $t = 0$ s). In the following period up to about 160 s, the cell voltage, power density and current of the hydrophobised part drops, whereas the current of the untreated part increases. This is due to a flooding of the right, hydrophobised part of the cell during that time and demonstrated by the neutron radiograph taken after 160 s, showing water droplets generated in the bottom, right part of the cathode flow field channels. A few seconds later, the water droplets start to disappear. Simultaneously, U_{cell}, p, p_{right} and j_{right} increase and j_{left} decreases. The enhanced performance can easily be explained by an improved oxygen supply of the right, hydrophobised part of the cell. However, it should be emphasised that the current in the hydrophobised part of the cell is always higher than that in the untreated part, independent on visible flooding effects in the flow field channels. This result again suggests that water management of GDLs can be even more important than the liquid distribution in the flow field channels. It further suggests future experiments to be performed not only in the through-plane mode, but also in the cross-sectional viewing direction to visualise the liquid distribution within the GDLs. This means neutron radiography as well as synchrotron X-ray investigations.

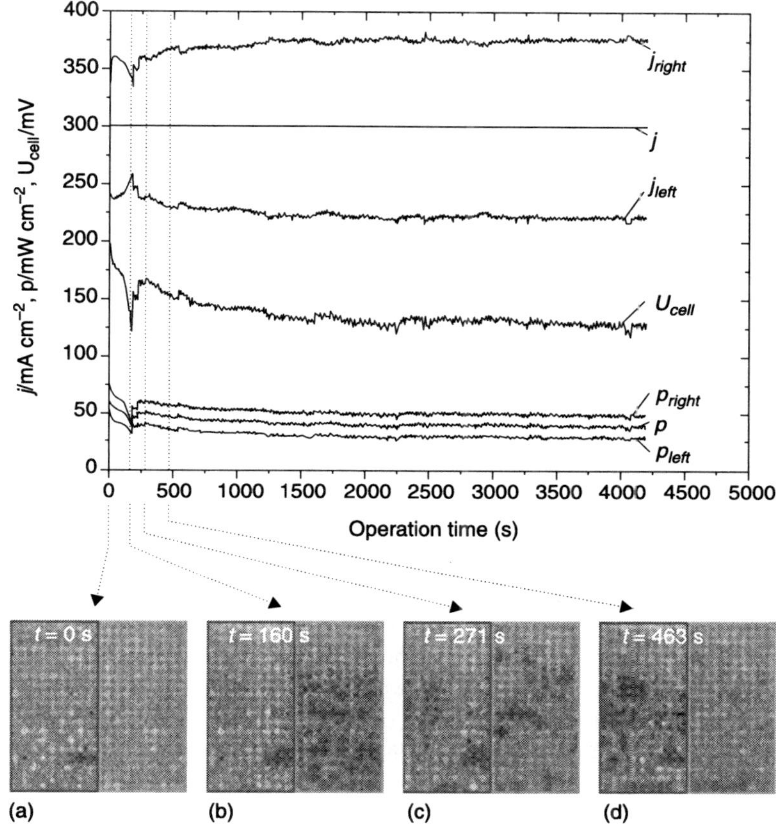

8.15 Time dependence of current density, power density and cell voltage during the experiment performed at a current density of 150 mA cm^{-2} with the MEA containing a vertically split cathode carbon cloth (see Fig. 8.3b).[26] In the case of current and power density, total values as well as data of untreated (left) part and hydrophobised (right) part are presented. Additionally, neutron radiographs indicating characteristic states of media distribution in the flow field channels are shown.

As the experiment continues, more and more water droplets appear in the left, untreated part of the cell (see neutron radiograph Fig. 8.15c). It takes about 460 s until steady-state conditions are achieved and the water droplets predominantly appear in the left, untreated part of the cell. No conclusive explanation for the observed 'shift' of water droplets in the cathode channels from the hydrophobised to the untreated part of the cell can be offered here. However, the time scale of this shift effect is current-dependent: e.g. if current density is enhanced from 150 to 300 mA cm^{-2}, the shift occurs three times faster. This suggests a mechanism where the untreated part of the GDL behaves like a sponge, taking up water in the beginning of the experiment after switching on the current. It means that oxygen transport in the untreated part of GDL is blocked from the beginning of the experiment. This effect explains the lower local current density in the left part of the cell throughout the experiment. Because the 'sponge' of untreated GDL fills with water, the current in the left part of the cell decreases and the current in the right part increases. Since most of the current is flowing in the right, hydrophobised part of the cell, the corresponding water production is higher. At the same time, the water uptake capacity of the hydrophobic GDL is smaller than that of the untreated part. It means that the water on the surface of the hydrophobic GDL cannot be removed solely in the gaseous state and water droplets appear on the surface of the hydrophobic GDL in the cathode channels. This effect is enhanced by the water droplets themselves, blocking the air flow in the right part of the flow field. When the water uptake capacity of the untreated GDL is exceeded ($t = 160$ s, see Fig. 8.15b), more and more water droplets are released on the surface of the untreated GDL, starting to block the air flow also in this part of the cell. Because the water droplets are more easily removed from the hydrophobic part of the GDL, their number decreases. Simultaneously, the air flow in the hydrophobic part increases and in the untreated part decreases. As a result, the water droplets 'shift' from the hydrophobic to the untreated part of the cell. The higher the current density, the more water is produced in the cathode gas diffusion electrode and the faster is the process of taking up water in the untreated part of the cathode GDL.

The MEA with divided cathode carbon cloth has a performance of less than 40 mW cm^{-2} (300 mA cm^{-2}) under steady-state conditions, which is only little more than one-third of the performance of undivided, standard MEAs under these conditions. This result can be explained by the restricted oxygen transport in the left, untreated part of the cell due to water accumulation in the cathode GDL and flow field channels, causing high cathode over-potential and a cell voltage drop. It means that the actual, low performance of the untreated part dominates the total power density of the cell.

8.5   Conclusions and future trends

Neutron radiography is a non-invasive method that is based on the high attenuation coefficient of hydrogen and allows observation of the fluid distribution in operating fuel cells.[16–26] In the case of direct methanol fuel cells (DMFCs), the method has been applied so far in the through-plane mode (neutron beam perpendicular to the active surface).[17, 25, 26] Depending on operating conditions and materials properties (e.g. wetting properties, porous structure), the following information about the fluid distribution in DMFCs can be obtained using this mode:

- Local distribution of water droplets in the cathode flow field channels.
- Carbon dioxide evolution in the anode flow field channels.

- In the case of special anode flow field geometries, local distribution of carbon dioxide bubbles.
- Combination of neutron radiography with local current distribution measurements: correlation of fluid and current distribution.

Neutron radiography in trough-plane mode offers the possibility to study the complete active area, even of large DMFCs, at one time. Figure 8.16 provides a look into the future, using neutron radiography technique for measuring large fuel cells for power DMFC applications. It is the first, normalised neutron radiograph of such a stack cell under DMFC operating conditions ($T = 70$ °C, $j = 150$ mA cm^{-2}, $\lambda_{air} = \lambda_{MeOH} = 4$).[36] The measurement was performed at the neutron tomography instrument CONRAD/V7 of Helmholtz Centre Berlin, where cell areas up to 20 cm × 20 cm can be investigated. The stack cell shown in Fig. 8.16 has an area of 315 cm^2 (21 cm × 15 cm). The design of this stack cell includes a multi-fold meander anode flow field and a multi-channel cathode flow field. Though distribution of the neutron beam intensity is not homogeneous and has to be further improved, the light grey meander channels and water droplets in parts of the cathode channels can be clearly identified. It should be emphasised that a simulation of the flow conditions in large stack cells by using small single cells is challenging. Hence, for future neutron radiography measurements in through-plane mode one should pay special attention to the investigation of practice-oriented, large stack cells.

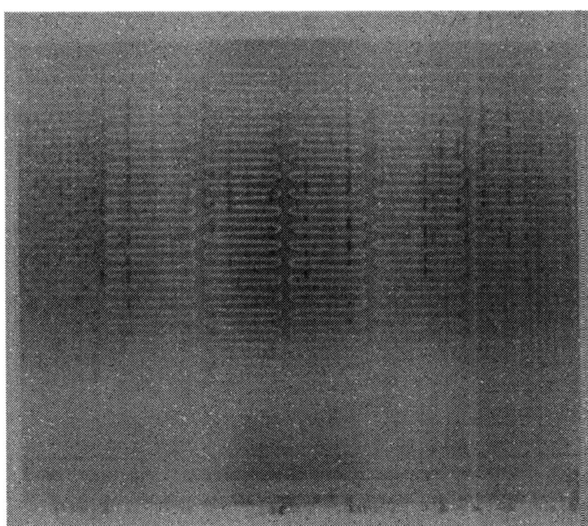

8.16 Normalised neutron radiograph of a large single stack cell with an area of 315 cm^2 (21 cm × 15 cm). Multi-fold meander anode and multi-channel cathode flow fields. Operating conditions: $T = 70$ °C, $j = 150$ mA cm^{-2}, $\lambda_{air} = \lambda_{MeOH} = 4$.

As a non-invasive technique, neutron radiography allows obtaining information without changing the design or materials of the flow fields. The end plates should be made of gold-plated aluminium instead of stainless steel, to improve the signal-to-noise ratio. However, even using steel end plates, sound results can be achieved. This is an advantage over other methods such as synchrotron X-ray radiography[12–15] or the use of transparent electrode covers.[4–11]

In the case of synchrotron X-ray radiography, the high absorption of X-rays by flow field and end plate materials requires the development of special cell designs including windows for the X-ray beam. Advantages of synchrotron X-ray radiography are:

- The higher spatial resolution (about 3 μm).
- The much higher temporal resolution (some hundred milliseconds) compared to neutron radiography (some 10 μm at a time resolution in the range of 15 s).

It should be stressed, that for both methods, temporal resolution is sacrificed for a gain of spatial resolution and *vice versa*. Because the spatial resolution of neutron radiography lies in a range similar to that of the thickness of catalyst or microporous layers, i.e. several tens of micrometres, distinguishing single water droplets or gas bubbles in these layers will hardly be possible, even in the in-plane mode. In contrast, synchrotron X-ray radiography allows the *in situ* observation of the dynamic fluid distribution on a micrometre scale, e.g. the dynamics of CO_2 bubble formation and detachment from the position of formation.[14]

Applying transparent covers requires the substitution of flow-field materials such as graphite or steel by, for example, acrylic glass. This is a disadvantage, because the exchange for acrylic glass also changes material properties such as wettability, electrical conductivity and thermal conductivity. For example, the wettability of the DMFC cathode flow field is crucial for the removal of liquid water in the cathode flow field channels.[37] Advantages of the method using transparent covers compared to neutron radiography are:

- Straightforward, simple method to be realised in a normal laboratory.
- Observation of dynamic processes on a sub-second time scale by means of a high-speed camera, e.g. movement of CO_2 bubbles in the anode channels or disruptive removal of water droplets in the cathode channels.

If neutron radiography is run in the through-plane mode, only information about the water and gas distribution in the flow field channels is obtained. However, the fluid distribution in the porous layers of the gas diffusion electrodes also appears to be crucial for the local cell performance and has to be visualised. It is therefore suggested that future neutron radiography experiments are performed on DMFCs in the in-plane mode (cross-sectional viewing direction) also, as successfully demonstrated for PEFCs.[16,22,23,25] Because the spatial resolution of neutron radiography lies in a range similar to that of the thickness of catalyst or microporous layers, i.e. several tens of micrometres, distinguishing single water droplets or gas bubbles in these layers will hardly be possible. Nonetheless, the following effects and processes could be observed:

- Overview of spatially and temporary resolved fluid distribution across the cell.
- Evaluation of water/methanol storage, transport and flooding mechanisms.
- Hydration/dehydration (swelling/shrinking) processes of membranes and ionomer-containing catalyst layers, e.g. swelling of electrodes into the flow field channels.
- Water distribution across the walls and on the back side of flow field channels. As a complementary method, synchrotron X-ray radiography both in the in-plane and through-plane mode would allow study of dynamic processes such as the fluid transport on a micrometre scale.

In conclusion, the method of high-resolution neutron radiography provides a detailed insight into the changing water and gas distribution in direct methanol fuel cells (DMFCs) under different operating conditions and surface properties of the cell components.[17,24,26] Future measurements should comprise both trough-plane and in-plane modes. As an *in situ* method, neutron radiography is especially valuable in combination with a simultaneous measurement of the local current distribution and impedance, because it allows the correlation of the local amount of liquid water/carbon dioxide and local performance.[24, 26] An intensified use of local impedance spectroscopy will also provide a correlation of local fluid distribution and kinetic parameters. The study of these interdependencies offers the possibility to optimise the water management and performance of DMFCs while respecting a homogeneous liquid (water/methanol), gas (carbon dioxide/air), current and temperature distribution. The homogeneity is essential to achieve high performance and durability of DMFCs.

8.6   Sources of further information

Advanced Tomographic Methods in Materials Research and Engineering, Edited by John Banhart, ISBN13: 9780199213245 ISBN10: 0199213240, 2008, Oxford University Press.

8.7   References

1. C. Hartnig, L. Jörissen, W. Lehnert, J. Scholta, *Direct Methanol Fuel Cells (DMFC), Materials for Fuel Cells*, Ed. M. Gasik, Cambridge, Woodhead Publishing Limited (2008) 185–208.
2. D. U. Sauer, T. Sanders, B. Fricke, T. Baumhöfer, K. Wippermann, A. A. Kulikovsky, H. Schmitz, J. Mergel, 'Measurement of the current distribution in a direct methanol fuel cell – Confirmation of parallel galvanic and electrolytic operation within one cell', *Journal of Power Sources* **176** (2008) 477–483.

3. T. Sanders, T. Baumhöfer, D. U. Sauer, A. Schröder and K. Wippermann, 'Spatially resolved impedance spectroscopy on direct methanol fuel cells considering fuel concentration oscillations in gas channels', *ECS Transactions* **25** (2009) 1719–1728.

4. P. Argyropoulos, K. Scott, W. M. Taama, 'Carbon dioxide evolution patterns in direct methanol fuel cells', *Electrochimica Acta* **44** (1999) 3575–3584.

5. H. Yang, T. S. Zhao, Q. Ye, '*In situ* visualization study of CO_2 gas bubble behavior in DMFC anode flow fields', *Journal of Power Sources* **139** (2005) 79–90.

6. C. W. Wong, T. S. Zhao, Q. Ye, J. G. Liu, 'Transient capillary blocking in the flow field of a micro-DMFC and its effect on cell performance', *Journal of The Electrochemical Society* **152** (2005) A1600–A1605.

7. Q. Liao, X. Zhu, X. Zheng, Y. Ding, 'Visualization study on the dynamics of CO2 bubbles in anode channels and performance of a DMFC', *Journal of Power Sources* **171** (2007) 644–651.

8. J. Inukai, K. Miyatake, Y. Ishigami, M. Watanabe, T. Hyakutake *et al.* '*In situ* and real-time visualisation of oxygen distribution in DMFC using a porphyrin dye compound', *Chemical Communications*, (2008) 1750–1752.

9. X. Liu, H. Guo, C. Ma, 'Water flooding and two-phase flow in cathode channels of proton exchange membrane fuel cells', *Journal of Power Sources* **156** (2006) 267–280.

10. X. Liu, H. Guo, F. Ye, C. Ma, 'Water flooding and pressure drop characteristics in flow channels of proton exchange membrane fuel cells', *Electrochimica Acta* **52** (2007) 3607–3614.

11. D. Spernjak, S. G. Advani, A. K. Prasad, 'Simultaneous neutron and optical imaging in PEM fuel cells', *Journal of The Electrochemical Society*, **156** (2009) B109–B117.

12. I. Manke, C. Hartnig, M. Grünerbel, W. Lehnert, N. Kardjilov, *et al.* 'Investigation of water evolution and transport in fuel cells with high resolution synchrotron x-ray radiography', *Applied Physics Letters* **90**, 174105 (2007).

13. C. Hartnig, I. Manke, R. Kuhn, N. Kardjilov, J. Banhart, W. Lehnert, 'Cross-sectional insight in the water evolution and transport in polymer electrolyte fuel cells', *Applied Physics Letters* **92**, 134106 (2008).

14. C. Hartnig, I. Manke, R. Kuhn, S. Kleinau, J. Goebbels, J. Banhart, 'High-resolution in-plane investigation of the water evolution and transport in PEM fuel cells', *Journal of Power Sources* **188** (2009) 468–474.

15. C. Hartnig, I. Manke, J. Schloesser, P. Krüger, R. Kuhn, *et al.* 'High resolution synchrotron X-ray investigation of carbon dioxide evolution in operating direct methanol fuel cells', *Electrochemistry Communications* **11** (2009) 1559–1562.

16. R. J. Bellows, M. Y. Lin, M. Arif, A. K. Thompson, D. Jacobson, 'Neutron Imaging technique for in situ measurement of water transport gradients within Nafion in polymer electrolyte fuel cells', *Journal of The Electrochemical Society* **146** (1999) 1099–1103.

17. D. Kramer, E. Lehmann, G. Frei, P. Vontobel, A. Wokaun, G. G. Scherer, 'An on-line study of fuel cell behavior by thermal neutrons', *Nuclear Instruments and Methods in Physics Research A* **542** (2005) 52–60.

18. N. Pekula, K. Heller, P. A. Chuang, A. Turhan, M. M. Mench, *et al.* 'Study of water distribution and transport in a polymer electrolyte fuel cell using neutron imaging', *Nuclear Instruments and Methods in Physics Research A* **542** (2005) 134–141.

19. I. Manke, C. Hartnig, M. Grünerbel, J. Kaczerowski, W. Lehnert, *et al.* 'Quasi–in situ neutron tomography on polymer electrolyte membrane fuel cell stacks', *Applied Physics Letters* **90**, 184101 (2007).

20. C. Hartnig, I. Manke, N. Kardjilov, A. Hilger, M. Grünerbel, *et al.* 'Combined neutron radiography and locally resolved current density measurements of operating PEM fuel cells', *Power Sources* **176** (2008) 452–459.

21. I. Manke, C. Hartnig, N. Kardjilov, M. Messerschmidt, A. Hilger *et al.* 'Characterization of water exchange and two-phase flow in porous gas diffusion materials by hydrogen–deuterium contrast neutron radiography', *Applied Physics Letters* **92**, 244101 (2008).

22. M. A. Hickner, N. P. Siegel, K. S. Chen, D. S. Hussey, D. L. Jacobson *et al.* '*In Situ* High-Resolution Neutron Radiography of Cross-Sectional Liquid Water Profiles in Proton Exchange Membrane Fuel Cells', *Journal of The Electrochemical Society*, **155** (2008) B427–B434.

23. P. Boillat, D. Kramer, B. C. Seyfang, G. Frei, E. Lehmann *et al.* '*In situ* observation of the water distribution across a PEFC using high resolution neutron radiography', *Electrochemistry Communications* **10** (2008) 546–550.

24. A. Schröder, K. Wippermann, J. Mergel, W. Lehnert, D. Stolten *et al.* 'Combined local current distribution measurements and high resolution neutron radiography of operating direct methanol fuel cells', *Electrochemistry Communications* **11** (2009) 1606–1609.

25 A. Turhan, S. Kim, M. Hatzell, M. M. Mench, 'Impact of channel wall hydrophobicity on through-plane water distribution and flooding behavior in a polymer electrolyte fuel cell', *Electrochim. Acta* **55** (2010) 2734–2745.

26 A. Schröder, K. Wippermann, W. Lehnert, D. Stolten, T. Sanders, *et al.* 'The influence of hydrophilic/hydrophobic properties of GDLs on DMFC performance: A combined local current distribution and high resolution neutron radiography study', *J. Power Sources* **195** (2010) 4765–4771.

27 N. Kardjilov, A. Hilger, I. Manke, M. Strobl, W. Treimer, J. Banhart, 'Industrial applications at the newcold neutron radiography and tomography facility of the HMI', *Nuclear Instruments and Methods in Physics Research* **A 542** (2005) 16–21.

28 N. Akhtar, A. Qureshi, J. Scholta, C. Hartnig, M. Messerschmidt, W. Lehnert, 'Investigation of water droplet kinetics and optimization of channel geometry for PEM fuel cell cathode', *International Journal of Hydrogen Energy* **34** (2009) 3104–3111.

29. A. A. Kulikovsky, H. Schmitz, K. Wippermann, J. Mergel, B. Fricke *et al.* 'DMFC: Galvanic or electrolytic cell?', *Electrochemistry Communications* **8** (2006) 754–760.

30. Q. Ye, T. S. Zhao, H. Yang, J. Prabhuram, 'Electrochemical reactions in a DMFC under open-circuit conditions', *Electrochem. Solid State Lett.* **8** (2005) A52.

31. M. Müller, H. Dohle, A. A. Kulikovsky, 'Comment on "Electrochemical reactions in a DMFC under open-circuit conditions" ', *Electrochem. Solid State Lett.* **9** (2006) L7.

32. A. A. Kulikovsky, '1D + 1D model of a DMFC: localized solutions and mixed potential', *Electrochem. Comm.* **6** (2004) 1259.

33. A. A. Kulikovsky, 'Analytical models of a direct methanol fuel cell', in: T. S. Zhao, K. D. Kreuer, T. Nguyen (Eds.), *Advances in Fuel Cells*, vol. 1, Elsevier, New York, 2007, pp. 337–418.

34. A. A. Kulikovsky, H. Schmitz, K. Wippermann, 'Direct methanol hydrogen fuel cell', *Electrochemical and Solid-State Letters* **10–8** (2007) B126–B129.

35. A. A. Kulikovsky, H. Schmitz, K. Wippermann, J. Mergel *et al.* 'Bifunctional activation of a direct methanol fuel cell', *J. Power Sources* **173** (2007) 420–423.

36. A. Schröder, K. Wippermann, T. Sanders, T. Arlt, Extended Abstract, 'Combined local current distribution measurements and high resolution neutron radiography of operating direct methanol fuel cells', *18th World Hydrogen Energy Conference, Essen, Germany, Proceedings of the WHEC*, May 16–21. 2010/Essen, Detlef Stolten, Thomas Grube (Eds.), 2010, Essen.

37. A. Schröder, K. Wippermann, G. Zehl, D. Stolten, 'The influence of cathode flow field surface properties on the local and time-dependent performance of direct methanol fuel cells', *Electrochemistry Communications* **12** (2010) 1318–1321.

9

Neutron tomography for polymer electrolyte membrane fuel cell characterization

I. MANKE AND N. KARDJILOV, Helmholtz Centre Berlin for Materials and Energy, Germany and C. HARTNIG, Chemetall GmbH, Germany

Abstract: Neutron imaging is a unique method for visualization of water distribution in low temperature polymer electrolyte membrane (PEM) fuel cells. As an expansion to two-dimensional (2D) radiography neutron tomography allows for three-dimensional quantification of the water amount and therefore a separation of the anodic and cathodic sectors of a fuel cell. We present tomographic studies of fuel cell stacks where the water content in each cell and each electrode has been analyzed and quantified individually. Furthermore, the influence of different flowfield designs and effects like electroosmotic drag and back-diffusion from cathode to anode on the water distribution was investigated. The prospects and the fundamental limitations of the technique in spatial and time resolution are discussed and demonstrated.

Key words: PEMFC, fuel cell stacks, water content, neutron imaging, water management.

9.1 Introduction

In the preceding chapters several aspects of radiographic imaging of fuel cells have been highlighted. However, radiographic imaging methods only provide a transmission picture of the inner structure of the investigated object.[1] In this way the exact localization and quantification of volumetric details cannot be accessed for bulky samples. The imaging of water distribution in 'real fuel cells', i.e. fuel cells stacks with more than one cell, is a scientific problem which requires radiation that transmits several centimeters of metal plates and provides sufficient contrast for small amounts of water and at the same time allows for an exact localization of the liquid water. This problem can be solved by using neutron tomography where the neutron radiation helps to obtain a three-dimensional (3D) image of the inner structure of fuel cell and to quantify the water distribution.

The origins of neutron imaging date back to 1935 when Kallmann and Kuhn performed the first experiments in Germany.[2–3] The progress of the imaging techniques with neutrons is related to the development and improvement of neutron sources worldwide.[4–6] At the time of writing nuclear research reactors and spallation sources provide neutron beams with sufficient intensities, i.e. flux, for a sufficient count rate in imaging investigations.[6] Additional improvements of detector technology, e.g. CCD cameras, and data processing and storage capabilities in the 1990s, strengthened the recognition of neutron imaging methods. It was shown for the first time by Bellows and colleagues in 1999 that neutron imaging is well suited for *in situ* investigations of the water distribution in fuel cells.[7] Today, neutron imaging is an important tool for non-destructive investigations of real fuel cell systems and their components.[8–21]

An example for this is the application of neutron imaging in the investigation of water management in polymer electrolyte membrane fuel cells (PEMFCs) where a lot of improvements have been obtained with regard to, e.g., durability, cost reduction and overall increase of power density.

The water management problem is related to the water production at the cathodic part of the fuel cell ($2H^+ + \frac{1}{2}O_2 + 2e^- \rightarrow H_2O$).[22–26] The amount of product water and water vapor originating from the humidification of the reactant gases (especially on the anode side) is crucial for fuel cell performance. On the one hand, water is needed for the fuel cell operation because only the wet membrane is proton conductive but, on the other hand, flooding is a significant source of power losses because the transport of reactant and product gases (O_2, H_2 and H_2O) is strongly hindered and the supply of the reactive spots is prohibited, i.e. the reactants cannot reach the active sites. The application of neutron imaging helps to optimize the flow field geometry in a way that a uniform distribution of reactants and water can be gained which in turn prevents local inactivities.

9.2   Complementarity of neutrons and X-rays

X-ray and neutron imaging are complementary non-destructive techniques for materials scientific investigations.[27] The two radiations interact differently with matter: X-rays interact with the electronic shell of atoms while neutrons as charge-neutral particles interact with the nuclei. Figure 9.1 shows a comparison between the interaction mechanisms of X-rays and neutrons with matter. The different interaction mechanisms provide different contrasts in the imaging techniques employing the two radiation sources. The attenuation coefficients of chemical elements for X-rays increase with their atomic number (electron number in the atomic shell) while the attenuation of neutrons is described by a rather complex nuclear interaction mechanism. Neighbor elements with similar numbers of electrons often have completely different attenuation coefficients for neutrons. This fact is in general used to distinguish material compositions that can hardly be separated by X-rays.

The nuclear interaction is one of the reasons that the neutrons distinguish between different isotopes.[1] A prominent example is the difference of the attenuation coefficient between 'normal' water H_2O and deuterated water D_2O or, more precisely, the different coefficients of hydrogen and heavy hydrogen: deuterium is basically transparent in the neutron beam whereas normal hydrogen is heavily attenuating the neutron radiation, a fact that can be used to visualize the exchange dynamics of light water (H_2O) and heavy water (D_2O) in operating fuel cells.[16]

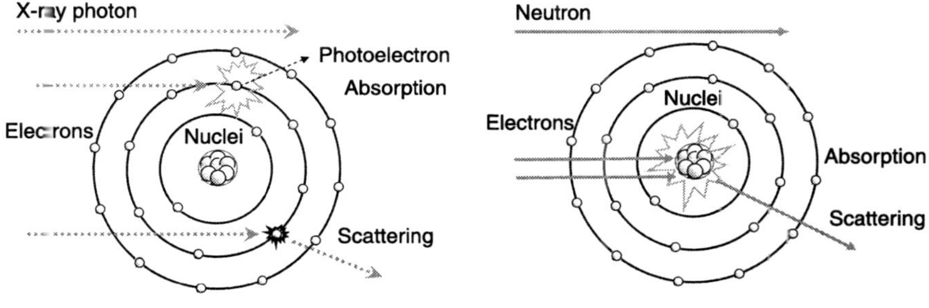

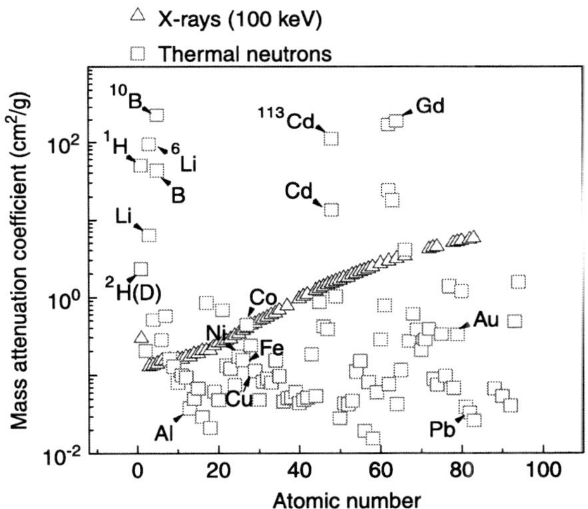

9.1 Comparison of the interaction of X-rays and neutrons with atoms.

The unique combination of a comparatively high transmission through metals and the high sensitivity to light elements such as hydrogen and lithium (high attenuation coefficients) opens a wide variety of opportunities in fuel cell and battery research to follow processes inside opaque objects in a non-destructive manner.

The interaction probability of the radiation with matter decreases with increasing energy of the used radiation. Neutrons used in radiographic and tomographic applications are so-called 'cold' and 'thermal' neutrons. They have energies typically in the range of a few to some tenths of meV. Cold neutrons allow for a higher contrast in imaging while their penetration depth is smaller compared to thermal neutrons. For fuel cell research purposes the higher sensitivity of cold neutrons to water has proven to be of great importance.

The advantages of using neutron imaging for fuel cell research are summarized in Table 9.1.

Table 9.1 Properties of neutron tomography in fuel cell research

	Neutron tomography
General properties	High contrast for light elements like H or Li in the presence of thick metallic components High penetration depth in metals Larger sample size up to several ten cm
Disadvantages	Low beam intensities (low time and spatial resolution) Polychromatic beam, reduced water quantification accuracy Large-scale facility necessary
Typical fluxes (intensity)	Typically 10^6–10^8 neutrons/cm^{2}s
Maximum spatial resolution	Small fuel cells (<2 cm): currently 20 µm Large fuel cells (>10 cm): several 100 µm
Time for a complete tomography	Small fuel cells at high flux source and low spatial resolution (several 100 µm): 20–300 s overall measurement time Large fuel cells at typical fluxes at most facilities: 1–10 hours overall measurement time

9.3   Principles of neutron tomography

9.3.1   Beam attenuation

Conventional neutron imaging is based on the attenuation of a neutron beam that penetrates a sample and the remaining beam intensity is detected by means of a neutron sensitive detector; attenuation is mainly caused by absorption and scattering.[1,28] The resulting contrast in the obtained images is therefore a result of the specific interaction of the neutrons with matter and depends strongly on the investigated material. Because of the low beam fluxes (intensities) available at neutron sources the whole available neutron spectrum (a 'white beam') is normally used for real-time imaging applications, i.e. a polychromatic beam is used especially for fast *in situ* radiography measurements on PEM fuel cells. The beam attenuation can be described by means of the basic rules of optics. The intensity $I_n(x, y)$ of a neutron beam passing a sample with a thickness of $d = n\Delta z$ in the z-direction is given by:

$$I_n(x,y) = I_0(x,y) \cdot \exp(-\mu_1(x,y,z_1) \cdot \Delta z) \cdot \exp(-\mu_2(x,y,z_2) \cdot \Delta z)...$$

$$\exp(-\mu_n(x,y,z_n) \cdot \Delta z) = I_0(x,y) \cdot \exp\left(\sum_{i=1}^{n} -\mu_i(x,y,z_i) \cdot \Delta z\right) \quad [9.1]$$

or

$$I_n(x,y) = I_0(x,y) \cdot \exp\left(\int -\mu(x,y,z)dz\right)$$

where $\mu(x, y, z)$ is the distribution of the local attenuation coefficient in the sample, Δz the width of the corresponding voxel (volume pixel) within the sample and $I_0(x, y)$ is the initial beam intensity distribution, see Fig. 9.2.

The attenuation coefficient $\mu(x, y, z)$ is a product of various physical events that contribute to the overall beam attenuation such as coherent and incoherent scattering and absorption.

9.3.2   Detection and spatial resolution

In neutron imaging, detector systems based on charge-coupled device (CCD) cameras are mainly used nowadays. The main advantages of state-of-the-art CCD-detectors and digital imaging are the high time resolution and a good linear dynamic range of up to 16 bits, which provides a high accuracy for data quantification. The conversion of neutron radiation in visible light occurs by scintillator screens. In order to protect the CCD from radiation damage a mirror setup is normally used together with an optical system to project the image from the scintillator on to the light sensitive chip. The properties, i.e. the 'quality', of a detector system are determined by the CCD, the scintillator screen used and the optics. Besides the lens system, the scintillator screen is in the focus of recent improvements.

Great progress has been made in detector development by achieving spatial resolutions down to 20 µm, which were used for the investigation of the water distribution in PEM fuel cells. This is the result of the development of new scintillator screens based on, e.g. crystalline ^{6}LiF/ZnS:Ag with thicknesses down to 0.05 mm or optimized GADOX (Gd_2O_2S). In addition, improved lens systems provide high optical spatial resolution and excellent light efficiency. More details can be found in Chapter 12.

Another factor that affects the spatial resolution is the beam collimation, which is best defined by the so-called L/D-ratio. The beam divergence causes a pixel blurring d, which depends on the distance between the scintillator screen and the sample because a point of the sample is projected to a spot by non-parallel rays (see Fig. 9.3). For this reason the divergence of a neutron beam has to be chosen to be as small as possible. This can be realized by a limitation of the effective neutron source size by a spatial limiter (e.g. a pinhole). The intensity of the neutron beam depends on the pinhole size and therefore a compromise between

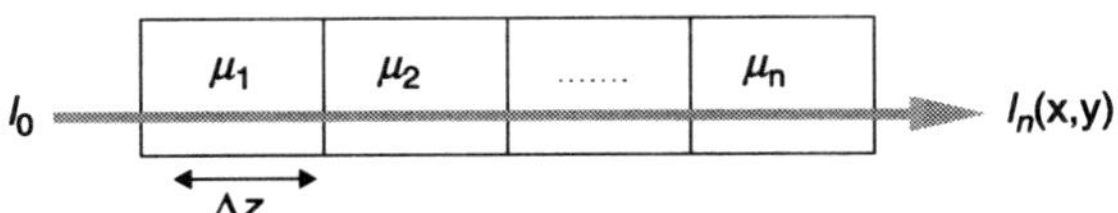

9.2 Schematic description of beam attenuation.

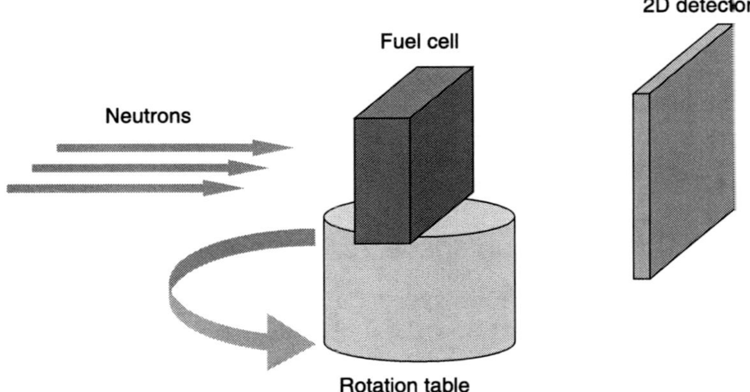

9.3 Standard setup for neutron radiography and tomography.

intensity and blurring d has to be found. For further quality improvements, the distance l between the scintillator screen and the sample should be as small as possible compared to the distance between the sample and the pinhole L (see Fig. 9.4).

A measure for the blurring d and the 'quality' of the neutron beam configuration is given by the L/D ratio:

$$\frac{L}{D} = \frac{l}{d} \qquad\qquad [9.3]$$

which is one of the main parameters that describes the performance of a given experimental setup.[1,6] Typical L/D values are between 200 for fast imaging (large pinhole) and 2000 (small pinhole) for high spatial resolution imaging.

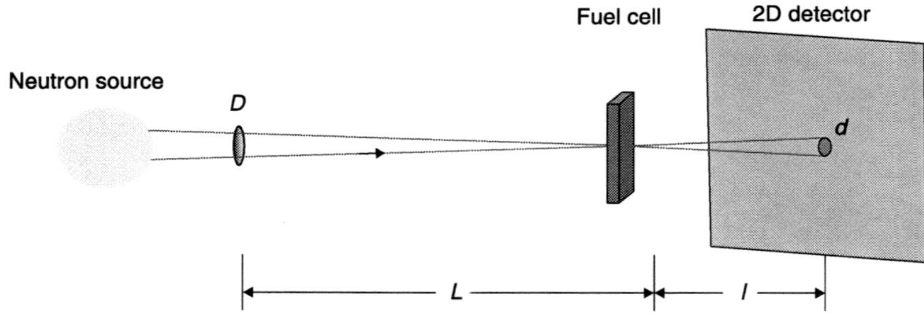

9.4 Schematic representation of neutron radiography and tomography experimental setup involving a pinhole setup (D = pinhole diameter).

9.3.3   Measuring procedure and reconstruction

The definition of tomography is comparatively simple. A tomogram of a sample is a set of radiographies taken under a series of different angular positions around a rotation axis that is preferably orientated vertically with respect to the beam. This set of radiographic projections is used to reconstruct a complete 3D image that represents the distribution of the local attenuation coefficient in the sample (see Fig. 9.5).[1, 29]

In order to prepare the radiographic images for the tomographic reconstruction algorithm a normalization procedure should be performed. The standard procedure for image normalization in neutron imaging of fuel cells is briefly described below.

The radiographic projections are corrected for two contributions, which both might lead to errors in the final image. One error is caused by electronic noise from the detection device (dark fields), and the second error is caused by non-uniform beam distribution (flat fields). Dark fields describe the electronic noise per pixel caused by the camera (the CCD chip). Cooling the CCD chip results in a reduced contribution of thermal pixel noise which cannot, however, be entirely neglected. These images are recorded with the radiation beam switched off in order to also identify contributions from the ambient light (which are also minor error sources) that reaches the camera from the outside. The dark field images (electronic noise) are subtracted as offset contributions from all recorded images.

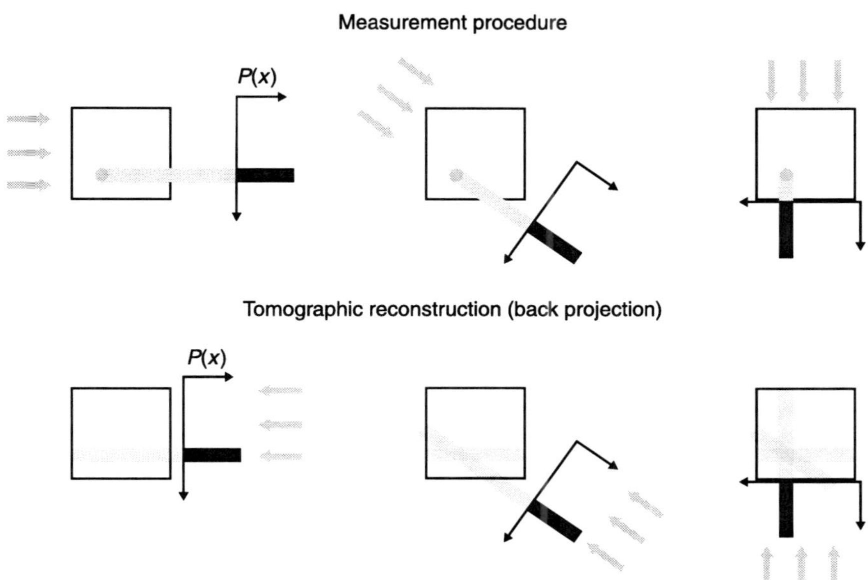

9.5 Tomographic measurement and reconstruction.

The flat fields are taken from the unperturbed beam, which means the object is moved out of the beam direction. The final normalization step is to divide all radiographic projections by the collected flat field images.

Different reconstruction algorithms can be used to reconstruct the tomogram of a sample from two-dimensional (2D) angular projections. The most widely used is the filtered back projection algorithm. The mathematical basis of the algorithm was introduced by Radon in 1917 (Radon transform). The high computational efficiency of the back-filtered algorithm and the high quality of the reconstructed images means that this is the most common algorithm used currently in neutron as well as in X-ray based tomographic techniques.

Using an up-to-date computer system, the computation times vary between a few minutes and several hours to reconstruct a tomogram depending on the amount of data and the desired image quality. The trend towards CCD-systems with more than 2048×2048 pixels causes longer reconstruction times and requires more memory space (RAM). A typical tomogram of a standard 2048×2048 pixel camera consists of about 32 Gbytes of data. Currently, 4096×4096 pixel CCDs are being introduced, which leads to an increase of the amount of data by a factor of eight.

9.4   Limitations and artifacts

In a radiographic or tomographic experiment some deviations from the ideal theoretical considerations can occur. These effects are observed as artifacts in the radiographic projections which should be interpreted and – if possible – corrected in order to perform quantitative data analysis. Typical artifacts observed in neutron imaging are described below:

- Neutron scattering by the sample, especially by liquid water, causes an additional background that depends on the sample-detector distance.
- Secondary radiation (e.g. gamma quants or γ-rays) scattered from the sample hits the scintillator materials and the detector and causes so-called 'white spots' in the images.
- The response of a pixel of an imaging detector to an incoming signal is not perfectly linear and varies between individual pixels, i.e. an increase in the beam intensity does not yield an increase in the measured intensity by the same value.
- Incident beam intensity fluctuations in space and time.
- Electronic noise.
- Small-angle scattering and refraction effects in the sample.

In tomographic reconstructed data additional artifacts occur that are related to the underlying reconstruction procedure:

- Crescent-like artifacts (in the case of a 180° tomogram) caused by a non-perfect rotation axis.

- Ring artifacts caused by defects or impurities on the scintillator crystals or imperfect detector elements.
- Beam hardening: the attenuation coefficient depends on the beam energy. When a polychromatic beam is used the beam spectrum changes with penetration depth in the sample. This results in higher attenuation coefficients at the sample edges of a reconstructed tomogram.
- Missing wedge/limited angles. Large fuel cells (active area above 100 cm^2) can hardly be penetrated using the complete 180° range. This causes blurring of all structures in the direction perpendicular to the missing projection directions.
- Motion artifacts due to slight position deviations of the fuel cell during measurement.

Many of these effects can be partially eliminated by appropriate adjustments, image normalization procedures, image filtering and optimized reconstruction algorithms, but still represent a source of error that has to be considered when evaluating the results.

For the water quantification in fuel cells a very important issue is the thorough correction of neutron scattering.[1, 30] During investigation of fuel cells, incoherent scattering caused by water strongly alters the measured transmission values through the cell. Neutrons that hit a hydrogen atom are scattered more or less isotropically in all directions. Some of the scattered neutrons hit the detector and contribute to a background signal, especially at locations with large amounts of water in the sample. This complicates the quantification of the amount of water in the cell volume and requires a thorough correction procedure. Because of the isotropic scattering the intensity of this background depends on the distance between the scattering center in the sample and the scintillator screen. Scattering effects can be reduced by increasing the distance between the sample and the detector. However, this results in larger spot blurring due to the beam divergence and limits the spatial resolution (see Fig. 9.4). This limitation has to be considered with the utmost care when aiming for measurements with very high spatial resolutions; in this case the setup requires a position of the scintillator screen as close as possible to the sample.

In order to analyse the contribution of scattering effects, the whole image formation process, including description of beam characteristics, sample transmission and detector response was simulated with the help of a Monte Carlo method using the MCNPX code. The results of the study are presented elsewhere.[31]

The spatial resolution has also been improved by mathematical methods: A correction algorithm named quantitative neutron imaging (QNI) based on the estimation of the so-called point scattered function (PScF) was developed, which describes the scattering contribution for each point of the object.[32] Due to the difficulty (better: mere impossibility) of calculating the corresponding PScF for all combinations of materials, beam lines and sample detector distances, a parameterised approximation was introduced based on the setup of a special data library. An example for the application of the QNI algorithm is shown in Plate V in the colour section between pages 252 and 253, where a model system of soil cubes with different moisture content was investigated by neutron tomography.

9.5   Examples of applications

9.5.1   Neutron tomography for structural investigations

The first neutron tomographic measurement of a fuel cell was reported by Satija *et al.* in 2004[33] at the National Institute of Standardization and Technology (NIST). A dry fuel cell stack was investigated in order to analyze the structure and possible physical defects within the cells. Figure 9.6 shows a 3D view of the cell's volume and a cross-section through the tomogram.[33] The shown data set is

9.6 Different viewing angles and perspectives of the tomogram of a PEM fuel cell as described in Fig. 9.7. Tomography proved successful in identifying many of the cell's features, yet the system lacked the resolution to distinguish the PEM membrane from the catalyst and diffusion layers.[33]

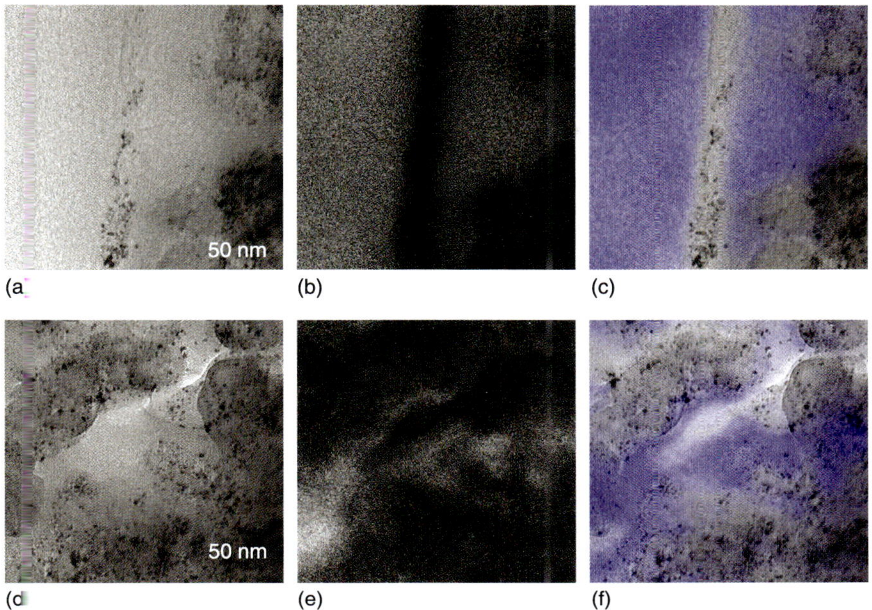

Plate I Unfiltered transmission electron micrographs of (a) membrane electrode interface and (d) a part of the electrode layer. (b, e) Fluorine maps of these regions obtained by energy filtered transmission electron microscopy. Bright parts in the image correspond to a high fluorine concentration. (c, f) A superposition of the fluorine map with the unfiltered images.

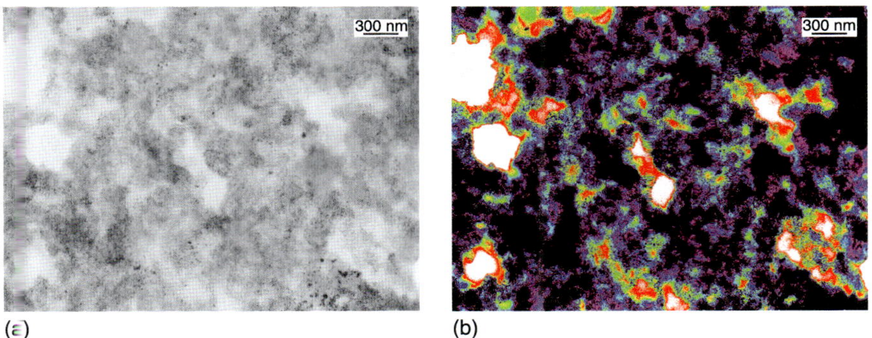

Plate II (a) Original grey-level TEM image of part of an electrode. (b) Same image as in (a), but with the grey-level information re-mapped to a rainbow colour scheme to improve the visibility of faint contrast images.

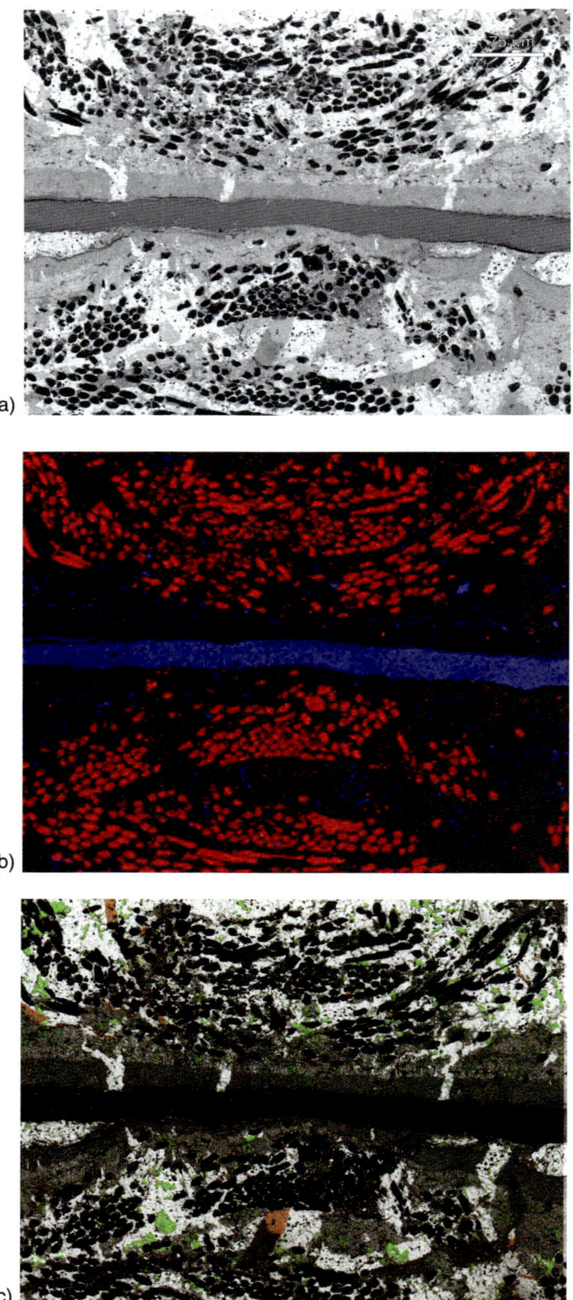

Plate III Part of the Wood's metal infiltrated MEA/GDL structure.
(a) BSE image. (b) EDX mappings of carbon (red) and fluorine (blue).
(c) Superposition of the BSE image and maps of tin (green) and
cadmium (red).

(a)

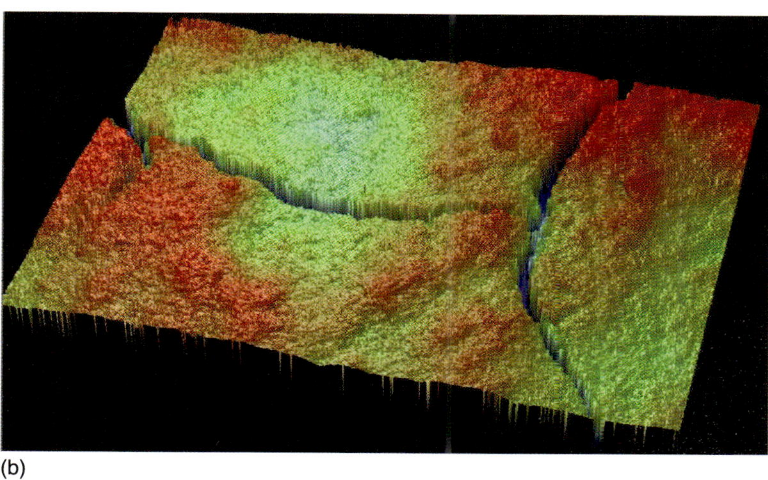

(b)

Plate IV Surface morphological features of (a) gas diffusion layer and (b) microporous layer obtained by optical profilometry scanning.

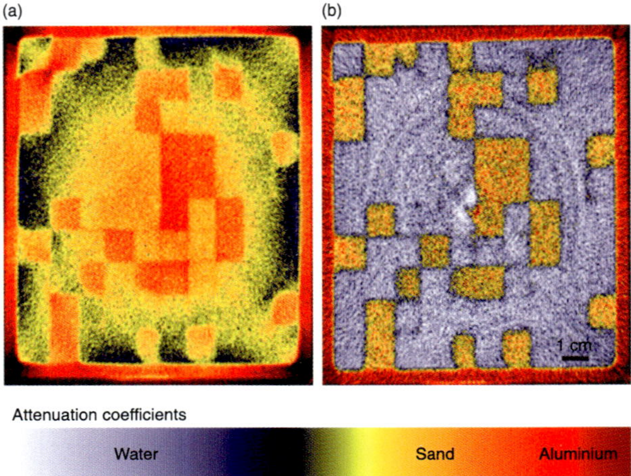

Plate V A tomographic slice through a model system of soil cubes with different moisture content. The direct reconstruction (a) shows artefacts due to neutron scattering of liquid water. The reconstruction from corrected projections using the QNI algorithm (b) clearly shows the inner structure of the system.[32]

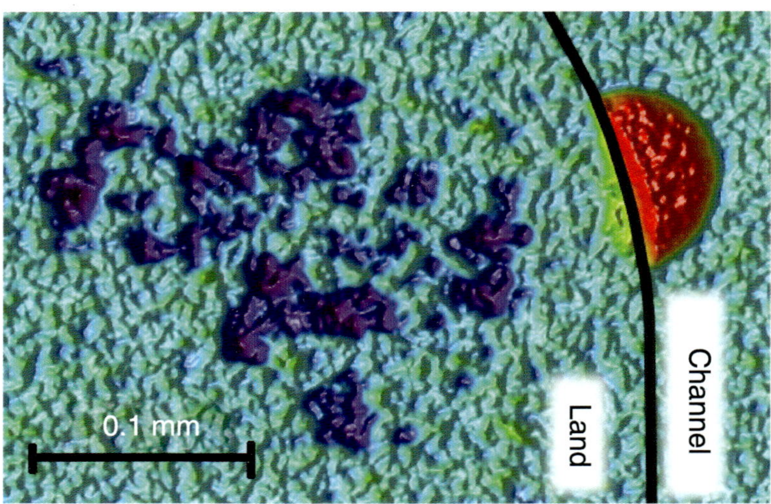

Plate VI Water transport mechanism of liquid water in an operating fuel cell. Blue areas under the land of the flow field (left side) represent empty pores, water stemming from these volumes forms the droplet (represented as red spot) in the gas channel (right side).

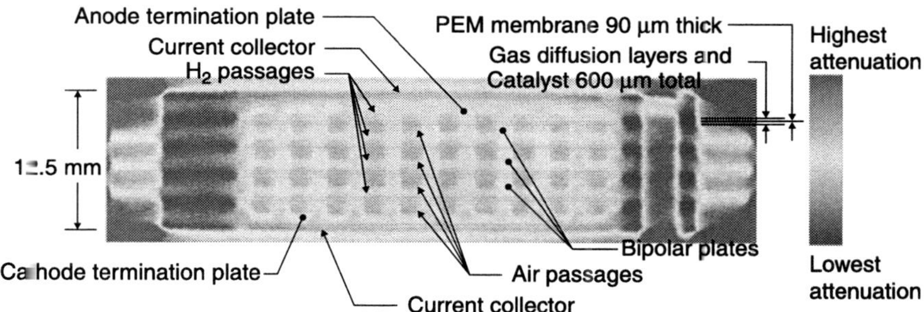

9.7 Structure of PEMFC superimposed onto the slice reconstruction shown in Fig. 9.6.[33]

reconstructed from 450 radiographic projection images. The virtual image of the stack represents a sound basis for further investigations as it can be sliced in various ways to display structural details within the cells (Fig. 9.7).

9.5.2   3D mapping of water distribution

The overall exposure time necessary for acquiring a full tomographic data set for such a large fuel cell is somewhere between 30 minutes and several hours. The most important prerequisite for a reliable interpretation of the obtained results is an unchanged water distribution on this time scale which, however, makes a tomogram of larger cells or even stacks extremely difficult and renders *in situ* measurements almost impossible. In the case of smaller fuel cells, tomographic measurements can be performed much faster.

In Fig. 9.8 a tomographic analysis of a small fuel cell performed at the NIST is displayed.[34–35] The tomogram was split into several slices and reveals the water distribution in all parts of the cell, i.e. the anode and cathode parts can be visualised separately. In this study, water distribution peaks in the anodic gas diffusion medium. Furthermore, interesting details on the water transport mechanism and the water content at different fuel cell temperatures have been addressed.

Another possibility to increase the speed of tomographic imaging is to use an increased neutron flux as available at some neutron facilites. In the case of the Helmholtz Centre Berlin such a special high flux position is available. It provides a cold neutron beam with an intensity of 2×10^8 neutrons/cm^2/s. This allows for comparatively fast measurements in addition to a high imaging quality. In Fig. 9.9 the tomogram of a fuel cell with a diameter of 3 cm is shown.[27] The power range of this fuel cell type is a few watts. Water agglomerations are colored in dark grey with the graphite composite components colored semi-transparent in light grey. The overall tomogram was performed within 15 min. This leads to a high image quality that is the precondition for sufficient water quantification and a comparison of different operating modes, for example variation of the applied electric load.

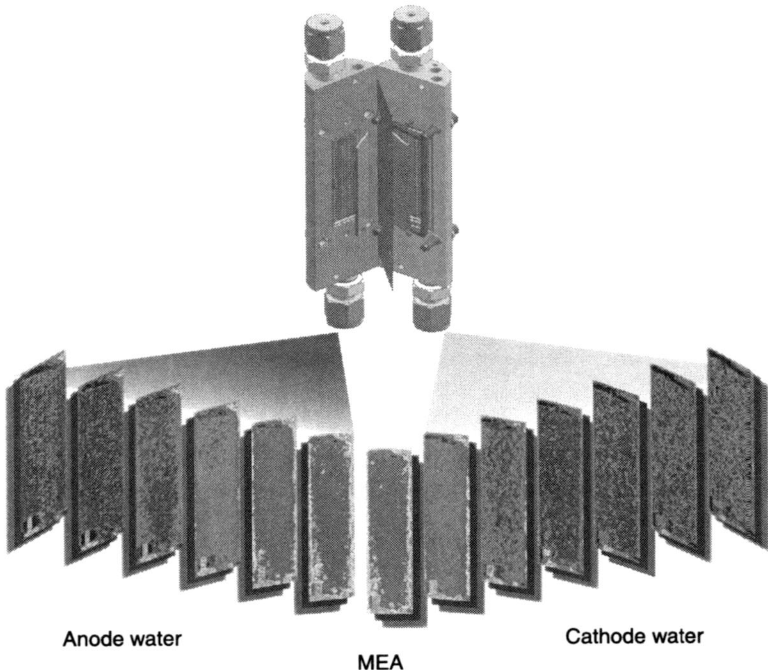

9.8 Images of the water distribution obtained from tomographic measurement of a test cell. Width of each slice: 0.125 mm.[34–36]

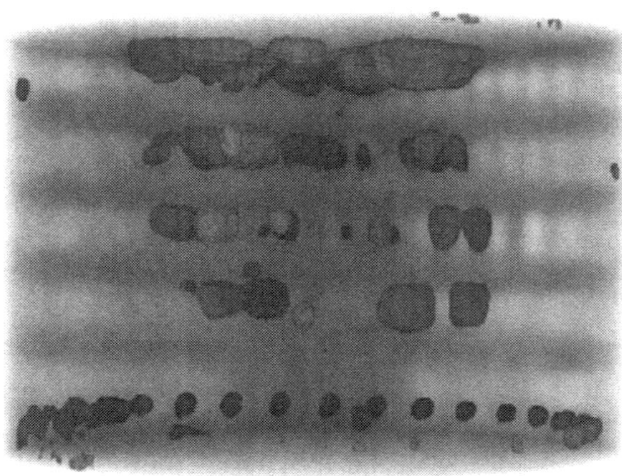

9.9 Neutron tomogram of water in a fuel cell (2–4 W_{el} performance) with a diameter of 3 cm. Liquid water agglomerations can be identified as dark grey spots; the structure of the flow fields is displayed as semi-opaque.[27]

9.5.3 *In situ* tomography

A further increase of measurement speed can be achieved at the cost of quality of the resulting images. The Neutrograph neutron tomography station at the ILL in Grenoble can provide a very high neutron flux[37] of about 10^9 neutrons/cm^2/s. At the same time, the beam spectrum is mainly composed of thermal neutrons, which are less sensitive to hydrogen but have a much greater penetration depth, i.e. the transmission signal is much higher than that of a cold neutron beam. This allows for exposure times for single radiographic projections[37] down to 50 ms, leading to a total imaging time for a complete tomogram of just 10–60 seconds. These short measurement times are suited for real-time *in situ* 3D investigations of fuel cells. Figure 9.10 shows different slices through the tomogram of a fuel cell taken with a total imaging time[37] of 30 s. Although the image quality is comparatively low and can be considered to be at the lower limit, it is still sufficient to analyze the water distribution. Besides a reduced spatial resolution the high flux leads to an activation of the fuel cell and other surrounding components (such as the mounting equipment). This activation in turn requires longer waiting times until the fuel cell set-up can be altered or exchanged.

9.5.4 Large fuel cell stacks

For many applications the possibility to study larger fuel cells or fuel cell stacks with typical sizes between 10–20 cm is challenging. Since the overall measurement time of 1–10 hours can hardly be reduced without impact on the image quality, a new approach has been developed that allows 'freezing' of the water distribution in the cell at a given condition. For this purpose the fuel cell is switched off at the previously defined operation point and the gas connections sealed. In this way the water is retained in the cell at the original spots as of the point of switching off the cell.[38]

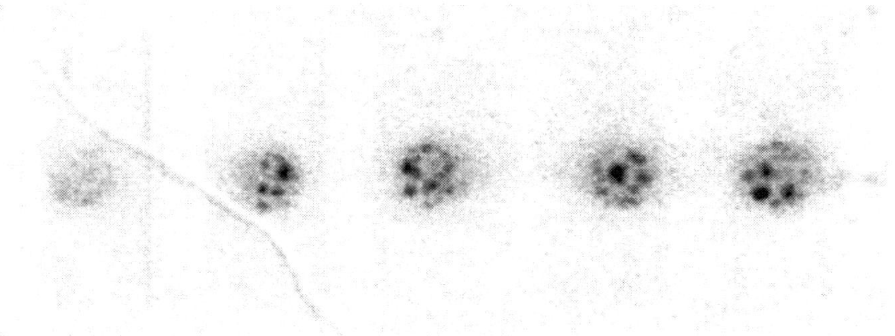

9.10 Water condensation in a micro fuel cell (water is displayed as black spots).[37]

An example for a successful application of this procedure is given in Fig. 9.11.[38] The investigated three-fold fuel cell stack had dimensions of 14×14 cm^2 with an active area of 10×10 cm^2. The serpentine flow fields with 1 mm wide ribs and channels were made of graphite composite. The overall measurement time for the complete tomogram was 5 hours, resembling 600 single radiographic projections with an exposure time of 30 s per projection. The voxel size was $113 \times 113 \times 113$ μm^3 and the physical spatial resolution was about 250–300 μm.

The tomographic measurement allowed not only for a quantification of the water amount in all three cells of the stack but also for a separation of the water content of the anode and cathode (Fig. 9.11) and for analysis of effects such as back-diffusion and electro-osmotic drag. Figure 9.11a displays the water distribution on the anode and cathode of the first of the three cells at two different current densities. At a current density of 300 mA/cm^2 about equal amounts of water were found on both sides. Although the cathodic gas stream was externally humidified and water is produced at the cathode, the back-diffusion from cathode to anode is sufficient to humidify the anode. Back-diffusion proves at this current density to be one of the key factors for a uniform humidification of the active area in a fuel cell. An increase of the current density to 500 mA/cm^2 leads to a strong decrease of the anodic water content. This can be assigned to an increase in the electro-osmotic drag that significantly outweighs the back diffusion in this case.

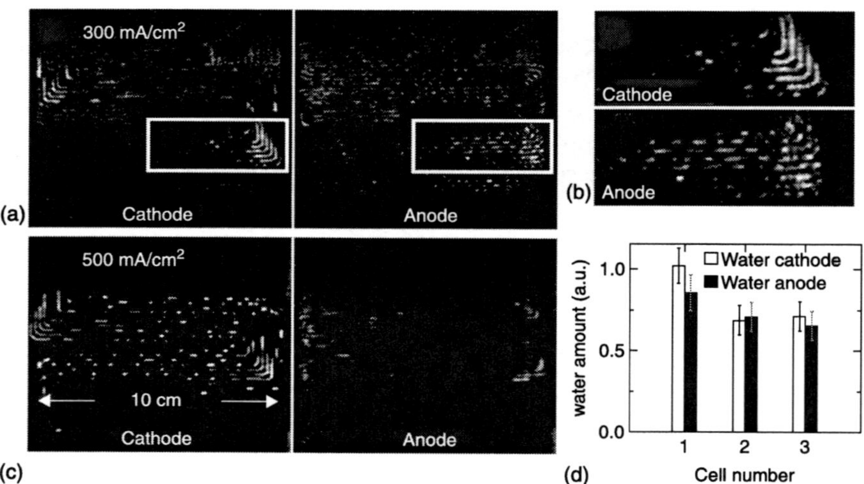

9.11 (a) Water distributions in the anodic (right) and cathodic (left) flow field channels of the first cell in a triple fuel cell stack ($i_0 = 300$ mA cm^{-2}). (b) corresponding enlargements of the marked areas. (c) Water distributions at higher current density ($i_0 = 500$ mA cm^{-2}). (d) Total water content in the anodic and cathodic flow fields of all three cells at $i_0 = 300$ mA cm^{-2}. (Source: Manke *et al.*[38] Copyright 2007, American Institute of Physics.)

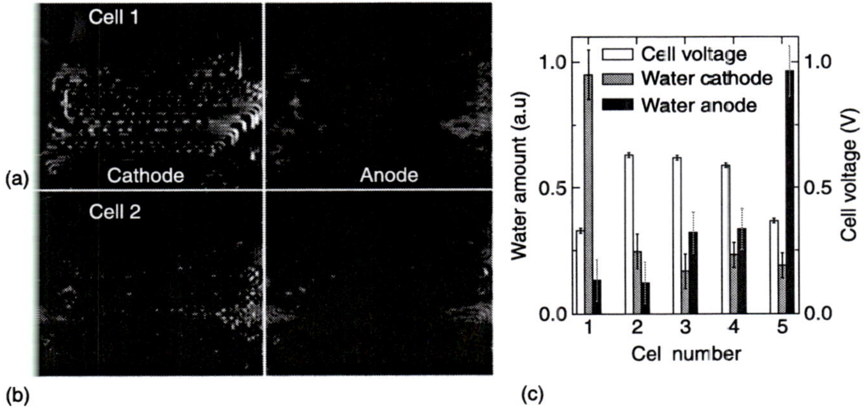

9.12 Water distribution and performance of a five-fold fuel cell stack (active area of 10 × 10 cm^2). (a, b) Water content in the outermost and the second cell. (c) Performance and water content of the respective cell at i_0 = 500 mA cm^{-2}. (Source: Manke *et al.*[38] Copyright 2007, American Institute of Physics.)

Membrane properties such as the thickness can be varied to balance electro-osmotic drag and back-diffusion.

In another example (Fig. 9.12) a five-fold stack has been examined. At the applied operating conditions (current density 500 mA/cm^2) only small amounts of liquid water were found in the cells. However, the first and the last cells within the stack have a significantly increased amount of water (Fig. 9.12c). Concurrently, cell voltages are strongly reduced by about 40% in these cells. Most of the liquid water in these two cells is located on the corresponding electrodes facing towards the end plates, i.e. the cathode of the first cell on one side and the anode of the last cell. This behavior has often been found in the outmost cells and is mainly caused by slight thermal deviations which can be as low as a few degrees but are obviously sufficiently high to cause enhanced condensation in the outermost electrodes.

9 5.5 Optimization of tomographic reconstruction methods

The reduced quality of tomograms taken from larger fuel cells requires the implementation of new or modified tomographic reconstruction algorithms. The main problems are caused by the fact that the fuel cell is only transparent up to a certain rotation angle. At an angle of 60° (taken between the electrochemically active area and the beam direction) the penetration path for neutrons is strongly increased and the transmission is reduced in parallel. The edges of the active areas especially contain several highly absorbing materials, mainly the gaskets. At increased rotation angles the edges are hiding the rest of the cell and the projection images contain hardly any information about the water distribution in the fuel cell.

For the mathematical reconstruction it might be even better to refrain from using the large-angle projection because this might impact the results negatively.[1] The standard tomographic reconstruction algorithms such as, for example, filtered back projection, are disturbed by the unclear information resulting from these angle positions. This in turn means only a small part of typically between 100–140 degrees can be used for the tomographic imaging and subsequent reconstruction which is usually called a 'limited angle' or 'missing wedge' problem.

Several approaches to enhance reconstruction quality have been developed and are already realized. Here, the authors would like to demonstrate an example of a recently introduced new algorithm that was tested with focus on the reconstruction of neutron tomograms of fuel cells. The DIRECTT algorithm was developed by the Federal Institute of Materials Research and Testing in Berlin, Germany and is especially suited for limited angle tomograms with varying pixel smearing caused by a limited aperture. The algorithm is best described in a short technical outline: this algorithm is based on the identification of sine graphs in the sinograms of the measurement.[39–40]

Figure 9.13 shows an application and the resulting improvement of resolution of the algorithm that has been used to enhance the reconstruction of a tomogram of a triple fuel cell stack.[40] Two slices through the neutron tomogram of this large (1 active area: 100 cm^2) fuel cell stack are compared with each other. Figure 9.13a was reconstructed with a conventional filtered back projection algorithm and Fig. 9.13b was reconstructed employing the novel algorithm. Already on the first glance an improved image quality becomes obvious; the enhanced quality also serves as the basis for a more accurate and reliable determination of the water

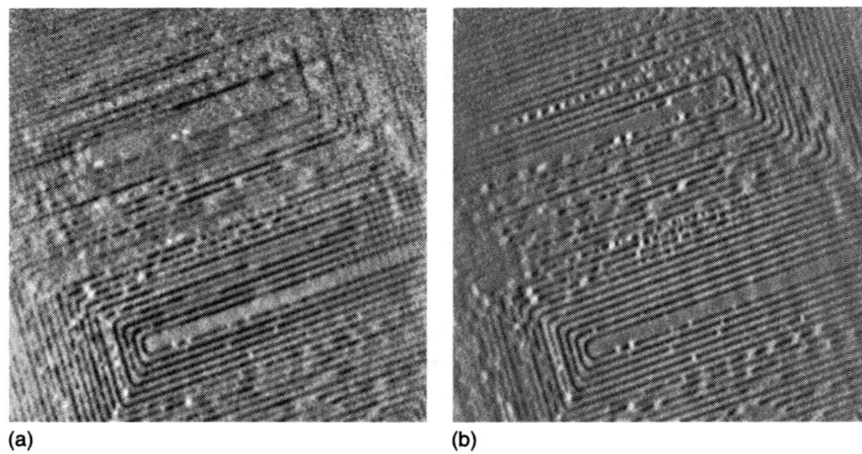

(a)   (b)

9.13 Comparison of (a) filtered back-projection (FBP) (b) DIRECTT reconstructions of the central cathode plane (active area about 10 × 10 cm^2, typical channel/land dimensions: 1 mm): different contrast and improved resolution (perceptibility) of water droplets.[40]

content of the respective compartments. The 'smearing' artifacts are especially reduced by the new algorithm, i.e. with a standard reconstruction algorithm the water distribution on, for example, the anode causes shadows in the cross-section through the cathode. In turn, separate quantification of anode and cathode water content becomes easier.

9.6   Outlook

Neutron-based methods are an interesting tool for non-destructive studies of operating fuel cells; the extension to larger cells and concurrently from radiographic to tomographic imaging makes neutron tomography and/or radiography valuable methods to study the 3D water distribution and content of the different components in minute detail.

There are still a number of improvements required covering all aspects from sample size up to adapted reconstruction algorithms, which not only improve the quality of the images, but at the same time lead to a reduction of the overall measurement time once the number of orientations can be clearly reduced.

9.7   References

1 Advanced Tomographic Methods in Materials Research and Engineering; J. Banhart, Ed.; Oxford University Press: Oxford, UK, 2008.
2 C.-O. Fischer In *Proceedings 4th World Conference on Neutron Radiography 1992*; J. B. Barton, Ed.; Gordon and Breach, Chemin de la Sallaz: San Francisco, USA, 1994, p 291.
3 H. Kallmann, E. Kuhn *Zeitschrift für Naturforschung* **1946**, 557.
4 N. Kardjilov, A. Hilger, I. Manke, M. Strobl, M. Dawson, J. Banhart Nuclear Instruments and Methods in Physics Research Section A: Accelerators, Spectrometers, Detectors and Associated Equipment **2009**, 605, 13.
5 E. Lehmann *Pramana Journal of Physics* **2008**, *71*, 653.
6 M. Strobl, I. Manke, N. Kardjilov, A. Hilger, M. Dawson, J. Banhart *Journal of Physics D-Applied Physics* **2009**, *42*.
7 R. J. Bellows, M. Y. Lin, M. Arif, A. K. Thompson, D. Jacobson *Journal of The Electrochemical Society* **1999**, *146*, 1099.
8 J. Zhang, D. Kramer, R. Shimoi, Y. Ono, E. Lehmann *et al. Electrochimica Acta* **2006**, *51*, 2715.
9 M. A. Hickner, N. P. Siegel, K. S. Chen, D. N. McBrayer, D. S. Hussey, *et al. Journal of The Electrochemical Society* **2006**, *153*, A902.
10 M. A. Hickner, N. P. Siegel, K. S. Chen, D. S. Hussey, D. L. Jacobson, M. Arif *Journal of The Electrochemical Society* **2008**, *155*, B427.
11 D. J. Ludlow, C. M. Calebrese, S. H. Yu, C. S. Dannehy, D. L. Jacobson, *et al. Journal of Power Sources* **2006**, *162*, 271.
12 D. Kramer, J. Zhang, R. Shimoi, E. Lehmann, A. Wokaun, *et al. Electrochimica Acta* **2005**, *50*, 2603.
13 P. Boillat, D. Kramer, B. C. Seyfang, G. Frei, E. Lehmann, *et al. Electrochemistry Communications* **2008**, *10*, 546.

14. P. Boillat, G. Frei, E. H. Lehmann, G. G. Scherer, A. Wokaun *Electrochemical and Solid-State Letters* **2010**, *13*, B25.
15. C. Hartnig, I. Manke, N. Kardjilov, A. Hilger, M. Grünerbel *et al. Journal of Power Sources* **2008**, *176*, 452.
16. I. Manke, C. Hartnig, N. Kardjilov, M. Messerschmidt, A. Hilger *et al. Applied Physics Letters* **2008**, *92*, 244101.
17. A. Schröder, K. Wippermann, J. Mergel, W. Lehnert, D. Stolten *et al. Electrochemistry Communications* **2009**, *11*, 1606.
18. A. Schröder, K. Wippermann, W. Lehnert, D. Stolten, T. Sanders *et al. Journal of Power Sources* **2010**, *195*, 4765.
19. J. Park, X. Li, D. Tran, T. Abdel-Baset, D. S. Hussey *et al. International Journal of Hydrogen Energy* **2008**, *33*, 3373.
20. M. M. Mench, Q. L. Dong, C. Y. Wang *Journal of Power Sources* **2003**, *124*, 90.
21. N. Pekula, K. Heller, P. A. Chuang, A. Turhan, M. M. Mench *et al.* Nuclear Instruments and Methods in Physics Research Section A: Accelerators, Spectrometers, Detectors and Associated Equipment **2005**, 542, 134.
22. *Encyclopedia of Electrochemical Power Sources*; J. Garche, C. K. Dyer, P. T. Moseley, Z. Ogumi, D. A. J. Rand, B. Scrosati, Eds.; Elsevier: Amsterdam, 2009; Vol. 3.
23. C.-Y. Wang In *Handbook of Fuel Cells – Fundamentals, Technology and Applications*; W. Vielstich, A. Lamm, H. A. Gasteiger, Eds.; John Wiley & Sons: Chichester, 2003; Vol. 3, p 337.
24. C.-Y. Wang *Chemical Reviews* **2004**, *104*, 4727.
25. *Handbook of Fuel Cells – Fundamentals, Technology and Applications*; W. Vielstich, A. Lamm, H. A. Gasteiger, Eds.; John Wiley & Sons: Chichester, 2003; Vol. 3.
26. L. Carrette, K. A. Friedrich, U. Stimming *Fuel Cells* **2001**, *1*, 5.
27. J. Banhart, A. Borbely, K. Dzieciol, F. Garcia-Moreno, I. Manke *et al. International Journal of Materials Research* **2010**, *101*, 1069.
28. B. Schillinger, E. Lehmann, P. Vontobel *Physica B: Condensed Matter* **2000**, *276-278*, 59.
29. A. C. Kak, M. Slaney *Principles of computerized tomographic imaging*; IEEE Service Center, Piscataway, NJ, 1988.
30. A. K. Heller, L. Shi, J. S. Brenizer, M. M. Mench Nuclear Instruments and Methods in Physics Research Section A: Accelerators, Spectrometers, Detectors and Associated Equipment **2009**, 605, 99.
31. L. S. Waters MCNPX User's Manual, version 2.4.0, Los Alamos National Laboratory Report, 2002.
32. E. Lehmann, N. Kardjilov; J. Banhart, Ed.; Oxford University Press: Oxford, UK, 2008.
33. R. Satija, D. L. Jacobson, M. Arif, S. A. Werner *Journal of Power Sources* **2004**, *129*, 238.
34. M. Arif, D. L. Jacobson, D. S. Hussey In *FY 2006 Annual Progress Report* 2006, p 875.
35. D. S. Hussey, J. P. Owejan, D. L. Jacobson, J. Gagliardo, T. A. Trabold, M. Arif In *International Conference on Neutron Scattering* Sydney, Australia, 2005.
36. D. S. Hussey, J. P. Owejan, D. L. Jacobson, T. A. Trabold, J. Gagliard *et al. Tomographic Imaging of an Operating Proton Exchange Membrane Fuel Cell*; Destech Publications, Inc: Lancaster, 2008.
37. A. Hillenbach, M. Engelhardt, H. Abele, R. Gähler Nuclear Instruments and Methods in Physics Research Section A: Accelerators, Spectrometers, Detectors and Associated Equipment **2005**, 542, 116.
38. I. Manke, C. Hartnig, M. Grunerbel, J. Kaczerowski, W. Lehnert *et al. Applied Physics Letters* **2007**, *90*, 184101.

39. Computed tomography reconstructions by DIRECTT-2D model calculations compared to filtered backprojection; A. Lange, M. P. Hentschel, A. Kupsch, Eds.; Carl Hanser Verlag, 2008; Vol. 50, 272–277.

40. A. Lange, A. Kupsch, M. P. Hentschel, I. Manke, N. Kardjilov *et al. Journal of Power Sources* **2010**, *196*, 5293.

10

Magnetic resonance imaging (MRI) techniques for polymer electrolyte membrane and direct alcohol fuel cell characterization

K. W. FEINDEL, National Research Council Canada, Canada

Abstract: This chapter introduces the fundamental concepts of magnetic resonance imaging (MRI) for the study of water in operating polymer electrolyte membrane fuel cells (PEMFCs). Practical aspects of investigating PEMFCs with MRI are covered, including adaptation of designs and materials, and choice of appropriate MRI experiment. Common image artefacts and experimental limitations are outlined. Key results from the literature are presented, covering distribution of water in the PEM and flow channels under various operating conditions, and methods for quantitative analysis of water content. The use of nuclear magnetic resonance spectroscopy to investigate direct alcohol fuel cells and future directions of MR techniques for the study of operating PEMFCs are also discussed.

Key words: magnetic resonance imaging, water distribution, water dynamics, quantitative water content, flow field flooding.

10.1 Introduction

In recent decades, numerous modalities have been applied or developed specifically to investigate and visualize the distribution of water in polymer electrolyte membrane fuel cells (PEMFCs). These methods include neutron radiography (Chapters 7, 8), neutron tomography (Chapter 9), Raman spectroscopy (Chapter 11), and magnetic resonance imaging (MRI) techniques. Both neutron and MRI techniques have enabled the visualization of areas within operating PEMFCs not observable by direct optical methods. Magnetic resonance (MR) techniques, such as spectroscopy and imaging, are based on the fundamental nuclear magnetic resonance (NMR) interactions. Herein, the terms NMR and MRI are used to refer to general- and imaging-specific principles, respectively. The achievable resolution of MRI is in the order of μm, ultimately limited by the frequency and decay of the NMR signal (Ciobanu *et al.*, 2002; Lee *et al.*, 2001). The resolution is comparable to that of optical microscopy and hence, the term 'NMR microscopy' is often used when referring to high resolution MRI of materials (Eccles and Callaghan, 1986). Although X-ray or neutron source techniques for materials characterization can achieve atomic scale resolution, the variety of image contrasts that can be obtained based on the sensitivity of NMR parameters to spin-density and molecular interactions (Xia, 1996) makes MRI extremely robust and the applications diverse. For example, image contrast can be gathered from NMR relaxation parameters, molecular motion (e.g. diffusion, flow), differences in chemical environment, and magnetic susceptibility variations (Callaghan, 1991; Kimmich, 1997; Haacke, 1999; Xia, 1996). This diverse sensitivity has facilitated the establishment of NMR techniques as some of the most powerful methods for molecular structural elucidation (Ernst, 1992), characterization of materials (Blümich, 2000; Codd and Seymour, 2009; Stapf and Han, 2006), and for diagnostic medical imaging (Lauterbur, 2005; Mansfield, 2004). Application of MRI outside of the life sciences realm is often technically challenging due to the non-aqueous or heterogeneous composition of the materials of interest (Codd and Seymour, 2009; Stapf and Han, 2006).

PEMFCs generally have a heterogeneous material composition, some which are favourable for MRI investigations (e.g. the water-containing components) and others which can be detrimental (e.g. metallic and conductive components). Prior to the application of ^{1}H MRI to study operating PEMFCs, no available experimental technique could monitor water in the polymer electrolyte membrane (PEM) between the operating catalyst layers separate from other regions of the fuel cell (FC). Other experimental techniques provide either an indirect measure of water content (e.g. PEM impedance measurements) or provide integral measurements of total water content (e.g. neuron imaging techniques). MRI allows one to select and investigate a well-defined region of interest, and as discussed in Section 10.9, the gas diffusion and catalyst layers of a PEMFC facilitate investigation of water within the PEM without interference from, e.g., liquid water in the gas flow channels.

This chapter covers the application of NMR techniques to the *in situ* investigation of operating PEMFCs, particularly to the distribution of water within the PEM and flow channels. The fundamental concepts and theory of NMR and MRI and typical laboratory hardware required for performing MRI experiments are introduced. Subsequently, adaptations to PEMFC materials required to facilitate MRI investigations and technical considerations will be outlined. The application of MRI for studying water distribution in operating PEMFCs will be reviewed, noting key results, developments, and advances. In addition, advantages, limitations, and future trends are discussed.

10.2   Concepts of nuclear magnetic resonance (NMR)

For MRI investigations of operating PEMFCs we are concerned primarily with determining the distribution of water, for which the NMR active nuclei are ^{1}H, ^{2}H, and ^{17}O (tables of nuclear spin properties are available (Harris *et al.*, 2001)). ^{1}H is the most suitable with the highest natural abundance (99.99%) and the largest magnetogyric ratio, γ (26.7522128 $\times$ 10^7 rad s^{-1} T^{-1}). NMR techniques perturb and detect the bulk nuclear magnetism of an ensemble of spins, which is created in the presence of a strong applied external magnetic field, B_0. With careful control of the external applied magnetic field (e.g. by introducing linear field gradients that either add to, or subtract from, B_0), experiments have been devised to obtain the spatial distribution of the ensemble of spins (Lauterbur, 2005; Mansfield, 2004). This section provides an introduction to the fundamental theory of NMR and to specific phenomena that are relevant for imaging water in PEMFCs.

10.2.1 The Zeeman interaction and the NMR response

The NMR response relies on the fundamental interaction of a magnetic field with the magnetic dipole moment associated with the orbital angular momentum of a nucleus (Zeeman, 1897). Many atomic nuclei possess half-integer or integer spin-angular momentum, $I\hbar$, co-linear with a magnetic dipole moment, $\mu = \gamma I\hbar$, where γ is the magnetogyric ratio, and $\hbar$ is the Planck constant h divided by 2π. As shown in Eq. 10.1, the Zeeman Hamiltonian describes the energy of interaction of $I\hbar$, with an external magnetic field, $\mathbf{B}_0$,

$$\mathcal{H}_z = -\gamma I\hbar\mathbf{B}_0 \tag{10.1}$$

Typically, only the largest component of the $\mathbf{B}_0$ column vector (B_x, B_y, B_z) is considered, which defines the z-axis of the laboratory frame of reference, $\mathbf{B}_0 = (0, 0, B_0)$.

A proton possesses a magnetic dipole moment and has a half-integer nuclear spin, $I = \frac{1}{2}$. In the presence of an external magnetic field, the magnetic dipole moment aligns either parallel or anti-parallel to the direction of B_0. As shown in Fig. 10.1, different energies are obtained for the two possible projections with respect to the applied magnetic field, and correspond to the non-degenerate nuclear spin eigenstates: $m_I = \frac{1}{2}$ (parallel) and $m_I = -\frac{1}{2}$ (anti-parallel). The frequency corresponding to the difference in energy between the two states, ΔE, is defined by the Larmor equation, Eq. 10.2:

$$v_L = \frac{|\gamma| B_0}{2\pi} = \frac{\Delta E}{h} \tag{10.2}$$

where v_L is referred to as the Larmor frequency. Both Eqs 10.1 and 10.2 show the magnetic field-dependent nature of the NMR frequency. The Larmor frequencies of protons in commercial superconducting NMR magnets are in the order of 10^8 Hz, or in the radio frequency (RF) range.

10.2.2 Creating and detecting the NMR response

The NMR signal generated by an ensemble of nuclei is governed by Boltzmann statistics, which determines the relative populations of the nuclear spin eigenstates. In thermal equilibrium the net magnetization vector of a sample under study is oriented with the external applied magnetic field, and is often denoted M_0. To detect an NMR signal, M_0 must be reoriented from the large external applied field. An oscillating magnetic field, B_1, is applied transverse to the z-axis, with frequency

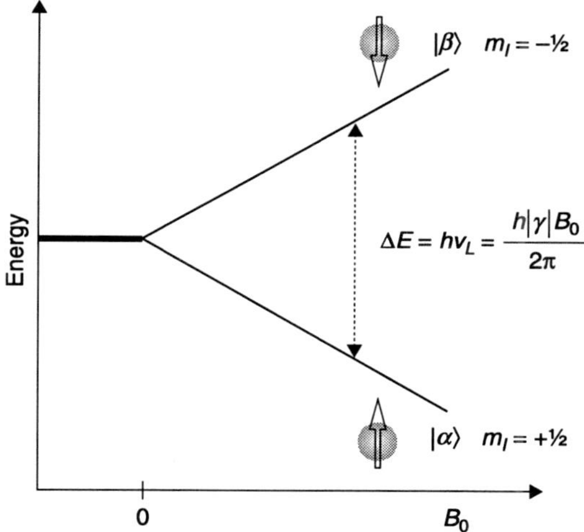

10.1 Zeeman energy levels for a nucleus with $I = \frac{1}{2}$ demonstrating the separation of the non-degenerate nuclear spin eigenstates $m_I = +\frac{1}{2}$ (parallel) and $m_I = -\frac{1}{2}$ (anti-parallel) as a function of applied magnetic field, B_0.

that matches v_L. The duration and amplitude of the B_1 field together determine the extent to which the net magnetization vector rotates about the B_1 field away from the z-axis. A projection of the magnetization vector is oriented along the plane transverse to the external applied field and is often denoted $M_{x,y}$. $M_{x,y}$ continues to process at the Larmor frequency, creating a detectable signal called the free-induction decay (FID), the amplitude and phase of which are measured as a function of time, this is the NMR signal.

10.2.3 Relaxation phenomena

NMR relaxation is a general term that is used to describe the phenomena of either the recovery of M_0 or the decay of $M_{x,y}$. The time that is required for the magnetic moments of an ensemble of spins in random orientation to align with an applied magnetic field depends on many variables including material density and molecular motion (e.g. solid, liquid, mobile), and applied field strength. Similarly, the components of $M_{x,y}$ 'dephase' or lose coherence, and the detectable response decays.

At thermal equilibrium M_0 is in the direction of B_0 along the z-axis. Application of an RF pulse that rotates the magnetization away from the z-axis and into the x,y plane disturbs the equilibrium of the spin system. Over time the equilibrium will be restored, and the magnetization along the z-axis at time t, $M_z(t)$ can be described in Eq. 10.3 using the classical equation of motion derived by Bloch (1946):

$$M_z(t) - M_0 = [M_z(0) - M_0]\exp\left(-\frac{t}{T_1}\right)$$

[10.3]

where $M_z(0)$ is the magnetization remaining along the z-axis after the RF pulse and T_1 is the first-order time constant that describes the recovery of M_0 after the RF pulse. This time constant is generally referred to as the spin-lattice or longitudinal relaxation time. As described by the mono exponential dependence, during a period T_1, ~63% of the original equilibrium magnetization recovers, after $3T_1$ ~ 95% recovers, and essentially complete relaxation occurs after $5T_1$. Thus, the number of times that M_z can be tipped (e.g. to signal average) in a particular time period (often called the repetition time, T_R) without saturating the signal is related to T_1 (i.e. a delay of several seconds may be required to reach $M_z = M_0$).

Pre-treatment and the amount of water in Nafion are known to affect the ^{1}H NMR relaxation times (Boyle et $al.$, 1983; Chen et $al.$, 1995; Fontanella et $al.$, 1996; MacMillan et $al.$, 1999). A summary of ^{1}H NMR T_1 values for pure $H_2O(l)$ at various temperatures and for H_2O in Nafion–117 is provided in Table 10.1. There is a large spread in the T_1 values measured for Nafion. Neither Chen et $al.$ (1995) nor Fontanella et $al.$ (1996) reported any membrane pre-treatment to remove possible paramagnetic contaminants. MacMillan et $al.$ (1999) completed a procedure to convert the Nafion to acid form, and to remove paramagnetics, raw polymer lightweight oligomers, and precursor fragments remaining from

Table 10.1 Summary of ^{1}H NMR spin-lattice relaxation times, T_1, for pure $H_2O(l)$ and H_2O in Nafion–117

	T (K)	Condition[†] (wt%)	T_1 (ms)	Reference
$H_2O(l)$				
	273		1 730 ± 35	(Krynicki, 1966)
	293		3 150 ± 63	(Krynicki, 1966)
	298		3 570 ± 71	(Krynicki, 1966)
	373		12 750 ± 255	(Krynicki, 1966)
Nafion-117				
	298	4.0	42	(Fontanella et $al.$, 1996)[‡]
	298	5.9	250	(MacMillan et $al.$, 1999)[c]
	297	9.8	57	(Chen et $al.$, 1995)[§]
	298	10.4	84	(Fontanella et $al.$, 1996)[‡]
	298	15.9	1300	(MacMillan et $al.$, 1999)[*]
	297	18.0	124	(Chen et $al.$, 1995)[§]
	298	20.8	110	(Fontanella et $al.$, 1996)[‡]

† 1.66 wt% ($\lambda = 1$, where $\lambda = nH_2O/SO_3H$
‡ Values measured graphically (from Fig. 6 Fontanella et $al.$, 1996)
* Values measured graphically (from Fig. 3 MacMillan et $al.$, 1999)
§ Values measured graphically (from Fig. 3 Chen et $al.$, 1995)

synthesis; however, their procedure differs from that commonly used for FC applications. Although specific values are not reported, Zawodzinski *et al.* (1991) report that for purified, acid-form Nafion–117, the ^{1}H T_1 values ranged from ~80–200 ms. In general, the PEM in an operating PEMFC is not saturated with water and the T_1 is reduced significantly when in contact with paramagnetics such as the catalyst layers or $O_2(g)$ (Hausser and Noack, 1965; Glasel, 1972). As well, hot-pressing is known to significantly reduce the ability of Nafion to take up water (Cappadonia *et al.*, 1995), and therefore will also affect T_1.

Once the magnetization vector is rotated away from B_0, and into the transverse plane (i.e. $M_z \rightarrow M_{x,y}$), $M_{x,y}$ will decay due to random homogeneous interactions and result in no net transverse magnetization. For molecules in solution this relaxation process and the transverse magnetization at time t, $M_{x,y}(t)$, can generally be treated using the theory of Bloembergen, Purcell, and Pound (Bloembergen *et al.*, 1948) and described by Eq. 10.4:

$$M_{x,y}(t) = M_{x,y}(0)\exp\left(-\frac{t}{T_2}\right)$$

[10.4]

where $M_{x,y}(0)$ is the transverse magnetization immediately following the RF pulse and T_2 is the spin-spin or transverse relaxation time. In general, $T_2 \leq T_1$. In the absence of paramagnetic materials, aqueous or semi-solid systems containing water with a homogeneous composition tend to exhibit ^{1}H T_2 values on the order of 10^{-1} s or longer. The longest ^{1}H T_2 reported for H_2O in Nafion–117 is 750 ms (15.9 wt%, 293 K) (MacMillan *et al.*, 1999); however, the T_2 values observed by Zawodzinski *et al.* (1991) were generally one-half the measured T_1 values, and the T_2 values reported by Wang *et al.* (2010) for water in the PEM of a prepared membrane electrode assembly (MEA) ranged from 20 ms (10.0 wt%; $\lambda = 6$) to 40 ms (19.1 wt%; $\lambda = 11.5$).

10.2.4 Spectroscopy

NMR is a very sensitive probe of local molecular and electronic environment. The dominant interactions for $I = \frac{1}{2}$ nuclei are the magnetic shielding interaction (which is typically scaled to the field independent, and referenced chemical shift (CS) scale) and the direct and indirect (i.e. *J*-coupling) dipolar interactions (Levitt, 2008). Exploiting these interactions, NMR spectroscopy has long been recognized as a powerful tool for identification of molecular species and elucidation of molecular structure (Ernst, 1992). The ^{1}H NMR CS of water in Nafion is related to the amount of water in the PEM; as the water content decreases, the ^{1}H CS of water increases, a result of interaction with the acidic sulfonyl groups (Bunce *et al.*, 1986). ^{2}H and ^{13}C NMR spectroscopy have proven useful for characterizing reaction intermediates in PEMs obtained from alcohol PEMFCs (see Section 10.12) (Paik *et al.*, 2008; Paik *et al.*, 2009).

10.2.5 Diffusion

In the presence of a magnetic field gradient, $M_{x,y}$ will dephase in a predicable manner, and can be refocused with an equal gradient of opposite sign. However, if the ensemble of spins is subject to diffusion, the two magnetic fields experienced by the spins will be of different magnitude, and an attenuation of the transverse magnetization will result. Stejskal and Tanner (1965) demonstrated that this principle could be applied to obtain the bulk rate of diffusion, D, of the molecular species giving rise to the NMR signal. In the absence of other relaxation mechanisms (i.e. T_2), the attenuated transverse magnetization due to unrestricted translational diffusion, as measured by a pulsed field gradient spin echo experiment, can be expressed by Eq. 10.5 (Kimmich, 1997; Callaghan, 1991):

$$M_{x,y}(t) = M_{x,y}(0)\exp\left[-\gamma^2 G^2 D\delta^2\left(\Delta - \frac{\delta}{3}\right)\right]$$

[10.5]

where G is the strength of the applied gradient, δ is the duration of the gradient pulse, and Δ is the time between application of the two gradient pulses. Table 10.2 provides a summary of the range of ^{1}H translation diffusion rates for pure $H_2O(l)$ and H_2O in acid form Nafion–117, as determined by NMR techniques at atmospheric pressure.

Table 10.2 Summary of ^{1}H translational diffusion rates for pure $H_2O(l)$ and H_2O in acid-form Nafion–117, as determined by NMR techniques at atmospheric pressure

	T (K)	Condition[‡] (wt%)	D (10^{-10} m^2 s^{-1})	Reference
$H_2O(l)$ Nafion–117	298		23.0 ± 0.2	(Harris and Woolf, 1980)
	303	3.3	0.6 ± 0.06	(Zawodzinski *et al.*, 1991; Zawodzinski, 1993a)
	303	5.0	1.2 ± 0.08	(Zawodzinski *et al.*, 1991; Zawodzinski, 1993a)
	288	6.6	0.83 ± 0.08	(Jayakody *et al.*, 2004)
Nafion–117	303	6.6	2.10 ± 0.06	(Zawodzinski *et al.*, 1991; Zawodzinski, 1993a)
	303	10.0	3.70 ± 0.15	(Zawodzinski *et al.*, 1991; Zawodzinski, 1993a)
	303	15.0	4.40 ± 0.07	(Zawodzinski *et al.*, 1991; Zawodzinski, 1993a)
	288	22.0	5.56	(Jayakody *et al.*, 2004)
	303	23.2	5.80 ± 0.21	(Zawodzinski *et al.*, 1991; Zawodzinski, 1993a)
	303	36.5	7.38[‡]	(Zawodzinski, 1993a)

† 1.66 wt % · λ = 1, where λ = nH_2O/SO_3H
‡ Value measured graphically from Fig. 5 (Zawodzinski, 1993a)

One must be aware that diffusive attenuation occurs whenever a field gradient exists. The effect is particularly noticeable in high-resolution MRI experiments during which long-duration and/or large field gradients are employed. Pulsed field gradient experiments can also be used to measure coherent motion such as flow, where rather than an attenuation of the signal, a phase-shift proportional to the velocity and gradient properties is introduced (Kimmich, 1997; Callaghan, 1991).

10.3   Introduction to magnetic resonance imaging (MRI)

MRI requires careful use of magnetic field gradients to encode spatial information into the NMR signal. This section introduces basic concepts that explain how an image is obtained from acquisition of a time-dependent NMR signal.

10.3.1 Magnetic field gradients

According to the Larmor equation (Eq. 10.2), the NMR frequency of a nucleus is related directly to the magnitude of the external applied magnetic field. The application of well-defined and controlled pulsed magnetic field gradients can make the frequency and phase of the NMR signal vary as a function of space. Figure 10.2a shows the relation between a magnetic field gradient, G_y, applied on to B_0, and the resulting well-defined spatial dependence of frequency. Note that the magnetic field created is in the z-direction, the subscript y denotes the axis along which the magnitude of the field varies. The corresponding change in resonance frequency, Δv, can be described as a function of the applied field gradient and therefore of position in space, y, by $\Delta v(y) = \gamma G_y y / 2\pi$. For example, Fig. 10.2b shows a sample experiencing a uniform magnetic field B_0, which results in a single peak at frequency v_0. In comparison, if a magnetic field gradient, G_y, is applied across the sample as shown in Fig. 10.2c, the total magnetic field experienced by nuclei varies with respect to their position within the applied gradient.

MRI affords the possibility to acquire an image from a plane of arbitrary thickness, a technique colloquially called slice selection. The slice selection process relies on a magnetic field gradient across the sample creating a spread in v_L. A slab or a slice of a sample can be selected for study by simultaneously applying an appropriately shaped band-selective RF pulse in combination with a magnetic field gradient (see Fig. 10.3a). The amplitude of the RF pulse used to generate the detectable NMR signal is often shaped in the time domain (e.g. sinc-modulated) such that the frequency profile of the B_1 field is approximately square. Multi-slice (MS) acquisitions are typically performed in which images from multiple slices can be acquired in one T_R, reducing total experimental time while maintaining a long repetition time for an individual slice. In practice, the frequency profiles of the RF pulses used for slice excitation are imperfect (i.e. are not rectangular) and for MS acquisition can lead to inadvertent perturbation of sample outside the intended image slice. To reduce this effect gaps can be left between

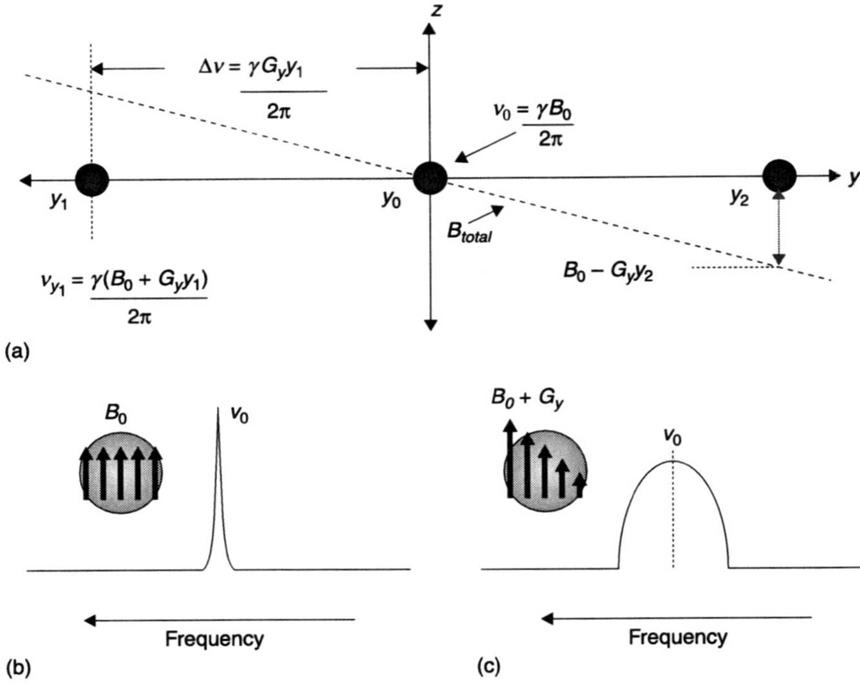

10.2 (a) A magnetic field gradient along the *y*-axis, G_y, modulates the magnitude of the total applied magnetic field along the *z*-axis, B_{total}. The change in magnetic field is related to the strength of the gradient and the distance from the origin of the axis along which the gradient is applied such that $B_{total} = B_0 + G_y y$. The resulting resonance frequency, v_y, can be related to position in space with respect to the gradient axis. (b) A sample placed in a uniform B_0 will exhibit a single peak with frequency centred at v_0. (c) Application of G_y across the sample introduces a spatial dependence of the frequency, and results in a peak shape representative of the projection of the sample profile.

neighbouring slices, and the slices can be interlaced to maximize the time elapsing between perturbation of adjacent slices.

10.3.2 NMR signal and **k**-space

The amplitude of the NMR signal is acquired as a function of time and is denoted $S(t)$. In MRI, the frequency of the signal from a volume of spins at position **r**, i.e., the spin density, $\rho(\mathbf{r})$, is a function of the position with respect to the magnetic field gradient vector, **G**. The signal integrated over volume has the form of a Fourier transform, FT, and assuming mixing with a reference frequency, can be expressed by Eq. 10.6 (Callaghan, 1991):

$$S(t) = \iiint \rho(\mathbf{r}) \exp[i\gamma \mathbf{G} \cdot \mathbf{r} t] d\mathbf{r} \qquad [10.6]$$

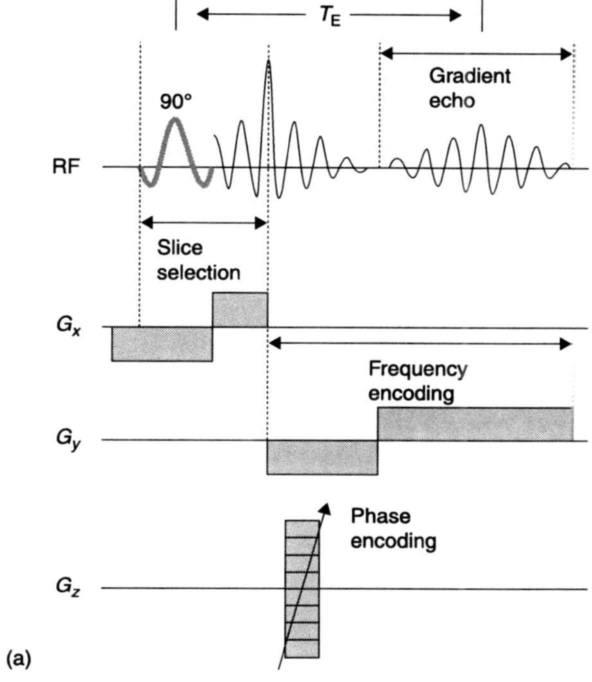

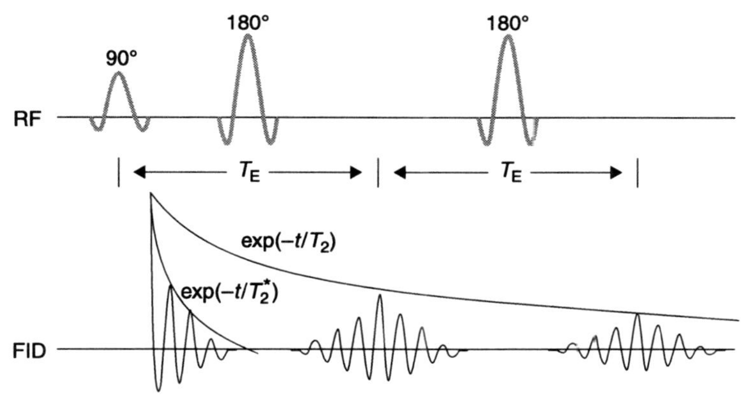

10.3 (a) A schematic of a 2D gradient echo imaging sequence with a slice-selective 90° RF pulse along the x-axis followed by frequency encoding along y, and phase encoding along z. The echo time, T_E, is the time between the RF pulse and the peak of the gradient echo. (b) A schematic of a basic spin-echo sequence. The decay of the FID occurs with characteristic time T_2^*, and the envelope formed by the peaks of the spin-echoes decays with characteristic time T_2. (c) A schematic of a 2D spin-echo imaging sequence with a multi-echo loop.

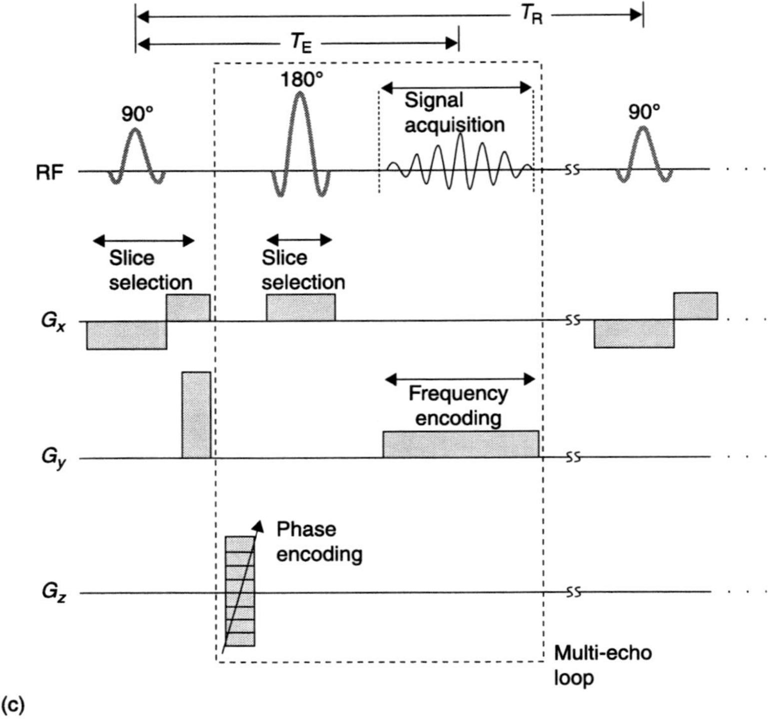

10.3 Continued.

Mansfield introduced the concept of a reciprocal space vector, **k**, of the form in Eq. 10.7 (Mansfield and Grannell, 1973; Mansfield and Grannell, 1975; Mansfield, 1976):

$$\mathbf{k} = \frac{\gamma \mathbf{G} t}{2\pi} \qquad [10.7]$$

which shows that **k**-space can be traversed by gradient magnitude or by time. Shown in Eqs 10.8 and 10.9 is the fundamental relationship of MRI, that the signal, $S(\mathbf{k})$, and $\rho(\mathbf{r})$ are a FT pair:

$$S(\mathbf{k}) = \iiint \rho(\mathbf{r})\exp[i2\pi\mathbf{k}\cdot\mathbf{r}]d\mathbf{r} \qquad [10.8]$$

$$\rho(\mathbf{r}) = \iiint S(\mathbf{k})\exp[-i2\pi\mathbf{k}\cdot\mathbf{r}]d\mathbf{k} \qquad [10.9]$$

Subjecting **k**-space to a multi-dimensional FT converts the data into an image, i.e., a map of signal intensity as a function of frequency, and therefore physical space.

10.3.3 1D imaging: frequency-encoding

As discussed above, a frequency-dependent spatial variation of an NMR signal can be imposed by applying a magnetic field gradient across a sample. This method of spatial encoding is typically called frequency encoding (FE). As shown in Fig. 10.3a, to obtain a one-dimensional (1D) image (i.e. a profile), the data points are sampled as the FID evolves under the influence of a gradient of fixed strength, G_y, applied across the sample (i.e. fix **G**, and vary t in Eq. 10.7). Equation 10.10 shows the relation between the field-of-view (FOV) and **k**-space:

$$FOV_{FE} = \frac{1}{\Delta k_y} = \frac{2\pi}{\gamma G_y \Delta t} \qquad [10.10]$$

The time between data points, Δt, is the reciprocal of the receiver bandwidth (BW), $\Delta t = 1/BW$. Following the RF pulse, successive magnetic field gradients of equal magnitude and opposite sign are applied to move across **k**-space and to create an echo at the middle of the acquisition period; hence, this signal is referred to as the gradient echo (GE). The period of time from application of the RF pulse to the peak of the echo is called the echo-time, T_E. In this manner, a line of **k**-space is traversed for each FID.

10.3.4 2D imaging: phase-encoding

To obtain the second (2D) (or third (3D)) dimension of an image, the spatial information is encoded prior to acquisition of the signal (see Fig. 10.3a). In this case, **k**-space is traversed by incrementing **G** over a short, fixed time period, τ, as described by Eq. 10.11:

$$FOV_{PE} = \frac{1}{\Delta k_z} = \frac{2\pi}{\gamma \Delta G_z \tau} \qquad [10.11]$$

During application of this magnetic field gradient, the phase evolution of all spins due to field heterogeneity is constant except for the phase-shift introduced by the incremented gradient. Hence, this method of spatial encoding is often called phase encoding (PE).

10.3.5 Gradient echo imaging

A GE imaging sequence combines the use of an RF pulse to tip the magnetization into the transverse plane, and both a PE gradient and a FE gradient to traverse **k**-space. The sequence is repeated for incremented values of the PE gradient to traverse the second and third dimensions of **k**-space. GE sequences are well suited to homogenous systems. If the local magnetic field is not homogeneous (e.g. susceptibility differences) the characteristic decay time-constant of the FID (i.e.

T_2^*) will be significantly less than T_2 (see Fig. 10.3b). Thus, GE experiments have limited application to the study of PEMFCs.

10.3.6 Spin echo imaging

In the presence of static magnetic field inhomogeneity, spin echo (SE) based imaging techniques are often used. After the initial RF pulse to tip the magnetization into the transverse plane, a RF refocusing pulse can be applied to flip $M_{x,y}$ in the transverse plane (i.e. rotate by 180°). This rotation of the magnetization reverses the direction of inhomogeneous dephasing, and results in the formation of a SE. A train of appropriate RF refocusing pulses will yield a series of SEs, the decay envelope of which can be described by a mono exponential function of T_2 (Fig. 10.3b). This method, referred to as the CPMG experiment, after Carr and Purcell (1954), and Meiboom and Gill (1958), is often used to measure T_2. In general, an SE imaging experiment incorporates a GE for FE, and the GE and SE are timed to coincide. Shown in Fig. 10.3c is a schematic of a slice-selective SE imaging sequence. The train of echoes (multi-echo loop) can be summed into a single FID to increase the signal-to-noise ratio (SNR), each echo can be used to traverse an additional line of k-space, or each echo can be used for a separate image to provide a series of T_2-weighted images (see Section 10.7). Typically, multi-slicing is performed in combination with SE imaging, yielding the acronym MSSE.

10.3.7 Pure phase encode or single-point imaging

Single-point imaging is a general term to describe a class of experiments by which all directions in k-space are traversed via PE, one point at a time (in contrast to FE experiments that traverse an entire line of k-space for each FID or echo). As one may expect, these experiments can require a comparatively long time to complete. However, as a single point is acquired at the top of each echo, SPI techniques do not suffer from resolution limitations or artefacts due to static field inhomogeneity which cause rapid decay of the FID or due to frequency offset effects (see Section 10.5). Zhang *et al.* (2008) have had excellent success with achieving high resolution (i.e. 6 μm) across the PEM of an operating FC with an SPI-based technique (see Section 10.9.1)

10.3.8 Image contrast

The contrast of an MR image is related to numerous factors including the volume and spin-density of the sample, and various relaxation mechanisms. To obtain an image with contrast that represents only one of the factors, the data must be collected in a manner to isolate the contribution, or corrected with knowledge of the other parameters. For example, T_1-weighted contrast can be obtained by acquiring images with different T_R values ($< 5T_1$) which results in different amounts of magnetization saturation. To obtain an image with T_2-weighted contrast only, T_1 weighting is removed by adjusting the T_R value to prevent saturation of the magnetization (i.e. $T_R > 5T_1$). Images with contrast related to T_2 can be obtained by acquiring images at a series of different echo times (see Fig. 10.3b). This is generally facilitated using multi-echo experiments. In most MRI investigations of water in FCs to date, the images have a combination of T_1 and T_2 weighting. As covered in Section 10.7 there are corrections that can be applied to the data to obtain image contrast that represents the true spin-density.

10.4   NMR and MRI hardware

This section outlines and briefly discusses the standard hardware required to perform modern high-field (i.e. > 1.5 T) MRI experiments for the study of operating PEMFCs. The major components of an MRI system include: superconducting magnet, console/spectrometer, magnetic field gradient unit, gradient current supplies, RF amplifier, and RF coils. An example hardware setup for MRI of PEMFCs, in addition to components for monitoring and controlling the FC is shown in Fig. 10.4.

10.4.1 Superconducting magnet

Commercial superconducting magnets are available in applied fields up to 23.5 T (1000 MHz). Imaging hardware is commonly fitted to vertical-bore magnets with

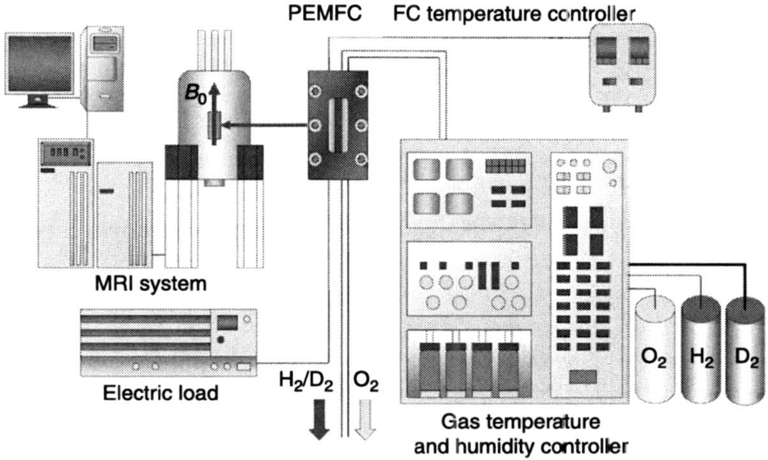

10.4 Schematic of a hardware setup for MRI studies of an operating PEMFC. (Source: Adapted from Fig. 2 in Kotaka *et al.* (2007) by permission of ECS – The Electrochemical Society.)

a room-temperature (RT) bore diameter of 89 mm, ranging in field from 7 T to 21.1 T. Some super-wide-bore magnets with 154 mm bore diameter are also available. Horizontal-bore magnets are better suited for imaging large objects, typically designed for small animal applications, in fields up to 16.4 T (RT bore = 260 mm). PEMFC studies have been performed at a range of fields including 2.35 T (Bedet *et al.*, 2008b), 2.4 T (Zhang *et al.*, 2008), 7.05 T (Feindel *et al.*, 2004; Teranishi *et al.*, 2002), 9.4 T (Shim *et al.*, 2009), 11.7 T (Minard *et al.*, 2006), and 14.1 T (Dunbar and Masel, 2007).

10.4.2 Magnetic field gradient unit

The image resolution attainable in MRI experiments is linked directly to the ability to create a spatially varying magnetic field such that nuclei at different locations resonate at different NMR frequencies. To generate the uniform, spatially-dependent magnetic fields a set of three actively shielded (Mansfield and Chapman, 1986) gradient coils surround the RF coil assembly. The magnetic field gradients either add to, or subtract from, B_0, and the gradient units are designed such that the pulsed magnetic field increases with distance from the centre of the device. The magnetic fields are traditionally generated along the z-axis (G_z) using a Maxwell pair (Tanner, 1965), and in the mutually orthogonal x- and y-directions (G_x, G_y) using concentric saddle coils (Hoult and Richards, 1975; Moore and Holland, 1980).

The magnetic field gradient unit nests within the RT bore of the magnet. The ID of gradient units range from 40 mm (typical for vertical bore magnets with 89 mm ID RT bore) to 260 mm (for horizontal bore magnets with 400 mm ID RT bore), with maximum available pulsed field strengths of 1.5 T m^{-1} and 0.23 T m^{-1} respectively. A variety of gradient units have been used for MRI studies of PEMFCs, including 1.0 T m^{-1} units (40 mm ID) (Feindel *et al.*, 2004; Minard *et al.*, 2006), 0.5 T m^{-1} (75 mm ID) (Zhang *et al.*, 2008), and 0.18 T m^{-1} (120 mm ID) (Bedet *et al.*, 2008a).

10.4.3 Spectrometer/console

The spectrometer or console refers to the instrumentation components that interpret instructions from the software and determines and controls the RF pulses, field gradients, signal acquisition, etc. Many commercially available software packages for materials imaging were derived from those for medical applications. The packages generally have very convenient routines for defining image geometry with respect to reference or localizer images. Virtually all commercial software packages include basic 2D and 3D GE and SE imaging sequences. As hardware and software become more advanced, specialized techniques for materials imaging are also emerging in standard commercial packages (e.g. ultra-short TE (Robson *et al.*, 2003), and sweep imaging with Fourier transformation (SWIFT) (Idiyatullin *et al.*, 2006)).

10.4.4 Radio frequency (RF) coil

An RF coil is the component designed to perturb the magnetization vector and to detect the NMR signal in the transverse plane (which induces a voltage in the coil). The pulsed RF field applied to generate the B_1 magnetic field transverse to B_0, is generally desired to be homogeneous over the sample volume. To achieve greater RF uniformity over larger volumes typically required for imaging, several designs have been introduced, such as the birdcage coil (Hayes *et al.*, 1985) and the Litzcage (Doty *et al.*, 2007). In addition, RF coils can be driven in quadrature which affords a $2^{\frac{1}{2}}$ increase in SNR and a reduced power deposition. Also, as discussed in Section 10.6.3, application-specific RF coil designs are feasible.

10.5   MRI technical considerations

In this section, some of the technical considerations of MRI experiments with respect to studies of PEMFCs are covered.

There are many factors that may influence the quality of images obtained in a MRI experiment, related to both the hardware used for the experiment and the materials contained in the object being imaged. A variety of artefacts can be imparted in MR images due to material properties, imaging techniques, RF interference, motion, etc. In this section technical considerations with respect to some of these factors are discussed.

10.5.1 Signal to noise ratio

To increase the signal to noise ratio (SNR) of an NMR experiment, one can either average the signal (SNR $\propto$ [number of FIDs acquired]$^{1/2}$), use a larger sample, or use a stronger applied magnetic field (in practice SNR $\propto B_0$(Hoult and Richards, 1976)). In addition, the SNR is related to the 'filling factor' of the RF coil, the volume from which signal is acquired relative to the volume of the RF coil. Each pixel in a two-dimensional image slice is a projection of the signal contained within the volume defined by the width and height of the pixel and the slice thickness. This three-dimensional volume element is called a voxel. As described by Eq. 10.12:

$$\text{SNR} \propto \frac{V_{\text{voxel}}}{V_{\text{coil}}} \qquad [10.12]$$

where V_{voxel} is the volume of a voxel, and V_{coil} is the volume of the RF coil. As a consequence, standard RF coils can present some limitations for the study of PEMFCs, in which the volume of interest is small relative to the volume of the coil. Zhang *et al.* (2008) have shown that an RF coil integrated into a PEMFC (in combination with an appropriate imaging sequence) can provide high resolution and high SNR without simply resorting to high B_0 (see Sections 10.6.3 and 10.9.1).

10.5.2 RF interference

At room temperature, the population difference of the eigenstates (determined by Boltzmann statistics) is very small (i.e. in the order of 1 in 10^5), which results in an inherently low amplitude NMR signal. As a result, extraneous signals detected by the RF coil can cause a significant decrease in the SNR. For MRI studies performed at $B_0 \sim 2.35$ T (with $v_L(^1H) \sim 100$ MHz), interference from radio stations can be particularly problematic. Often these instruments are housed in RF shielded rooms (e.g. covered in Cu foil to block high frequency interference). RF interference can also result in stripes in images called zipper artefacts (Mitchell and Cohen, 2004), which commonly occur along the PE axis at the point of zero frequency (see Fig. 10.5e and f), but may also occur elsewhere. RF interference is particularly problematic for studies of operating PEMFCs; cables used to monitor the power output of the PEMFC to correlate with the results of the MRI studies can introduce noise directly into the RF coil. To reduce or eliminate RF interference, in several MRI studies on operating PEMFCs, the researchers opted to place the FC load within the bore of the magnet and not monitor the current and voltage (Zhang *et al.*, 2008; Bedet *et al.*, 2008b).

10.5.3 Motion

In operating PEMFCs, one must also consider the dynamic behaviour of water, particularly in the flow fields. Water is generated at the cathode and may be transported into the PEM, collect in the gas diffusion layer (GDL), accumulate in the flow channels, and subsequently be removed by the gas flow. MR images are obtained from a finite volume of samples over a finite period of time. In general, MRI experiments are in the order of seconds or minutes, and thus the image is an average representation. Sample motion can introduce blurring due to misregistration of position. In particular, periodic motion along the PE direction can lead to superposition of images, termed Nyquist ghosting. As mentioned in Section 10.2.5, velocity or flow can be encoded into the MRI signal; however, dynamic changes in the volume of water could result in an error in the calculated velocity.

10.5.4 Magnetic susceptibility

The extent to which magnetization can be induced in a material by a large external magnetic field is termed magnetic susceptibility, χ. The magnetic susceptibility of materials can vary widely; PEMFCs are heterogeneous structures, and contain materials with magnetic susceptibilities different from that of water. Such differences introduce inhomogeneity in the local magnetic field, proportional to both the difference in magnetic susceptibility and the external applied magnetic field, and can lead to image distortion due to the resulting shifts in the local Larmor frequency (Blümich, 2000) and a rapid relaxation of transverse magnetization. Off resonance water can contribute to a variety of image artefacts, including intensity

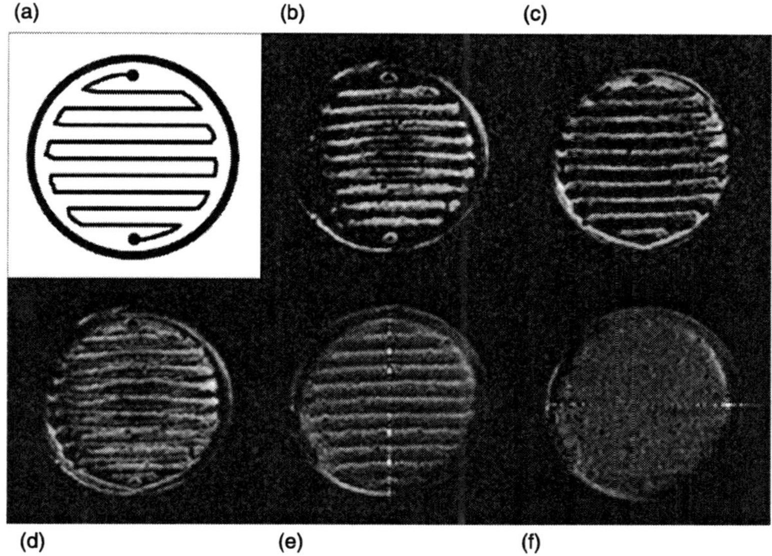

10.5 Optimization of imaging parameters and characterization of static field heterogeneity: (a) Schematic showing the pattern of milled channels in one of the gas manifolds. When sandwiched between two gas manifolds the Nafion–117 is located either adjacent to air (black) or PEEK (white) and differences in the magnetic susceptibility shift the NMR frequency of water depending on its location. (b) MR image of water in a 500 Hz band near the frequency anticipated when molecules are sandwiched between apposing air interfaces. (c) MR image of water in a 500 Hz band near the frequency anticipated when molecules are sandwiched between apposing PEEK interfaces. (d) Overlay of (b) and (c). (e) MR image acquired when the bandwidth of applied RF is greater than the NMR frequency range spanned water in the PEM, with frequency encoding applied in the vertical direction. (f) Same as (e) except frequency encoding applied in the horizontal direction. (Source: Fig. 3 in Minard *et al.* (2006) with permission from Elsevier.)

variations caused by excitation and refocusing errors (Zweckstetter and Holak, 1998), poor slice localization (Lei and Dunn, 2001), and spatial misregistration along the frequency-encoded axis (Callaghan, 1991). Understanding the origin of artefacts introduced into MR images by conductors and susceptibility differences is of interest for both materials and medical imaging (Lauer *et al.*, 2005; Ogawa *et al.*, 1990). Except for simple geometries such as a sphere or cylinder, the distortion imparted on a magnetic field by susceptibility effects must be solved numerically (Mitchell and Cohen, 2004). Artefacts can often be reduced by using a SE technique with a short T_E or a pure PE technique; however, if the field distortions become very large, susceptibility inhomogeneities can result in complete loss of signal.

To optimize image quality Minard *et al.* (2006) used a 2D MSSE experiment with a frequency-selective RF excitation pulse to evaluate the field heterogeneity in the plane of a hydrated PEM. The NMR signal from water in the PEM was found to span ~4500 Hz or ~9.0 ppm at 11.7 T (i.e. 4500 Hz/500 MHz). For the experiment parameters, the differences in the NMR frequency of water were predicted to cause a misregistration of ~6 pixels (400 μm) in the FE direction. As shown in Fig. 10.5, the light and dark regions of image intensity (Fig. 10.5b and c) that mimic the pattern of the flow field (Fig. 10.5a) resulted because the frequencies of water adjacent to air and adjacent to the FC housing result in a spatial shift relative to each other. Similar imaging artefacts were also recognized by Bedet *et al.* (2008a; 2008b). To visualize all water inside the PEMFC an RF pulse with a BW of 10 kHz was used to minimize errors in slice localization and artefacts caused by off-resonant RF pulses (Fig. 10.5e), and misregistration artefacts were minimized by placing the FE axis of the image parallel with the gas channels (Fig. 10.5f) (Minard *et al.*, 2006).

10.5.5 RF attenuation

Several groups have noted the effects of RF attenuation due to conductive materials in PEMFC components (Feindel *et al.*, 2006a; Minard *et al.*, 2006; Bedet *et al.*, 2008a). The GDLs are one component in particular that generally remain unmodified in PEMFCs for MRI studies (see Section 10.6.2). High-frequency electromagnetic radiation, such as RF waves in the MHz range, is effectively attenuated by conductors (Cheng, 1989). Feindel *et al.* (2006a) noted that the through-plane conductivity of the TGP-H-060 carbon paper used in their experiments was 1250 S m^{-1} (Barbir, 2005) which, for a thickness of 190 μm, will attenuate a 300 MHz electromagnetic plane-wave by approximately one-fifth. They noted that the SNR was related to the orientation of the FC within the RF coil. A quadrature-driven birdcage resonator generates a circularly polarized B_1 field, which can be represented by two plane-polarized fields oscillating perpendicular to each other and to B_0. Thus, the minimum RF attenuation is expected to occur with the plane of the PEM aligned parallel with the B_1 field. Minard *et al.* (2006) constructed their FC such that the PEM was perpendicular to B_0. With this design, significant screening of the RF field by the gold-coated manifolds was prevented and the SNR would be independent of the the rotational alignment of the FC within the RF coil.

Bedet *et al.* (2008b) investigated the potential of imaging FCs with RF field gradient experiments (Canet, 1997). Such experiments are useful for investigating samples with internal field gradients, which decrease the local homogeneity of B_0, and for samples with very short T_2 values that could not be imaged with the more common B_0 field gradient approach (Maffei *et al.*, 1992). The presence of the GDLs was found to significantly affect the RF pulses preventing adequate images from being obtained. Several different materials were investigated as replacements for the GDLs, including copper, brass, and stainless steel meshes, but these materials were found to result in larger artefacts. The presence of

GDLs and other conductors precludes the utility of RF field gradient experiments for imaging FCs.

10.5.6 Receiver bandwidth

The size of the receiver BW used for signal acquisition affects the rate of data acquisition (i.e. BW = $1/\Delta t$) and the frequency range of each image pixel (e.g. 100 kHz/128 pixels ~780 Hz/pixel). According to Eq. 10.10, assuming a constant FOV, a larger receiver BW will require a stronger FE gradient due to the decreased time between sampled points. A larger BW facilitates faster data acquisition and reduces artefacts due to an increase in Δv_L (i.e. each pixel has a larger frequency range), magnetic field inhomogeneity, and motion, but also results in lower SNR. A smaller BW increases the time required to sample each echo, which significantly increases the echo time, T_E, and may introduce unwanted T_2 or T_2^* contrast. In practice, the signal strength is proportional to B_0, whereas the SNR increases only as the square root of BW reduction; therefore, one may effectively compensate for the increased severity of Δv_L artefacts at higher fields with an increased BW and still achieve greater signal overall. However, with a very large FE gradient, significant diffusive attenuation of the signal may occur (i.e. G^2 contribution in Eq. 10.5).

10.5.7 Physical resolution limitations

For experiments using FE, the achievable spatial resolution, $1/\Delta y$, as described by Eq. 10.13, is limited by the spread in observed frequencies or linewidth at half-height of the NMR peak, $v_{1/2}$, and the strength of the applied gradient, G_y (Blümich, 2000):

$$\frac{1}{|\Delta y|} = \left| \frac{\gamma G_y}{2\pi v_{1/2}} \right| \qquad [10.13]$$

In practice, the maximum strength of the applied gradient is limited (e.g. the maximum commercially available gradients for NMR microscopy are approaching 3 T m^{-1}). The linewidth is related to the time required for the FID to decay (i.e. T_2^*, see Fig. 10.3b); as a result of inhomogeneity in the magnetic field experienced, $v_{1/2}$ is related to the shorter apparent spin-spin relaxation time, T_2^*, such that $v_{1/2} = (\pi T_2^*)^{1/2}$. In an operating PEMFC the ^{1}H T_2^* is dependent on temperature, the water content of the PEM, and the homogeneity of the magnetic field. Thus, T_2^* will vary depending upon the conditions under which the PEMFC is operated, the materials used, and the subsequent ability to maintain a homogeneous B_0. In addition, if the total signal readout is long compared to T_2^*, the ends of the FE axes in k-space contain mostly noise which results in image blurring. To obtain $\Delta y \approx 10$ μm with a typical NMR microscopy gradient strength of 1.0 T m^{-1} a linewidth of less than $v_{1/2} = 425$ Hz would be required, and therefore T_2^* would have to be 7.4 ms or longer. For systems with short T_2 values, small diffusion constants, or, if weak gradients are used, relaxation broadening, will limit the achievable physical resolution (Callaghan, 1991). Also, as noted by Dunbar and Masel (2007), partial volume averaging within each voxel can introduce ambiguity in the MR images for the location of the interface of the PEM, catalyst layer, GDL, and flow field.

10.6   Adaptation of polymer electrolyte membrane fuel cell (PEMFC) design and materials

Many of the PEMFCs for MRI studies were designed to investigate achievable imaging performance and assess limitations imposed by FC materials. As a result, designs were neither intended to be representative of a commercial device nor were performance characteristics rigorously optimized. Instead, data was sought to examine the feasibility of exploiting MRI for the diagnostics of commercial devices. Numerous designs of MRI compatible PEMFCs can be found in the literature, several of which are outlined below. Further details can be found in the respective original literature.

10.6.1 Size restrictions

The first step in designing an MRI-compatible PEMFC is determining the size restrictions imposed by the MRI hardware to be used. The operating FC must be able to reside within the RF coil. Commercial RF coils for imaging typically have cylindrical sample spaces, and are sometimes accessible only from one end. MRI-compatible PEMFCs of various sizes have been reported in the literature for use in RF coils of 10 mm ID (Feindel *et al.*, 2006a), 30 mm ID (Feindel *et al.*, 2004; Feindel *et al.*, 2006a; Dunbar and Masel, 2007; Minard *et al.*, 2006), and 40 mm ID (Shim *et al.*, 2009).

10.6.2 Material selection

A fundamental requirement for the materials used in construction of PEMFCs for MRI investigations is that they be non-magnetic. Ferromagnetic materials must be avoided (anything containing iron, nickel, or cobalt). Commercial FCs are generally constructed from conductive materials, such as graphite or stainless steel, into which the gas flow channels are machined. In addition to the attenuation of RFs in the MHz region, pulsed RF or magnetic fields can induce eddy currents (disrupting the local magnetic field) or large voltages (10^2 V) in conductors. Therefore only those conductive materials essential to the operation of the FC are typically retained (i.e. GDL/current collector) and are used in small quantities.

As discussed below, for MRI studies PEMFCs have been machined from acrylic, polyoxymethylene (POM, Delrin®), polyetheretherketone (PEEK), polymethylmethacrylate (PMMA), polytetrafluoroethylene (PTFE, Teflon®), and fibreglass-supported printed circuit (PC) board. A variety of PEMs have been used in preparation of MEAs including Nafion (115, 117, 1110), Aciplex (S–1112, S–1104, S–1102), and hydrocarbon. The geometric area of the MEA catalyst layers has ranged from ~0.5 cm^2 (0.8 cm diameter) (Feindel *et al.*, 2004) to ~10 cm^2 (3.2 cm × 3.2 cm) (Zhang *et al.*, 2008). In addition, designs have been employed that align with the MEA either parallel with, or perpendicular to, B_0.

Shown in Fig. 10.6 are several of the PEMFCs that have been designed for use in MRI investigations. The FC design of Tsushima and Teranishi has varied, but

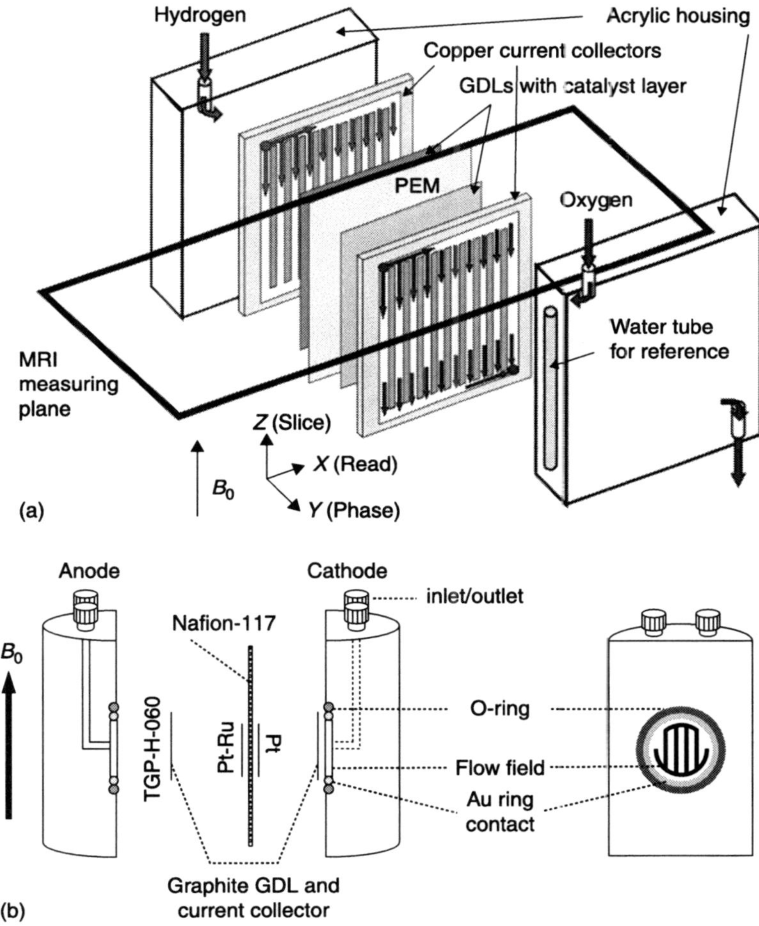

10.6 Examples of PEMFCs designed for MRI studies: (a) (Source: Fig. 1 in Tsushima *et al.* (2007a) with permission from Springer). (b) (Source: adapted from Fig. 1 in Feindel *et al.* (2007a).) (c) (Source: Fig. 1 in Dunbar and Masel (2008a) with permission from Elsevier). (d) (Source: adapted from Fig. 2 in Bedet *et al.* (2008a) with permission from Elsevier.)

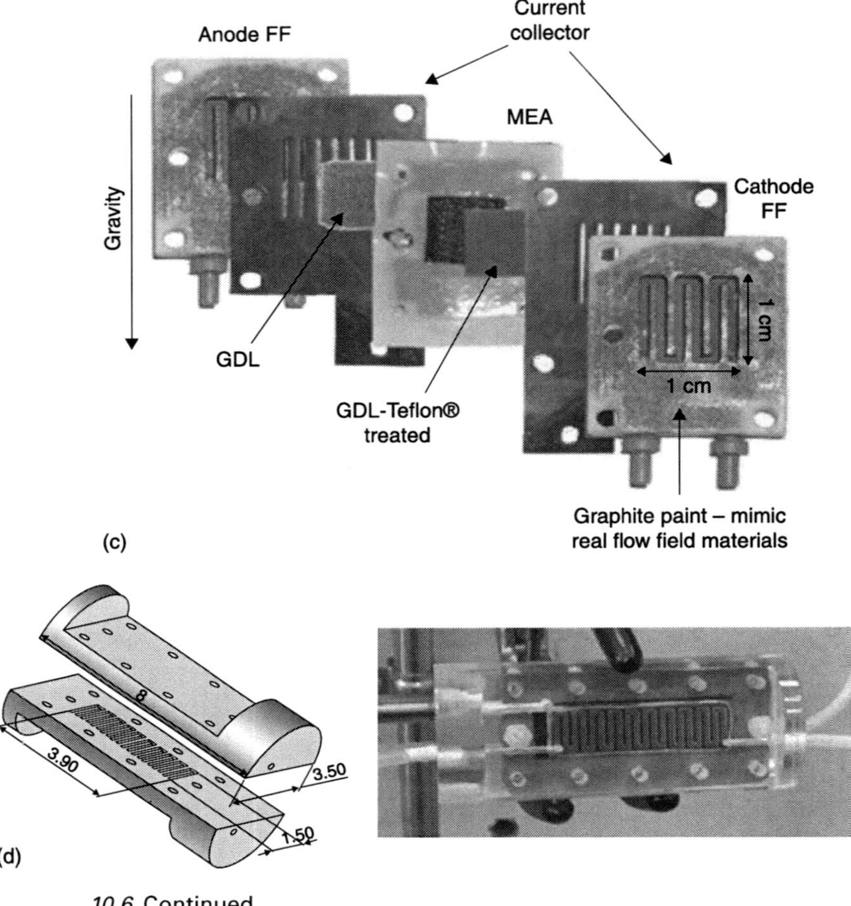

10.6 Continued.

the majority of the essential components remain consistent (Fig. 10.6a). Acrylic was used for the FC housing, and the MEA aligned parallel with B_0, was composed of a PEM (Aciplex, Nafion, or hydrocarbon) sandwiched by GDLs with dispersed Pt particles. An early design used a Cu current collector in combination with Au mesh (Teranishi *et al.*, 2002); however, Cu is subject to acidic degradation by the PEM, and latter designs have used an Au-coated Cu current collector (Tsushima *et al.*, 2004).

The research reported by Feindel and coworkers (Feindel *et al.*, 2004; Feindel *et al.*, 2006a) also used a design in which the planes of the MEA and gas flow fields were oriented parallel with B_0. A schematic of the design of the PEMFC is shown in Fig. 10.6b. Non-conductive POM was selected to machine a cylindrical

FC housing (28 mm OD) for its hard, durable, and acid resistant properties. The MEA was produced using a modified decal transfer method, with Nafion–117 as the PEM, and PtRu/C and Pt/C as the anode and cathode catalysts, respectively. The diameter of each catalyst decal was ~0.8 cm yielding an area of ~0.5 cm^2. The face of each flow field was pressed against a GDL to sandwich the MEA. A Simriz (copolymer of TFE and perfluorovinyl ether) o-ring was seated around each flow field to create a gas-tight seal against the PEM when the two halves of the FC assembly were screwed together. Au wire was used as the electrical circuit contact to form an unclosed ring around the flow field. Shielded cables were used to monitor the operating PEMFC. The authors note that several cables tested were found to be slightly magnetic.

Minard *et al.* (2006) constructed a PEMFC with a 25.4 mm OD cylindrical housing from PEEK with the MEA oriented perpendicular to B_0. The PEEK gas manifolds were Au-coated to act as the current collector. A commercial triple-layer MEA using Nafion–117 and 0.3 mg Pt cm^{-2} (geometric area of catalyst decal was 1.7 cm^2) was sandwiched between wet-proofed carbon cloth GDLs.

The PEMFC used by Dunbar and Masel (2007) (Fig. 10.6c) was machined from PTFE, with 1 mm wide and 3 mm deep serpentine flow channel flow fields (either uncoated or with a thin graphite coating painted onto the flow channels to simulate the hydrophobicity of a commercial solid graphite flow field (Dunbar and Masel, 2008a)) adjacent to current collectors fabricated from Au. The MEA was composed of Nafion–115 and catalyst inks consisting of Pt on carbon and Nafion solution were applied using a direct paint technique, with a geometric area of 1 cm^2. The GDLs were Teflon-treated carbon cloth for the cathode and plain carbon cloth for the anode. Water-filled capillaries were attached to the FC flow fields to act as calibration standards for analyzing the water signal. The current and voltage were measured using multimeters and controlled via a variable resistor.

Transparent PMMA was used by Bedet *et al.* (2008a; 2008b) to construct a PEMFC with 1 mm wide and 1 mm deep machined gas flow channels (Fig. 10.6d). A commercial Paxitech MEA composed of Nafion–115 and low platinum density electrodes with a geometric area of 6 cm^2 (4 cm × 1.5 cm) was sandwiched by GDLs in contact with current collectors constructed from a ring of 0.5 mm diameter Au wire.

10.6.3 Integrated PEMFC and RF coil

To achieve high resolution profiles of water content across the PEM of an operating PEMFC, Zhang *et al.* (2008) developed a PEMFC integrated with the RF circuit of a parallel plate resonator. Two G10 fibreglass substrates were machined to create flow channels and supported gold plated PC boards that were used as the electrodes of the FC and the resonating plates of the RF probe. The MEA was composed of two catalyst-sprayed GDLs (3.2 cm × 3.2 cm) sandwiching

Nafion–1110 (3.5 cm × 3.5 cm). The design of the RF probe is specific to investigating the water content distribution across the PEM; the B_1 fields within the gas channels are very small and will not effectively elicit an NMR response for water or vapour in the channels. A pure phase encoding MRI technique designed for thin-film depth imaging (Ouriadov *et al.*, 2004) was used; the double half **k**-space (DHK) spin echo (SE) single point imaging (SPI) technique yields images free from distortions due to B_0 inhomogeneity, susceptibility variations, and chemical shift.

10.6.4 Operating temperature

Most MRI investigations of operating PEMFCs have been conducted with unregulated FC temperature. In general, the area of the MEA is small (e.g. several cm^2) and minimal heat is expected to be generated. The gradient units for imaging are often water-cooled, and the temperature of the sample within the RF coil is assumed to be close to the temperature of the gradient unit (e.g. 20 °C) (Feindel *et al.*, 2006b). Tsushima *et al.* (2006a) incorporated a thermocouple on the current collector, and subsequently developed an MRI compatible FC that utilizes glass-plate heaters (Kotaka *et al.*, 2007; Shim *et al.*, 2009).

10.7   Quantification of water content

Quantifying the water content in an operating PEMFC requires careful consideration of influences on the ^{1}H NMR signal. Investigations in the literature tend to report MRI signal intensities in arbitrary units and assume that the signal is a good representation of water content, or convert the signal intensities to values of λ; however, few studies provide adequate (if any) experimental details on how the values of λ were determined. Inclusion of a reference tube in the FC housing with a known amount of $H_2O(l)$ (Dunbar and Masel, 2007; Tsushima *et al.*, 2006a) seems suitable for quantifying $H_2O(l)$ in the flow channels; however, quantifying the water in the PEM is less trivial.

In theory, as shown by Eq. 10.8 the signal intensity of a ^{1}H MRI image is directly proportional to the ^{1}H spin-density, which in the operating PEMFC arises almost exclusively from water. Correlation between ^{1}H MRI signal intensity and power density have been shown (see Section 10.9); however, the T_2 value of water in a PEM is well known to vary with water content, which invalidates the assumption that the MRI signal intensity is representative of the true water content (i.e. different T_2 values result in different image intensity). Regardless, qualitative correlations between image intensity and water content appear valid as long as the caveats are recognized. Recently, several reports have detailed how image intensity can be corrected to obtain true water content in the PEM (Wang *et al.*, 2010; Zhang *et al.*, 2008).

10.7.1 Internal calibration standards

Dunbar and Masel (2007) attached water-filled capillaries to each of the principle axes of their FC to act as calibration standards to quantify the amount of water. The MRI signal from water in the capillaries was compared to that from the water-filled flow fields of the PEMFC. As the signal intensities were within experimental error of each other, the method was deemed sufficient for quantification. Average signal intensity in the reference capillary was calculated and assumed to be the intensity for the known volume of pure water. Similarly, the average intensity was determined for each MR image and the mass of water calculated. Using this method the authors reported the concentration for water (g cm^{-3}) for both the MEA and flow field regions (see Section 10.10).

10.7.2 MRI signal for known values of λ

One approach to relate the MRI signal intensity to the PEM water content is to acquire a set of images from PEMs with known values of λ. The mass of water in a PEM can be determined through fastidious weight measurements relative to a dry PEM, and in conjunction with the equivalent weight of the PEM, the value of λ can be determined (Zawodzinski *et al.*, 1993a). PEMs are known to exhibit different water uptake behaviour depending on the relative humidity (RH) and temperature to which it is exposed. Thus, by exposing the PEM to a known temperature and RH, a relation could be established between MRI signal intensity and λ (Tsushima *et al.*, 2009). The reliability of such calibrations would be related to the reproducibility of the MEA preparation and assembly of the PEMFC (e.g. catalyst loading, position of GDL, clamping pressure of the FC assembly). In addition to the effects of NMR relaxation, the integrated MRI signal is related to experimental parameters such as slice thickness and orientation, pixel size, the number of signal averages, receiver gain, etc.; therefore, in practice a calibration would have to be performed for each different set of MRI parameters.

10.7.3 Correction of MRI signal for T_2 relaxation

Early research using MRI to investigate operating PEMFCs focused on qualitative explorations of water distribution and the effects of operating conditions. The success of these early studies is demonstrated by the desire and recent efforts to pursue high resolution and quantitative measurements of water in operating PEMFCs (Wang *et al.*, 2010; Zhang *et al.*, 2008). The importance of considering the effects of T_2 relaxation on the measured MRI signal and the relation to PEM water content was nicely demonstrated by Zhang *et al.* (2008). With a known value of T_2, the MRI signal intensity can be corrected to yield true water content maps. For each experiment, multiple echoes were acquired to obtain spatially resolved T_2 maps. The water content profiles measured as a function of T_E (i.e.

T_2-weighted) were fit to determine the true water content profile on a point-by-point basis via the relation shown in Eq. 10.14:

$$S(y) = \rho_0(y) \cdot \exp\left(-\frac{nT_E}{T_2(y)}\right)$$

[10.14]

where $S(y)$ is the MRI signal intensity from the spin density at location y, $\rho_0(y)$. Shown in Fig. 10.7 is an example of a T_2 fitting and a T_2 map across the PEM

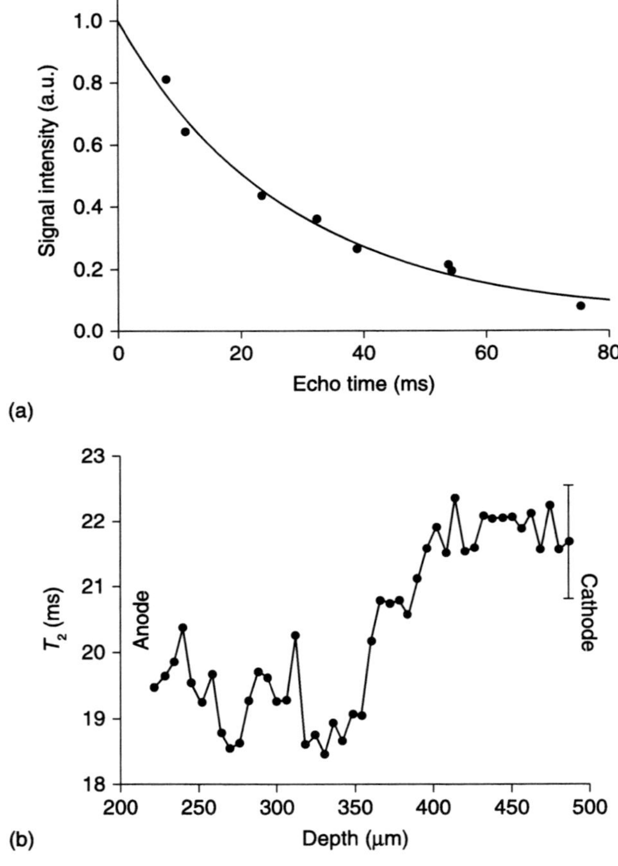

10.7 (a) Representative T_2 fitting of data obtained with an integrated PEMFC and RF coil using the double half k-space spin-echo single point imaging technique. The data of the pixel at 400 µm depth, and 1 h experiment time is shown. The single exponential decay is clear. The T_2 is ~27 ± 2 ms. (b) Composite T_2 map averaged over a 6 h time window. The T_2 values do not vary widely, but those close to the anode have a lower value than those close to the cathode. The standard error of the mean is illustrated by the representative error bar at the far right. (Source: Fig. 9 in Zhang *et al.* (2008) with permission from Elsevier.)

averaged over a 6 hr time window, showing that the T_2 values were higher at the cathode side of the PEM.

Recently, Wang *et al.* (2010) demonstrated that a calibration curve relating the T_2 corrected intensity of MR images to the λ value of Nafion–117 could be used to obtain the true water content of the PEM in an operating PEMFC. For the calibration curve the values of λ were determined for a PEM with heat-pressed catalyst decals. The maximum value of λ was 11.5 ± 0.5, which corresponds well with that expected for Nafion dried at a temperature above 105 °C (Zawodzinski *et al.*, 1993b). For a range of λ values, multi-echo images were acquired and the image intensities were corrected for T_2 relaxation. The T_2 corrected image intensities were plotted as intensity relative (RI) to the maximum level of hydration (i.e. RI = 1 for λ = 11.5) versus λ value to yield the calibration curve for the PEM in their FC. The authors note that the calibration curve determined is not universal for all MEAs using Nafion–117; various pre-treatment and preparation methods reported in the literature, in addition to the MEA history can affect both the maximum water uptake of the PEM and T_2 values.

10.8   General water distribution

The use of MRI for investigating the distribution of water in an operating PEMFC was first presented at a conference in 2001 (Tsushima *et al.*, 2005a). H–1 MR images of water in Nafion–117 (thickness 178 µm) were obtained with a resolution of 25 µm × 780 µm every 15 s during FC start-up, and revealed that the MRI signal rapidly decreased during the first 100 s, to a plateau after ~150 s. The change in signal intensity was well correlated with the decrease in output current density. A subsequent MRI study demonstrated that the distribution of water across the PEM (Aciplex S–1112, thickness 340 µm, EW 1025) from anode to-cathode was not uniform, and that the amount of water in the PEM was found to vary with operating conditions (Teranishi *et al.*, 2002).

These proof-of-concept results encouraged research groups internationally to pursue the study of water distribution in operating PEMFCs using MRI. In the past decade significant advances have been made in the application of MRI techniques to investigate and understand the amount and distribution of water in PEMFCs.

10.8.1 Reference/localizer images

Prior to starting an MRI study of an operating PEMFC, the geometry of the region to be investigated is first identified. Typically, a localizer or reference image is obtained which enables the image slices or volume to be positioned at regions of interest (ROI) in the FC. The ability to align slices accurately depends on both the resolution and SNR of the reference image. The resolution needs to be high enough (i.e. image pixels small enough) such that the boundaries between the

MEA and the flow fields are well defined. As mentioned previously, each pixel represents the average signal within the originating voxel. The resulting blurring may misrepresent the location of an object by approximately one voxel. Shown in Fig. 10.8 is an example reference image obtained from a PEMFC filled with water, with some physical dimensions labelled (Feindel *et al.*, 2007a). Figure 10.8a indicates the orientation of the image slices relative to B_0 and the PEMFC. Fig. 10.8b and c show example images obtained with the PEMFC filled with water. Note the region void of signal between the PEM and the flow channels, where the GDLs reside.

Water contained in the catalyst layers or GDLs is difficult to observe due to the effects of the properties of the material on the ^{1}H NMR signal (see Section 10.5). The GDL is one of the few components for which a fully MRI-compatible substitute has yet to be found. With the GDLs, however, MRI can provide a unique view of water within the PEM between the operating catalysts, separate from other regions of the PEMFC. The dimensions of the MEA components and flow channels provided on the schematic in Fig. 10.8c demonstrate that a slice of 500 μm thickness can be aligned such that the image will contain signal from only the PEM.

10.9   Water in the PEM

Ultimately, information regarding the three-dimensional distribution of water in the PEM is sought. In general, due to the small thickness of PEMs used and the resulting MRI limitations with respect to achieving adequate SNR with a temporal resolution on the order of seconds or minutes, the distribution of water is generally investigated in one of two orientations: across the thickness of the PEM (i.e. transverse direction) or in the plane of the PEM. As the MRI results discussed below will show, the distribution of water in the PEM of an operating PEMFC is very dynamic, and understanding the capabilities of MRI and potential artefacts is critical for interpretation of results. This section reviews the literature and highlights key results. The specific PEMFC operating conditions and MRI parameters vary widely, and therefore for complete details the reader is referred to the original literature.

10.9.1 Water content across the thickness of the PEM

Teranishi *et al.* (2002) commenced investigations using MRI to measure water distribution across the PEM from anode to cathode during FC operation. The majority of their studies have focused on the Aciplex S–1112 PEM with a thickness of 340 μm (Teranishi *et al.*, 2002; Teranishi *et al.*, 2004; Teranishi *et al.*, 2006), although the use of Aciplex S–1104 and S–1102 PEMs (thickness 117 and 56 μm, respectively) (Teranishi *et al.*, 2003; Tsushima *et al.*, 2005a), Nafion–117, and Flemion PEMs has been reported (Tsushima *et al.*, 2005a). In the first few reports of the use of MRI to study operating PEMFCs (Teranishi *et al.*, 2002;

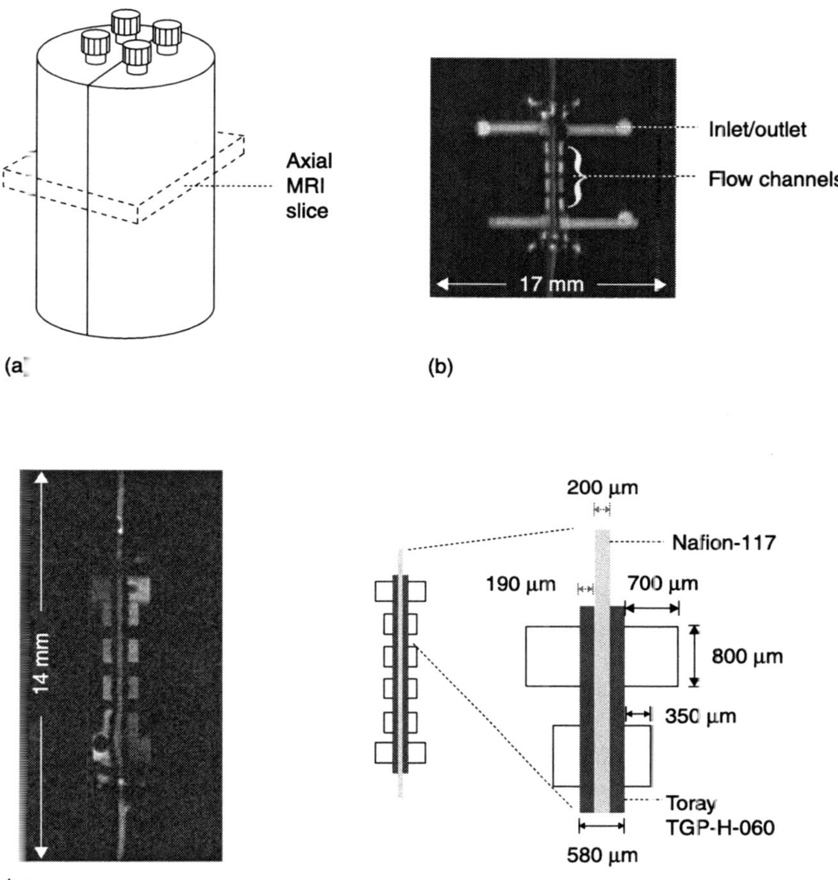

10.8 (a) Schematic of a PEMFC showing the orientation of the assembly with respect to the external applied magnetic field, B_0, and an example of a slice defined for MRI. (b) An example MR image acquired from the PEMFC filled with water. (c) A high resolution MR image acquired from the PEMFC filled with water and a schematic showing the approximate thickness of the Nafion–117 (200 µm), the Toray GP-H–060 (190 µm), and dimensions of the flow channels. Image acquisition parameters: (b) SE, slice = 1.0 mm, T_R = 3.0 s, T_E = 5.2 ms, FOV = 17 mm × 30 mm (h × v; FE × PE), pixels = 128^2, pixel size = 134 µm × 234 µm, 8 echoes, 1 image, 32 averages; (c) SE, slice = 500 µm, T_R = 1.5 s, T_E = 4.7 ms, FOV = 14 mm × 14 mm (h × v; PE × FE), pixels = 256^2, pixel size = 55 µm × 55 µm, 16 averages. (Source: Fig. 2 in Feindel *et al.* (2007b).)

Teranishi *et al.*, 2003; Teranishi *et al.*, 2004; Tsushima *et al.*, 2004), the water content and distribution across the PEM was found to be affected by the FC power output.

Using a SE imaging experiment to obtain signal from a large 50 μm × 800 μm × 2.5 mm voxel, changes in water content in the S–1112 PEM during FC start-up were monitored with a temporal resolution of 50 s (Tsushima *et al.*, 2004). A drop in FC voltage was correlated with an overall decline of water in the PEM during the first 200 s of operating with a small circuit resistance, and the anode side of the PEM dehydrated more rapidly than the cathode (Tsushima *et al.*, 2004). Numerous studies have found that the PEM was partially dehydrated at the anode side due to electro-osmotic drag under FC operation (Tsushima *et al.*, 2005a; Tsushima *et al.*, 2005b; Teranishi *et al.*, 2005; Teranishi *et al.*, 2006; Tsushima *et al.*, 2004; Tsushima *et al.*, 2006a), and that the overall water content in the PEM decreases with increasing current (Teranishi *et al.*, 2002; Teranishi *et al.*, 2003; Teranishi *et al.*, 2004; Tsushima *et al.*, 2005a; Tsushima *et al.*, 2004).

Teranishi *et al.* (2003) also investigated the effect of PEM thickness on FC performance using three different Aciplex membranes (S–1112, S–1104 and, S–1102). Water content profiles were obtained with a MSSE sequence over 32 minutes with a pixel size of 25 μm × 400 μm. The water content of each PEM was low when the FC was operated at high current density, which implies that electro-osmotic drag was the dominant effect determining the water content of the PEM. In a separate study using either a 50 μm-thick hydrocarbon or perfluorinated PEM, higher current densities (ranging from 0–0.45 A cm^{-2}) were found to increase the level of hydration (Tsushima *et al.*, 2007a).

In an attempt to maintain a high water content in the PEM of an operating PEMFC, Tsushima *et al.* (2003) implemented a direct supply of water to the anode side of the PEM and noted an improvement in performance (at 0.375 A cm^{-2}, voltage increased from 0.18 to 0.25 V). Achieving a pixel size of 25 μm × 800 μm with a slice thickness of 5 mm, the region of the PEM in contact with $H_2O(l)$ was found to be swollen relative to other areas of the PEM (Tsushima *et al.*, 2005b). This demonstrated that a direct supply of water to the PEM effectively increases the level of hydration; however, the localized swelling may indicate that the rate of transport of water within the PEM is insufficient to hydrate neighbouring regions to the same extent.

Mechanical confinement

Teranishi *et al.* (2003) have also used the results from their MRI experiments to parameterize a one-dimensional model for water transport in a PEM (Teranishi *et al.*, 2004; Teranishi *et al.*, 2006). The absorption of water by a PEM in an assembled FC was investigated, and they noted that at high vapour activity the PEM water content was smaller than measured previously on free PEMs (Hinatsu *et al.*, 1994). This result suggested that an upper limit of water uptake resulting from mechanical confinement of the PEM by the FC housing. To reproduce the water distribution in the operating FC, a parameter capping the maximum water content at $\lambda \sim 6$ was required. This value corresponds well with the relation between water content and PEM swelling; at $\lambda < 6$ the PEM thickness does not increase with water content, but at larger water content, the PEM thickness increases roughly linearly with increasing water content. Shown in Fig. 10.9 are plots of the experimental and modelled water content profiles across the PEM with current densities of 0, 89, and 178 mA cm^{-2}, in Fig. 10.9a, b, and c, respectively (details were not provided regarding determination of λ values). Similarly, Bedet *et al.* (2008a; 2008b) noted alternating regions of high and low water content in the same pattern as the flow channels which was also attributed to mechanical compression.

Gas humidity

Determining the appropriate level of supply gas humidification to maintain a high water content in the PEM, while not leading to accumulation of $H_2O(l)$ in the flow channels, is critical for efficient FC operation.

In an investigation of hydrogen cross-leak behaviour of PEMs, Tsushima *et al.* (2007b) used MRI to visualize the water content distribution in the PEM under FC operation. The FC was operated at a temperature of 80 °C with a current density of 0.1 A cm^{-2} with supply gases that were dry or were humidified to either RH = 13% or 85%, and the PEM hydration was found to be greater with increased RH. Ikeda *et al.* (2008) followed up with a study to investigate the effect of RH and current density on the transverse water content profile in Nafion–1110 (thickness 254 µm) under FC operation. Images were obtained with a MSSE sequence with a pixel size of 50 µm × 500 µm over a FOV of 12.8 mm × 64 mm, with an experiment time of 1 hr. The FC was operated at a temperature of 70 °C at each RH until the current became constant, and then for a further 3 hrs to ensure a steady state had been reached. Shown in Fig. 10.10a–c are the water content profiles across the PEM, measured with the FC operating with current densities of 0.1 and 0.2 A cm^{-2}, and with RH of 92%, 80%, and 40%. Values for λ were reported; however, details regarding the calibration to obtain λ were not provided. The water content in the PEM increased with greater RH. For RH = 80% and 92% higher current density also increased water in the PEM. In all cases partial dehydration at the anode was observed. The highest water content profile was observed for RH = 92% with a current density of 0.2 A cm^{-2}, and the value of λ approached that expected for Nafion in contact with $H_2O(l)$. The relatively flat water content profile measured, in comparison to that at a current density of 0.1 A cm^{-2}, was postulated to result from the high level of water at the cathode, which would facilitate back-diffusion to the anode.

The water content profile across a PEM in a FC operated under conditions to dehydrate the cathode was investigated by Zhang *et al.* (2008). While maintaining

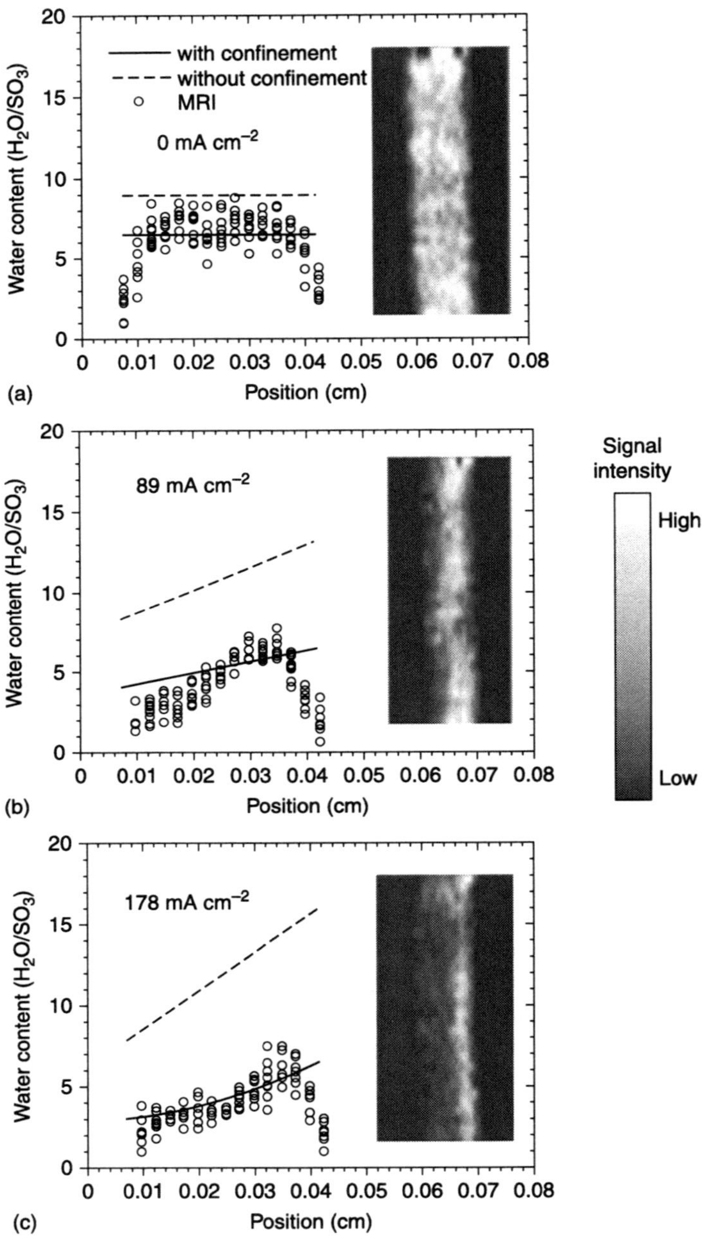

10.9 Water content profiles in a PEM under various FC current densities. The lines are the results from a model and the open circles are measurements from experiment. In the MR image, the left side is the anode. The output current density is (a) 0, (b) 89, and (c) 178 mA/cm². (Source: adapted from Fig. 7 in Teranishi *et al.* (2006) by permission of ECS – The Electrochemical Society.)

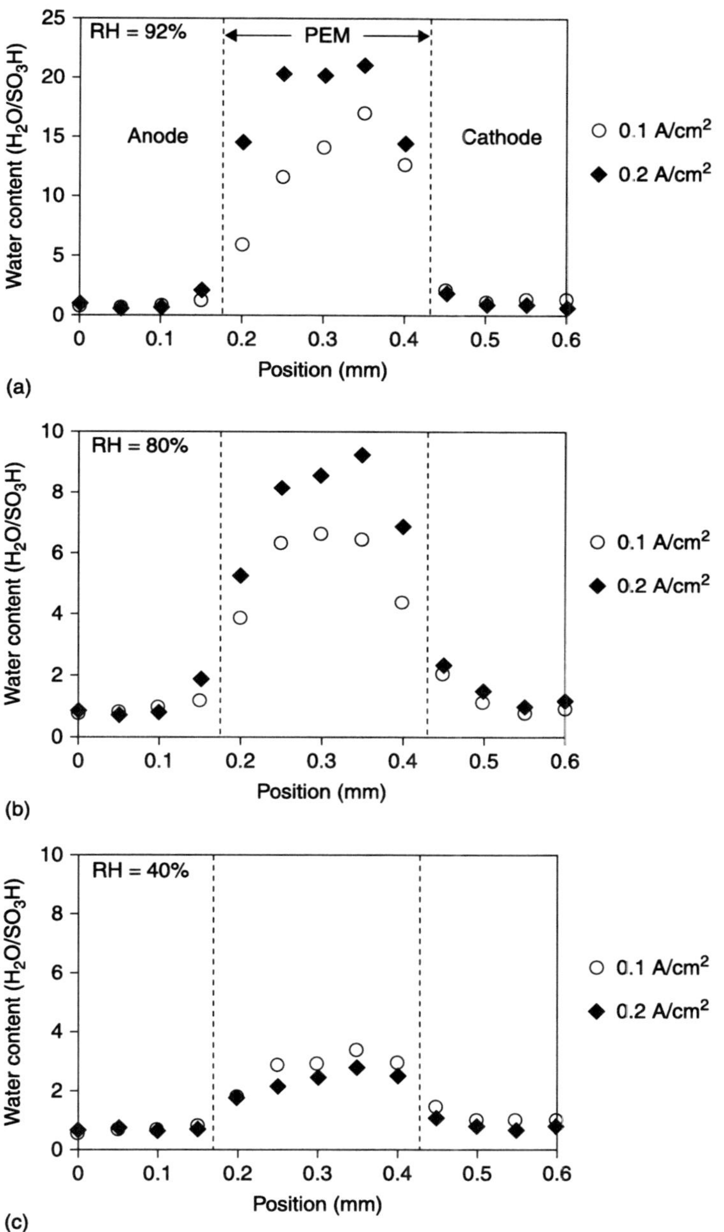

10.10 Water content profiles in a PEM under FC current densities of 0.1 A cm^{-2} and 0.2 A cm^{-2} and RHs of (a) 92%, (b) 80%, and (c) 40%. (Source: adapted from Fig. 3 in Ikeda *et al.* (2008) with permission from Elsevier.)

the gas flow rate, the RH of the incoming air was reduced to 12% and the temperature decreased to 20 °C. Immediately the dry air flow began to dehydrate the PEM as demonstrated by both decrease in MRI signal and in T_2 values, and the Nafion was dehydrated within 6 min.

High resolution across PEM

Zhang *et al.* (2008) performed some exquisite research to obtain one-dimensional water content profiles across Nafion–1110 with 6 μm nominal resolution in an operating FC. Profiles were acquired at $B_0 = 2.4$ T using a multi-echo DHK SE SPI experiment (Ouriadov *et al.*, 2004). Using this approach the FC was studied during various stages of operation: activation, steady-state operation, and cathode dehydration. Shown in Fig. 10.11 are plots of the signal intensity profile across the PEM at 6 min. intervals during FC activation. Nafion was initially hydrated by the humidified reactant gas flow, but within the first 6 min. of operation, the MRI signal decreased at the anode and increased significantly at the cathode (essentially reaching the peak level of hydration), a feature suggesting the action of electro-osmotic drag. As the FC continued to operate the anode became hydrated and the MRI signal across the PEM became increasingly linear. Shown in Fig. 10.12a are plots of the MRI signal profiles across the PEM for various steady-state operating times. The signal was corrected for T_2 relaxation to obtain water content profiles and replotted as shown in Fig. 10.12b. The water content on the anode side was stable,

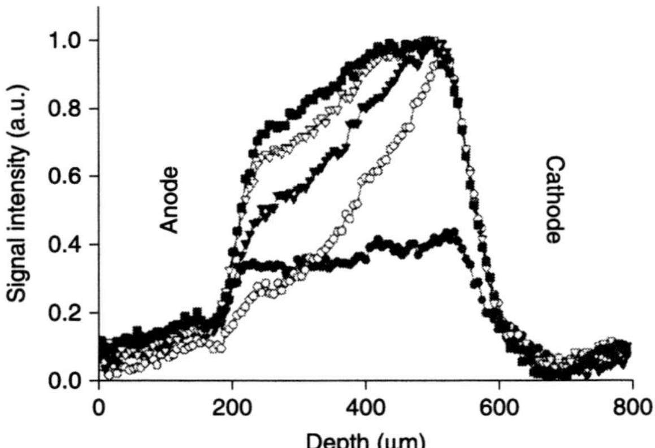

10.11 Water content profile across Nafion–1110 during fuel cell activation. Each profile was acquired in 6 min, and is resolved by 128 pixels with 6 μm nominal resolution. Data shown for experiments at times of 0 (●), 6 min (o), 12 min (▼), 18 min (∇), and 24 min(■) after commencing gas flow. The signal intensity close to the cathode is much higher than that close to the anode at early times. (Source: Fig. 6 in Zhang *et al.* (2008) with permission from Elsevier.)

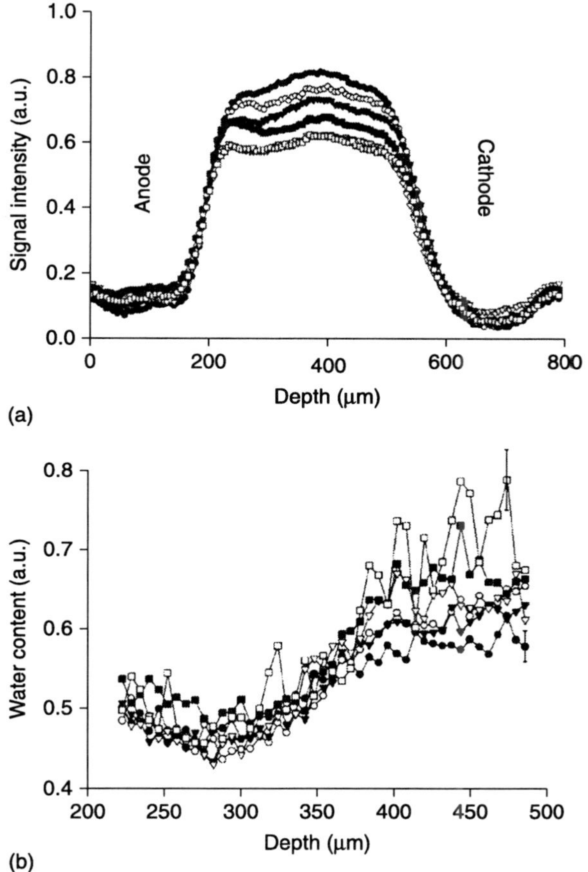

10.12 (a) Water content profiles across Nafion–1110 during quasi-steady state fuel cell operation, starting at approximately 1.2 h and lasting 6 h. Ten first-echo MRI profiles were averaged over 1 h intervals to produce the profiles shown in (a). Profiles (•),(o),(▼),(∇),(■), and (□) correspond to time windows of, 1.1–2.1, 2.1–3.1, 3.1–4.1, 4.1–5.1, 5.1–6.1, 6.1–7.1 h. (b) True water content profiles determined by relaxation time mapping. The profiles displayed are also averaged. The profiles in (b) show stable water content close to the anode, with a higher water content near the cathode. Experimental uncertainties, due to fitting, are illustrated by the representative error bars at the far right. (Source: Fig. 8 in Zhang *et al.* (2008) with permission from Elsevier.)

but increased with operating time on the cathode side. The authors often noted that although the MRI signal intensity showed variation with time, the change was due to variation in T_2, and the water content remained relatively constant.

Recently, Shim *et al.* (2009) reported results from a high-resolution study of the water content distribution across Nafion–115 (thickness 127 µm). The experiments were performed at 9.4 T with a SE sequence, and images were obtained with pixel sizes of 156 μm along the PEM and either 25 μm, 10 μm or 5 μm across the thickness of the PEM. The authors note that there was little difference between the 10 μm and 5 μm images. The direction of PE was not provided; to approach such high resolution in an operating PEMFC, the PE direction would need to be across the thickness of the PEM.

10.9.2 Water content in the plane of the PEM

The distribution of water in the plane of the PEM has been investigated by numerous research groups (Tsushima *et al.*, 2007b; Feindel *et al.*, 2006a; Minard *et al.*, 2006; Bedet *et al.*, 2008a). Feindel *et al.* (2004) noted that with no ^{1}H MRI signal from the GDL, water in the PEM could be investigated separately from $H_2O(l)$ in the gas flow channels. As shown in Fig. 10.8c a slice thickness of 500 μm was well suited to isolate the MEA from the flow fields, and two neighbouring slices each of 500 μm thickness encompassed each flow field. After starting with a PEM that had been dehydrated by dry gases, as the FC was operated for 6 hrs the amount of water in the PEM surrounding the MEA and outside the o-ring seals increased (Feindel *et al.*, 2004). This observation demonstrated that water formed at the cathode, and brought into the PEM by electro-osmotic drag from the anode, diffuses radially down the concentration gradient through the PEM away from the MEA. Such transport of water may reduce flooding at cathode sites near the edges of the catalyst stamp and decrease the amount of water in the PEM between the catalysts. Over the course of a 96 hr period, the PEMFC was operated under constant resistive load, and the distribution of water in the PEM was not uniform, with more water in the PEM that was in direct contact with water pooled around the gold ring (similar to observation by Tsushima *et al.* (2003; 2005b) with direct supply of water to PEM).

Progressive PEM dehydration

Both Minard and Tsushima noted an interesting pattern of PEM dehydration when operating a FC with a serpentine flow field. Tsushima *et al.* (2006a) measured the effect of typical parallel and serpentine gas flow channel designs on the in-plane distribution of water in a 340 μm Asahi PEM in an operating FC. Humidified gases were supplied in co-flow and a 2D MSSE sequence was used to obtain images with 100 μm × 600 μm nominal resolution from 2.5 mm-thick slices in an experiment time of 18 min. In the case of no power generation, both flow channel designs afforded a uniform distribution of water in the PEM. With increasing current density, PEM water content decreased and dehydration of the PEM occurred from inlet to outlet. In the serpentine flow, dehydration of the PEM did not occur along the direction of the flow pattern; a dehydration front progressed from inlet to outlet. The authors postulate that an

H$_2$O(*l*) blockage in the serpentine flow channel may lead to gas passing through the GDL to an adjacent channel.

Minard *et al.* (2006) visualized water inside a PEMFC with a serpentine flow channel pattern during 11.4 hrs of continuous operation with humidified gases and a constant load. Images were correlated to the current output and operating characteristics of the FC. A total of 320 images were acquired with a 2D MSSE sequence, each requiring 128 s. The water distribution was not stable during prolonged operation, and the series of images revealed that a dehydration front propagated over the surface of the PEM from gas inlets and progressed toward gas outlets (see Fig. 10.13). After only several minutes of operation, a distinct front of dehydration in the MEA became visible. As shown in Fig. 10.13b, after 66 min. the front was located about midway between gas inlet and outlets (and the current

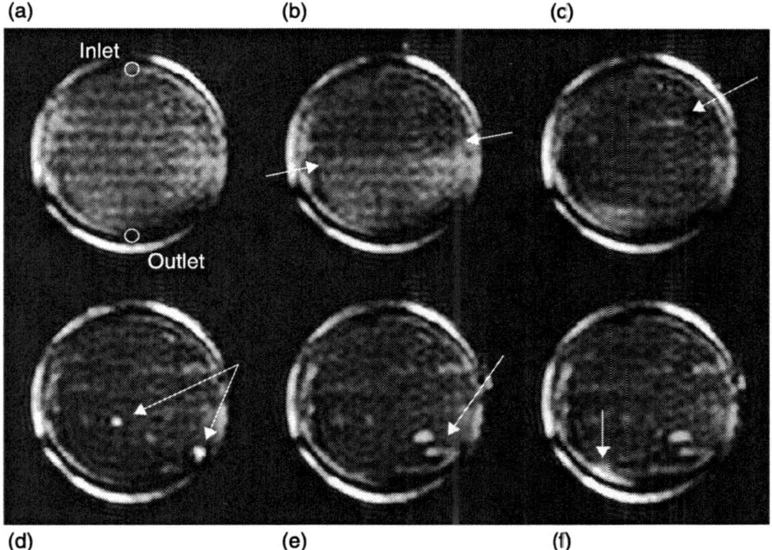

10.13 Selected MR images showing water in an operating fuel cell over an 11.4 h period. Images were acquired at: (a) the start of the experiment and after (b) 66 min, (c) 164 min, (d) 290 min, (e) 473 min, (f) 678 min. The fuel and oxidant inlet and outlet positions are labelled and the arrows in (b) highlight the position of the dehydration front that moved from the inlet towards the outlet. The grey scale was chosen to highlight the dehydration front formation and the actual signal intensity varied by approximately 10%. Regions showing flooding are also highlighted by white arrows. Image acquisition parameters: MSSE sequence, receiver BW = 200 kHz, slice = 4 mm (containing MEA and flow channels), T_R = 500 ms, T_E = 1.6 ms, FOV = 35 mm × 35 mm, 64 × 64 pixels (zero-filled to 512 × 512 prior to FT), 320 images, 128 s per image. (Source: Fig. 6 in Minard *et al.* (2006) with permission from Elsevier.)

density had diminished). The front remained clear and moved steadily within the plane of the MEA. After 164 min. (Fig. 10.13c) the dehydration front had moved from the inlet to the outlet. Subsequently $H_2O(l)$ appeared in the gas channels near the outlets and behind the dehydration front, but adjacent areas of the PEM did not rehydrate. After 290 min. (Fig. 10.13d) the dehydration front completely disappeared and the current began to decrease more rapidly. At 473 min. (Fig. 10.13e) further $H_2O(l)$ had pooled in the gas manifold and the current density had decreased further. During the final hour of imaging, water accumulated near the edge of the gas manifold, close to the gas outlets (Fig. 10.13f). The obstructions were not cleared indicating that gas may preferentially flow around the blocked areas through the GDL. The authors postulated that the dehydration front may be related to depletion of $H_2(g)$ from inlet to outlet, limiting both the reaction and electro-osmotic drag.

Influence of gas inlet/outlet configuration

Feindel *et al.* (2006b) investigated the influence of the gas inlet and outlet configuration on the in-plane distribution of water in the PEM, on the distribution of $H_2O(l)$ in the cathode flow field, and on performance of the FC. MR images with a nominal in-plane resolution of 234 µm × 234 µm were acquired with a MSSE sequence from five parallel slices of 500 µm thickness, using a 30 mm × 30 mm FOV, and requiring 128 s per image. By operating the FC with dry gas streams, the only source of water is that produced by the reduction of oxygen at the cathode. The authors found that water could readily be depleted from the PEM by operating with a high $O_2(g)$ flow rate and by increasing the current drawn from the FC. Subsequently, to study correlation between signal intensity in the MEA and FC performance, in each configuration the FC was operated starting with a water-depleted MEA (Feindel *et al.*, 2007a). The power output was adjusted to allow water in the PEM to reach a constant level and facilitate accumulation of $H_2O(l)$ in the cathode flow field. The co-flow gas configurations were found to require longer operating times and/or increased FC currents to facilitate $H_2O(l)$ accumulation in the cathode flow field. Fig. 10.14 shows a schematic of the four possible gas flow configurations, the distribution of water in the PEM and in the cathode flow field. The MEA difference image was obtained by subtraction of the image of the low water-content PEM from the image obtained at the denoted operating time. As shown in Fig. 10.14, the counter-flow configurations (Fig. 10.14a and b) were found to lead to a relatively uniform in-plane distribution of water in the PEM. Co-flow configurations (Fig. 10.14c and d) did not lead to a uniform in-plane distribution of water in the PEM, rather the PEM near the gas inlets contained less water than the region of the PEM near the gas outlets. Integrated signal intensities from a ROI that encompassed the entire MEA were found to be larger for counter-flow configurations than for co-flow configurations. In addition, the integrated signal intensity for the anode inlet/

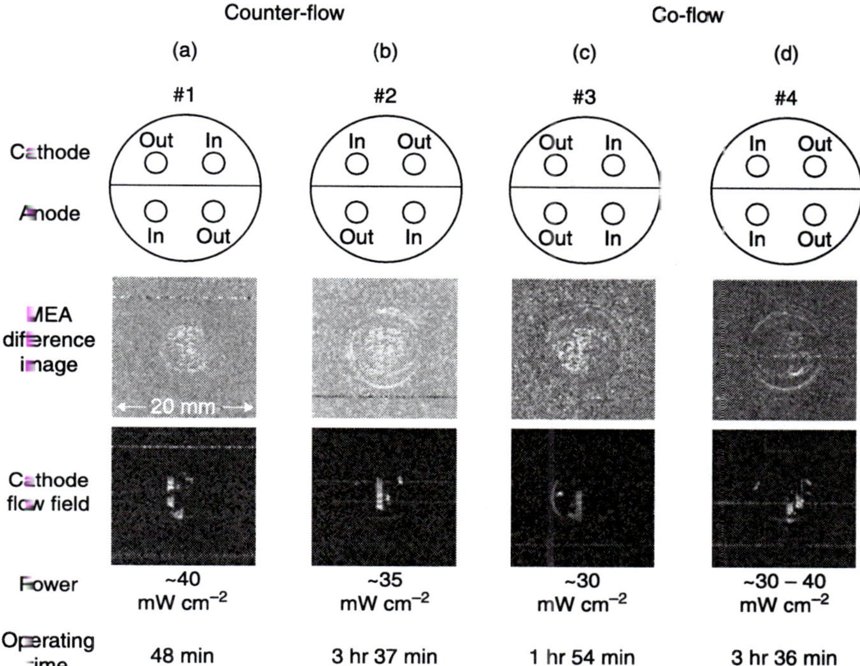

10.14 The four possible gas inlet/outlet configurations and a representative example of the resulting water distribution for each. The different images show the distribution of water in the PEM relative to a low water content PEM. The cathode flow field images show the locations of $H_2O(l)$ accumulations. (Source: Adapted with permission from Fig. 7 in Feindel *et al.* (2006b). Copyright 2006 American Chemical Society.)

cathode outlet half of the MEA was less than that for the anode outlet/cathode inlet. Such signal intensity variations can be easily rationalized qualitatively, for example, by considering the relative gas flow rates on the anode and cathode inlets; however, further quantitative investigation of subtle variations is warranted.

Bedet and coworkers found similar results when operating their FC with gases in co-flow configuration (Bedet *et al.*, 2008a, 2008b). The signal intensity in the PEM increased near the gas outlets, and was associated with $H_2O(l)$ in the gas flow channel (visual confirmation). Water was also found to pool around the gold ring current collector, resulting in an increased level of PEM hydration in that area.

Interdependence of operating conditions and power output

Several studies have found that FC power peaks as $H_2O(l)$ appears in the cathode flow field, but sustained accumulation of $H_2O(l)$ results in decline of FC power

(Feindel *et al.*, 2006b; Feindel *et al.*, 2007a). For example, under a fixed external resistive load, the influence of the $O_2(g)$ flow rate on the PEM water content and power output of a self-humidifying PEMFC was investigated (Feindel *et al.*, 2007a). As shown in Fig. 10.15, the FC power output and MRI signal intensity from a ROI encompassing the MEA appear closely related (shape of response curves are similar). The plot shows the variation in power density and MRI signal intensity from water in the MEA and $H_2O(l)$ in the cathode flow field as a function

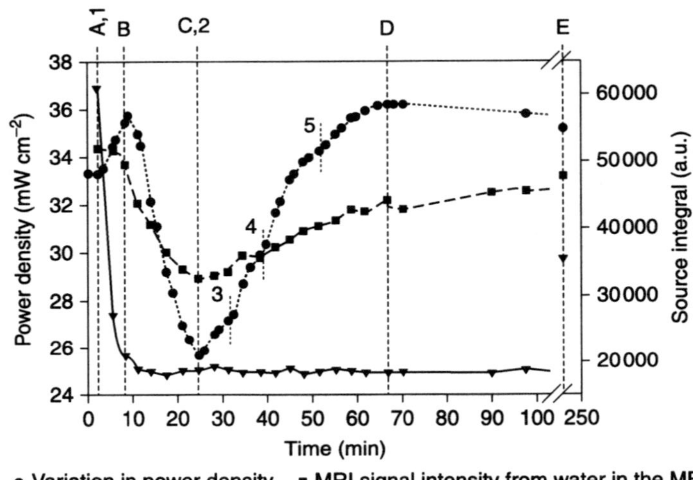

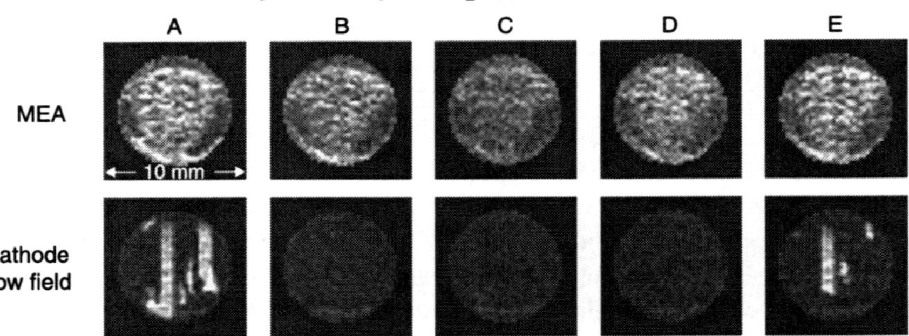

10.15 Top: Plot showing the sensitivity of the power density (•), MEA water content (■), and $H_2O(l)$ in the cathode flow field (▼) to changes in the $O_2(g)$ flow rate. The $O_2(g)$ flow rate was: 50 mL min⁻¹ from $t = 0$ (A,1) to $t = 25$ min (C,2); 20 mL min⁻¹ from $t = 25$ min to $t = 32$ min (3); 10 mL min⁻¹ from $t = 32$ min to $t = 39$ min (4); 5 mL min⁻¹ from $t = 39$ min to $t = 52$ min (5); and 2.5 ml min⁻¹ from $t = 52$ min onward. Bottom: ¹H MR images (A-E) of the MEA and cathode flow field regions of interest from which the intensities were integrated. The PEMFC was operating in a counter-flow gas configuration. (Source: Fig. 7 from Feindel *et al.* (2007a).)

of $O_2(g)$ flow rate. At point A, 1 on the plot, the MEA water content was high and a large amount of $H_2O(l)$ was in the cathode flow field. The $O_2(g)$ flow rate was then increased from 2.5 to 50 mL min^{-1}. Fig. 10.15, points A,1 through B, show that the $H_2O(l)$ was quickly purged from the flow field and resulted in an increase in power. A small decrease in the MEA water content was also observed. As shown in Fig. 10.15, points B–C, under the high $O_2(g)$ flow rate the ROI source integrals for both the MEA and cathode flow field, and the FC power output declined. At point C,2 on the plot, the $O_2(g)$ flow rate was reduced to 20 mL min^{-1} and resulted in increased power output and MEA water content. The $O_2(g)$ flow rate was then decreased to 10, 5, and 2.5 mL min^{-1} at points 3, 4, and 5, respectively. Evident from the plot and comparison of MR images C and D, upon each reduction in $O_2(g)$ flow rate both the FC power output and the MEA water content increased. After the $O_2(g)$ flow rate was reduced to 2.5 mL min^{-1}, the MEA water content continued to increase; however, the FC power output began to decline as $H_2O(l)$ accumulated in the cathode flow field (Fig. 10.15 points D–E). When the power output of the FC peaks as $H_2O(l)$ appears in the cathode flow field, some flooding of the catalysts and GDL must be occurring. However, the increased ionic conductivity of Nafion–117 when in contact with $H_2O(l)$ may initially offset any detrimental effect on power output. Considering both theoretical models and experimental observations, the authors inferred that for a low-humidity PEMFC operating at low current densities, any detrimental effect on performance due to minimal accumulation of $H_2O(l)$ at the catalyst layer or GDL is superseded by the improved ionic conductivity of the PEM when in contact with $H_2O(l)$ (Feindel *et al.*, 2006b).

Feindel *et al.* (2007a) also investigated the effects of changing both the power output of the FC and the gas flow rates on water in the PEM and $H_2O(l)$ in the cathode flow field. Higher power densities were found to increase water in the PEM, which indicates that the PEM was hydrated by back-diffusion of water produced at the cathode.

10.10 Flow channels

Numerous studies have demonstrated the utility of MRI for investigating the distribution and dynamics of water in the flow channels of operating PEMFCs. Of particular interest is the variability in results obtained, dependent on both FC design and operating conditions.

10.10.1 Water build-up and movement

Feindel *et al.* (2004) demonstrated that MRI was an effective technique for monitoring the distribution of $H_2O(l)$ in the flow channels. When their PEMFC was operated under constant load for 96 hrs and imaged intermittently, $H_2O(l)$ was noted in the cathode flow field and pooled around the current collector surrounding the flow field. Water in the cathode flow field created transient obstructions, and $H_2O(l)$ was also present in the anode flow field if the humidity of the $H_2(g)$ was too high. After operating for several hours, or when the FC was operated with the PEM saturated with water immediately before operating the FC, $H_2O(l)$ typically began to accumulate in the outlet half of the cathode flow field (Feindel *et al.*, 2006b). $H_2O(l)$ was not observed in the anode flow field when using dry gas feeds (even with $H_2O(l)$ in the cathode flow field). In contrast, Dunbar and Masel (2007) noted that when operating their FC at low currents with a dry anode gas stream, water was visible in the anode flow field, corresponding to a location directly across the PEM from the area of highest water concentration in the cathode. When running at high currents, there was no water visible in the anode flow field (Dunbar and Masel, 2008a). The presence of water at low currents implies that diffusion and backpressure induced by the hydrophobic GDL play a significant role in the transport of water into the anode.

MRI has also been used to study the flow and accumulation of $H_2O(l)$ and to analyse the dynamics of $H_2O(l)$ transport in a PEMFC operating in steady state (Dunbar and Masel, 2007). The region of the MR image corresponding to the location of the GDL was void of signal; however, water droplets existed on the surface of the GDL and were proposed to be quickly transported into the moving water waves in the cathode flow field, with more water accumulating as the reactant gas progressed from inlet to outlet. The total water content of the PEMFC was analysed, finding that the cathode contained the largest proportion of water (> 60%), while the remainder was split between the anode and the MEA. In some cases the water was not in contact with the surface of the cathode GDL, but instead was observed along the bottom wall of the flow field, away from the MEA and GDL. Figure 10.16 shows an image of the water overlaid on a photo of the FC and a plot of the water content resolved from the MEA to the cathode flow field. Water motion did not look like slug flow, but wavy-stratified flow, where waves of water move slowly through the flow field, transported by friction with the moving gas, but at a slower rate. The waves did not have a fixed wavelength or wave spacing.

Dunbar and Masel (2008a) used a FC with flow fields painted with a graphite layer (Aquadag) to more accurately mimic the hydrophobicity of commercial flow fields. Initially large $H_2O(l)$ droplets were present, clinging to the Teflon-treated GDL. After operating for ~15 min. large water waves accumulated along the bottom of the flow channels. The waves became stagnant and the volume increased. Slow movement of the waves was observed, but the location of their accumulation was consistent between images. This observation lead to the postulation that the location of the water was related to defects or non-uniformities in the surface of the flow channel. Defects will exert an additional frictional force on the moving water wave, and in some cases would be large enough to stall the wave. However, as water accumulates, the gap between the water surface and the GDL decreases, and the viscous and shape forces increase until they are sufficient

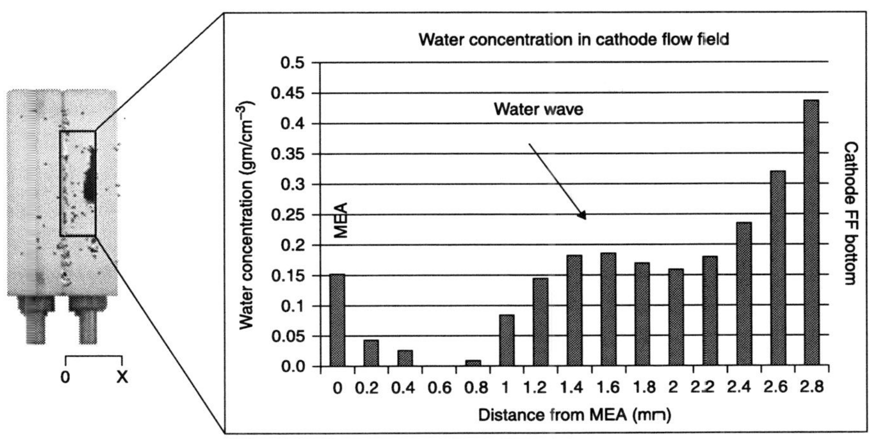

10.16 Average water concentration distributions of the cathode flow fields at various distances from the MEA. Concentration is highest along the back wall of the cathode flow field. (Source: Fig. 6 in Dunbar and Masel (2007) with permission from Elsevier.)

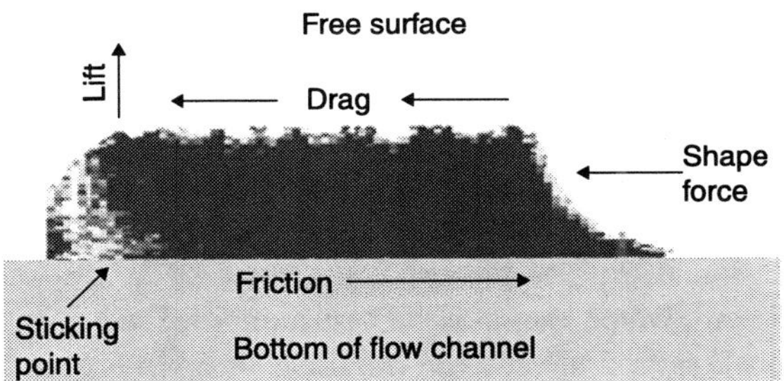

10.17 A diagram of the forces on a water wave in the cathode flow field just before it slips. The moving air creates a shape (drag) force pushing along the front surface, and a viscous drag force because of the air blowing across the top surface. Friction from the bottom wall slows the motion down. There is also a lift force near the front edge that produces the Jeffries waves. (Source: Adapted from Fig. 6 in Dunbar and Masel (2008a) with permission from Elsevier.)

to push the water along. Shown in Fig. 10.17 is an image of a water wave, labelled with the various forces influencing the shape and movement.

Dunbar and Masel (2008b) have aimed to increase the temporal resolution of the MRI experiments by focusing on smaller regions of interest. By decreasing the FOV and number of pixels in the PE direction, images with 200 μm resolution can be obtained in < 16 s for a FOV covering two channels, and in < 5 s for images of a single channel. Preliminary results showed the increase in volume of water from a single drop to a larger drop or wave over the course of 80 s.

10.11 Hydrogen–deuterium contrast

The use of hydrogen–deuterium (H–D) exchange as a method to introduce contrast in ^{1}H MRI images and to investigate the distribution of water throughout an operating H_2/O_2 PEMFC was demonstrated by Feindel *et al.* (2006a; 2007b). D_2O readily exchanges with H_2O in the MEA under non-operative conditions. Subsequently supplying $H_2(g)$ as fuel, recovery of the ^{1}H NMR signal can be monitored. Similarly, alternating between $H_2(g)$ and $D_2(g)$ was effective to provide image contrast without saturating the PEM. The use of H–D exchange is useful for a qualitative perspective of the dynamic distribution of water within a PEMFC operating at steady state.

Under normal operating conditions at ambient pressure, no change in FC performance was found when using $D_2(g)$ as fuel, or when hydrating the PEM with $D_2O(l)$. However, a kinetic-isotope effect was reported when using $D_2(g)$ as fuel at a low partial pressure (0.7 kPa) resulting in a four-fold decrease in power output (Katsaounis *et al.*, 2005; Tsampas *et al.*, 2006). This effect was attributed to proton-tunnelling between adjacent sulfonate groups at the surface of the PEM and becomes less important with increasing partial pressure of $D_2(g)$.

10.11.1 Saturated PEM

Flowing $D_2O(l)$ through the cathode flow field of a FC results in a fully hydrated PEM with negligible contribution to the ^{1}H MRI signal (Feindel *et al.*, 2006a). Subsequent operation of the FC with dry $H_2(g)$ as fuel allowed monitoring of recovery of the ^{1}H signal, due to the oxidation reaction at the anode. After saturating the MEA with water, the in-plane distribution of water in the Nafion remained uniform for a period of time (on the order of hours) regardless of gas flow configuration. Using the PEMFC shown in Fig. 10.6b, after approximately 10 min. with $H_2(g)$ as fuel, recovery of the ^{1}H MRI signal in the MEA region was evident and continued to increase over the course of ~90 min. to reveal a relatively uniform distribution of water in the MEA (Feindel *et al.*, 2007b); the high water content in the PEM that resulted from cycling $D_2O(l)$ through the flow channels was sufficient to compensate for water removed by the dry gases entering the flow channels while operating the PEMFC over a period of 90 min. This indicates that the amount of water in the PEM had yet to reach a steady-state; in a co-flow gas configuration, the region of the PEM near the gas inlets would typically be dehydrated (Feindel *et al.*, 2006b). The MR images also showed that as the FC was operated, D_2O in the PEM was transported from the MEA to surrounding regions, similar to the radial diffusion outward from the MEA observed during FC start-up with a dry membrane (Feindel *et al.*, 2004). The MR images indicated that D_2O extended ~7 mm beyond the MEA ~100 min. after operating the FC on $H_2(g)$. This distance is greater than expected due to translational diffusion alone; the authors inferred that the distribution of D_2O in the PEM was increased by either ionic conductivity or a water concentration gradient between the area of the PEM within and outside the o-rings (Feindel *et al.*, 2007b).

10.11.2 Dry $H_2(g)$ or $D_2(g)$ as fuel

Alternating between dry $H_2(g)$ and dry $D_2(g)$ introduced contrast that enabled investigation of changes in the water in the PEM when the PEMFC operated under steady-state conditions (Feindel *et al.*, 2007b). After reaching a steady state of operation with $H_2(g)$ as fuel, when $D_2(g)$ was supplied as fuel, there was a significant delay (ranging from ~30 to 45 min.) before the ^1H MRI signal in the MEA started to decline (Fig. 10.18b). This is in contrast to experiments in which the PEM was saturated with water first (signal started to decline within 10 min.). As shown in Fig. 10.18d, little ^1H MRI signal remained in the MEA or in the flow field after operating on $D_2(g)$ for 69 min., and after 10 hrs significant decline of the ^1H MRI signal was evident outside of the MEA in the surrounding PEM (Fig. 10.18e).

Alternating between $H_2(g)$ or $D_2(g)$ as fuel may facilitate approximation of the proportion of ion transport channels utilized by the PEM (Feindel *et al.*, 2007b). The water uptake of Nafion is known to depend on whether the PEM is in contact

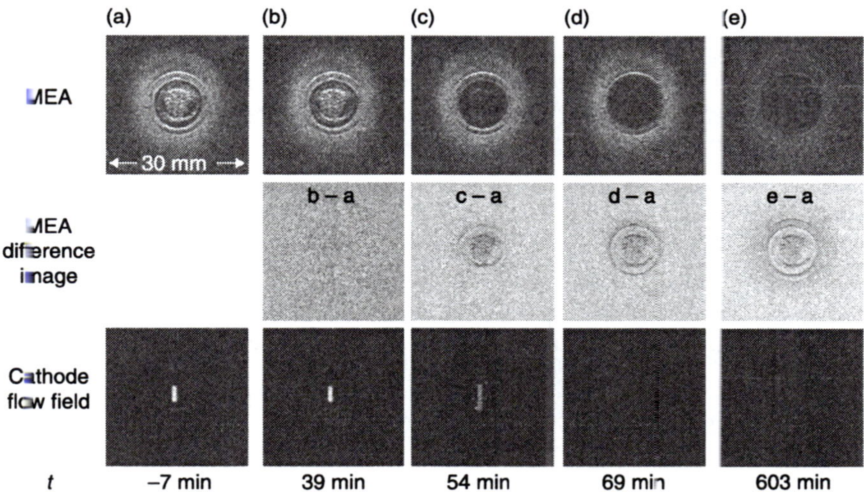

10.18 ^1H MR images from 500 µm slices containing the MEA (top row), the cathode flow field (bottom row), and the MEA difference images (middle row) showing the decrease in ^1H MRI signal while operating the PEMFC on $D_2(g)$ as fuel with a power output of ~30 mW cm^{-2} (22 mA, 0.67 V). At $t = 0$, the fuel was switched from $H_2(g)$ to $D_2(g)$ using counter-flow gas configuration #1 as shown in Fig. 10.14a. (Source: Adapted from Fig. 7 in Feindel *et al.* (2007b) with permission from Elsevier.)

with water vapour or liquid water and is thought to result from the hydrophobic fluorine-rich surface morphology of Nafion. Atomic force microscopy (AFM) studies have shown an increase in the size of the conductive hydrophilic domains of Nafion in high RH environments or those in contact with liquid water. Clusters of hydrophilic domains of 4–10 nm in diameter at ambient humidity were found to swell to 7–15 nm in diameter when the Nafion was in contact with liquid water (McLean *et al.*, 2000), and clusters up to 40 nm have been reported (Aleksandrova *et al.*, 2007). The increased size of the hydrophilic clusters improves significantly the ion transport properties of Nafion. In MRI studies, Feindel *et al.* (2007b) noted that H–D exchange began sooner over a large portion of the PEM between catalyst layers when the PEM was saturated with D_2O (~10 min.) than when operating under low RH (~40–60 min.), despite that in both cases the total ionic current passing through the PEM was similar. The authors therefore proposed that a smaller volume of the hydrophilic domains are involved in ion transport and that the current per conductive area is higher when the PEM is operated at low RH than when saturated with water.

In addition, the period of time required for H–D exchange to be evident with water in the PEM was longer when operating on $H_2(g)$ (non-saturated D_2O containing PEM) versus $D_2(g)$ (non-saturated H_2O containing PEM), consistent with expected isotope effects; however, no effect on FC operating performance was noted. After the FC was operated with $D_2(g)$ and subsequently fuelled with $H_2(g)$, ~60 min. of operating was required before the ^{1}H MRI signal was evident in the MEA and cathode flow field, and the signal continued to increase over a period of 120 min. Deuteron exchange rate for D_2O is slower than proton exchange rate for H_2O; therefore one might expect the recovery of the ^{1}H NMR signal in a D_2O-containing PEM to take longer than the signal decline in a H_2O-containing PEM.

The variation in the distribution of water in the PEM for the co-flow versus counter-flow gas configurations observed using H–D exchange (Feindel *et al.*, 2007b) was consistent with the results reported when starting the PEMFC with low water content (Feindel *et al.*, 2006b); in co-flow gas configurations the area of the PEM near the gas inlets contained less water than near the outlets, and in counter-flow gas configurations the in-plane distribution of water appeared uniform.

10.11.3 Across the thickness of the PEM

Tsushima *et al.* (2006b) investigated the change in ^{1}H MRI signal across a 340 μm Aciplex PEM as the FC was operated on $D_2(g)$ as fuel at constant current and voltage. A SE MRI experiment was used to obtain a resolution of 40 μm across the thickness of the PEM. When $D_2(g)$ was supplied to the FC, decrease in the ^{1}H MRI signal was evident at the anode within 150 s. After operating for > 300 s, there was little remaining ^{1}H MRI signal from the PEM. In addition, the decrease in ^{1}H MRI signal was expedited when operating the FC at a higher current density. The authors interpreted these results with respect to the various proton-transport mechanisms in a PEM and concluded that the Grotthus mechanism dominated proton (and deuteron) conduction; Grotthus transport facilitates more rapid net proton flow to the cathode than direct vehicular and surface transport of protons obtained from oxidation of the fuel. This interpretation is consistent with their observation that upon using $D_2(g)$ as fuel, the ^1H MRI signal decreases rapidly at the anode, but only after a significant delay at the cathode (Tsushima *et al.*, 2006b).

The PEM hydration path (i.e. anode humidification, cathode humidification, or generated water) was investigated by using various combinations of $H_2(g)$ or $D_2(g)$ as fuel with $H_2O(l)$ or $D_2O(l)$ to humidify the anode and cathode gas streams (Kotaka *et al.*, 2007; Hirai and Tsushima, 2008). A FC with a single channel (3 mm wide, 30 mm long) flow field, imbedded glass heaters, and Nafion–1110 with a 1 cm × 4 cm catalyst decal was used. Gas temperature and humidity were controlled to 60 °C and 84% RH with a flow rate of 100 mL min^{-1}. Prior to MRI experiments, the FC was operated for 2 hrs to reach a steady state. A SE imaging sequence was used to obtain a resolution of 50 μm × 400 μm (FOV = 6.4 mm × 51.2 mm) along the thickness of the PEM in 34 min. Profiles of the ^1H MRI signal across the PEM for each condition are shown in Fig. 10.19. The effects of operating the PEMFC with $D_2(g)$ as fuel and humidifying either the anode or cathode gas stream with H_2O are shown in Fig. 10.19, plots a and b, respectively. Operating with $H_2(g)$ as fuel (both gas streams humidified with $D_2O(l)$) versus $D_2(g)$ as fuel

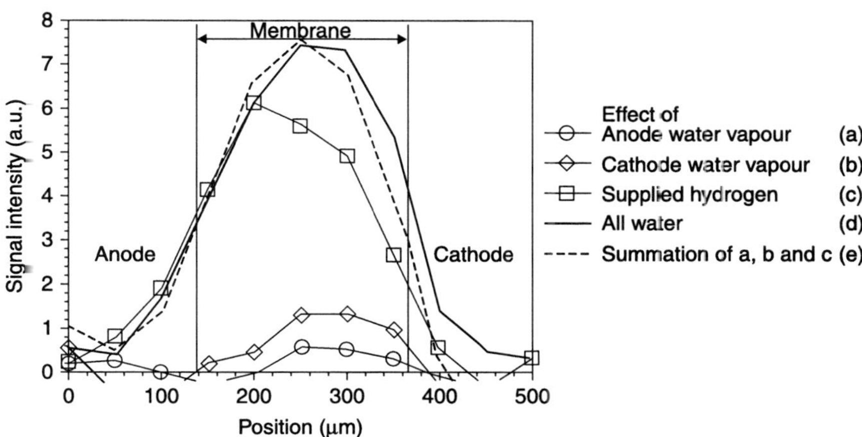

10.19 Through-plane one dimensional water distribution across the PEM resulting from different experiments using H–D contrast. (a) $D_2(g)/H_2O(l)$ and $O_2(g)/D_2O(l)$. (b) $D_2(g)/D_2O(l)$ and $O_2(g)/H_2O(l)$. (c) $H_2(g)/D_2O(l)$ and $O_2(g)/D_2O(l)$. (d) $H_2(g)/H_2O(l)$ and $O_2(g)/H_2O(l)$. (e) sum of a, b, and c. The lines denote the edge of the PEM and the anode is on the left side. (Source: Fig. 4 in Kotaka *et al.* (2007) by permission of ECS –The Electrochemical Society.)

(both gas streams humidified with $H_2O(l)$), demonstrated that water in the PEM resulted from the electrochemical reaction, and not from the humidified gases (Fig. 10.19 plot c). The authors concluded that under these operating conditions, water vapour from the anode and the cathode had little effect on the PEM hydration.

10.12 Application to direct alcohol fuel cells

The majority of investigations that used MRI to study PEMFCs focused on the distribution of water in hydrogen fuel cells. However, the inherent sensitivity of NMR to different chemical species has been used to investigate the chemical reactions in direct alcohol fuel cells as reviewed in this section.

10.12.1 NMR spectroscopy

As noted in the introduction, NMR has found use in a variety of applications, and is particularly important for chemical species identification and molecular structure analysis. NMR spectroscopy investigations have been used to improve the understanding of the morphology of PEMs (Mauritz and Moore, 2004), the state and behaviour of water within them, and the mechanisms of proton conduction (Traer and Goward, 2010; Ye *et al.*, 2006; Akbey *et al.*, 2009). These studies utilize NMR experiments to act as molecular-level probes of bulk or average material properties. The ^{1}H NMR chemical shift of water in Nafion has long been known to be dependent on the amount of water in the PEM (Bunce *et al.*, 1986).

Recent studies have shown that insight about the chemical reactions in direct alcohol FCs can be obtained by using NMR spectroscopy. ^{2}H and ^{13}C solid-state NMR spectroscopy were used to study the chemical species present in the PEM and exhaust of a DMFC (Paik *et al.*, 2008). Subsequently, ^{13}C solution and solid-state NMR spectroscopy were utilized to investigate direct ethanol FCs (Paik *et al.*, 2009). In each study, the MEA was composed of a hot-pressed triple-layer of Nafion–117 sandwiched between catalyst-sprayed carbon cloth acting as the electrodes (2.5 cm × 1.8 cm). As the PtRu/C anode and Pt/C cathode catalysts were applied to the outer PEMs, the middle PEM could be extracted free from the electrode components (e.g. catalysts, carbon black, and carbon cloth).

^{2}H and ^{13}C NMR studies of a DMFC

The natural abundance of ^{2}H and ^{13}C are 0.015% and 1.1%, respectively. Thus, to facilitate the NMR experiments, a conditioned FC was operated with an isotopically enriched fuel (i.e. 2 M CD_3OH or 2 M $^{13}CH_3OH$) (Paik *et al.*, 2008). When operating the DMFC on CD_3OH, the authors found that the power density was reduced by about 10% due to the deuterium isotope effect observed during the cleavage of methanol on the anode. After operating at 50 mA cm^{-2} for 15–30 min. the middle PEM layer of the MEA was removed and freeze-milled at 77 K to obtain a powder for magic angle spinning (MAS) ^{2}H and ^{13}C NMR experiments at 9.4 T (^{2}H NMR frequency ~60.6 MHz, ^{13}C NMR frequency ~100 MHz). As shown in Fig. 10.20, apart from the dominant ^{13}C NMR signal from methanol, the NMR spectrum indicates the presence of formic acid (HCOOH) and methyl formate (HCOOCH$_3$). In addition, the authors infer that formaldehyde (HCOH) is formed during DMFC operation, but reacts readily with methanol to form a small amount of methoxyhydroxymethane (H$_2$C(OH)OCH$_3$). A DMFC was also operated with Pt/C catalysts at both the anode and the cathode. The same reaction intermediate species were present; however, the Pt/C catalyst at the anode produced three times more formic acid than PtRu/C (Paik *et al.*, 2008).

The dominant ^{2}H MAS NMR signals were from CD$_3$OH and HOD or D$_2$O (Paik *et al.*, 2008). A small signal corresponding to DCOO$^-$ was also evident. The ^{2}H chemical shift for D$_2$O was located at 5.6 ppm, shifted significantly from pure water (4.8 ppm), reflecting the acidic environments of the ion pores in Nafion (Bunce *et al.*, 1986). The chemical shift of the methyl groups was not shifted, indicating that they do not interact directly with the sulfonyl protons of the Nafion side chains.

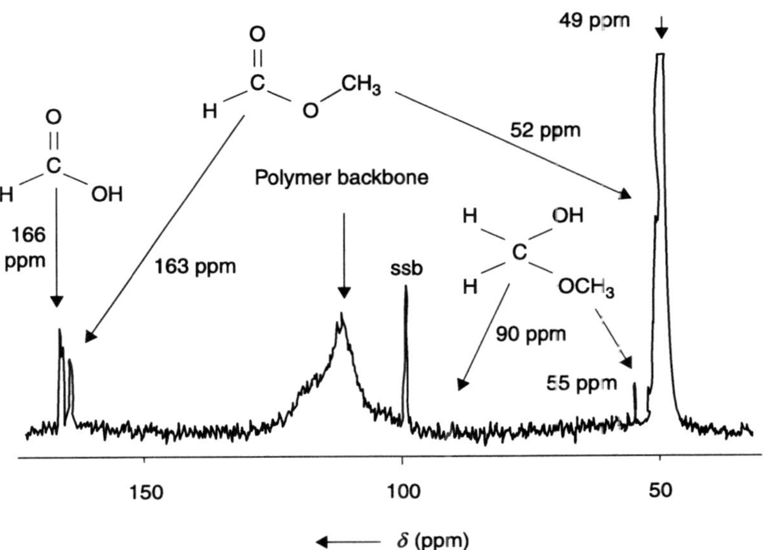

10.20 ^{13}C MAS NMR spectrum of the middle PEM layer extracted from the MEA of a DMFC after operation on 2M ^{13}CH$_3$OH. The species assigned to the peaks are shown in the spectrum (see text for more details). The methanol peak (at δ = 49 ppm) is truncated to 10% of the height to achieve an appropriate display of other peaks. The sample was spun at the magic angle in a 4 mm rotor at a rate of 5 kHz, and the spinning sideband (ssb) of the methanol peak is marked in the spectrum. (Source: Fig. 3 in Paik *et al.* (2008). Copyright Wiley-VCH Verlag GmbH & Co. KGaA. Reproduced with permission.)

^{13}C NMR studies of a DEFC

The electro-catalytic oxidation of ethanol is more complex than that of methanol. Paik *et al.* (2009) have used both solution and solid-state ^{13}C NMR methods to identify and quantify the distribution of reaction intermediates and the fuel in the PEM and fuel pathways of a DEFC. A conditioned DEFC was operated for 15 min. with 2 M $CH_3$$^{13}CH_2OH$ at 80 °C and 50 mA cm^{-2} and exhaust was collected in 1 mL samples for analysis with ^{13}C solution NMR experiments at 11.7 T (^{13}C NMR frequency ~125 MHz). A known amount of $^{13}CH_3OH$ was added as an internal standard for quantification. From the NMR spectrum (Fig. 10.21), the authors estimated that $17 \pm 1\%$ of ethanol in the fuel was consumed. Numerous reaction products were found in the exhaust, including acetic acid (CH_3COOH) (17 mM) and acetaldehyde (CH_3COH) and its derivatives acetaldehyde gem-dihydroxyethane, methoxyhydroxyethane, and ethoxyhydroxyethane (23.2 mM total).

The dominant ^{13}C solid state NMR signal was from ethanol in the PEM (110 μmoles). The reaction intermediates amounted to 43% of the ethanol signal in the PEM, and was almost of an order of magnitude higher than observed in the anode exhaust (in contrast to what they observed in DMFCs (Paik *et al.*, 2008)). The authors attribute the discrepancy between the reaction intermediates in the PEM and in the exhaust to differences in the electro-catalytic reactions at the electrodes and in the cross-over rates of the chemical species under the FC operating conditions (Song *et al.*, 2005).

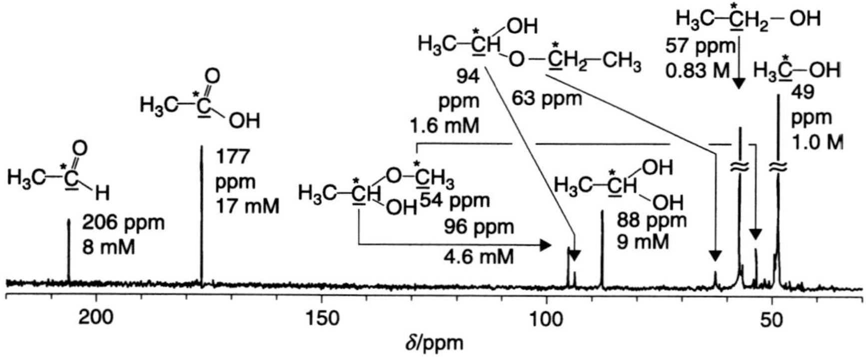

10.21 ^{13}C solution NMR spectrum of the exhaust from a DEFC containing an MEA with a triple-layer PEM, operated on 2 M $CH_3$$^{13}CH_2OH$ solution at 80 °C. The chemical species assigned to the peaks are shown in the spectrum. The ^{13}C-labelled sites are marked by asterisks and the carbons responsible for the ^{13}C NMR signal are underlined. (Source: Fig. 2 in Paik *et al.* (2009) with permission from Elsevier.)

10.13 Advantages and limitations

Each technique used to study the distribution of water in an operating PEMFC has advantages and limitations. The information gained from each modality is complementary, either providing unique insight or corroborating other measurements. Understanding the limitations and determining the utility of a technique is critical for obtaining reliable data and achieving meaningful outcomes.

There are several advantages of using MRI to investigate the distribution of water in operating PEMFCs. MRI affords the capability to define ROIs in arbitrary dimensions and locations within the PEMFC. This has proven useful for isolating water in specific PEMFC components (e.g. water in the PEM separate from water in the flow channels). MRI shares an advantage with neutron imaging, in that the distribution of water can be obtained through optically opaque materials. In addition, MRI can be used to investigate water in hydrocarbon membranes.

10.13.1 Spatial and temporal resolution

The dilemma of spatial versus temporal resolution in an MRI study of a PEMFC was discussed by Minard *et al.* (2006). The distribution of water in an operating PEMFC is inherently a four-dimensional function of space and time. In general, an increase in spatial resolution requires an increase in experimental time. In a 2D MRI sequence a decrease in slice thickness while maintaining planar resolution will reduce SNR, proportional to the change in slice thickness (i.e. change in volume of voxel). For example, to achieve the same SNR with one-fifth the slice thickness requires 25 times more averages, since SNR increases with the root of the number of averages. Thus, MRI protocols often establish a compromise between spatial and temporal resolution.

10.13.2 NMR/MRI hardware size and cost

The infrastructure for performing NMR spectroscopy experiments exists in essentially all major universities with a science-based research program. For most vertical wide-bore superconducting magnets used for solid-state NMR experiments, commercial solutions to enable MRI experiments can be easily obtained. Although most major hospitals have large horizontal-bore superconducting magnets for human medical imaging, MRI facilities for materials research are far less common. Medical MRI scanners typically have limitations on gradient slew rates, lower maximum gradient strengths, and limitations on RF power deposition compared to MRI scanners intended for materials imaging. To date, the majority of MRI investigations on operating PEMFCs have been performed on small purpose-built FCs, designed to operate within the small space of already available MRI systems.

The cost of superconducting magnets scales approximately linearly with increasing B_0, and with the square of bore radius. Beyond the initial investment in the NMR/MRI hardware (in the order of \$ millions), a facility also needs to commit substantial funds annually to maintain the infrastructure. Annual operating costs include cryogens (i.e. $N_2(l)$, $He(l)$), inevitable repairs, additional hardware (e.g. RF coils), in addition to salaries for well-trained technical staff.

As noted by Minard *et al.* (2006), low-cost MRI alternatives are possible, relying on low-field permanent, rather than superconducting, magnets. To obtain the same SNR at lower B_0, either an increased voxel size with the same experimental time, or vice versa, is required. Or, if improved spatial resolution is required, temporal resolution can be sacrificed. For example if B_0 was reduced from 11.7 T to 0.2 T, a 2D SE image with the same SNR could be obtained in 128 s by increasing the pixel size from 0.55 mm × 0.55 mm to 3.8 mm × 3.8 mm (Minard *et al.*, 2006). Spatial resolution of a few millimetres may be sufficient for diagnostics on larger FCs with larger unobstructed regions. In addition, in exchange for reduced SNR at lower B_0, RF penetration increases inversely with B_0^2, and field variations decrease linearly with B_0, which reduces artefacts resulting from off-resonance water in close proximity to FC materials. Low-field MRI may reduce the extent to which standard PEMFC components are substituted with non-conductive materials, and open the possibility of visualizing water gradients inside of the GDLs.

10.14 Future trends

MRI has been demonstrated to be effective for investigating the distribution of water in the PEM and flow fields of operating PEMFCs. Understanding the interplay of $H_2O(l)$ accumulation, the amount of water in the PEM, and FC power output is critical for optimization of operating conditions. Research to date has demonstrated that widely different results can be obtained depending on PEMFC design and operating conditions. Although this exemplifies the complicated interdependence of PEMFC operating conditions, power output, and design, in the absence of standardization of design or PEM preparation, correlation of the results from independent MRI investigations is difficult.

Significant advances have been made toward quantification of the amount of water in operating PEMFCs. However, further research needs to be completed to understand the effects of PEM pre-treatment, MEA preparation, and mechanical compression. MRI studies are particularly well suited to study the effects of flow field materials and design on the accumulation and transport of $H_2O(l)$.

Recently developed sequences for imaging materials with ultra-short T_E (Idiyatullin *et al.*, 2006; Robson *et al.*, 2003) remain to be applied to operating PEMFCs. In addition, a foray into the application of low-field MRI for investigating commercial PEMFC designs is certainly warranted.

10.15 Sources of further information

There is a large variety of monographs and reviews available on the subjects of NMR and MRI, with applications to both materials science and medicine. The *Concepts in Magnetic Resonance* journals published by Wiley are a great source of concise presentations of many fundamental aspects of magnetic resonance. The books mentioned below are far from an exhaustive list, and are simply intended to serve as a brief guide to material I have found informative.

Abragam A (1961), *The Principles of Nuclear Magnetism*, Oxford, Oxford University Press.
Blümich B (2000), *NMR Imaging of Materials*, Oxford, Oxford University Press.
Callaghan PT (1991), *Principles of Nuclear Magnetic Resonance Microscopy*, Oxford, Clarendon Press.
Codd SL and Seymour JD (eds) (2009), *Magnetic Resonance Microscopy Spatially Resolved NMR Techniques and Applications*, Chichester, Wiley-VCH.
Fukushima E (1986), *Experimental Pulse NMR: A Nuts and Bolts Approach*, Reading, Addison-Wesley.
Haacke E (1999), *Magnetic Resonance Imaging: Physical Principles and Sequence Design*, New York, Wiley.
Kimmich R (1997), *NMR: Tomography, Diffusometry, Relaxometry*, Berlin, Springer.
Levitt M (2008), *Spin Dynamics: Basics of Nuclear Magnetic Resonance* 2nd ed., Chichester, John Wiley & Sons.
McRobbie D (2007), *MRI from Picture to Proton* 2nd ed., Cambridge, Cambridge University Press.
Mitchell DG and Cohen MS (2004), *MRI Principles* 2nd ed., Philadelphia, Saunders.
Price W S (2009), *NMR Studies of Translational Motion: Principles and Application*, New York, Cambridge University Press.
Slichter C (1990), *Principles of Magnetic Resonance* 3rd ed., Berlin, Springer-Verlag.
Stapf S and Han S (eds) (2006), *NMR Imaging in Chemical Engineering*, Weinheim, Wiley-VCH.

10.16 References

Abbey Ü, Granados-Focil S, Coughlin EB, Graf R and Spiess HW (2009), '^{1}H solid-state NMR investigation of structure and dynamics of anhydrous proton conducting triazole-functionalized siloxane polymers', *J Phys Chem B*, **113**, 9151–9160. doi: 10.1021/jp9030909
Aleksandrova E, Hiesgen R, Eberhard D, Friedrich KA, Kaz T *et al.* (2007), 'Proton conductivity study of a fuel cell membrane with nanoscale resolution', *ChemPhysChem*, **8**, 519–522. doi: 10.1002/cphc.200600704
Barbir F (2005), *PEM Fuel Cells: Theory and Practice*, San Diego, Elsevier Academic.
Bedet J, Mutzenhardt P, Canet D, Maranzana G, Leclerc S *et al.* (2008a), 'Étude du comportement de l'eau dans une pile à combustible à membrane échangeuse d'ions (PEMFC): étude par RMN et IRM', *C R Chim*, **11**, 465–473. doi: 10.1016/j.crci.2007.07.004
Bedet J, Maranzana G, Leclerc S, Lottin O, Moyne C *et al.* (2008b), 'Magnetic resonance imaging of water distribution and production in a 6 cm^2 PEMFC under operation', *Int J Hydrogen Energy*, **33**, 3146–3149. doi: 10.1016/j.ijhydene.2008.01.053

Bloch F (1946), 'Nuclear induction', *Phys Rev*, **70**, 460–474. doi: 10.1103/PhysRev.70.460.

Bloembergen N, Purcell E and Pound RV (1948), 'Relaxation effects in nuclear magnetic resonance absorption', *Phys Rev*, **73**, 679–712. doi: 10.1103/PhysRev.73.679

Blümich B (2000), *NMR Imaging of Materials*, Oxford, Oxford University Press.

Boyle NG, McBrierty VJ and Douglass DC (1983), 'The behavior of water in Nafion membranes', *Macromolecules*, **16**, 75–80. doi: 10.1021/ma00235a015

Bunce NJ, Sondheimer SJ and Fyfe CA (1986), 'Proton NMR method for the quantitative determination of the water content of the polymeric fluorosulfonic acid Nafion-H', *Macromolecules*, **19**, 333–339. doi: 10.1021/ma00156a015

Callaghan PT (1991), *Principles of Nuclear Magnetic Resonance Microscopy*, Oxford, Clarendon Press.

Canet D (1997), 'Radiofrequency field gradient experiments', *Prog Nucl Magn Reson Spectrosc*, **30**, 101–135. doi: 10.1016/S0079–6565(96)01031-X

Cappadonia M, Erning J, Niaki S and Stimming U (1995), 'Conductance of Nafion 117 membranes as a function of temperature and water content', *Solid State Ionics*, **77**, 65–69. doi: 10.1016/0167–2738(94)00289–5

Carr HY and Purcell EM (1954), 'Effects of diffusion on free precession in nuclear magnetic resonance experiments', *Phys Rev*, **94**, 630–638. doi: 10.1103/PhysRev.94.630

Chen RS, Stallworth PE, Greenbaum SG, Fontanella JJ and Wintersgill MC (1995), 'High pressure NMR and electrical conductivity studies in acid form NAFION membranes', *Electrochimica Acta*, **40**, 309–313.

Cheng D (1989), *Field and Wave Electromagnetics*, Reading, Addison-Wesley.

Ciobanu L, Seeber D and Pennington C (2002), '3D MR microscopy with resolution 3.7 μm by 3.3 μm by 3.3 μm', *J Magn Reson*, **158**, 178–182. doi: 10.1016/S1090–7807(02)00071-X

Codd SL and Seymour JD (eds) (2009), *Magnetic Resonance Microscopy Spatially Resolved NMR Techniques and Applications*, Chichester, Wiley-VCH.

Doty FD, Entzminger G, Kulkarni J, Pamarthy K and Staab JP (2007), 'Radio frequency coil technology for small-animal MRI', *NMR Biomed*, **20**, 304–325. doi: 10.1002/nbm.1149

Dunbar Z and Masel RI (2007), 'Quantitative MRI study of water distribution during operation of a PEM fuel cell using Teflon® flow fields', *J Power Sources*, **171**, 678–687. doi: 10.1016/j.jpowsour.2007.06.207

Dunbar ZW and Masel RI (2008a), 'Magnetic resonance imaging investigation of water accumulation and transport in graphite flow fields in a polymer electrolyte membrane fuel cell: Do defects control transport?', *J Power Sources*, **182**, 76–82. doi: 10.1016/j.jpowsour.2008.03.057

Dunbar ZW and Masel RI (2008b), 'Magnetic resonance imaging as an *in-situ* diagnostic method to characterize water flooding', *ECS Trans*, **16**, 1001–1008. doi: 10.1149/1.2981940

Eccles CD and Callaghan PT (1986), 'High-resolution imaging. The NMR microscope', *J Magn Reson*, **68**, 393–398. doi: 10.1016/0022–2364(86)90261–1

Ernst RR (1992), 'Nuclear magnetic resonance Fourier transform spectroscopy (Nobel Lecture)', *Angew Chem Int Ed*, **31**, 805–823. doi: 10.1002/anie.199208053

Feindel KW, Bergens SH and Wasylishen RE (2006a), 'The use of ^{1}H NMR microscopy to study proton-exchange membrane fuel cells', *ChemPhysChem*, **7**, 67–75. doi: 10.1002/cphc.200500504

Feindel KW, Bergens SH and Wasylishen RE (2006b), 'Insights into the distribution of water in a self-humidifying H_2/O_2 proton-exchange membrane fuel cell using ^{1}H NMR microscopy', *J Am Chem Soc*, **128**, 14192–14199. doi: 10.1021/ja064389n

Feindel KW, Bergens SH and Wasylishen RE (2007a), 'The influence of membrane electrode assembly water content on the performance of a polymer electrolyte membrane fuel cell as investigated by ^{1}H NMR microscopy', *Phys Chem Chem Phys*, **9**, 1850–1857. doi: 10.1039/b617551a

Feindel KW, Bergens SH and Wasylishen RE (2007b), 'Use of hydrogen–deuterium exchange for contrast in ^{1}H NMR microscopy investigations of an operating PEM fuel cell', *J Power Sources*, **173**, 86–95. doi: 10.1016/j.jpowsour.2007.04.079

Feindel KW, LaRocque LP-A, Starke D, Bergens SH and Wasylishen RE (2004), '*In situ* observations of water production and distribution in an operating H_2/O_2 PEM fuel cell assembly using ^{1}H NMR microscopy', *J Am Chem Soc*, **126**, 11436–11437. doi: 10.1021/ja0461116

Fontanella JJ, Edmondson CA, Wintersgill MC, Wu Y and Greenbaum SG (1996), 'High-pressure electrical conductivity and NMR studies in variable equivalent weight NAFION membranes', *Macromolecules*, **29**, 4944–4951. doi: 10.1021/ma9600926

Glasel JA (1972), 'Nuclear magnetic resonance studies of water and ice', in Franks F (ed), *Water: A Comprehensive Treatise*. New York, Plenum Press, 215–254.

Haacke E (1999), *Magnetic Resonance Imaging: Physical Principles and Sequence Design*, New York, Wiley.

Harris RK, Becker ED, de Menezes SMC, Goodfellow R and Granger P (2001), 'NMR nomenclature. Nuclear spin properties and conventions for chemical shifts (IUPAC Recommendations 2001)', *Pure Appl Chem*, **73**, 1795–1818. doi: 10.1351/pac200173111795

Hausser R and Noack F (1965), 'Kernmagnetische relaxation und korrelation im system wasser – sauerstoff', *Z Naturforsch*, **20A**, 1668–1675.

Hayes C, Edelstein W, Schenck J, Mueller O and Eash M (1985), 'An efficient, highly homogeneous radiofrequency coil for whole-body NMR imaging at 1.5 T', *J Magn Reson*, **63**, 622–628. doi: 10.1016/0022–2364(85)90257–4

Hietatsu JT, Mizuhata M and Takenaka H (1994), 'Water uptake of perfluorosulfonic acid membranes from liquid water and water vapor', *J Electrochem Soc*, **141**, 1493–1498.

Hiai S and Tsushima S (2008), 'Water Transport and Degradation Analysis in PEMFC by *in-situ* MRI Visualization', *ECS Trans*, **16**, 1337–1343. doi: 10.1149/1.2981974

Holt DI and Richards RE (1975), 'Critical factors in the design of sensitive high resolution nuclear magnetic resonance spectrometers', *Proc R Soc A*, **344**, 311–340. doi: 10.1098/rspa.1975.0104

Holt DI and Richards RE (1976), 'The signal-to-noise ratio of the nuclear magnetic resonance experiment', *J Magn Reson*, **24**, 71–85. doi: 10.1016/0022–2364(76)90233-X

Idiyatullin D, Corum C, Park J-Y and Garwood M (2006), 'Fast and quiet MRI using a swept radiofrequency', *J Magn Reson*, **181**, 342–349. doi: 10.1016/j.jmr.2006.05.014

Ikeda T, Koido T, Tsushima S and Hirai S (2008), 'MRI investigation of water transport mechanism in a membrane under elevated temperature condition with relative humidity and current density variation', *ECS Trans*, **16**, 1035–1040. doi: 10.1149/1.2981944

Kasaounis A, Balomenou S, Tsiplakides D, Brosda S, Neophytides S *et al.* (2005), 'Proton tunneling-induced bistability, oscillations and enhanced performance of PEM fuel cells', *Appl Catal B-Environ*, **56**, 251–258. doi: 10.1016/j.apcatb.2004.08.018

Kimmich R (1997), *NMR: Tomography, Diffusometry, Relaxometry*, Berlin, Springer.

Kotaka T, Tsushima S and Hirai S (2007), 'Visualization of membrane hydration path in an operating PEMFC by nuclei-labeling MRI', *ECS Trans*, **11**, 445–450. doi: 10.1149/1.2780958

Lauer U, Graf H, Berger A, Claussen C and Schick F (2005), 'Radio frequency versus susceptibility effects of small conductive implants — a systematic MRI study on aneurysm clips at 1.5 and 3 T', *Magn Reson Imaging*, **23**, 563–569. doi: 10.1016/j.mri.2005.02.012

Lauterbur PC (2005), 'All science is interdisciplinary – from magnetic moments to molecules to men (Nobel Lecture)', *Angew Chem Int Ed*, **44**, 1004–1011. doi: 10.1002/anie.200462400

Lee S, Kim K, Kim J, Lee S, Yi J *et al.* (2001), 'One micrometer resolution NMR microscopy', *J Magn Reson*, **150**, 207–213. doi: 10.1006/jmre.2001.2319

Lei H and Dunn J (2001), 'The effects of slice-selective excitation/refocusing in localized spectral editing with gradient-selected double-quantum coherence transfer', *J Magn Reson*, **150**, 17–25. doi: 10.1006/jmre.2001.2304

Levitt M (2008), *Spin Dynamics: Basics of Nuclear Magnetic Resonance* 2nd ed., Chichester, John Wiley & Sons.

MacMillan B, Sharp AR and Armstrong RL (1999), 'An n.m.r. investigation of the dynamical characteristics of water absorbed in Nafion', *Polymer*, **40**, 2471–2480. doi: 10.1016/S0032–3861(98)00484–4

Maffei P, Kiene L and Canet D (1992), 'Application of NMR microimaging by radio-frequency field gradients to the observation of solvent penetration in polymeric materials', *Macromolecules*, **25**, 7114–7118. doi: 10.1021/ma00052a007

Mansfield P (1976), 'Proton spin imaging by nuclear magnetic resonance', *Contemp Phys*, **17**, 553–576.

Mansfield P and Chapman B (1986), 'Active magnetic screening of gradient coils in NMR imaging', *J Magn Reson*, **66**, 573–576. doi: 10.1016/0022–2364(86)90205–2

Mansfield P and Grannell PK (1973), 'NMR "diffraction" in solids?', *J Phys C: Solid State Phys*, **6**, L422-L426.

Mansfield P and Grannell PK (1975), ' "Diffraction" and microscopy in solids and liquids by NMR', *Phys Rev B*, **12**, 3618–3634. doi: 10.1103/PhysRevB.12.3618

Mansfield P (2004), 'Snapshot magnetic resonance imaging (Nobel Lecture)', *Angew Chem Int Ed*, **43**, 5456–5464. doi: 10.1002/anie.200460078

Mauritz KA and Moore RB (2004), 'State of understanding of Nafion', *Chem Rev*, **104**, 4535–4586. doi: 10.1021/cr0207123

McLean RS, Doyle M and Sauer BB (2000), 'High-resolution imaging of ionic domains and crystal morphology in ionomers using AFM techniques', *Macromolecules*, **33**, 6541–6550. doi: 10.1021/ma000464h

Meiboom S and Gill D (1958), 'Modified spin-echo method for measuring nuclear relaxation times', *Rev Sci Instrum*, **29**, 688–691. doi: 10.1063/1.1716296

Minard KR, Viswanathan VV, Majors PD, Wang LQ and Rieke PC (2006), 'Magnetic resonance imaging (MRI) of PEM dehydration and gas manifold flooding during continuous fuel cell operation', *J Power Sources*, **161**, 856–863. doi: 10.1016/j.jpowsour.2006.04.125

Mitchell DG and Cohen MS (2004), *MRI Principles* 2nd ed., Philadelphia, Saunders.

Moore WS and Holland GN (1980), 'Experimental considerations in implementing a whole body multiple sensitive point nuclear magnetic resonance imaging system', *Phil Trans R Soc B*, **289**, 511–518. doi: 10.1098/rstb.1980.0070

Ogawa S, Lee T-M, Nayak AS and Glynn P (1990), 'Oxygenation-sensitive contrast in magnetic resonance image of rodent brain at high magnetic fields', *Magn Reson Med*, **14**, 68–78. doi: 10.1002/mrm.1910140108

Ouriadov AV, MacGregor RP and Balcom BJ (2004), 'Thin film MRI–high resolution depth imaging with a local surface coil and spin echo SPI', *J Magn Reson*, **169**, 174–186. doi: 10.1016/j.jmr.2004.04.015

Paik Y, Kim SS and Han OH (2009), 'Spatial distribution of reaction products in direct ethanol fuel cell', *Electrochem Commun*, **11**, 302–304. doi: 10.1016/j.elecom.2008.11.036

Paik Y, Kim SS and Han OH (2008), 'Methanol behavior in direct methanol fuel cells', *Angew Chem Int Ed*, **47**, 94–96. doi: 10.1002/anie.200703190

Robson MD, Gatehouse PD, Bydder M and Bydder GM (2003), 'Magnetic resonance: an introduction to ultrashort TE (UTE) imaging', *J Comput Assist Tomogr*, **27**, 825–846.

Shim JY, Tsushima S and Hirai S (2009), 'High resolution MRI investigation of transversal water content distributions in PEM under fuel cell operation', *ECS Trans*, **25**, 523–528. doi: 10.1149/1.3210602

Song S, Zhou W, Liang Z, Cai R, Sun G *et al.* (2005), 'The effect of methanol and ethanol cross-over on the performance of PtRu/C-based anode DAFCs', *Appl Catal B-Environ*, **55**, 65–72. doi: 10.1016/j.apcatb.2004.05.017

Stapf S and Han S-I (eds) (2006), *NMR imaging in chemical engineering*, Weinheim, Wiley-VCH.

Stejskal EO and Tanner JE (1965), 'Spin diffusion measurements: Spin echoes in the presence of a time-dependent field gradient', *J Chem Phys*, **42**, 288–292. doi: 10.1063/1.1695690

Tanner JE (1965), 'Pulsed field gradients for NMR spin-echo diffusion measurements', *Rev Sci Instrum*, **36**, 1086–1087. doi: 10.1063/1.1719808

Teranishi K, Tsushima S and Hirai S (2002), 'Measurement of water distribution in polymer electrolyte fuel cell by MRI', *Therm Sci Eng*, **10**, 59–60.

Teranishi K, Tsushima S and Hirai S (2004), 'Study on mass transfer in a polymer electrolyte fuel cell by using magnetic resonance imaging', *Therm Sci Eng*, **12**, 91–92.

Teranishi K, Tsushima S and Hirai S (2006), 'Analysis of water transport in PEFCs by magnetic resonance imaging measurement', *J Electrochem Soc*, **153**, A664–A668. doi: 10.1149/1.2167954

Teranishi K, Tsushima S and Hirai S (2003), 'Membrane thickness effect on PEFC's performance by measurement of water distribution', *Therm Sci Eng*, **11**, 35–36.

Teranishi K, Tsushima S and Hirai S (2005), 'Study of the effect of membrane thickness on the performance of polymer electrolyte fuel cells by water distribution in a membrane', *Electrochem Solid-State Lett*, **8**, A281–A284. doi: 10.1149/1.1897343

Traer JW and Goward GR (2010), 'The proton dynamics of imidazole methylphosphonate: an example of cooperative ionic conductivity', *Phys Chem Chem Phys*, **12**, 263–272. doi: 10.1039/b917360a

Tsampas M, Pikos A, Brosda S, Katsaounis A and Vayenas C (2006), 'The effect of membrane thickness on the conductivity of Nafion', *Electrochim Acta*, **51**, 2743–2755. doi: 10.1016/j.electacta.2005.08.021

Tsushima S, Takita S, Hirai S, Kubo N and Aotani K (2009), 'Magnetic resonance imaging of a polymer electrolyte membrane under water permeation', *Expt Heat Transfer*, **22**, 1–11. doi: 10.1080/08916150802530484

Tsushima S, Teranishi K and Hirai S (2005a), 'Water diffusion measurement in fuel-cell SPE membrane by NMR', *Energy*, **30**, 235–245.

Tsushima S, Teranishi K, Nishida K and Hirai S (2005b), 'Water content distribution in a polymer electrolyte membrane for advanced fuel cell system with liquid water supply', *Magn Reson Imaging*, **23**, 255–258.

Tsushima S, Hirai S, Kitamura K, Yamashita M and Takase S (2007a), 'MRI application for clarifying fuel cell performance with variation of polymer electrolyte membranes: Comparison of water content of a hydrocarbon membrane and a perfluorinated membrane', *Appl Magn Reson*, **32**, 233–241. doi: 10.1007/s00723–007–0009–0

Tsushima S, Nanjo T and Hirai S (2007b), 'Investigation of hydrogen cross-leak behaviors in a polymer electrolyte membrane by direct gas mass spectroscopy and magnetic resonance imaging', *ECS Trans*, **11**, 435–443. doi: 10.1149/1.2780957

Tsushima S, Nanjo T, Nishida K and Hirai S (2006a), 'Investigation of the lateral water distribution in a proton exchange membrane in fuel cell operation by 3D-MRI', *ECS Trans*, **1**, 199–205. doi: 10.1149/1.2214490

Tsushima S, Teranishi K and Hirai S (2004), 'Magnetic resonance imaging of the water distribution within a polymer electrolyte membrane in fuel cells', *Electrochem Solid-State Lett*, **7**, A269–A272. doi: 10.1149/1.1774971

Tsushima S, Teranishi K and Hirai S (2003), 'Direct supply of water into a polymer electrolyte membrane for improvement of cell performance', *Therm Sci Eng*, **11**, 31–32.

Tsushima S, Teranishi K and Hirai S (2006b), 'Experimental elucidation of proton conducting mechanism in a polymer electrolyte membrane of fuel cell by nuclei labeling MRI', *ECS Trans*, **3**, 91–96. doi: 10.1149/1.2356127

Wang M, Feindel KW, Bergens SH and Wasylishen RE (2010), '*In situ* quantification of the in-plane water content in the Nafion® membrane of an operating polymer-electrolyte membrane fuel cell using ^{1}H micro-magnetic resonance imaging experiments', *J Power Sources*, **195**, 7316–7322. doi: 10.1016/j.jpowsour.2010.05.029

Xia Y (1996), 'Contrast in NMR imaging and microscopy', *Concepts Magn Reson*, **8**, 205–225. doi: 10.1002/(SICI)1099–0534(1996)8:3<205::AID-CMR4>3.0.CO;2–2

Ye G, Janzen N and Goward GR (2006), 'Solid-state NMR study of two classic proton conducting polymers: Nafion and sulfonated poly(ether ether ketone)s', *Macromolecules*, **39**, 3283–3290. doi: 10.1021/ma0523825

Zawodzinski TA, Derouin C, Radzinski S, Sherman RJ, Smith VT *et al.* (1993a), 'Water uptake by and transport through Nafion® 117 membranes', *J Electrochem Soc*, **140**, 1041. doi: 10.1149/1.2056194

Zawodzinski TA, Springer TE, Davey J, Jestel R, Lopez C *et al.* (1993b), 'A comparative study of water uptake by and transport through ionomeric fuel cell membranes', *J Electrochem Soc*, **140**, 1981–1985. doi: 10.1149/1.2220749

Zawodzinski TA, Neeman M, Sillerud LO and Gottesfeld S (1991), 'Determination of water diffusion coefficients in perfluorosulfonate ionomeric membranes', *J Phys Chem*, **95**, 6040–6044. doi: 10.1021/j100168a060

Zeeman P (1897), 'The effect of magnetism on the nature of light emmited by a substance', *Nature*, **55**, 347–351. doi: 10.1038/055347b0

Zhang Z, Martin J, Wu J, Wang H, Promislow K and Balcom BJ (2008), 'Magnetic resonance imaging of water content across the Nafion membrane in an operational PEM fuel cell', *J Magn Reson*, **193**, 259–266. doi: 10.1016/j.jmr.2008.05.005

Zweckstetter M and Holak TA (1998), 'An adiabatic multiple spin–echo pulse sequence: Removal of systematic errors due to pulse imperfections and off-resonance effects', *J Magn Reson*, **133**, 134–147. doi: 10.1006/jmre.1998.1437

11

Raman spectroscopy for polymer electrolyte membrane fuel cell characterization

H. BETTERMANN and P. FISCHER,
Heinrich-Heine-University of Düsseldorf, Germany

Abstract: This contribution shows how Raman spectroscopy can be used to characterize processes inside polymer electrolyte membrane (PEM) fuel cells. Two experimental setups have been designed for this purpose. In the first, a Raman microscope was constructed that allows hydrogen, oxygen, nitrogen, gaseous and liquid water to be detected. This device was used to detect changes in gas composition and water distribution induced by an artificial pinhole within the membrane electrolyte assembly (MEA). In the second setup, a multiple-fiber Raman spectrometer was developed to refine *in situ* fuel cell investigation. This setup allows gas composition, liquid water and water vapor to be recorded simultaneously and also reveals how these substances change locally during fuel cell operation.

Key words: Raman spectroscopy, PEM fuel cells, pinholes, *in situ* measurements, gas and water vapor distribution.

11.1   Introduction

In recent years, a number of different physical and physico-chemical methods have been adopted in the examination of fuel cells in operation (Borup *et al.*, 2007). These methods are primarily focused on changes in electrical parameters (Wu *et al.*, 2008a, 2008b). For instance, electrochemical impedance spectroscopy is a suitable tool for investigating electrochemical processes in greater detail and also allows the electrical resistance of the membrane unit to be determined (Yuan *et al.*, 2010, Kuhn *et al.*, 2006). Further insights into fuel cell operations, particularly water evolution (Tsukada *et al.*, 1999, Kramer *et al.*, 2005) and the distribution of specific substances (Geiger *et al.*, 2000) have been achieved through neutron tomography. In addition to the global setups, locally resolved methods involving segmented cells were used to monitor local current densities or the distribution of temperatures (Mench *et al.*, 2003).

This contribution intends to explain how Raman spectroscopy can be used to examine fuel cells from the inside, and to demonstrate the advantages that this technique can offer over the usual methods in terms of the information it can provide.

Raman spectroscopy is a spectroscopic tool that is currently used with success in a wide range of applications, in fields including pharmaceutical chemistry, environmental protection, material science and geoscience, and is thus well-

321

established in academic and industrial research laboratories. The main task of Raman spectroscopy is to evaluate the molecular structures of chemical compounds as well as their concentrations in gaseous, liquid and solid states. The main principle behind the technique is the recording of vibrations and rotations of sample molecules. This effect was discovered in 1928 by the Indian physicist Chandrasekhara Venkata Raman (1888–1970) (Raman and Krishnan, 1928), who in 1930 received the Nobel prize for his fundamental work on molecular light scattering.

One striking advantage of Raman spectroscopy is the simplicity of sample design and preparation. This enables measurements to be taken in locations that would otherwise be barely accessible (Chalmers and Griffiths, 2001). Furthermore, Raman signals generally experience less interference from light scattering by water. Raman spectroscopy is also one of the few methods able to quantitatively analyze homo-nuclear diatomic gases or mixtures of them (Ogilvie, 1998). The last two properties make Raman spectroscopy particularly suitable for the investigation of fuel cell processes.

As well as being used for *in situ* measurements focused on polymer electrolyte membrane (PEM) fuel cell operations, Raman spectroscopy also allows structural defects within membranes after different operation stages to be studied, and permits the identification and tracking of reaction products. This can be carried out by, for example, analyzing the changing patterns of molecular vibrations, and has been used in the investigation of degradation processes of membrane materials (Mattsson *et al.*, 2000) and in the detection of intermediates formed on solid oxide fuel cell (SOFC) cermet anodes (Pomfret *et al.*, 2006, 2007).

This chapter shows how Raman spectroscopy can be used to record the distribution of gas within flow field channels. Two experimental setups were designed for this purpose. First, a microscope objective was inserted into one bipolar plate. This allowed the detection of hydrogen, oxygen, nitrogen, water vapor and liquid water and proved that following gas distributions inside PEM fuel cells is indeed possible. This setup made possible detection of gases, and was used to record changes in gas compositions and water distribution induced by an artificial pinhole within the membrane electrolyte assembly (MEA).

The second setup employed a multiple-fiber Raman spectrometer with seven twin fibers, developed to refine the technique for examining fuel cells from the inside. This setup allows gas composition, liquid water and water vapor along a flow field channel to be recorded simultaneously, and provides answers to the question of how these species change locally and temporally during cell operations. The monitoring of local gas distributions and their time-dependent changes are primarily used as a means of identifying nascent degradation processes and the early stages of membrane damage.

Before presenting the Raman experiments, this chapter begins with a brief theoretical survey of the Raman effect and of the experimental conditions necessary to obtain spectral data.

11.2 Raman fundamentals

The basic principle behind the Raman effect is the inelastic scattering of monochromatic light. Inelastic light scattering means that the sample molecules emit light whose frequency differs from that of the excitation light. The basic experimental setup for recording scattered light is presented in Fig. 11.1.

The sample molecules are excited by light from a strong monochromatic light source, usually a laser. The excited sample molecules emit scattered light that is collected by a lens system. The lenses then image the scattered light to a monochromator unit in which it is dispersed spectrally. A typical spectrum of scattered light consists of three parts (Fig. 11.2).

The strong emission in the center of the spectrum is called Rayleigh scattering. The Rayleigh emission is spectrally unshifted, has the same frequency as the excitation light, and is accompanied by the Stokes- and anti-Stokes-lines of Raman emission. Stokes-lines are red-shifted and anti-Stokes-lines are blue-shifted in comparison to the excitation frequency.

11.2.1 Origin of Raman scattering

Molecular light scattering essentially originates from the direct interaction of a propagating light wave with a molecule. The electric field vector $\underline{E}$ of the oscillating electromagnetic wave acts on the electronic system of the sample molecule, deforming its electronic shell.

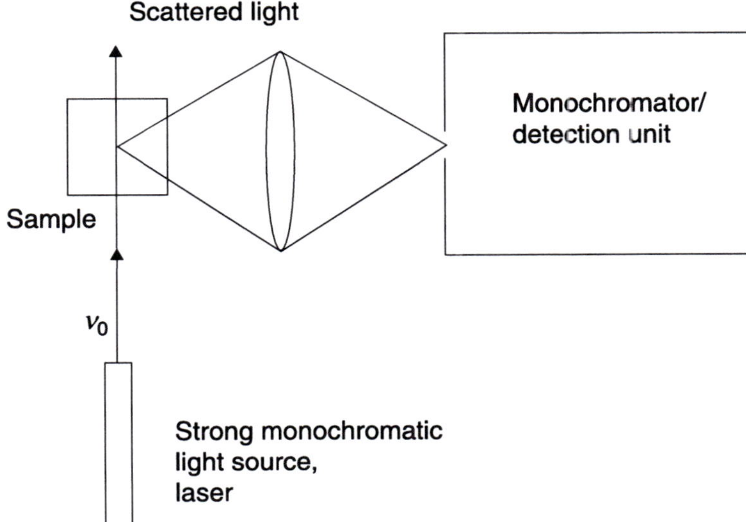

11.1 Basic experimental setup for the recording of Raman spectra.

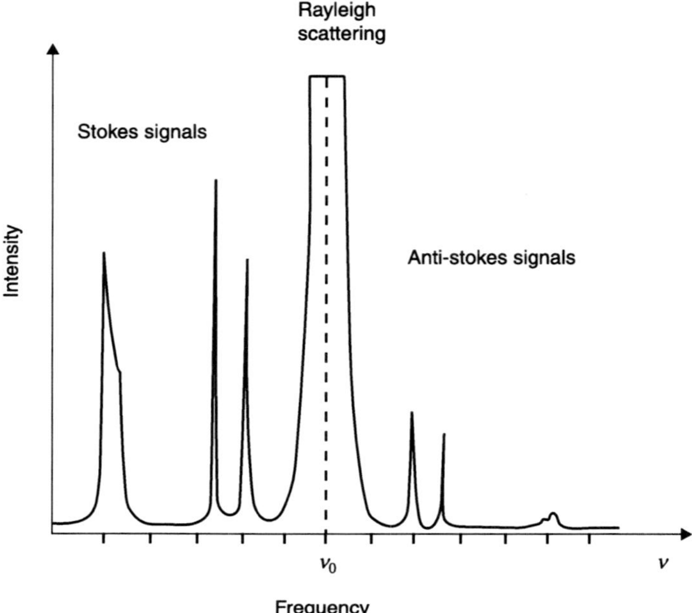

11.2 Schematic spectrum of molecular light scattering.

$$\underline{E} = \underline{E}_0 \cos 2\pi v_0 t \qquad [11.1]$$

where $\underline{E}_0$ is the amplitude of the electric field vector, v_0 is the frequency of the excitation light, t is time.

The deformation induces an electric dipole moment μ' within the molecule, which is directly proportional to the propagating electric field vector. Its magnitude depends on the size of the polarizability α. Polarizability is a specific molecular property that describes the tendency of a charge distribution to be distorted from its equilibrium disposition.

$$\underline{\mu}' = \alpha \cdot \underline{E} \qquad\qquad \underline{\mu}' = \alpha \cdot \underline{E}_0 \cos 2\pi v_0 t \qquad [11.2]$$

Because of the periodic oscillations of the attacking electromagnetic field, the molecular electronic shell also oscillates, and acts like a Hertz dipole, capable of emitting light. The total power P of the dipole radiation of one molecule depends on the magnitude of the squared induced dipole moment and the fourth power of the emission frequency v.

$$P = \frac{32\pi^4 v^4 \mu'^2}{3\varepsilon_0 c^3} \qquad [11.3]$$

where c is the speed of light, ε_0 is the electric field constant.

The deformation of the electronic shell includes temporal changes of the binding conditions within the molecule, causing nuclei to be displaced from their former equilibrium positions. This process can be expressed by a change in polarizability induced by the atom motion within the molecule. Mathematically, this dependence is expressed by a term that covers the changes in polarizability caused by the nuclear displacements along a selected normal mode Q_k. This contribution to the polarizability is added to the polarizability α_0 of the rigid molecular framework.

$$\alpha = \alpha_0 + \left(\frac{\partial \alpha}{\partial Q_k}\right)_0 Q_k \qquad [11.4]$$

Normal modes are harmonic vibrations in which all nuclei of a molecule move through their equilibrium positions simultaneously. Each normal mode can be expressed by its amplitude Q_k^0 and its vibrational frequency v_k (Nakamoto, 1986).

$$Q_k = Q_k^0 \cos 2\pi v_k t \qquad [11.5]$$

The insertion of Eq. 11.4 and 11.5 into Eq. 11.2 yields

$$\underline{\mu}' = \alpha_0 \cdot \underline{E}_0 \cos 2\pi v_0 t + \left(\frac{\partial \alpha}{\partial Q_k}\right)_0 Q_{k0} \underline{E}_0 (\cos 2\pi v_0 t) \cdot (\cos 2\pi v_k t) \qquad [11.6]$$

The second term in Eq. 11.6 is then split into two terms by applying a simple addition theorem.

$$\underline{\mu}' = \alpha_0 \cdot \underline{E}_0 \cos 2\pi v_0 t + \tfrac{1}{2}\left(\frac{\partial \alpha}{\partial Q_k}\right)_0 Q_{k0} \underline{E}_0 (\cos 2\pi (v_0 - v_k)t + \cos 2\pi (v_0 + v_k)t) \qquad [11.7]$$

The term containing $(\cos 2\pi(v_0 - v_k)t)$ specifies the partial induced dipole moment which describes the Stokes emission. As already shown in Fig. 11.3, a Stokes

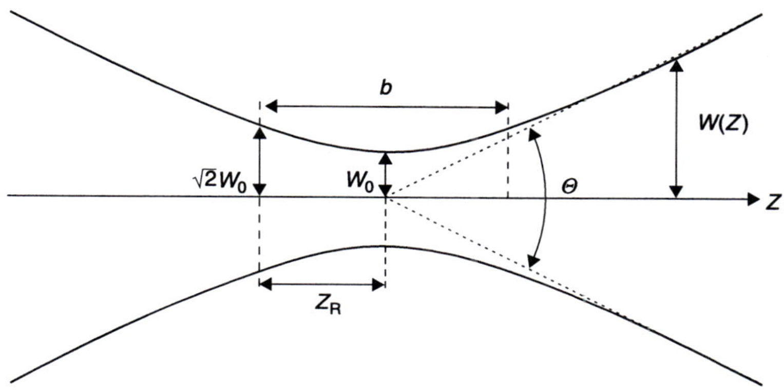

11.3 Zone of the laser beam waist, w_0 radius of the waist, z_R: Rayleigh length, Θ: angle of divergence, z: direction of propagation.

signal has a smaller emission frequency $(v_0 - v_k)$ than the excitation light and this emission is therefore red-shifted with respect to the excitation frequency. The second term containing $(\cos 2\pi(v_0 + v_k)t)$ has a higher emission frequency $(v_0 + v_k)$ and describes the part of the induced dipole moment that belongs to the anti-Stokes emission. The last term contains an unshifted emission frequency and characterizes the Rayleigh emission.

The power of Stokes- and anti-Stokes emission is obtained by inserting the corresponding parts of the total induced dipole moment into Eq. 11.3. Since Eq. 11.3 describes the power P which is emitted by one molecule, the scattered light power of an ensemble of molecules is achieved by multiplying P by the number n of molecules that are exposed. The squared induced dipole moment is proportional to the squared electric field vector and this quantity is in turn proportional to the power density I_{ex} (power per unit area) of the excitation light source. All equations derived previously can then be combined to give a final expression of the Raman intensity I_{Raman} that is proportional to the number of molecules exposed, the power density of the light sources and the scattering cross-section $\sigma_{scattering}$

$$I_{Raman} = n \cdot \sigma_{scattering} \cdot I_{ex} \qquad [11.8]$$

As seen in Fig. 11.2, anti-Stokes signal intensities are considerably smaller than those of Stokes signals. This is because anti-Stokes signals are generated from a smaller number of molecules. Anti-Stokes signals are emitted from molecules that are vibrationally excited. The population of an upper vibrational state occurs through thermal excitation of the sample molecules.

11.2.2 Experimental realization

Spontaneous Raman scattering generates very weak intensities, particularly when measuring trace-like samples such as gases in small sample volumes, as a result of the small values of Raman cross-sections. To express the intensity in photon numbers, as a rule of thumb, one needs one thousand photons to generate one Rayleigh photon but one million photons to obtain one Raman photon. Every attempt must therefore be made experimentally to avoid losses of Raman intensity. The intensity of Raman emission that can be achieved in experimental conditions is principally dependent on the quality of optical imaging that is expressed by the instrument constant γ. The dimensionless constant relates the theoretically possible Raman intensity to the experimentally obtainable light intensity.

$$I_{Raman} = \gamma \cdot n \cdot \sigma_{scattering} \cdot I_{ex} \qquad [11.9]$$

The constant γ can be determined by a combination of factors. Each factor describes a special attenuation of the initially generated Raman signal. A collection of possible loss factors are listed in Eq. 11.10.

$$\gamma = \gamma_{NA} \cdot \gamma_F \cdot \gamma_M \cdot \gamma_Q \cdot \gamma_A \qquad [11.10]$$

where γ_{NA} is losses related to numerical apertures, γ_F is losses related to notch filters and dichroitic mirrors, γ_M is losses related to the monochromator and the dispersion of light, γ_Q is losses related to the quantum efficiency of the detector, γ_A is losses related to the absorption of the scattered light.

As a result of the small scattering cross-sections and the numerous attenuation factors, a Raman experiment requires strong light sources, units with high throughputs of light, well-chosen optic imaging conditions, and very sensitive detectors.

Due to Eq. 11.9, an increase in the power density of laser irradiation leads to a concomitant increase in the Raman intensity. Considering a gaussian beam (Yariv, 1975), the highest power density is achieved inside the zone of the focused laser beam with its waist w_0 and length $b = 2z_R$ (Fig. 11.3). The largest number of Raman photons is generated within that zone. The volume from which Raman photons are observed is then the convolution of the waist zone with the volume that is defined by the depth of field of the chosen microscope objective.

If all factors of Eq. 11.9 are known, along with the exact volume from which the scattered light is collected, the concentration or the number of species can be directly established on the basis of the recorded Raman signal intensities. Since the size of the excitation volume cannot be established by the experiment itself, and since the instrument constant is heavily dependent on the actual optical adjustment of the outcoupled laser beam, the most practical solution is to examine the relation between the investigated signals and reference signals.

11.2.3 Available quantities derived by *in situ* Raman spectroscopy

With regard to measurements inside flow field channels, some parameters that characterize gases can be established. The initial quantity measured by Raman spectroscopy is the time-averaged signal intensity of a selected vibration.

$$\bar{I} = \frac{1}{\tau} \int_0^\tau I(t)\,dt \qquad [11.11]$$

where τ is integration time. Because of the small sample volume, from which signals of gaseous species are detected, Raman signals have to be integrated over a time interval to increase the signal-to-noise ratio.

According to Eq. 11.9 the averaged signal intensity corresponds directly to the average number of molecules within the excitation volume. When this number is multiplied by the molecular weight, the average intensity is directly proportional to the total mass of molecules within the excitation volume. This means that the averaged signal intensity is also proportional to the local density ρ of the gas at the chosen measuring port.

$$\rho = \kappa \cdot \bar{I} \qquad [11.12]$$

Assuming that in the nearest surroundings of the measuring port the gas flows are stationary within the flow field channel, the ideal gas law is valid.

$$p\dot{V} = \dot{m}R_sT \tag{11.13}$$

where p is pressure, $\dot{V}$ is volume flow rate, $\dot{m}$ is mass flow rate, R_s is specific gas constant, T is absolute temperature.

Since $\rho = \dfrac{\dot{m}}{\dot{V}} = \dfrac{m}{V}$, the local pressure p can be derived from the density by

$$p = \rho R_sT \tag{11.14}$$

if the local temperature is known at the observation port. As shown later in Section 11.4.2, local temperatures can also be derived by Raman spectroscopy. Under this condition the intensity I corresponds directly to the local pressure p.

When several observation ports are installed along the flow field, it is possible to measure the pressure difference between two ports and to determine volume flow rates under further limiting conditions. Assuming the incompressibility of the gases (surely confirmed by the small mass flows adjusted), stationary gas flows and the constancy of channel cross-sections, the local volume flow rate can be estimated using the Hagen-Poiseuille equation

$$\dot{V} = \frac{\pi r^4}{8\eta}\frac{\Delta p}{L} \tag{11.15}$$

where r is radius of a cylindrical channel (there are equivalent expressions for noncylindrical forms), L is distance between two observation ports, Δp is pressure drop between two ports, η is dynamic fluid viscosity.

This quantity additionally requires reliable values to be established of the dynamic fluid viscosity for different degrees of humidity within the channel. The usefulness of this calculation depends to a large extent on the constant cross-section of the channel, which is surely affected by the presence of droplets within the chosen channel segment. Considering all limitations, the bordering conditions can only be suitably fulfilled if the distance L is kept small.

11.3   Experimental setup

The application of Raman spectroscopy to the analyses of fuel cell processes was carried out in two steps. First the concept was proven by developing a Raman microscopy setup. The experiences which were gained by this setup were then used to construct a multi-fiber setup.

11.3.1 Raman microscopy setup

The experimental setup is sketched in Fig. 11.4. The Raman signals were generated by light from an argon ion laser (488 nm, Spectra Physics, model 2085). In order to remove the spontaneous plasma tube emission, the laser light was first passed through an Amici-prism as a pre-monochromator and was then directed via several apertures to a tilted optical notch filter (tilt angle: 45°). This notch-filter

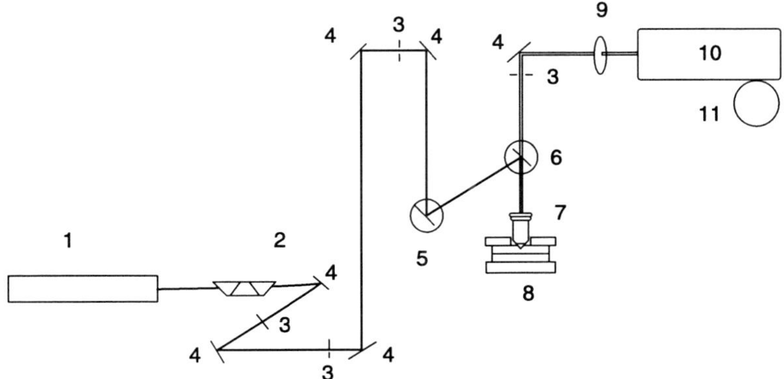

11.4 Schematic setup of the Raman microscope assembly for recording fuel cell processes. 1: Argon ion laser, 2: Pre-monochromator, 3: Aperture, 4: Folding mirrors, 5: Adjustable mirror, 6: Notch filter as dichroitic mirror, 7: Microscope objective, 8: Fuel cell, 9: Incoupling objective, 10: Spectrograph, 11: CCD camera.

(cut-off wavelength 514 nm, Andover) acted as a dichroitic mirror and reflected the beam to the microscope objective (Nikon, CF-Plan 50x working distance 13.4 mm) that focused the laser beam into the flow field channel. An observation port (diameter: 1 mm) was therefore inserted into one bipolar plate. The port passes into a conical channel which was adapted to the aperture of the microscope objective (Fig. 11.5). The flow field channel was sealed by a thin sapphire glass.

11.5 Sectional view of the fuel cell with the embedded microscope objective.

The scattered light generated within the flow field was collected by the microscope objective and directed to the spectrograph (HoloSpec f/1.8i, Kaiser Optical Systems) via the tilted notch filter. The spectrograph was equipped with different volume-phase holographic gratings. The choice of the gratings depends on which part of the Raman-spectrum (high frequency region/low frequency region) is of interest (VPE-gratings: HSG-488-LF/HSG-488-HF/HSG-532-HF). The spectrograph was equipped with a slit of constant width (100 µm). This caused a spectral resolution of about 8 cm^{-1}. Because of the nonlinear characteristic of the dispersion of VPE-gratings, the wavelength positions within the spectra were corrected by a polynomial of sixth order. The Raman scattering was detected by a back-illuminated CCD-camera (1340 × 400 pixels, cooled by liquid nitrogen, temperature kept at −100 °C, Princeton Instruments, model Spec-10).

Model fuel cell

The bipolar plates of the model fuel cell (diameter about 5 cm) were made at ZBT (Fuel Cell Research Center, Duisburg, Germany) from a hot compression-moulded polypropylene-graphite composite. The flow field was milled into the plates and has a segmented layout (Fig. 11.6). The MEA consists of a commercially available sulphonated PTFE membrane loaded equally with Pt as the catalyst (0.4 mg·cm^{-2}). During Raman measurements, the fuel cell was maintained at room temperature and was not humidified externally. For the measurements of current voltage characteristics, the standard mass flow of the gases was set to 60 cm^3·min^{-1} for hydrogen (N50) and 200 ccm·min^{-1} for synthetic air (Air Liquide, hydrocarbon free). The experiments were carried out in galvanostatic mode. Currents were controlled by a potentiostat (PP200, Zahner-Elektrik).

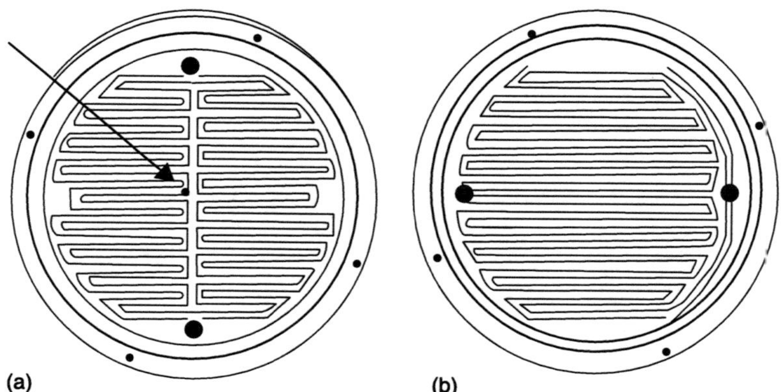

11.6 The flow fields of the model fuel cell. (a) Anode. (b) Cathode. The arrow indicates the measuring port.

11.3.2 The multi-fiber setup

The Raman microscope setup was then extended to look inside the meander structure of fuel cells at multiple observation ports simultaneously, through use of a multi-fiber setup. Along a flow field channel, several observation ports were placed at locations thought to be relevant to the study of the local properties of fuel cell operation. Ports at the entrance and the exit of the channel as well as at both straight and curved parts of the channels are of course considered.

The general setup is sketched in Fig. 11.7. The multi-fiber arrangement consists of an excitation laser, twin-fiber units and a spectrograph-detector assembly. In the setup presented here, the system is dimensioned for seven measuring ports along the flow field. The fuel cell and the optical components are connected by seven twin fibers, housed in SMA connectors, to allow the excitation and collection of Raman scattering. The SMA connectors are directly plugged into mountings located on the bipolar plate/current collector.

The beam of a powerful laser (frequency-doubled thin-disk Yb:YAG laser, 515 nm, up to 14 W) was first directed onto a beam homogenizer consisting of a micro-lens assembly that provides an almost uniform uncoupling of light into the seven fibers.

The excitation fibers are multimode-step-index hard-coated quartz fibers with a low content of OH functions. The last of these properties is fundamental to the study of water inside a flow field. In fibers with a high content of OH functions the passing laser beam generates a large number of OH Raman signals. These signals are reflected by the lens assembly and transferred to the detection unit by the second fiber. They then interfere with the detection of water vapor signals.

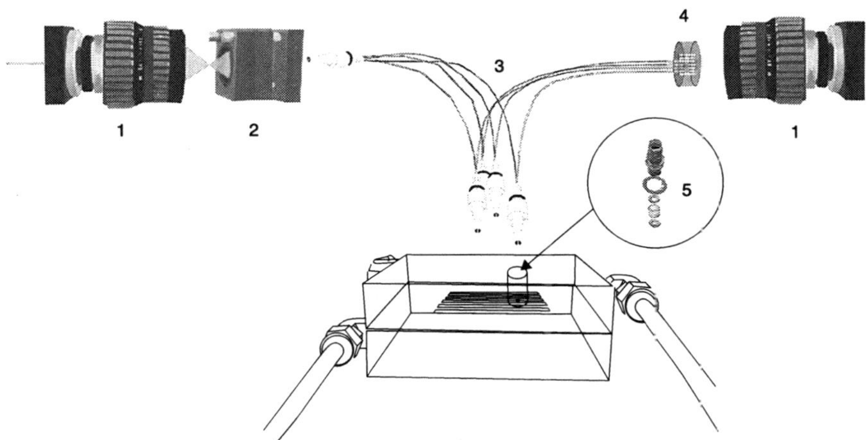

11.7 Sketched setup of the multi-fiber Raman spectroscopy, 1: Objective, 25 mm-Schneider Xenon 1:0.95, 2: Beam homogenizer, Suss-MicroOptics, 3: Twin fiber with SMA-connector, 4: Fiber array assembly, 5: SMA-screw coupling with gasket and micro-lenses.

At the opposite end of each excitation fiber a micro-lens assembly consisting of two aspheric lenses guides the excitation light through a small hole (1 mm in diameter) and focuses it into the flow field channel. The second task of the lens assembly is to collect the back scattered light which was generated within the flow field. The scattered light is directed to the detection unit by a second fiber that accompanies each excitation fiber. The detection unit consists of a CCD-camera and a bespoke spectrograph, containing a collection of photo-objectives which direct the scattered light through the entrance slit to the volume-phase grating. A notch filter within the spectrograph rejects the Rayleigh scattering. The high luminosity of the spectrograph is one property that enables spectra to be recorded in a short space of time.

Fuel cell parameters

The PEM fuel cell investigated had an active surface area of about 25 cm^2. The membrane electrode assembly (MEA) consisted of a commercially available sulphonated PTFE membrane that was equally loaded with platinum (0.6 mg·cm^{-2}) as a catalyst for both electrodes. The gas diffusion layer was made from carbon paper. Both flow fields were constructed as single serpentine channels, milled into stainless steel plates, which were completely gold-plated. Anodic and cathodic gases flowed in a co-flow arrangement.

The fuel cell was connected with a test station that was designed and manufactured by the German Aerospace Centre (DLR, Stuttgart, Germany). The test station controlled the humidification of the anode and cathode flow field separately, the temperature of the fuel cell and the humidifiers, gas flows and pressures on both the anode and cathode side. For the Raman measurements presented in this work, the fuel cell was operated in a constant voltage mode. Gas flows were set to fixed flow rates due to a stoichiometric factor λ of 1.5 for a current density of 1 A·cm^{-2}. In most measurements, the cathode was replaced by a segmented device developed at DLR. This device allowed locally resolved current densities to be recorded, and local temperatures along the meander to be measured. The segmented device is described in detail in (Schulze *et al.*, 2007).

The positions of the observation ports along the single serpentine meander of the anodic flow fields are shown in Fig. 11.8.

11.3.3 Thermal heating by laser excitation

Before using Raman spectroscopy to investigate fuel cell processes, it was verified that the chosen laser power settings only interfere in fuel cell processes to a very small extent. Possible interferences are a local increase in the temperature of the flow field gases and material damage to the GDL and the flow field.

As mentioned in Section 11.3.1, for the Raman microscopy experiment the bipolar plates in which the flow field channels were integrated were made of a

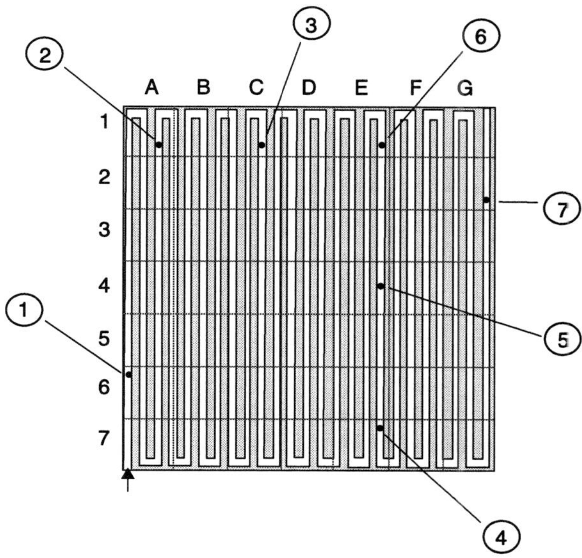

11.8 Positions of the observation ports along the serpentine of the anode flow field; the ports are marked by dark spots and denoted by numbers, arrow indicates the direction of gas flow; the grid width is denoted by 1 to 7 and A to G indicates the segments of the local current density measurements.

polypropylene-graphite composite. Their surfaces did not show any visual indication of damage after irradiation by focused laser light.

The flow field of the multi-fiber setup made of gold-plated stainless steel has a high reflectivity of light and proper heat conductivity. Thus, in both experimental setups the absorption of light by the black GDL is the only source of interference in the experimental results.

The possible local temperature increase of the GDL was investigated by a sensitive thermographic camera equipped with a close-up lens. In this procedure, the cell was demounted and a piece of the GDL was fixed at one bipolar plate. The laser light was directed to the GDL by the fiber via the lens set. To record the temperature increase induced by irradiation, the thermographic camera was installed on the opposite side of the GDL.

In the multi-fiber setup the GDL temperature increase depends on the arrangements of the twin fibers. The twin fibers can be positioned in one of two ways (Fig. 11.9). In Fig. 11.9a the fibers are arranged parallel to the channel path. The complete laser beam hits the GDL surface.

If the fibers are positioned perpendicular to the channel grooves (see Fig. 11.9b), the diverging laser light is reflected at the channel wall and is further diverged within the channel. The expansion of the beam causes a reduction in

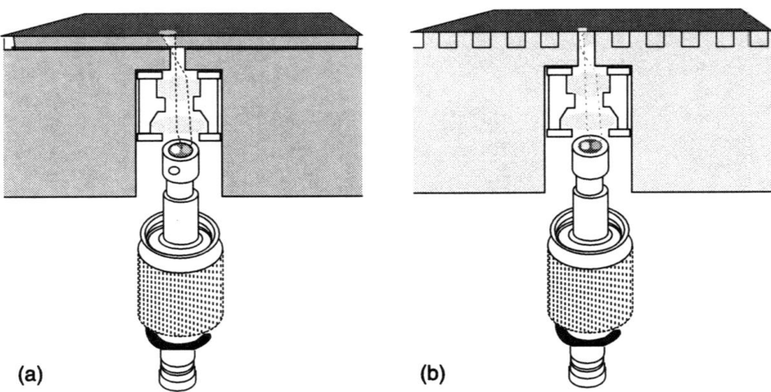

11.9 Arrangements of the twin fibers relative to the orientation of the flow field channels. (a) Parallel to the channel grooves. (b) Perpendicular to the channel grooves.

power density at the surface of the GDL. The distribution of laser light along the channel path mitigates the temperature increase. In this case, the signal of the thermographic camera exhibited a minimal value. The temperature rose linearly to 8.5 K for an increase of 100 mW output power of the fiber. This arrangement of twin fibers has less effect on fuel cell operation and was consequently adopted for the experiments. When fibers are positioned parallel to the channel grooves, an increase of 22 K per 100 mW laser output power was measured.

Similar results have been obtained for the GDL material of the microscope setup.

Finally, a potential increase in the temperature of gaseous samples can be directly verified by measuring the Raman signals of water vapor quantitatively or alternatively by measuring the signal intensity changes of pure rotational Raman transitions of hydrogen (see Section 11.4.2). In the case of the Raman microscope setup, the measurements showed that even when the laser power was set to 1 W, no temperature increase could be identified.

11.4   Raman spectroscopic investigations on polymer electrolyte membrane (PEM) fuel cells

11.4.1 Raman spectra of gaseous species and liquid water

Diatomic molecules have only one vibration. This single vibration of homo-nuclear molecules can be detected only through using Raman spectroscopy. Since the vibrational frequency reflects the specific binding conditions within diatomic molecules such as hydrogen, oxygen and nitrogen, their Raman signals are distributed over a large spectral range (Table 11.1).

Table 11.1 Relative normalized Raman cross-sections for Q-branches of PEM fuel cell gases for different excitation wavelengths in the emission range of argon ion laser and frequency-doubled thin disk Yb:YAG laser

Molecule	Transition/cm^{-1}	Excitation wavelength (nm)			
		435.8	457.9	488	514.5
N_2	2331	1.0	1.0	1.0	1.0
O_2	1555	1.0	1.0	1.0	1.0
H_2	4156	4.1		3.6	3.4

Source: Murphy, 1969; Fouche, 1971; Fouche, 1972; Fenner, 1973; Hochenbleicher, 1977.

In the gaseous state, vibrational motions are generally coupled with molecular rotations. This does affect the value of the pure vibrational transition to a small extent, but has a considerable effect on the shape of the spectrum thanks to the appearance of rotational-vibrational transitions. Due to quantum mechanic selection rules, the rotational-vibrational transitions are classified by their rotational transitions and the accompanying changes of rotational quantum numbers. Three types of rotational-vibrational transitions are possible: those in which the rotational quantum number J remains constant (Q-branch); those in which it is decreased (O-branch); and those in which it is lifted (S-branch) by two units (Weber, 1979). For instance, the most intense rotational-vibrational transition within the spectrum of molecular hydrogen (4156 cm^{-1}) corresponds to the Q-branch for which both the initial and the final J is 1 (Fig. 11.10) (Veirs and Rosenblatt, 1987). The prominent Q-branch of the rotational-vibrational transition of oxygen is recorded at 1555 cm^{-1} and that of nitrogen is detected

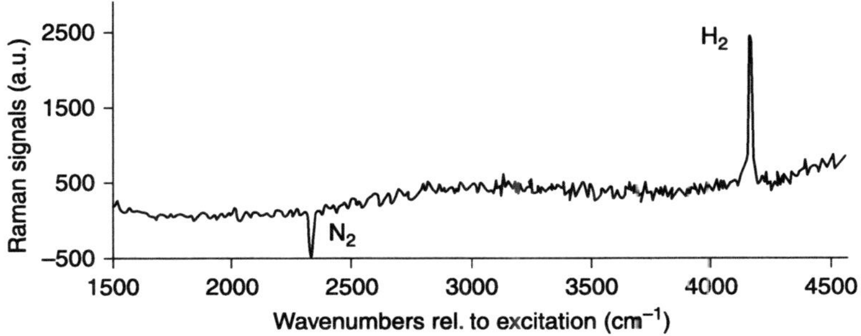

11.10 Raman spectrum of hydrogen (4156 cm^{-1}) recorded from the anodic flow field by the first experimental setup; the Raman signal of nitrogen (2331 cm^{-1}) is used as reference; the negative signal of nitrogen is caused by subtraction of spectra.

at 2331 cm^{-1}. Because of the different scattering cross-sections, the Raman signal intensity of O$_2$ is around one fourth of that of H$_2$ (Danichkin *et al.*, 1981). This implies that under the same conditions of laser excitation, studies on oxygen require exposure times four times longer than those required for measuring hydrogen.

In addition, Q-branches themselves consist of several transitions, as shown for the H-H stretching vibration of hydrogen (see Fig. 11.11). These transitions result from different occupations of rotationally excited states, from which rotational-vibrational transitions begin. The population of an upper rotational state occurs through thermal excitation of the sample molecules.

Water as a three-atomic molecule has three normal vibrations. These are a symmetric OH stretching mode, an asymmetric OH stretching mode and a totally symmetric HOH bending vibration. In the symmetric stretch vibration the hydrogen atoms simultaneously move towards and away from the oxygen atom while in the asymmetric stretch vibration one H-atom moves towards and the second away from the oxygen and vice versa. In the bending mode the H-O-H angle changes periodically. All three vibrations of water are Raman-active (Murphy, 1978). As in the case of diatomic molecules, a sequence of rotational-vibrational transitions of

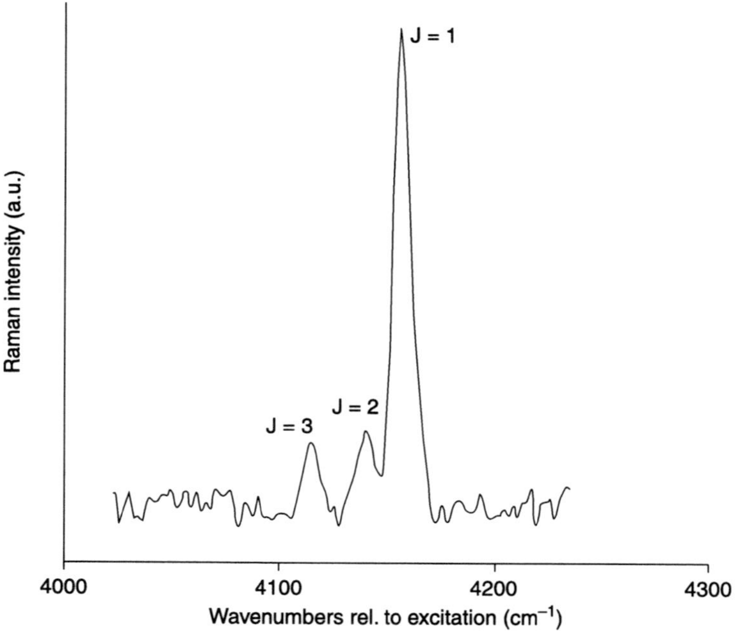

11.11 Q-branch transitions of the H-H stretching vibration of molecular hydrogen, extended wavenumber scale, the transitions are assigned by the rotational quantum number J.

water is detectable in the gaseous state. When measuring inside flow fields, only one Q-branch transition of the symmetric stretch vibration (3652 cm^{-1}) could be detected. Recording the water vapor inside fuel cell channels generally proves more difficult than, for example, the measurement of hydrogen or oxygen. In contrast to the partial gas concentrations, water vapor concentrations are low at temperatures near room temperature. For example, the molar water content of water-saturated air at 25 °C amounts to 4% (Keenan, 1978). For this reason, water vapor at room temperature could only be detected with a strongly powered laser (close to 1 W) and long exposure times. Figure 11.12 shows the evolution of the water vapor signal for some temperatures measured by the Raman microscopy setup.

In its condensed state, water forms intermolecular H-bridges. The formation of intermolecular bonds allows the vibrational frequency to shift and the line shape of the vibrational transition to broaden. Moreover, intermolecular bonds inhibit the rotation of single molecules so that a rotational substructure of vibrations is missing. These effects cause the merging of both Raman-active OH stretch vibrations to form one broad non-resolved transition with its center at about 3400 cm^{-1} (see Fig. 11.13).

Thanks to its condensed state and the increased particle density, liquid water inside a channel can be easily detected. Thus, small water droplets passing the detection port were recorded during cell operation. However, no continuous adhesion of liquid water to the observation port was observed.

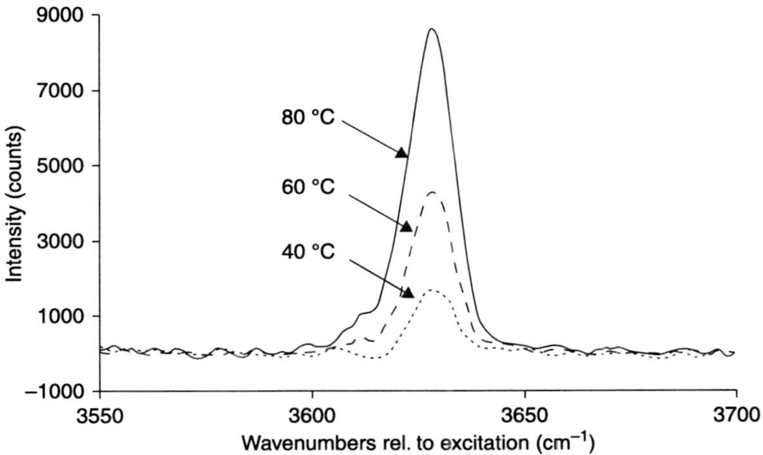

11.12 The water vapor signal at different temperatures, Q-branch of the totally symmetric OH stretch mode, centre of signal: 3652 cm^{-1}; excitation wavelength: 477 nm, laser power about 1 W, exposure time: 10 min.

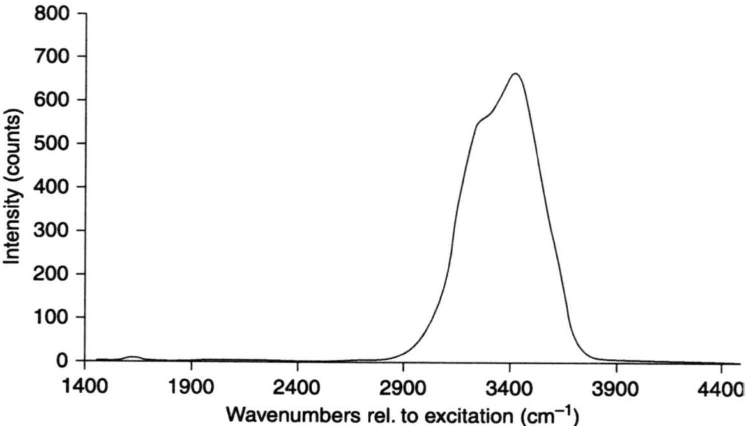

11.13 The Raman spectrum of the OH-stretching vibrations of liquid water obtained from a fuel cell channel, the shoulder at about 3650 cm^{-1} indicates the presence of water vapor.

11.4.2 Passage of air through a defect in the membrane electrode assembly (MEA)

The Raman microscopy setup was used to study a common membrane defect (Bettermann and Fischer, 2010). Since pinholes have been the subject of various investigations (Weber, 2008), an artificial hole was generated within the membrane. To achieve this, the GDL on the anode side was removed and the laser beam was directly focused on the membrane for a few seconds. The absorption of laser irradiation led to local membrane damage. After the experiments, an inspection of the MEA showed that the GDL on the cathode side remained intact, which was proven visually by examining the damage under a microscope.

As a result of this defect, nitrogen and oxygen Raman signals were measured on the anode side. The amount of nitrogen flowing through the defect was determined by comparing the nitrogen signals to signals obtained from nitrogen/hydrogen mixtures. For these purposes, all hydrogen/nitrogen mixtures had a constant flow of 60 ccm·min^{-1}.

As seen from Table 11.1, oxygen and nitrogen have very similar Raman cross-sections; consequently, the signal intensities can directly be used to compare the concentrations of the two gases. It was found that the ratio between oxygen and nitrogen amounted to 29:71. The ratio revealed that air that was nearly unconverted was recorded on the anode side of the defect.

In Fig. 11.14, three measurements are shown in which signals of both gases are plotted against the cathodic air flow. The lowest signal level was measured first, followed by further measurements on subsequent days, which showed an increase in signal intensities. The nitrogen and oxygen signals were obtained by reducing

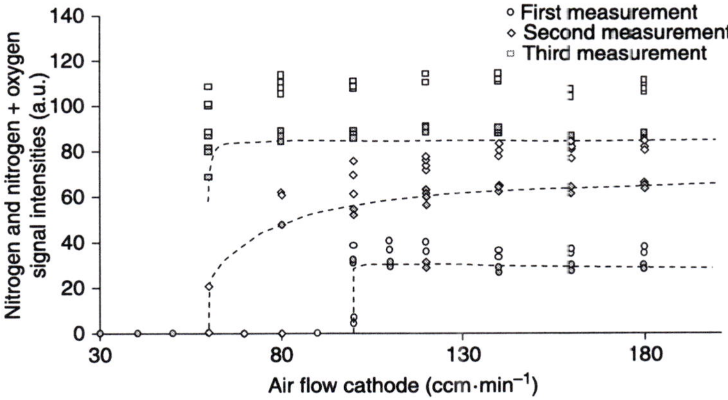

11.14 The evolution of nitrogen and oxygen on the anode side versus the cathodic air flow. Filled marks: nitrogen signals; unfilled marks: sum of nitrogen and oxygen signals; lines indicate the course of the nitrogen signals of each measurement.

the cathodic air flow in steps of 20 ccm·min^{-1}, starting from a maximum air flow of 200 ccm·min^{-1}. The measurements show a general trend: the signals of air remain at a constant level over a wide range of cathodic air flows, but at certain air flows the flows broke down instantly.

The constancy of the gas signals over a wide range of flow rates shows that the passage of gases is unaffected by the cathodic flow rate, and is therefore independent of the cathodic air pressure. In most setups that demonstrate the crossover of gases through pinholes, the observed passage of air is attributed to diffusive transports (Weber, 2008). Both normal molecular diffusion and Knudsen diffusion depend on variations in gas concentration or density between the cathode side and the anode side. Because of the incompressibility of gases under the chosen gas flow conditions, the density of gases on the cathode side remains constant for all rates of cathodic flow tested.

The abrupt breakdown of the air flow on the anode side indicates that the diffusive flux is instantly interrupted. The interruption of the air passage was accompanied by a strong liquid water signal. This characteristic was observed in all three measurements. The presence of liquid water was attributed to the catalytic combustion of oxygen. Since liquid water clogs the porous structure of the GDL, the passage of air is restricted by the presence of liquid water. At high air flows the water droplets were transported from the defect to the exit of the cathodic meander, but at a certain threshold of the air flow rate, the transport of droplets is interrupted. The GDL at the defect is then soaked with water and the diffusive process is stopped. When measured over time, breakdown occurs at increasingly low air flow rates and the air flow on the anode side increases. This is explained by an enlargement of the hole as the measurement sequence progresses.

Hydrogen signals could not be detected at the measuring point over a wide range of cathodic air flows. When the air flow was stopped, a significant increase in hydrogen temperature was also observed. This was proven by measuring pure rotational Raman transitions (Fig. 11.15). The spectrum exhibits three rotational transitions (583, 807, 1031 cm^{-1}) that are assigned as S_0 branch transitions. This notation means that the molecules are in the vibrationless state denoted by the subscript 0, and that Raman excitation changes their rotational states by two quantum numbers. Indicating the transitions by the initial rotational quantum number J, the signal at 583 cm^{-1} is attributed to a transition between J = 1 and J = 3. The subsequent signals belong to transitions between J = 2 and J = 4 as well as J = 3 and J = 5 (Veirs and Rosenblatt, 1987). The intensity distribution of these three signals allows the temperature of hydrogen at the measuring point to be determined simply by applying Boltzmann's distribution law. Due to this relation, the local temperature in the direct surroundings of the defect was estimated to be about 98 °C, while the global temperature of the fuel cell remains near room temperature.

11.4.3 Fuel cell investigations using a multi-fiber Raman setup

Evolution of water vapor along a flow field channel

Distributions of liquid water along flow field meanders up to the final flooding point have been successfully demonstrated by neutron tomography. In addition,

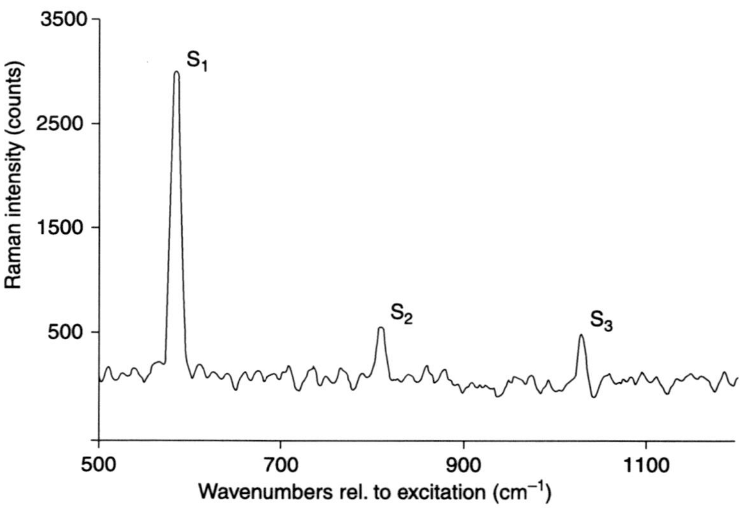

11.15 Pure rotational Raman transitions of hydrogen. S(index 1) J = 1 → J = 3, S(index 2) J = 2 → J = 4, S(index 3) J = 3 → J = 5 (Veirs and Rosenblatt, 1987); as identified by the CCD image.

Raman spectroscopy allows water vapor distribution along channels to be detected, making a complete description of the local and temporal humidification possible.

Figure 11.16 presents the changes in water vapor signals at three different ports along the anode meander. The temperature of the cell was set to 75 °C. The graphs first reveal that an increase in water vapor corresponds to an increase in channel length. This can be directly observed from the average values of water vapor obtained from the selected measuring ports.

The graphs also show that the time periods in which changes of water concentration take place become shorter as channel length increases. The apparent fluctuations of water vapor are attributed to condensation, which occurs primarily at the end of the meander, as this is the region in which the atmosphere of the channel is enriched with water vapor. In addition, condensation leads to a temporary contraction or complete blockage of the cross-sectional channel area. Both obstructions cause changes in water vapor pressure at the observation ports.

The highest track which resembles the water vapor concentration at the end of the flow field exhibits relative humidity values above 100%. This anomalous result is certainly caused by the difficult measuring conditions at the end of the flow field. Although the water vapor signal at the observation port next to the exit (fiber 6) was largest, it proved difficult to obtain exact measurements of its signal intensity, because of the large number of irrelevant high intensity signals in the

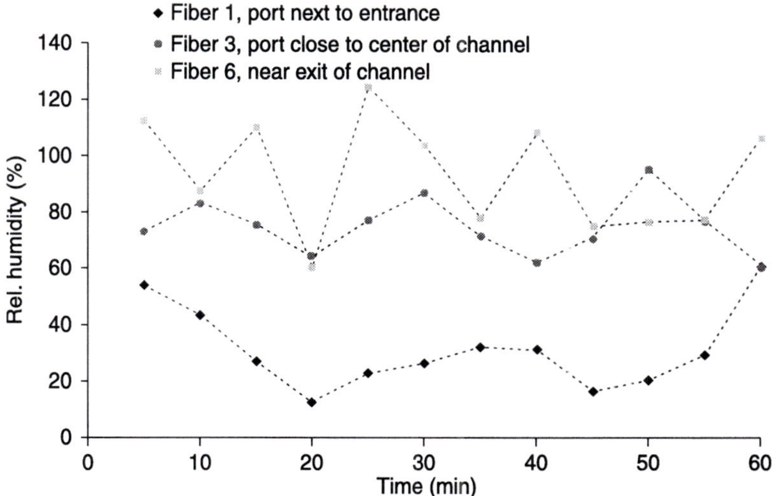

11.16 Evolution of relative humidity as a function of time at three different ports along the anode meander; the relative humidity was derived from the water vapor signals, cell temperature: 75 °C. For the notation of fibers see Fig. 11.8.

immediate surroundings of the water vapor signal. These noise-like signals are most likely caused by the high number of liquid water droplets, which generate non-molecular light scattering.

The low humidity value at fiber 1 (see Fig. 11.8) of the flow field is attributed to the high consumption of hydrogen by the electrochemical reaction at the entrance, since the transport of H^+ ions requires a considerable number of water molecules (electro-osmotic drag).

Local and temporal losses of hydrogen along the anode flow field

In this section the experimentally obtained hydrogen concentrations are compared to theoretically estimated hydrogen concentrations. The differences that appear between the two values are indicators of local hydrogen losses along the anode flow field. Moreover, the chronological evolution of the deviations between experimental and theoretical values shows that local losses at the selected measuring points are time-dependent.

The figures in this subsection illustrate a comparison of the relative hydrogen concentrations that were recorded with those predicted by calculations of the local consumption of hydrogen by the electrochemical reaction. The relation between the consumed hydrogen and the generated current is obtained by Faraday's law:

$$n_{H_2} = \frac{t_{measure} \sum_{k=1}^{N} j_k \cdot A_k}{2F} \qquad [11.16]$$

where n_{H_2} is the molar number, j_k is the current density at segment k, A_k is the effective area of segment k, F is the Faraday constant, $t_{measure}$ is measuring time.

In calculations of the consumption of hydrogen it is assumed that an exponential decay of the hydrogen concentration occurs along the meander. This setup is based on a model that assumes that the decrease of hydrogen along a small path of meander is constant and additionally proportional to the actual concentration. The local consumption of hydrogen is directly related to the local current density due to Eq. 11.16. Local current densities are provided by the segmented cell. At specific locations along the meander it is possible to draw a direct correspondence between the effective areas and the meander pathway (Fig. 11.8). These special areas are located at the points at which the meander enters a new row of segments. The constant of the exponential decay can then be derived from the calculated hydrogen consumption at these special locations. Following this model, the theoretical hydrogen consumption could be calculated for each point at which Raman signals were recorded. The expected hydrogen concentration at the measuring points is then obtained by subtracting the consumed hydrogen concentration from the initial concentration at the entrance. The values of the theoretically consumed hydrogen and the measured hydrogen concentration were both related to the initial concentration of hydrogen. For the measured signals, an intensity of 100 % at each measuring port is related to open circuit voltage (OCV) conditions.

The presence of water vapor means that these calculated values need to be corrected, taking into account the fact that hydrogen is mixed with water vapor within the volume from which signals are obtained. Thus, the percentage decrease is related to the relative humidity as shown in Fig. 11.16.

At fiber 1 (see Fig. 11.17) the predicted concentrations of hydrogen are mostly around 14% higher than the experimental values. This relatively large gap indicates local losses of the hydrogen that does not contribute to the cell current. These losses are likely to be largely the result of diffusive processes. The greatest degree of hydrogen diffusion must occur near the entrance of the meander, since this is the point at which the hydrogen density is highest. In addition, diffusion through the GDL decreases as the water content of the GDL increases. As seen in Fig. 11.15, the water vapor content at the entrance is remarkably low during the measurements. Because of the existing equilibrium between the water inside the GDL and the water vapor content of the flow field, it can be stated that the water content of the GDL was also low. It thus seems plausible that a high amount of hydrogen is lost by diffusion.

The time-dependent changes in the experimental hydrogen concentration at fiber 1 are attributed to the changing water content, which in turn causes changes in the diffusion coefficients of the GDL and the membrane.

At fiber 3 the differences between the calculated and measured hydrogen concentrations are in a range of about 7% (see Fig. 11.18). Given that a difference

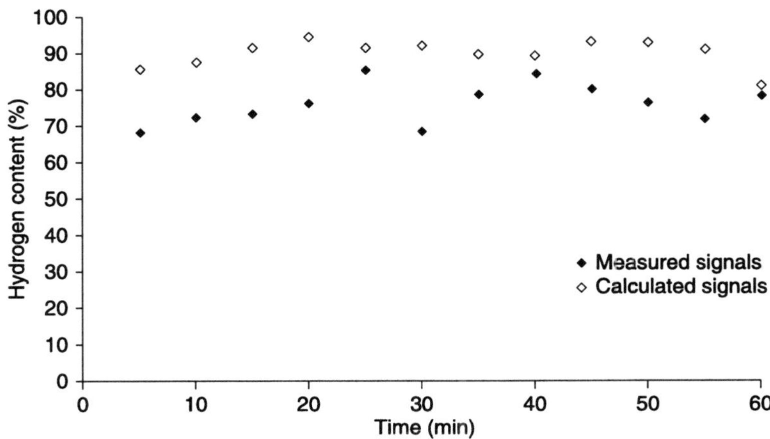

11.17 The relation of recorded hydrogen signals to expected signal intensities at fiber 1. The calculated values are related to the loss of hydrogen due the electrochemical reaction; cell temperature 75 °C, average current: 21.24 ± 0.9 A, voltage: 0.6 V (potentiostatic).

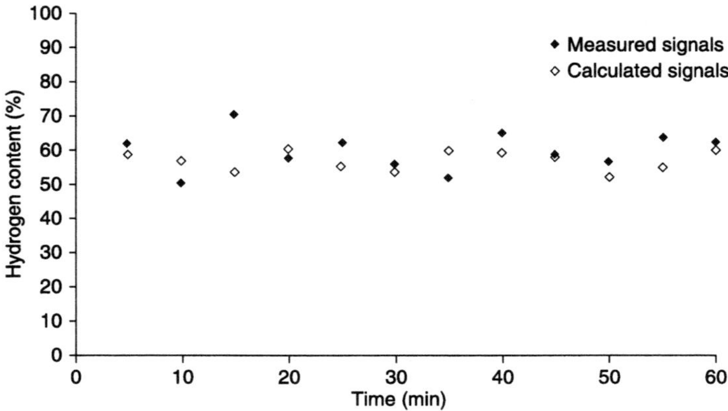

11.18 The relation of recorded hydrogen signals to expected signal intensities at fiber 3. The calculated values are related to the loss of hydrogen due the electrochemical reaction; cell temperature 75 °C, average current: 21.24 ± 0.9 A, voltage: 0.6 V (potentiostatic).

of around 5% can be attributed to uncertainties in the measurement of hydrogen signals at that measuring point, the deviation between the suggested theoretical hydrogen consumption and the experimentally obtained signals is very small. This is consistent with the fact that in the central region of the flow field the hydrogen concentration has become smaller as a result of the electrochemical reaction, and thus the diffusion rate also decreases.

At fiber 6, the values of hydrogen concentration obtained experimentally are higher than those theoretically estimated (Fig. 11.19). The theoretical values should in fact be higher than or equal to the experimental signal intensities. At present, the best explanation for this deviation is once again the uncertainty in determining the relative humidity near the end of the flow field.

Fluctuations of hydrogen concentration in relation to fluctuations of cell power

At a certain stage of aging, cell power is known to undergo considerable fluctuations. The same behavior was found in Raman signals of hydrogen at different measuring points (Bettermann and Fischer, 2010). Figure 11.20 and Fig. 11.12 show the relation between the changes of hydrogen density and the global power of the cell as functions of time.

Figure 11.20 shows the signal alterations at a measuring point next to the entrance of anode channel while Fig. 11.21 shows similar relations at the end of the meander. As seen in Fig. 11.20, the change in hydrogen density is inversely proportionate to the changes in the power. This inverse relation is in accordance with the chemical oxidation of hydrogen and the formation of the electric current. If higher currents are generated, the density of hydrogen should decrease, and vice

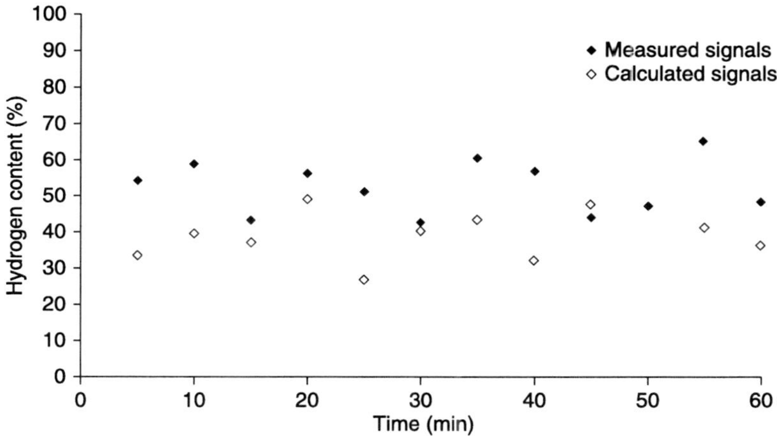

11.19 The relation of recorded hydrogen signals to expected signal intensities at fiber 6. The calculated values are related to the loss of hydrogen due the electrochemical reaction; cell temperature 75 °C, average current: 21.24 ± 0.9 A, voltage: 0.6 V (potentiostatic).

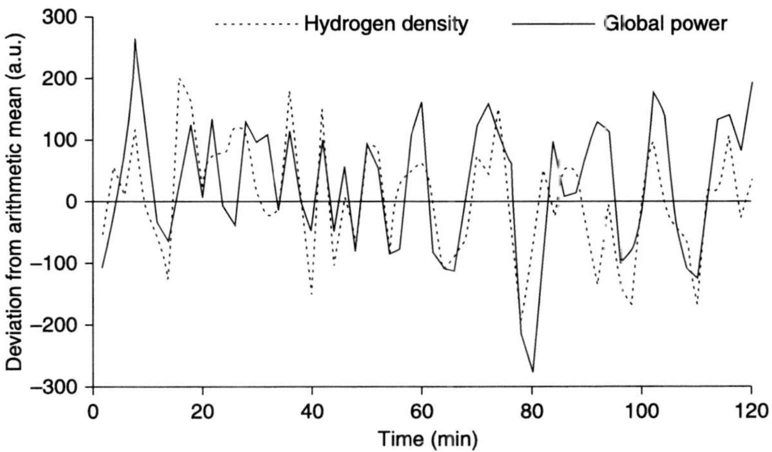

11.20 Changes of the hydrogen density and the global power as a function of time at the entrance of the anode meander (cell temperature: 55 °C). The hydrogen signals are inverted to visualize the correspondence of both signals better. The signal alterations are related to the mean values (electrical power: 7.8 W, Raman: peak area 63.5 in a.u.) respectively, which are set to 0. The ordinate scale is related to the standard deviation (power: 0.8 W, Raman: peak area 5.4 in a.u.) as a percentage.

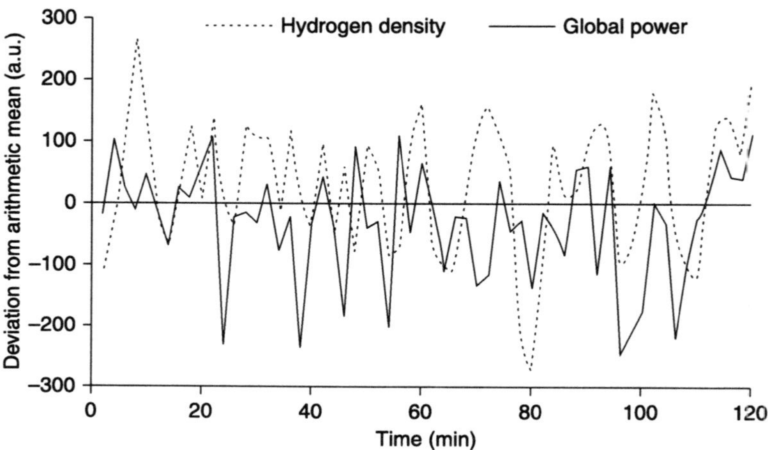

11.21 Changes of the hydrogen density and the global power as a function of time at the end of the anode meander (cell temperature: 55 °C). The hydrogen signals are inverted to visualize the correspondence of both signals better. The signal alterations are related to the mean values (el. power: 7.8 W, Raman: peak area 42 in a.u.) respectively, which are set to 0. The ordinate scale is related to the standard deviation (power: 0.8 W, Raman: peak area 5.2 in a.u.) as a percentage.

versa. However, this direct relation could only be observed next to the entrance of the anode flow field. At the end of the anode meander a stronger incoherence appeared between the changes in hydrogen signal and changes in power signal.

We currently attribute the fluctuations in hydrogen concentrations to the changing flow conditions within the flow field as a result of the humidification of the anode. At the chosen temperature the humidification of the anode channel may induce condensation of water. The droplets inside the channel are then moved by the gas flow, and can additionally change in size. The cross-section of the channel is therefore continuously altered locally and over the course of time.

11.5   Outlook and future prospects

In the first series of Raman experiments discussed in this contribution we collected experimental data from species that are present in flow field channels. Current measurements are concerned with back diffusion and electro-osmotic drag as well as CO poisoning of the catalysts.

These experiments have us provided insight into how the experimental setup can be further improved from a technical perspective, mainly with the aim of obtaining a better quantification of experimental data. With regard to the complex situations near the end of the flow field meanders, one task to be carried out in the near future is the inclusion of aerosol scattering. With this in mind, it seems logical to build on experiences from light detection and ranging (LIDAR) experiments that have been successfully used to investigate spatially resolved temperature and concentration gradients in the atmosphere.

Future goals are:

- The construction of test stations in which a Raman diagnostic tool is permanently implemented.
- With the use of spatially resolved Raman data for evaluating stages of degradation processes in semi-empirical models.
- The extension of Raman spectroscopic analysis to other types of fuel cells.

Investigations into the last of these have already begun, through examination of the reaction and degradation products of an alkaline ethanol fuel cell during operation.

11.6   Acknowledgments

The authors thank the German Aerospace Centre (DLR, Stuttgart, Germany) for providing them with a PEM fuel cell and a complete fuel cell test station as part of a collaboration within the project 'Online Diagnostik und Regelung für PEFC'. The financial support of this project by the German Federal Ministry of Education and Research (BMBF) is gratefully acknowledged. The authors are grateful for the helpful technical assistance received from Erich Gülzow and Dr Mathias Schulze (DLR).

Dr Can Kreutz (ZBT, Center for Fuel Cell Technology (ZBT GmbH), Duisburg, Germany) is thanked for constructing and manufacturing the fuel cell for the Raman microscope assembly as part of a collaboration within the project 'Entwicklung einer neuen Methode zur *In-situ*-Analyse der Eigenschaften von Brennstoffzellen durch laser-spektroskopische Verfahren' which was granted by the Ministry of Innovation, Science, Research, and Technology of the Federal State North Rhine-Westphalia.

11.7   References

Betermann, H., Fischer, P., 2010, 'On-line *in-situ* diagnostics of processes within PEM fuel cells by Raman spectroscopy', In *Proceedings of the ASME 2010 Eighth International Fuel Cell Science, Engineering and Technology Conference, FuelCell 2010–33320*, Brooklyn, N.Y.

Borup, R. et al., 2007, 'Scientific aspects of polymer electrolyte fuel cell durability and degradation', *Chemical Reviews*, **107**(10), 3904–51. DOI: 10.1021/cr050182l.

Chalmers, J., Griffiths, P., 2001, *Handbook of Vibrational Spectroscopy*, John Wiley & Sons, Vol. 4, Chichester. DOI: 10.1002/0470027320.

Danichkin, S.A., Eliseev, A.A., Popova, T.N., Ravodina, O.V., and Stenina, V.V., 1981, 'Raman scattering parameters for gas molecules (survey)', *Journal of Applied Spectroscopy*, **35**(4), 1057–1066. DOI: 10.1007/BF00605646.

Fenner, W.R., Hyatt, H.A., Kellam, J.M, and Porto, S.P., 1973, 'Raman cross-section of some simple gases', *Journal of the Optical Society of America*, **63**, 73–77.

Fouche, D.G., Chang, R.K, 1971, 'Relative Raman cross section for N_2, O_2, CO, CO_2, SO_2, and H_2S', *Applied Physics Letters*, **18**, 579–580.

Fouche, D.G, Chang, R.K, 1972, 'Relative Raman cross section for O_3, CH_4, C_3H_8, NO, N_2O, and H_2', *Applied Physics Letters*, **20**, 256–257.

Geiger, A., Lehmann, E., Vontobel, P. and Scherer, G.G., 2000, 'Direct methanol fuel cell – *In situ* investigation of carbon dioxide patterns in anode flow fields by neutron radiography', *Paul Scherrer Institut Scientific Report*, **V**, 84–85.

Hochenbleicher, J.G., 1977, 'Bestimmung von Molekülparametern aus Messungen von Linienlagen', Dissertation, University of Munich.

Keenan, J.K., 1978, *Steam Tables, Thermodynamic Properties of Water including Vapor, Liquid, and Solid Phase*, 2nd ed. Wiley, New York.

Kramer, D., Lehmann, E., Frei, G., Vontobel, P., Wokaun, A. and Scherer, G.G., 2005, 'An on-line study of fuel cell behavior by thermal neutrons', *Nuclear Instruments and Methods in Physics Research*, **A 542**, 52–60. DOI: 10.1016/j.nima.2005.01.011.

Kuhn, H., Andreaus, B., Wokaun, A. and Scherer, G.G., 2006, 'Electrochemical impedance spectroscopy applied to polymer electrolyte fuel cells with a pseudo reference electrode arrangement', *Electrochimica Acta*, **51**, 1622–1628. DOI: 10.1016/j.electacta.2005.02.108.

Mattsson, B., Ericson, H., Torell, L.M. and Sundholm, F., 2000, 'Degradation of a fuel cell membrane as revealed by micro-Raman spectroscopy', *Electrochimica Acta*, **45**, 1405–1408.

Mench, M.M., Wang, C. Y. and Ishikawa, M., 2003, '*In situ* current distribution measurements in polymer electrolyte fuel cells', *Journal of the Electrochemical Society*, **150**, A1052–A1059. DOI: 10.1149/1.1584440.

Murphy, W.F., Holzer, W., Bernstein, H.J., 1969, 'Gas phase Raman intensities – A review of pre-laser data', *Applied Spectroscopy*, **23**, 211–218.

Murphy, W. F., 1978, 'The rovibrational Raman spectrum of water vapour – v_1 and v_3', *Molecular Physics*, **36**(3), 727–732.

Nakamoto, K., 1986, *Infrared and Raman Spectra of Inorganic and Coordination Compounds*, 4th ed., Wiley, New York.

Ogilvie, J.F., 1998, *The Vibrational and Rotational Spectrometry of Diatomic Molecules*, Academic Press, San Diego.

Pomfret, M.B., Owrutsky, J.C. and Walker, A.W., 2006, 'High temperature Raman spectroscopy of solid oxide fuel cell materials and processes', *Journal of Physical Chemistry B*, **110**, 17305–17308. DOI: 10.1021/jp0639521.

Pomfret, M.B., Owrutsky, J.C. and Walker, A.W., 2007, '*In-situ* studies of fuel oxidation in solid oxide fuel cells', *Analytical Chemistry*, 79, pp. 2367–2372. DOI: 10.1021/ac062189o.

Raman, C.V., Krishnan, K.S., 1928, 'A new type of secondary radiation', *Nature*, **121**(3048), 501–502. DOI: 10.1038/121501c0.

Schulze, M., Gülzow, E., Schönbauer, S.T., Knöri, T., and Reissner, R., 2007, 'Segmented cells as tool for development of fuel cells and error prevention/prediagnostic in fuel cell stacks', *Journal of Power Sources*, **173**, 19–27. DOI: 10.1016/j.jpowsour.2007.03.055.

Tsukada, A., Lehmann, E., Vontobel, P., and Scherer, G.G., 1999, '*In situ* observation of water condensation in an operating polymer electrolyte fuel cell by means of neutron imaging at the spallation neutron source (SINQ)', *Paul Scherrer Institut Scientific Report*, **V**, 84–85.

Vers, D.K. and Rosenblatt, G.M., 1987, 'Raman line positions in molecular hydrogen: H$_2$, HD, HT, D$_2$, DT, and T$_2$', *Journal of Molecular Spectroscopy*, **121**, 401–419.

Weber, A., 1979, *Raman Spectroscopy of Gases and Liquids*, Springer-Verlag, Berlin, Heidelberg, New York, Chap. 2–3.

Weber, A.Z., 2008, 'Gas-crossover and membrane-pinhole effects in polymer-electrolyte fuel cells', *Journal of the Electrochemical Society*, **155**, B521–B531. DOI: 10.1149/1.2898130.

Wu, J., Yuan, X.Z., Wang, H., Blanco, M., Martin, J.J. and Zhang, J., 2008a, 'Diagnostic tools in PEM fuel cell research: Part I – Electrochemical techniques', *International Journal of Hydrogen Energy*, **33**(6), 1735–1746.

Wu, J., Ziyuan, X., Wang, H., Blanco, M., Martin, J., Zhang, J., 2008b, 'Diagnostic tools in PEM fuel cell research: Part II – Physical/chemical methods', *International Journal of Hydrogen Energy*, **33** (6), 1747–1757.

Yariv, A., 1975, *'Quantum Electronics'*, 2nd ed., John Wiley & Sons, New York.

Yuan, X.Z., Song, S., Wang, H., and Zhang, J., 2010, *Electrochemical Impedance Spectroscopy in PEM Fuel Cells*, Springer, London. DOI: 10.1007/978-1-84882-846-9.

Part III
Locally resolved methods for polymer electrolyte membrane and direct methanol fuel cell characterization

12

Submillimeter resolved transient techniques for polymer electrolyte membrane fuel cell characterization: local *in situ* diagnostics for channel and land areas

I. A. SCHNEIDER, M. H. BAYER and
S. VON DAHLEN, Paul Scherrer Institut (PSI), Switzerland

Abstract: Key results of submillimeter resolved local in situ diagnostics in channel and land areas of polymer electrolyte fuel cells (PEFCs) are presented. Cyclic voltammetry (CV) and electrochemical impedance spectroscopy (EIS) are employed in channel and land areas with submillimeter resolution to demonstrate how lateral inhomogeneities affect the value of the electrochemically available surface area as obtained from CV experiments and how their contribution can lead to misinterpretation of EIS results. The correlation between liquid water accumulation and cell performance in channel and land areas is demonstrated by using simultaneous submillimeter resolved current density distribution measurement and in-plane neutron radiography in transient experiments. Finally, the reverse current phenomenon is assessed in channel and land areas during start-up and shutdown of a PEFC.

Key words: flow field, channel, rib, gas diffusion layer, segmented cell, current distribution, mass transport, cyclic voltammetry, electrochemical impedance spectroscopy, neutron radiography, water management, flooding, start current, stop current, reverse current.

12.1 Spatially resolved characterization of polymer electrolyte fuel cells (PEFCs)

Polymer electrolyte fuel cells (PEFCs) are efficient electrochemical power sources. In practical PEFCs flow field plates are used to distribute the reactant gases over the active area and to remove the reaction products. The flow fields exhibit channel and land areas. The reactant gases are forced through the gas channels, whereas the current is collected in the land area from the gas diffusion layer (GDL) underneath the ribs. Evidently, the use of flow fields in PEFCs entails inhomogeneities in current generation over the active electrode area. Here, mass transport limitations play an important role. In-plane inhomogeneities occur on the one hand, in the down-the-channel direction, as a result of the finite gas flow rate and reactant conversion along the gas channel and, on the other hand, in the perpendicular-to-the-flow channel direction; in particular, due to the diffusion and conversion of reactants in a lateral direction underneath the ribs. Different

353

compressions of the porous GDL, temperature gradients, and anisotropic charge and mass transport properties must also be taken into account in this context.

Much effort has been made to monitor effects of down-the-channel inhomogeneities in PEFCs under both steady-state and transient operation conditions. The segmented cell approach[1–8] is a convenient tool for measuring the current distribution in flow direction. No obvious restrictions exist for a segmentation of the flow fields in the down-the-channel direction and, therefore, segmented cells find widespread use in a broad variety of applications, from fundamental dc current density and high-frequency resistance (HFR) distribution measurements (e.g., Refs 1–8), over cell characterization via spatially resolved transient techniques (e.g., Refs 9–15), to studies of local cell degradation phenomena (e.g., Refs 16–18). The required level of sophistication for local current measurement depends on the application and can vary significantly. In particular, the measurement of fast local current transients is challenging, where a low inductance of the current measuring circuitry is vital, as we have pointed out in our earlier work.[13]

Evidently, the effect of down-the-channel inhomogeneities[19, 20] can be excluded from the discussion by providing virtually homogeneous reaction conditions along the gas channels.[21] For this purpose small active area PEFCs are used. These small cells can easily be operated at a very high gas stoichiometry; the change in reactant composition in flow direction is kept at a differentially small level. These so-called 'differential cells'[22] can be regarded as quasi-two-dimensional cells, where in-plane inhomogeneities are restricted to the perpendicular-to-the-flow channel, i.e. lateral, direction. Employment of differential cells forms the basis for the evaluation of lateral inhomogeneities in channel and land areas of PEFC flow field structures. However, the assessment of the significance of these inhomogeneities requires current density distribution measurements in the perpendicular-to-the-flow channel direction. Here, several obstacles to the application of the segmented cell approach exist and, due to the difficulty of collecting the local cell current, only few reports exist in the literature, which scrutinize the effect of the flow field structure on the local current density in the channel and land area of PEFCs experimentally.

To the best of our knowledge, the first report dates back just to the year 2006.[23] In their work, Freunberger *et al.* demonstrated the importance of lateral inhomogeneities in PEFCs by measuring the potential distribution at the anode catalyst layer (CL)/GDL interface over the electrode area of a quasi-two-dimensional PEFC. From the potential distribution they also estimated the respective lateral current distribution. This estimation depends, however, on the results of finite element modeling of the GDL as a two-dimensional resistor.

Recently, our group at Paul Scherrer Institut presented a method that allows access to the lateral current distribution in PEFCs directly with submillimeter resolution.[24] This approach opens up unique insights into the local phenomena that occur on the millimeter scale of channels and lands. In this chapter we reveal a variety of key phenomena. Beyond submillimeter resolved steady-state current density distribution measurements, we focus primarily on the application of spatially resolved transient techniques and on the monitoring of transient phenomena in channel and land areas of PEFCs. We address aspects in the three basic fuel cell research areas, namely

- Membrane electrode assembly (MEA) development (Section 12.4).
- Water management (Section 12.5).
- Cell degradation phenomena (Section 12.6).

With regard to MEA development, we demonstrate how lateral inhomogeneities in gas composition affect the value of the electrochemically available surface area (ECA) as obtained from cyclic voltammetry (CV) experiments,[25–29] and how the contribution of in-plane processes such as reactant depletion underneath the ribs can lead to misinterpretation of results in electrochemical impedance spectroscopy (EIS) experiments[30–32] if classical one-dimensional impedance models[33] are used for the interpretation of results.

With regard to water management in PEFCs,[34–36] we shed light into the correlation between liquid water accumulation and cell performance[37–40] in channel and land areas. By using submillimeter resolved current density distribution measurement and high-resolution in-plane neutron radiography[41] simultaneously in voltage step experiments, the limitations of the local performance, which are inherent to the system and which must be attributed to the cell design, can be separated from losses due to liquid water accumulation by means of the time constant.

Finally, we put our attention to the reverse-current phenomenon,[42, 43] which has been identified as a main source of catalyst layer degradation[20, 44, 45] in PEFCs. It has been proposed that this phenomenon can occur even within small regions of local hydrogen starvation of a size of fractions of a millimeter.[43] An assessment of this phenomenon in channel and land areas with submillimeter resolution is therefore crucial. Here, we analyze the effect of the flow field design, the chosen operation conditions, and proposed carbon corrosion mitigation strategies during start-up and shutdown of a PEFC.

12.2 Approaches for the evaluation of the lateral current distribution in PEFCs

The segmented cell approach[1–8] has become an indispensable tool for the evaluation of down-the-channel inhomogeneities and their impact on local current generation. Unfortunately, this approach cannot straightforwardly be applied for the measurement of the lateral current distribution in the direction perpendicular to the flow channels. Here, several obstacles to the application of the segmented cell approach exist. This is illustrated in Fig. 12.1. The small scale of technical PEFC flow field structures, with a channel and land width of the order of 0.5–2 mm, requires a segmentation of the flow field within distances

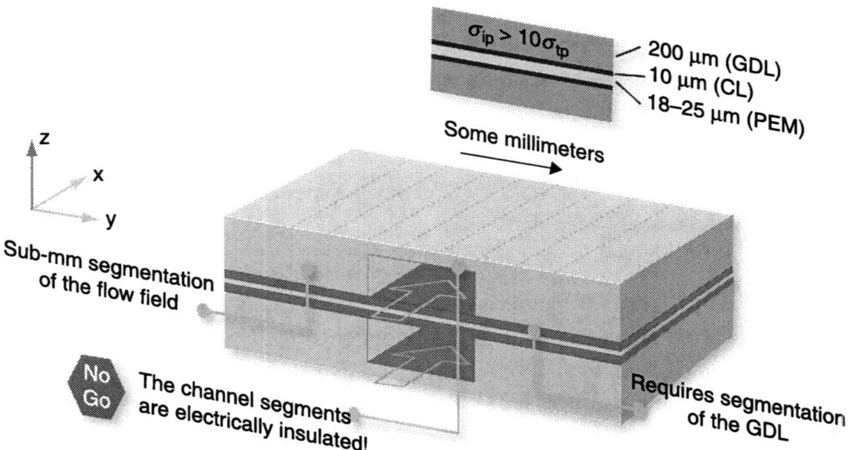

12.1 Obstacles to the application of the segmented cell approach for the measurement of the lateral current distribution in PEFCs (*y*-direction).

of at least some hundred micrometers to achieve sufficient resolution. This is challenging; yet, it is undoubtedly still technically feasible. However, the same requirement holds for the gas diffusion layer (GDL). Here, a segmentation is inevitable, as a result of the high in-plane conductivity of typical diffusion media $(\sigma_{ip} > 10\sigma_{tp})$.[47] To fulfill this requirement is all but impossible. The approach completely fails in the channel area, where no current can be collected from the GDL (Fig. 12.1).

Nevertheless, three experimental approaches for the evaluation of the current distribution in channel and land areas of PEFC flow field structures are reported in the literature:

- The partly catalyzed electrode approach.[48]
- The *in situ* shunt approach.[23]
- The segmented microstructured flow field approach.[24]

All these approaches make use of differential cells. Only the microstructured flow field approach allows direct access to the lateral current distribution in a PEFC, which employs a commercial catalyst coated membrane (CCM).

12.2.1 Partly-catalyzed electrode approach

The partly-catalyzed electrode approach was published by Liu *et al.* in 2008.[48] In this approach either the total channel current or the total rib current is measured separately in two different cells, where either the cathode area under the land or under the channel is loaded with catalyst as shown in Fig. 12.2. Strictly speaking, this method does not allow locally resolved current measurements in PEFCs, since only the integral cell current is obtained in a measurement.[48, 49]

The coarse spatial resolution of the reported method of 2 mm could, in principle, be improved by narrowing the active catalyst area and, therefore, even locally resolved data within the channel and land area of a PEFC could be obtained. However, any interaction of these portions, which undoubtedly exist in a fully catalyzed cell, e.g. due to oxygen depletion underneath the ribs, is denied by this method and no commercial standard CCM can be used when this method is employed.

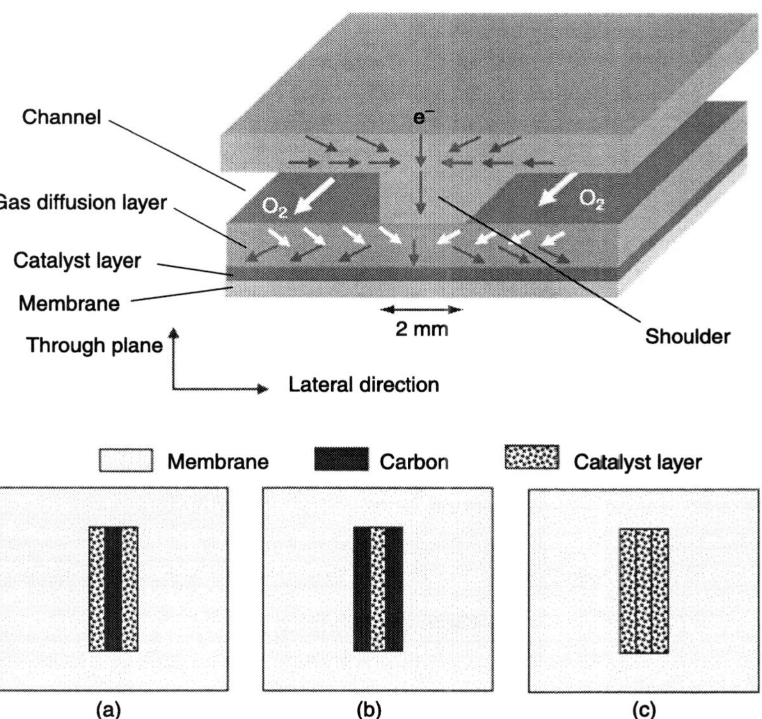

12.2 The partly catalyzed electrode approach: In separate measurements, the current generation at the cathode is either restricted to the channel region (a), or to the land region (b), no fully catalyzed (c) CCM can be used. The separate measurement does not account for any interaction between these regions, which exist if a fully catalyzed CCM is used. (Source: Wang and Liu[48]. Copyright 2008 Elsevier.)

12.2.2 The *in situ* shunt approach

The so-called *in situ* shunt method was published by Freunberger *et al.* in 2006.[23] This method can be used with a fully catalyzed membrane electrode assembly (MEA); yet, it is an indirect method, where the estimation of the current distribution relies on finite element modeling of the GDL. By the insertion of several local potential probes at the catalyst layer/GDL interface of the anode, just the respective potential distribution vs. the end plate potential can be obtained *in situ* with submillimeter resolution (0.2–0.5 mm). The obtained potential difference is attributed to the ohmic drop over the electron conductors of the cell. For the calculation of the respective current density distribution at the catalyst layer/GDL interface of the anode (Fig. 12.3), the characteristics of the resistance formed by the GDL and the flow field plate must be known, including all contact resistances, which are strongly dependent on compact pressure and cause a substantial fraction of the total ohmic drop.[50, 51] Estimation of the resistor characteristics is complex and relies on comprehensive *ex situ* material characterization procedures and numerical structural mechanics calculations.[50, 51] Because the approach assumes ideal material properties, the reliability and reproducibility depends on multiple factors.[50] Application of a smoothing spline or averaging of the same experiment

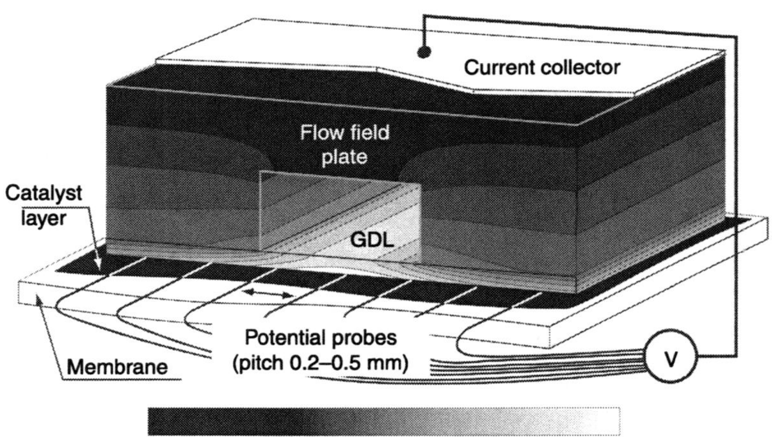

12.3 The *in situ* shunt approach: This method allows the measurement of the local potential difference between the GDL/CL interface and the current collector with submillimeter resolution via potential probes. However, the characteristics of the shunt formed by the GDL and the flow field plate is unknown and cannot be obtained *in situ* experimentally. Estimation of the current density distribution relies, therefore, on numerical structural mechanics calculations of the resistor characteristics. (Source: Freunberger *et al.*[23] Copyright 2006 The Electrochemical Society, Inc.)

over different independent cells is necessary to alleviate fluctuations and to avoid misinterpretation of the results.[50] In addition to all these restrictions, in particular, the poor signal-to-noise ratio of this indirect method[51] impedes the successful application of locally resolved transient techniques such as CV or EIS, where a low distortion measurement of small perturbation currents in the range of some mA/cm^2 is mandatory.

12.2.3 Segmented microstructured flow field approach

The segmented microstructured flow field approach was published by Schneider *et al.* in 2010.[24] This approach overcomes the obstacles to the application of the segmented cell approach for the measurement of the lateral current distribution in PEFCs (Fig. 12.4), at the expense of a reasonable simplification of the experimental system.

The cathode is responsible for the majority of PEFC voltage loss and degradation. Cathode losses comprise kinetic, proton transport, and oxygen transport contributions, whereas anode contributions are usually excluded from the discussion,[21] considering fast hydrogen oxidation reaction kinetics, surplus platinum catalyst loading, and rapid transport of molecular hydrogen. If we restrict our attention to phenomena related to the cathode flow field structure, a cell design must be chosen where the anode acts as a quasi-homogeneous counter electrode across the cell area. Therefore, the following precautions must be taken. The anode must be operated at a high gas flow rate to ensure low residence time and to exclude a notable change in anode gas composition between the hydrogen inlet and outlet. Inhomogeneities, which are induced by the anode flow field structure, must be insignificant on the length scale of the technical cathode flow field. To meet these requirements, a microstructured flow field may be used at the anode.

The segmented microstructured flow field approach for submillimeter-resolved local current measurement in PEFCs is presented in Fig. 12.4. At the anode of a differential cell, a segmented microstructured pin-type flow field (760 pins/cm^2) is directly attached to the catalyst layer. Since no GDL is used, the direct measurement of the local cell current with submillimeter resolution becomes possible in a fully catalyzed MEA, as a result of the finite in-plane conductivity of the CCM. The distance between the pins of the microstructured flow field is 0.2 mm (0.4 mm pitch). Note that the distance between carbon fiber bundles at the surface of typical carbon paper GDL material can be of an order of 100–200 μm,[47] as well. The membrane also acts as the anode seal.[52] The microstructured flow field is 19-fold segmented in the direction perpendicular to the cathode gas channels. The current density distribution in channel and land areas of the cathode flow field can directly be accessed with a resolution of 0.4 mm by the measurement of the respective branch currents. In this configuration, the anode is operated in cross-flow mode (Fig. 12.4). A high anode gas stoichiometry ensures a quasi-homogeneous anode gas composition in the direction perpendicular to the cathode

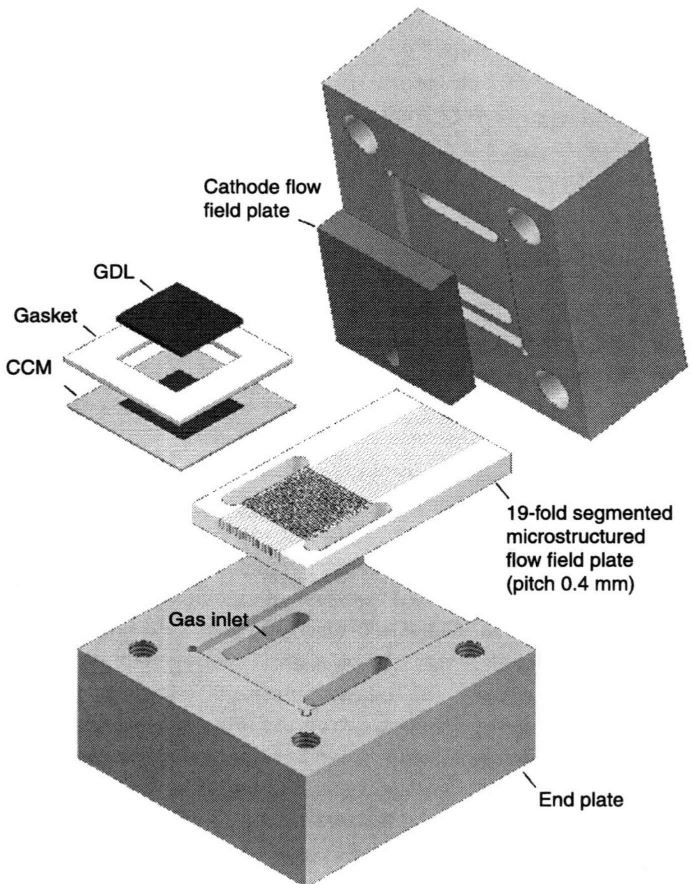

12.4 The segmented microstructured flow field approach: This method allows the direct measurement of current distribution in the direction perpendicular to the cathode flow channel with submillimeter-resolution (0.4 mm). A high cathode gas flow rate ensures a quasi-homogeneous gas composition along the cathode flow channel (differential flow conditions). The anode is operated as a quasi-homogeneous counter electrode and cannot be chosen arbitrarily. Here, a 19-fold segmented microstructured flow field is directly attached to the anode catalyst layer. (Source: Schneider *et al.*[24] Copyright 2010 The Electrochemical Society, Inc.)

gas channel. For current measurement high-precision shunt resistors are used. The voltage drop over these resistors is very low (approximately 0.5 mV at 1 A/cm^2). The voltage signals are amplified before acquisition. Due to the direct approach, no restrictions exist on using this method even in applications that require low noise, wide bandwidth and high-precision current measurement.

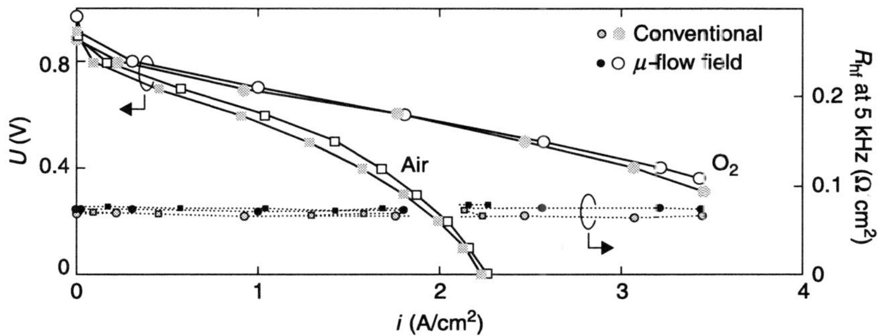

12.5 Integral i/U and i/R_{hf} characteristics of a conventional differential cell, which employs a 4 × 1 mm channel flow field at both anode and cathode (pitch 2 mm), and a cell, which is identical except for the anode, which employs the segmented microstructured flow field (Fig.12.4). T_{cell} = 70 °C, ambient pressure, V_{H2} = 100 mL/min, $V_{O2} = V_{air}$ = 200 mL/min, Gore Primea 5710 CCM (0.1 mg Pt/cm^2 (anode), 0.4 mg Pt/cm^2 (cathode)), Toray TGP-H-090 GDL (10% PTFE), 200 μm thick PTFE gasket. (Source: Schneider *et al.*[24] Copyright 2010 The Electrochemical Society, Inc.)

At the same time, the approach is able to handle high current densities. Figure 12.5 shows the integral i/U characteristics and the corresponding high-frequency resistance (HFR) R_{hf} (f = 5 kHz) of a conventional differential cell and of a cell that is identical, except for the anode, which employs the segmented microstructured flow field (Fig. 12.4). High performance is achieved at ambient pressure in both H_2/O_2 and H_2/air operation modes. Under the same operating conditions, both cells show a virtually equal performance. Evidently, the integral i/U curve is governed by several factors. However, if we take into account the similar HFR values of around R_{hf} = 0.07 Ω cm^2 (Fig. 12.5) in all measurements, the result clearly demonstrates that the employment of the segmented microstructured flow field at the anode has no notable detrimental effect on the cell performance. All results shown in the following sections were obtained with the segmented microstructured flow field approach.

12.3 Submillimeter-resolved local current measurement in channel and land areas

The result of submillimeter-resolved current density distribution measurements in channel and land areas of a 2 mm wide channel cathode flow field is shown in Fig. 12.6 for fully humidified operating conditions (Fig. 12.6a) and for operation of the cell under dry conditions (Fig. 12.6b). The results for cathode operation on fully humidified pure oxygen and fully humidified air are compared in Fig. 12.6a. The current density distribution is quasi-homogeneous for operation on pure oxygen (I_{cell} = 1 A/cm^2). Under fully humidified conditions, no significant

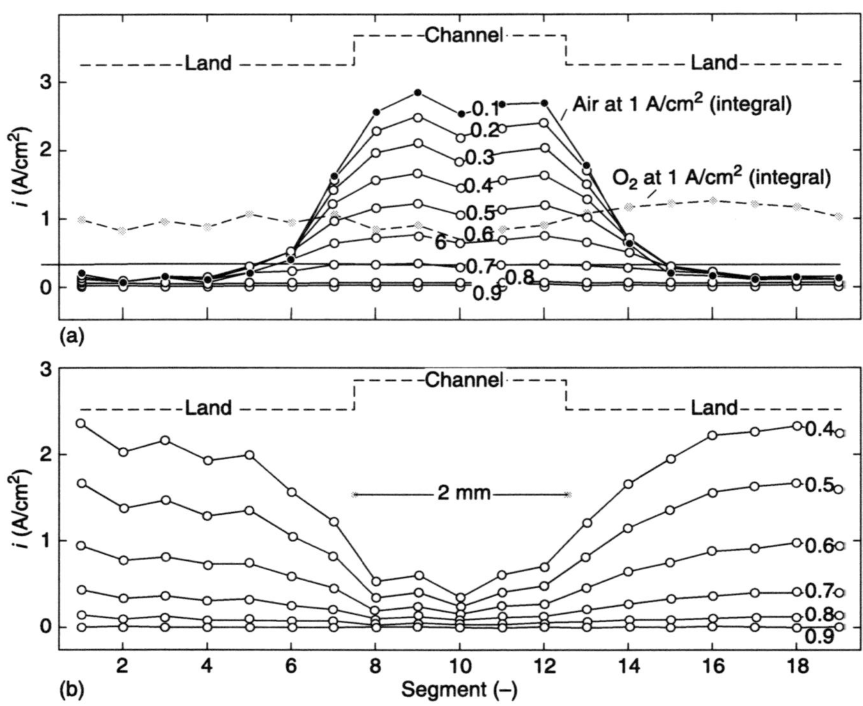

12.6 Current density distribution in channel and land areas of a 1 × 2 mm channel cathode flow field (number denotes cell voltage): (a) Operation on fully humidified H$_2$/O$_2$ (only data for U_{cell} = 0.6 V at I_{cell} = 1 A/cm^2 are shown) vs operation on fully humidified H$_2$/air (data were taken from U_{cell} = 0.9 V down to U_{cell} = 0.1 V at I_{cell} = 1 A/cm^2) and (b) operation on dry O$_2$ and H$_2$ (50% RH). T_{cell} = 70°C, ambient pressure, V_{H2} = 100 mL/min, V_{O2}= V_{air} = 200 mL/min, Gore Primea 5710 CCM (0.1 mg Pt/cm^2 (anode), 0.4 mg Pt/cm^2 (cathode)), Toray TGP-H-090 GDL (10% PTFE), 200 µm thick PTFE gasket. (Source: Schneider et al.[24] Copyright 2010 The Electrochemical Society, Inc.)

difference in the water partial pressure in channel and land areas of the cell can be expected and, consequently, the oxygen partial pressure is kept at a constant level in the binary gas mixture. This provides virtually homogeneous reaction conditions throughout the oxygen electrode area, unless liquid water starts to block pores and thereby inhibits the access of reactants to the active sites at higher cell currents. The slightly decreasing performance in the channel region (Fig. 12.6a, segments 8–12) is related to a higher contact resistance.[48]

This is different for air operation, where mass-transport-related inhomogeneities exist even at a rather low cell polarization (Fig. 12.6a). As most of the total cell current is generated in the channel area, the local current density approaches maximum values of more than 2.5 times the average value (I_{cell} = 1 A/cm^2). This results in a high cell polarization and a low cell voltage of U_{cell} = 0.1 V vs.

U_{cell} = 0.6 V for an operation on pure oxygen (Fig. 12.6a). Within the channel area, the oxygen concentration is kept high by the convective gas flow and, consequently, all parts of the channel (Fig. 12.6a, segments 8–12) show virtually the same performance. A limiting current density is still not reached at high local current density values beyond 2.5 A/cm^2. However, already within a distance of 1 mm from the channel, the cell performance drops below a tenth of this maximum value, as a consequence of the increasing diffusion path length for oxygen and its depletion in diluting nitrogen in the land area. The inverse characteristics are observed for an operation of the cell on dry oxygen (50% relative humidity (RH) H$_2$) as a result of water formation (Fig. 12.6b). The rising water partial pressure under the rib manifests in a strong performance gain in the land area due to the improving hydration of the ionomer.

12.4   Local transient techniques in channel and land areas

Transient techniques such as electrochemical impedance spectroscopy (EIS) or cyclic voltammetry (CV) are important tools in PEFC diagnostics. Employment of these techniques focuses primarily on a characterization of the most elemental part of a PEFC, namely the MEA,[22, 53–58] and therefore the effect of the flow fields and the associated in-plane inhomogeneities on transient response is often excluded from the discussion. However, these techniques are generally used *in situ* in technical cells and, consequently, the contribution of in-plane processes such as reactant depletion along the flow channels or underneath the ribs must seriously be taken into account.

Our earlier work focused on the effect of down-the-channel inhomogeneities on PEFC transient response.[9] These results have clearly demonstrated that simply concentrating on the MEA, while excluding in-plane processes from the discussion, runs the risk of focusing on the wrong phenomena and can lead to wrong conclusions.[14, 15, 59–66] Following up our results, we continue our discussion with an assessment of the importance of lateral inhomogeneities, when transient techniques are used *in situ* for MEA characterization. Therefore, submillimeter-resolved EIS and CV measurements were employed in channel and land areas of a differential cell.

12.4.1 Submillimeter-resolved cyclic voltammetry

CV is employed in PEFC diagnostics[25] for the determination of the electrochemically available surface area (ECA). In general, the ECA value is used as a measure for catalyst utilization in PEFC electrodes. In CV experiments, the area under the hydrogen underpotential deposition (H$_{upd}$) peaks is used for the determination of the ECA value.[26–29] These peaks appear in the potential range between 0.3 and 0.05 V (Fig. 12.7) vs. hydrogen electrode potential (HEP).

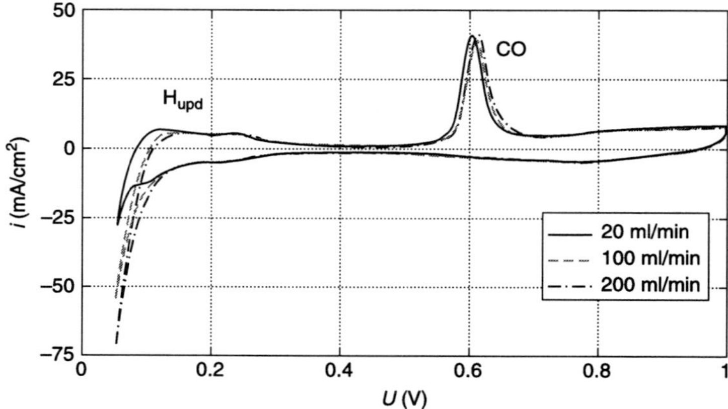

12.7 Effect of the nitrogen gas flow rate (V_{N2} = 20–200 ml/min) at the working electrode (3 × 1.2 mm channel flow field) on integral H_{upd} and CO stripping peaks (T_{cell} = 70 °C). The counter/reference electrode is operated on fully humidified pure hydrogen (V_{H2} = 100 ml/min). The working electrode was exposed to 1% CO in N_2 to adsorb CO on the platinum catalyst particles at E = 0.125 V and the working electrode was purged with fully humidified nitrogen (V_{gas} = 200 ml/min) prior and after adsorption. The first sweep (v = 10 mV/s) was then actually started at a potential of E = 0.125 V (vertex potential 1: E = 1 V, vertex potential 2: E = 0.05 V), the data is displayed starting from potentials above E > 0.35 V. CCM: N115, 0.5 mg Pt/cm^2 (Paxitech), Toray TGP-H-090 GDL (10% PTFE), 200 µm thick PTFE gasket.

Our recent work has focused on the effect of the convective inert gas flow, which is usually employed in CV measurements in PEFCs to exclude traces of oxygen in the flow field of the working electrode.[9, 14, 27, 66] Our results have clearly demonstrated that the convective inert gas flow leads to high cathodic hydrogen reduction currents in both the cathodic and the anodic sweep at lower positive potentials, as molecular hydrogen, which is removed from the system with the inert gas during the sweep, must be compensated by the formation of additional hydrogen.[9, 14] The effect hampers an accurate determination of the ECA from the H_{upd} area because a portion of this area is hidden by the strong cathodic hydrogen evolution current, as illustrated in Fig. 12.7.[9, 14, 24]

Note that the ECA value, as obtained via carbon monoxide (CO) stripping voltammetry is virtually unaffected by the inert gas flow rate, as we have shown earlier.[66] Here, the area of the CO oxidation peak is used for the determination of the ECA value. This peak appears at higher positive potentials of around 0.6 V vs. HEP. An increased inert gas flow rate has a pronounced effect on the H_{upd} area, whereas the area of the CO oxidation peak is virtually unaffected as demonstrated in Fig. 12.7. The CO oxidation peak potential is shifted however slightly toward more positive potentials with increasing inert gas flow rate.

Locally resolved CV measurements in the down-the-channel direction have clearly shown that the cathodic hydrogen evolution current and its effect on the H_{upd} area is most pronounced in the local CV at the nitrogen inlet. The effect levels off toward the gas outlet, as molecular hydrogen, which is formed during the sweep, accumulates in the nitrogen stream along the flow channels.[9, 14]

The results presented in Fig. 12.8 demonstrate that a similar effect also occurs on the scale of channel and lands in the direction perpendicular to the flow channel. The application of locally resolved CV in PEFCs is demonstrated here with submillimeter resolution. Fig. 12.8 shows local CVs as obtained in channel and land areas of a 2 mm wide channel cathode flow field.

The convective flow of inert gas strongly enforces the hydrogen reduction currents in the channel region. The respective local voltammograms exhibit a virtually congruent shape (Fig. 12.8, segments 8–12). Within the channel area, the hydrogen concentration is kept at a low level and, consequently, the cathodic currents approach high values of more than 100 mA/cm^2 at the vertex potential (Fig. 12.8, segments 8–12). As a result, a large portion of the local H_{upd} area is hidden by the strong cathodic hydrogen evolution current (Fig. 12.8). This effect entails a severe underestimate of the local ECA values in the channel region. The effect levels off in the land area (Fig. 12.8), where the cathodic hydrogen evolution current decreases, as a consequence of increasing diffusion path length for molecular hydrogen toward the channel.

Mitigation strategies should therefore address the effective diffusivity of the porous materials. A strong compression or a high water content of the GDL would be beneficial in this context. However, the best strategy is simply to stop the convective inert gas flow during the CV measurement.[9, 14, 24, 66] As molecular hydrogen is allowed to accumulate in the channel region, the cathodic hydrogen reduction current within both channel and land areas becomes much less pronounced at low potentials (figure not provided here).

12.4.2 Submillimeter-resolved electrochemical impedance spectroscopy

The effect of the flow fields on PEFC impedance response is often neglected. Most of the published ac impedance models for PEFCs take into account only those processes that occur across the MEA.[22, 25, 31–33, 53, 54, 58, 67–73] They are one-dimensional. Strictly speaking their use is therefore restricted to the interpretation of impedance spectra of so-called quasi-one-dimensional cells. This kind of cell provides quasi-homogeneous reaction conditions over the electrode area. Nevertheless, one-dimensional PEFC impedance models are often used for the interpretation of impedance spectra of technical (three-dimensional) single cells or of fuel cell stacks (e.g. Refs 69, 71, 73). Evidently, this simplifying approach, which focuses solely on the MEA, rules out any contribution of the flow fields to PEFC impedance response; thereby, it runs the risk of considering the wrong

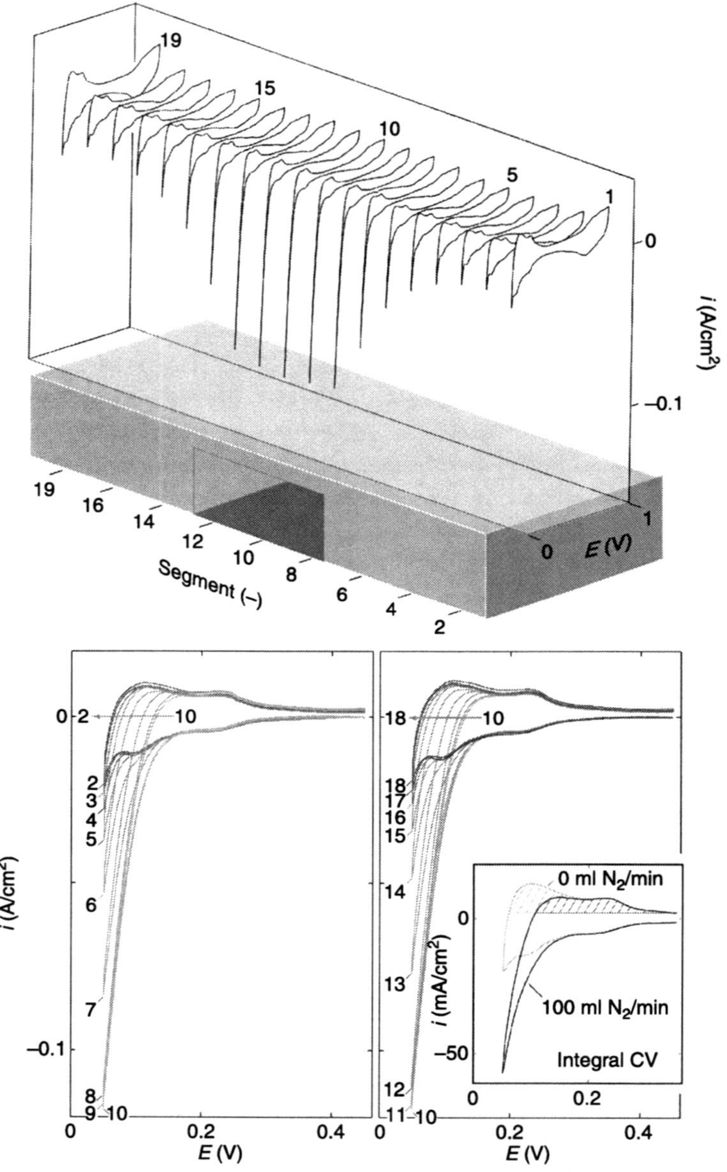

12.8 Effect of inert gas flow on integral (inset) and local CVs in channel and land areas of a 1 × 2 mm channel flow field. The local CVs are shown here for V_{N2} = 100 mL/min (number denotes the segment). CCM: N115, 0.5 mg Pt/. cm² (Paxitech), Toray TGP-H-090 GDL (10% PTFE), 200 μm thick PTFE gasket. (T_{cell} = 70 °C, v = 10 mV/s) (Source: Schneider et al.[24] Copyright 2010 The Electrochemical Society, Inc.)

phenomena.

Our recent work focused on the effect of down-the-channel inhomogeneities on PEFC impedance response.[9] Here, locally resolved EIS has clearly shown that the use of one-dimensional across-the-MEA impedance models is inadequate for the interpretation of results and can lead to wrong conclusions.[9, 15, 21, 59–65] This has been demonstrated, in particular, for PEFC air cathodes.[9, 15, 59–62, 64] In our experiments, a low-frequency capacitive loop appeared in both integral and local spectra only at lower air stoichiometry. Our results have proven that the appearance of this loop is primarily indicative of a limitation of cell performance, due to depletion of oxygen in the down-the-channel direction at finite gas flow rates.[9, 59, 60, 62, 64] The use of a one-dimensional across-the-MEA impedance model for the interpretation of the results would therefore undoubtedly take into account the wrong phenomena. A key feature of the local spectra, namely the appearance of negative low-frequency resistance values in the outlet region at low air stoichiometry is a direct consequence of high oxygen depletion[9, 10, 59, 62, 74] in flow direction. As an important consequence, this feature is only captured by models, which take into account the down-the-channel dimension e.g., 1D + 1D impedance models.[60, 64, 65]

Here, we focus on the assessment of the importance of inhomogeneities in the direction perpendicular to the flow channel on the millimeter scale of channel and lands on integral and locally resolved PEFC impedance response. This assessment is crucial, since it is often assumed in EIS measurements that small cells, which are operated at high gas flow rates (differential cells) provide quasi-homogeneous reaction conditions all over the active area. Therefore, locally resolved EIS[10, 12, 13] was employed for the first time in channel and land areas of a differential cell with sub-millimeter resolution. In addition to locally resolved impedance measurements in channel and land areas, we present a simple two-dimensional impedance model for a PEFC air cathode. This model accounts for lateral effects in PEFC impedance measurements. The simple model is used here to further support the discussion of the experimental results. It captures the experimentally obtained key effects of GDL mass transport losses on the impedance spectra.

Experimental approach – The segmented microstructured flow field approach[24] was used in the measurements (Fig. 12.4)). At the cathode a flow field (three channels, each channel 1.2 mm wide, channel-to-land ratio 1:1) was employed. The cell was operated on fully humidified H_2 and air at a high gas flow rate at ambient pressure. The locally resolved impedance spectra were obtained via calculation of the ratio of the Fourier transformed ac voltage and the Fourier transformed local ac current signals for all characteristic areas of the flow field at any given frequency of the sweep.[13] The respective integral cell impedance and the integral cell current were calculated from the locally resolved data.

Two-dimensional PEFC air cathode impedance model – The model considers mass transport in the GDL by binary diffusion of oxygen in nitrogen according to Fick's law (Eq. 12.1). For the sake of simplicity we assume an isotropic effective diffusion coefficient[75] of D_{eff} = 0.05 cm²/s. The calculated element is a single repetitive unit of the air cathode consisting of one channel and rib (1.2 mm wide each) with periodic boundary conditions in lateral direction (in-plane direction). In the across-the-MEA direction (through-plane direction), the model considers the GDL (t_{GDL} = 200 μm, ε = 0.6) and the catalyst layer, which is just treated here as an additional 10 μm thick GDL element. The electrochemical reaction, which causes a mass flux of oxygen due to the Faradaic current, Eq. 12.1 is assumed to take place in the center of the catalyst layer. The oxygen concentration in the gas channels is assumed to be constant (c_{ch} = 5.17 mol/m³) and uniform due to the very high air flow rate.

$$I_D(y,z,t) = -D_{eff} \cdot \nabla c(y,z,t), \quad I_F(y,t) = -\frac{j_F(y,t)}{4F} \qquad [12.1]$$

The electrochemical reaction is described by the Tafel equation (Tafel slope b = 69 mV/dec, j_0 = 1 mA/cm², c_{atm} = 35.52 mol/m³). Here, a reaction order of 0.79 is used,[76] which already considers the effect of the oxygen concentration on the Nernst potential. The current is calculated by the equation system in Eq. 12.2. The underlying local electrode model is a Randles cell equivalent circuit, which accounts for an ohmic series resistance R = 0.1 Ω cm² and for the capacitance of the electrochemical double layer[77] C_{dl} = 150 mF/cm². In Eq. 12.2, η_{el} is the local voltage drop across the double layer and c_{el} is the local oxygen concentration in the catalyst layer.

$$j_F(y,t) = j_0 \cdot \left(\frac{c_{el}(y,t)}{c_{atm}}\right)^{0.79} \cdot \exp\left(\frac{\eta_{el}(y,t)}{b}\right)$$

$$\frac{\partial}{\partial t}\eta_{el}(y,t) = \frac{1}{C_{DL}}(j(y,t) - j_F(y,t)) \qquad [12.2]$$

$$j(y,t) = \frac{\eta(t) - n_{el}(y,t)}{R}$$

The set of equations is discretized on a rectangular grid and solved using a low-order finite volume method. Conservation of species is the only conservation law considered, as given by Eq. 12.3.

$$\frac{\partial}{\partial t}c(y,z,t) = -\text{div}\,I(y,z,t), \quad I(y,z,t) = I_D(y,z,t) + I_F(y,z,t) \qquad [12.3]$$

For the steady-state calculations the time derivatives in Eqs 12.2 and 12.3 are set to zero. The resulting algebraic equations are solved in Mathematica. The steady-state solution of Eq. 12.2 can be expressed in terms of the Lambert W function $W\{x\}$ (Eq. 12.4).

$$j^{(ss)}(y)j_F^{(ss)}(y) = \frac{b}{R}W\left\{\frac{R \cdot j_0 \cdot (c_{el}^{(ss)}(y)/c_{atm})^{0.79} \cdot \exp(\eta^{(ss)}/b)}{b}\right\} \qquad [12.4]$$

For the calculation of the locally resolved impedance spectra, the local current response to a 1 mV step in the over-potential η is calculated at two different steady-state cell currents $j_1 = 0.6\,\text{A/cm}^2$ and $j_2 = 1.3\,\text{A/cm}^2$ within four characteristic sections of channel and land: channel center (CC), channel edge (CE), land center (LC), and land edge (LE). The results are transformed from time domain into frequency domain via discrete Fourier transform.[64]

Effect of lateral inhomogeneities on PEFC impedance response – Experimental results of steady-state and ac impedance measurements are presented in Fig. 12.9. Integral data and results of locally resolved measurements in channel and land areas of the cathode flow field are shown. The integral i/U curve of the cell is displayed in Fig. 12.9a (black curve). This curve appears to approach limiting current density at higher cell polarization. The higher slope of the i/U curve at high current densities above 1 A/cm^2 (Fig. 12.9a, black curve) is consistent with the observed strong increase of the integral impedance spectrum toward lower cell voltage (Fig. 12.9b, black curve). An approach based on one-dimensional across-the-MEA PEFC models[33] would probably take into account here an increasing mass transport limitation in the across-the-MEA direction (e.g. across the GDL) to account for these phenomena at lower cell voltage.

However, this conclusion is clearly put into question by the results of spatially resolved current density and impedance measurements in channel and land areas (Fig. 12.9). The i/U characteristics of channel and land regions are shown in Fig. 12.9a (grey curves). The current production increases, virtually unhindered, with increasing cell polarization in the channel area; yet, in the land area, the current production stops to increase already at cell voltages below $U = 0.4$ V. This result demonstrates that the apparent onset of limiting current conditions in the integral i/U curve (Fig. 12.9a, black curve) at higher cell polarization is actually more a change of the slope of this curve, as the current production stops increasing underneath the ribs and continues to rise only in half of the electrode area within the channel region.

An interesting phenomenon becomes apparent in the i/U characteristics of the rib center (Fig. 12.9c). At higher total cell polarization, the local current production actually decreases despite the increasing total cell polarization. A similar effect was observed and analyzed earlier in the down-the-channel direction of H$_2$/air PEFCs.[3, 9, 10, 60–62, 64, 74] The negative slope of the local polarization (i/η) curve in the rib center is a result of oxygen concentration polarization losses due to oxygen depletion in the lateral direction. Oxygen is consumed in the oxygen reduction reaction, while it diffuses under the rib. As a result of the 'ever' increasing oxygen consumption with increasing total cell polarization in the rib edge and channel area, the total increase in cell polarization is finally overcompensated in the rib center by increasing oxygen concentration polarization losses,[9, 61, 62, 64] and thereby the local current decreases despite an increasing total cell polarization. This phenomenon has an important impact on the locally resolved impedance spectra in channel and land areas.

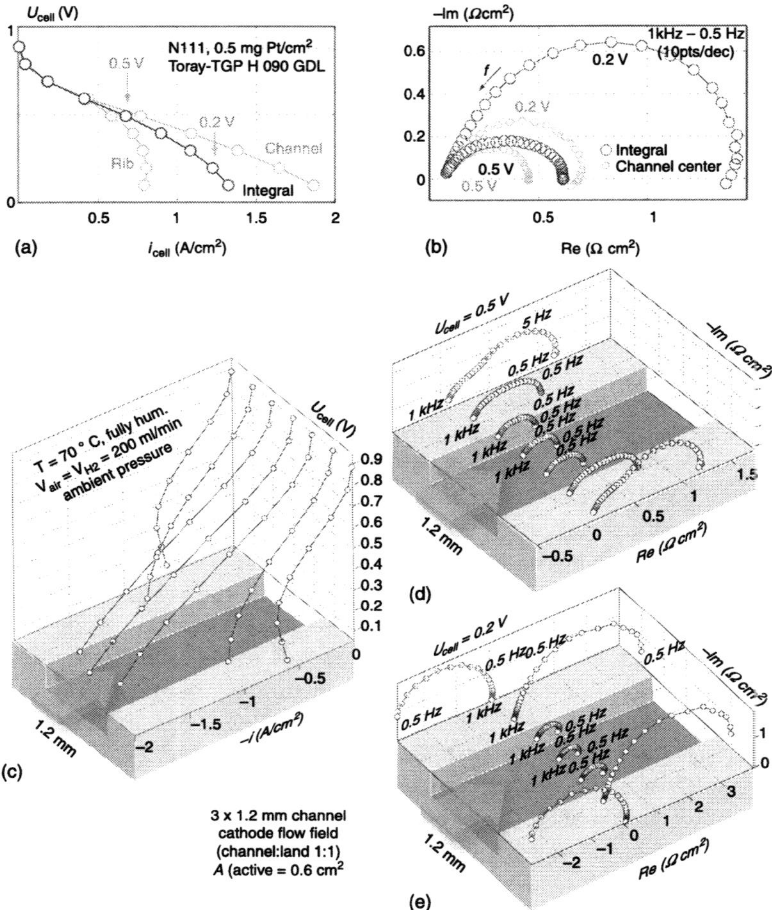

12.9 Integral and locally resolved measurements in a differential H₂/air PEFC: (a) integral i/U characteristic (black curve) and i/U characteristics of channel and land regions (grey curves) (b) integral impedance spectra (black curve) and local spectra in the channel center (grey curves) at a cell voltage of $U = 0.5$ V and $U = 0.2$ V (c) local i/U curves in channel and land areas of a repetitive unit of the 3 × 1.2 mm channel flow field and, (c, e) locally resolved impedance spectra (perturbation voltage amplitude $U_{mod} = 15$ mV, $f = 500$ mHz – 1000 Hz at 10 pts/dec) at $U = 0.5$ V (d) and $U = 0.2$ V (e). For the determination of the current density distribution and the locally resolved impedance spectra at a given cell voltage, the local dc and ac current signals were averaged within four characteristic areas of the cathode flow field: channel center (CC), channel edge (CE), land center (LC), and land edge (LE). Segments 1 and 19 were not considered here. CCM: N111, 0.5 mg Pt/ cm² (Paxitech), Toray TGP-H-090 GDL (10% PTFE), 200 μm thick PTFE gasket, $T_{cell} = 70$ °C, $V_{air} = V_{H2} = 200$ ml/min. (Source: Schneider *et al.*[114] Copyright 2011 The Electrochemical Society, Inc.)

Locally resolved impedance spectra as obtained at a cell voltage of $U = 0.5$ V and $U = 0.2$ V are presented in Fig. 12.9d and Fig. 12.9e. Already at a cell voltage of $U = 0.5$ V, a second low-frequency loop develops in the local spectra underneath the ribs. This loop becomes more strongly pronounced, with increasing distance from the channel and an increasing in-plane diffusion path length for molecular oxygen (Fig. 12.9d). This indicates that the appearance of the low-frequency loop must be associated with the depletion of oxygen in the land area. The conclusion is supported by the appearance of a negative low-frequency intercept[9, 60–64] in the local spectrum of the rib center at higher total cell polarization ($U = 0.2$ V) (Fig. 12.9e), where the slope of the respective local polarization ($i'\eta$) curve exhibits a negative value (Fig. 12.9c).

These unique features, which appear in the rib center at low cell voltage and high oxygen consumption, are key indicators for a limitation of cell performance, due to depletion of oxygen underneath the ribs. The spatially resolved steady-state and ac impedance measurements demonstrate that the strongly increasing loop diameter of the integral impedance spectrum toward lower cell voltage and higher current density (Fig. 12.9b, black spectra) is not primarily indicative of a mass transport limitation in the across-the-MEA direction. It is instead a direct consequence of depletion of oxygen underneath the ribs in the direction perpendicular to the flow channels. This conclusion is emphasized by the weak impact of the increasing current density on the local spectrum in the channel center when the cell voltage is decreased (Fig. 12.9b, grey curve). Strictly speaking, only this part of the MEA can be regarded as a quasi-one-dimensional cell. Consequently, use of differential cells does not generally justify a one-dimensional treatment of data by using across-the-MEA impedance models, in particular at higher cell currents where mass transport limitations underneath the ribs become important.

The experimental results clearly demonstrate that a change of the oxygen consumption with ac current in the rib edge or channel region has an impact on the impedance response in the rib center (Fig. 12.9c, 9e). An appropriate ac impedance model must take into account this coupling between different sections of the electrode area. The appearance of negative low-frequency resistance values in the local spectra underneath the ribs is therefore only captured by impedance models that take into account lateral effects in the direction perpendicular to the flow channel.

The results of two-dimensional steady-state and transient calculations for a PEFC air cathode are presented in Fig. 12.10 and Fig. 12.11. The calculated two-dimensional, steady-state, oxygen concentration profile in a single repetitive unit of the air cathode is shown for two loads $j_1 = 0.6$ A/cm^2 and $j_2 = 1.3$ A/cm^2 in Fig. 12.10.

The shape of the oxygen concentration profile at the air electrode governs the characteristics of the respective lateral current distribution at steady state as expressed by Eq. 12.4. The calculated oxygen concentration profile (Fig. 12.10) is consistent with the experimental observation of a local current maximum in the channel region at higher cell current (Fig. 12.9c).[24] In the channel area, the oxygen

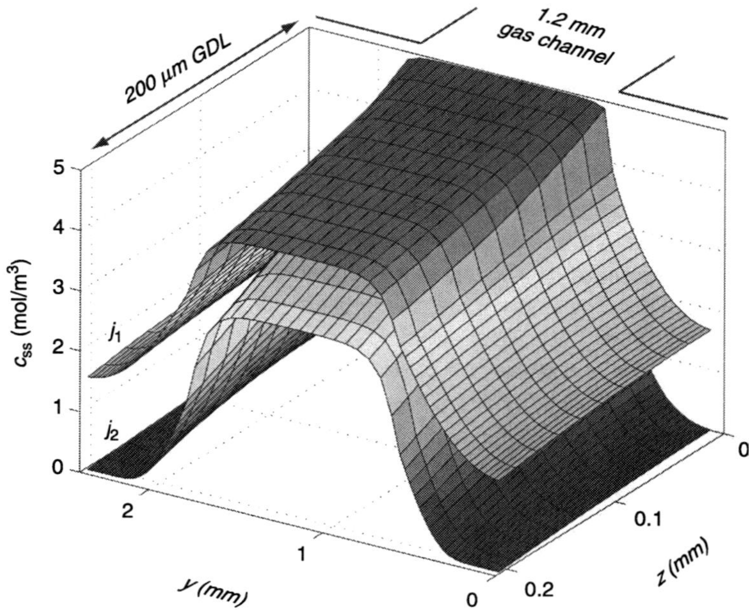

12.10 Calculated two-dimensional steady state oxygen concentration profile at a total current density of j_1 = 0.6 A/cm² and j_2 = 1.3 A/cm² for the repetitive unit of a 1.2 mm channel cathode flow field (channel:land ratio 1:1). (Source: Schneider *et al.*[114] Copyright 2011 The Electrochemical Society, Inc.)

concentration c_{ss} is kept at a high level in the through-plane direction across the whole GDL thickness. The oxygen concentration c_{ss} decreases, however, severely underneath the ribs in the in-plane direction. In particular, at higher current density, the local oxygen concentration c_{ss} approaches values near zero in the rib center (Fig. 12.10). At a low steady-state oxygen concentration c_{ss}, the local cell current in the rib center is highly sensitive to changes in the oxygen concentration Δc, since the associated change of the oxygen concentration polarization depends on the ratio $\Delta c/c_{ss}$.[9, 60, 61, 64] This effect becomes important in the discussion of the local transient response of the cell.

Calculated local current transients to a 1 mV step perturbation are shown in Fig. 12.11a, 11b for two steady-state cell currents j_1 = 0.6 A/cm² and j_2 = 1.3 A/cm². After the voltage step, the observed change in current density is limited initially by the ohmic series resistance (R = 0.1 Ω cm²) to a maximum value of 10 mA/cm² throughout the entire electrode area. The local current transients exhibit a capacitive behavior (Fig. 12.11a, 11b). The characteristic decay is governed by double layer charging and a changing oxygen concentration in through-plane and in-plane directions. The time to reach steady state is therefore, virtually equal, and at a minimum in the channel area (CC, CE). Underneath the ribs (LC, LE), however, this

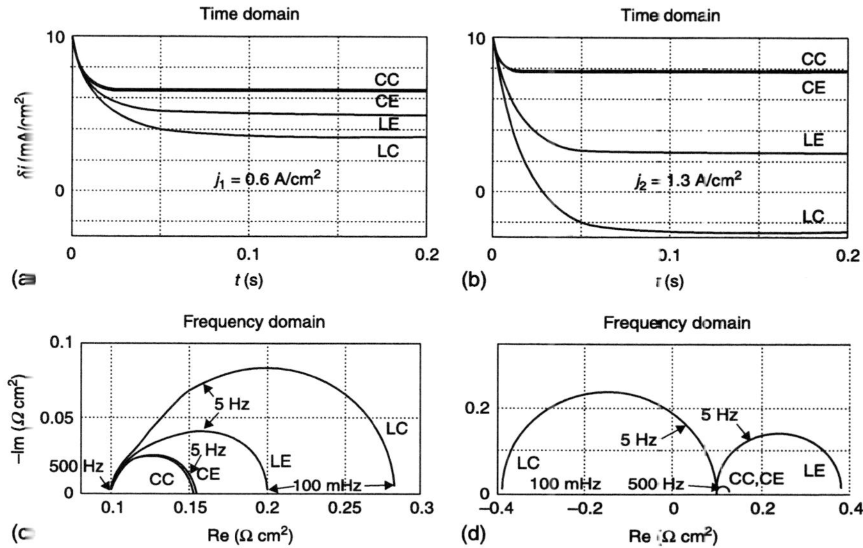

12.11 Calculated local current transients (1 mV step perturbation) for a steady state cell current of (a) $j_1 = 0.6$ A/cm^2 (b) $j_2 = 1.3$ A/cm^2 in four characteristic areas of the flow field (Fig. 12.10): channel center (CC), channel edge (CE), land edge (LE) and land center (LC), and (c, d) locally resolved impedance spectra as obtained via discrete Fourier transform of the time domain data at (c) j_1 (d) j_2. (Source: Schneider *et al.*[114] Copyright 2011 The Electrochemical Society, Inc.)

time increases gradually in lateral direction with increasing diffusion path length for molecular oxygen as a consequence of increasing distance from the channel (Fig. 12.11a, 11b).

The opposite trend is observed for the total increase of the local steady-state current density (Fig. 12.11a, 11b) after the step. Here, a decreasing value is observed toward the rib center (LC) as a result of increasing oxygen concentration polarization losses in lateral direction, as oxygen is consumed while it diffuses in-plane underneath the ribs. At lower cell current ($j_1 = 0.6$ A/cm^2) the obtained local change in current density is still positive in all four characteristic areas of the flow field (Fig. 12.11a). At higher cell current ($j_2 = 1.3$ A/cm^2) and low oxygen concentration c_{ss} underneath the ribs (Fig. 12.11b), the increase in oxygen concentration polarization in the rib center (LC) overcompensates the total perturbation voltage amplitude upon reaching steady state after the step.[62, 64] As a result, the local steady-state current density actually decreases despite an overall increasing cell polarization (Fig. 12.11b (LC)). Consequently, after Fourier transform of the time domain data, a negative low-frequency intercept is obtained in the local spectrum[64] of the rib center (Fig. 12.11d) at high load current ($j_2 = 1.3$ A/cm^2), while the local spectra still exhibit positive low-frequency resistance values at a lower load current ($j_1 = 0.6$ A/cm^2) in all characteristic areas of the cathode flow field (Fig. 12.11c).

With the set of parameters used here, the three captured processes of (i) double layer charging, (ii) through-plane diffusion, and (iii) in-plane diffusion and consumption of oxygen cannot be separated, and only one loop is distinguishable. Nevertheless, the purely diffusive model is able to capture the key experimental finding, namely the negative low-frequency intercept in the local spectrum underneath the ribs at high load.

The results demonstrate that care must be taken if one-dimensional across-the-MEA PEFC impedance models are used to analyze the impedance response of two-dimensional, i.e. differential, cells. Operation of small active area PEFCs at high gas stoichiometry (differential operation), is an efficient way to exclude inhomogeneities in the down-the-channel direction. However, at higher loads mass transport limitations underneath the ribs become significant in the direction perpendicular to the flow channels. In differential cells, only the channel center can be regarded as a quasi-one-dimensional cell. The contribution of in-plane processes to PEFC impedance response must seriously be taken into account. Simply focusing on through-plane processes in the across-the-MEA direction runs the risk of considering the wrong phenomena and can lead to wrong conclusions.

12.5   Combined use of local transient techniques and neutron radiography

Electrochemical *in situ* transient techniques such as EIS or potential step experiments are not stand-alone techniques. This applies in particular to the use of transient techniques for the characterization of complex systems, such as fuel cells. Here, several processes occur in different directions of the cell and contribute to the voltage or current response during cell perturbation. A description of the system solely based on the current, voltage or impedance response is therefore challenging. Information from complementary *in situ* techniques on distinct parameters can be crucial for the analysis of results. Evidently, in the case of PEFCs, the combined use of electrochemical transient techniques and methods for static and dynamic *in situ* monitoring of water plays a key role. In PEFCs water coexists, upon reaching saturated vapor pressure, in both the liquid and the vapor phase. In this context, our earlier work has successfully demonstrated the combination of locally resolved EIS in the down-the-channel direction with through-plane neutron radiography[38–41] or with tunable diode laser spectroscopy (TDLAS)[78–80] for liquid and gaseous water detection.[65, 81, 82] Here we focus on the effect of liquid water on local cell performance in channel and land areas of a H$_2$/air PEFC.

Much effort has been spent to understand two-phase flow phenomena by both experimental and modeling investigations. In the face of improved physical models[35, 36] and emerging diagnostic tools, e.g. advanced imaging techniques,[83] the picture of liquid water transport and its accumulation in the porous cell components is still nebulous. The same holds for flooding related phenomena. In particular, the demonstration to correlate the local water content to performance loss[38, 39, 84] remains challenging, if we consider that local mass transport limitations are not necessarily governed by the effect of liquid water flooding. In this context, the use of transient techniques provides an avenue to separate flooding related losses from the system-inherent limitation due to mass transport in the porous layers by means of the time constant. The results presented in the previous section have clearly demonstrated that time constants for the diffusive transport of oxygen in PEFCs are in the order of only fractions of a second. In contrast, water saturation profiles develop within tens of seconds, while the rearrangement of water may take minutes. These processes are the slowest at all, if effects due to changing membrane hydration can be excluded.[35, 36, 85–91] A separation of flooding related losses should therefore be possible by means of the time constant.

For this purpose, submillimeter-resolved current density distribution measurements and high-resolution in-plane neutron radiography have been used simultaneously in voltage step and voltage sweep experiments. Thereby, the simultaneous examination of the temporal evolution of local water accumulation and local cell performance in cathode channel and land areas becomes possible.

12.5.1 Experimental approach for high time resolution

In transient measurements, a series of neutron images must be taken synchronously with the voltage and current readings. The micro setup with tilted scintillator (Fig. 12.12a) and optical adaptor, employed[92] for neutron beam detection at the ICON beam line at the Paul Scherrer Institut, provides a full width at half maximum (fwhm) resolution of 20 μm in the through-plane direction and of 200 μm in the in-plane direction at a low exposure time[93] of 10 s.

A temporal resolution equal to the required exposure time of 10 s is appropriate in voltage sweep experiments ($v = 10$ mV/s). Here, no special measures must be taken. This is different in the voltage step experiments, where the cell voltage is changed quasi-instantaneously from OCV to 0.1 V. Here, a temporal resolution in the order of 1–2 s is required. A temporal resolution better than the total exposure time can be achieved if the required total exposure time of 10 s is accumulated by the averaging of multiple series of neutron images as received upon repetitive cell perturbation from steady state.[94] The approach requires a precise synchronization of the potentiostatic cell perturbation, the data acquisition and the image capture sequence with the CCD shutter signal as illustrated in Fig. 12.12.[94] The CCD shutter signal is used for synchronization. During an image capture sequence, a number ($n = 32$) of images is taken at an interval of ~2 s after cell perturbation. The cell is then disconnected from the potentiostat and, after gas purge at OCV, the next sequence is triggered. In total, $N = 5$ series of neutron images were averaged to achieve a temporal resolution of ~2 s at a spatial resolution of 20 μm (total exposure time: ~5.2s). Here,

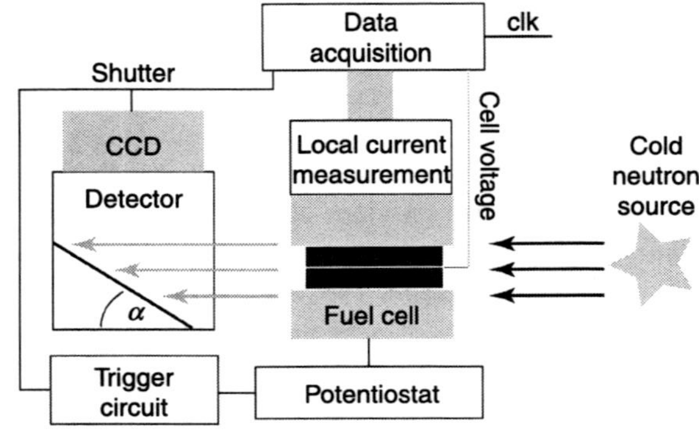

(a)

12.12 (a) Experimental setup[94] used in the voltage step experiments. Exposure time is accumulated by the averaging of multiple series of neutron images as received onto repetitive cell perturbation from OCV to 0.1 V. The first current measurement is taken after 3 ms at $i_{cell} < 3.5$ A/cm^2, and (b) characteristic positions of the cathode flow field used for the evaluation of (i, ii) the local water accumulation (the full line is a guide for the eye (fit $y(t) = k*(1 - e^{-at})$)) and (iii) the local current density. In both voltage step and voltage sweep experiments, the local current transients were averaged before evaluation within each of the four characteristic areas of the flow field. $A = 0.6$ cm^2, $T_{cell} = 40$ °C, $T_{hum\_air} = T_{hum\_}H_2 = 40$ °C, $V_{air} = V_{H2} = 200$ ml/min, ambient pressure, cathode flow field: 3 × 1.2 mm channel (channel:land ratio 1:1), GDL: untreated Toray TGP-H-090 (78% porosity, 280 μm thickness), 200 μm thick PTFE gasket, CCM: Nafion 112, 0.5 mg Pt/cm^2, 70% Pt/C. (Source: Schneider *et al.*[94] Copyright 2010 The American Chemical Society.)

the reproducibility of the observed current transients during the flooding period after cell perturbation from OCV to 0.1 V is considered as a prerequisite for the averaging of image series. The accumulation of liquid water after cell perturbation was determined from the neutron image sequence. Values are stated in percent of the total volume, which includes the pore space and the solid space.

Evidently, the liquid water distribution at the cathode will depend on the anode diffusor characteristics, which cannot be chosen arbitrarily in our cell design (Fig. 12.4). However, we focus primarily on a correlation of the liquid water distribution at the cathode and the local cell performance.

12.5.2 Voltage step experiments

In the voltage step experiments the cell voltage is changed quasi-instantaneously from OCV to 0.1 V. Local transients of the liquid water accumulation and the current density are shown in Fig. 12.12b (i–iii) for the four characteristic areas of the cathode flow field (three channels, each 1.2 mm wide, channel:land ratio 1:1): channel center (CC), channel edge (CE), land edge (LE), and land center (LC).

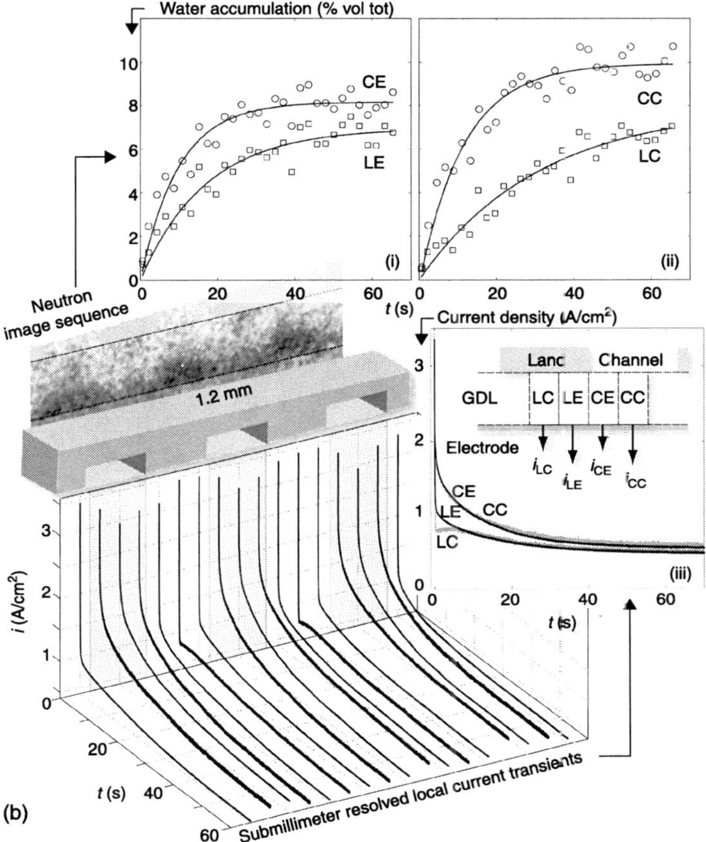

12.12 (Continued)

Liquid water accumulates in the cathode GDL within some tens of seconds, after cell perturbation (Fig. 12.12b (i–ii)). The weaker increase of the total water content in the land center (Fig. 12.12b (ii)) cannot solely be explained by the lower current density (Fig. 12.12b (iii)). A higher compression and lower permeability of the GDL under the ribs might favor water permeation through the membrane.[95] However, the average saturation change approaches comparable values of around 13% in channel (CC) and land (LC, LE) areas within a time frame of 60 s (Fig. 12.12b (i, ii)), considering 78% porosity of the uncompressed Toray paper GDL and ~30% compression of this material in the land area. The same characteristic time constant is observed in the local current transients (Fig. 12.12b (iii)). The associated performance loss is most strongly pronounced in the channel area. In the land area, the local current already drops to low values before a notable liquid water accumulation in the porous GDL. This phenomenon is illustrated in more detail in Fig. 12.13.

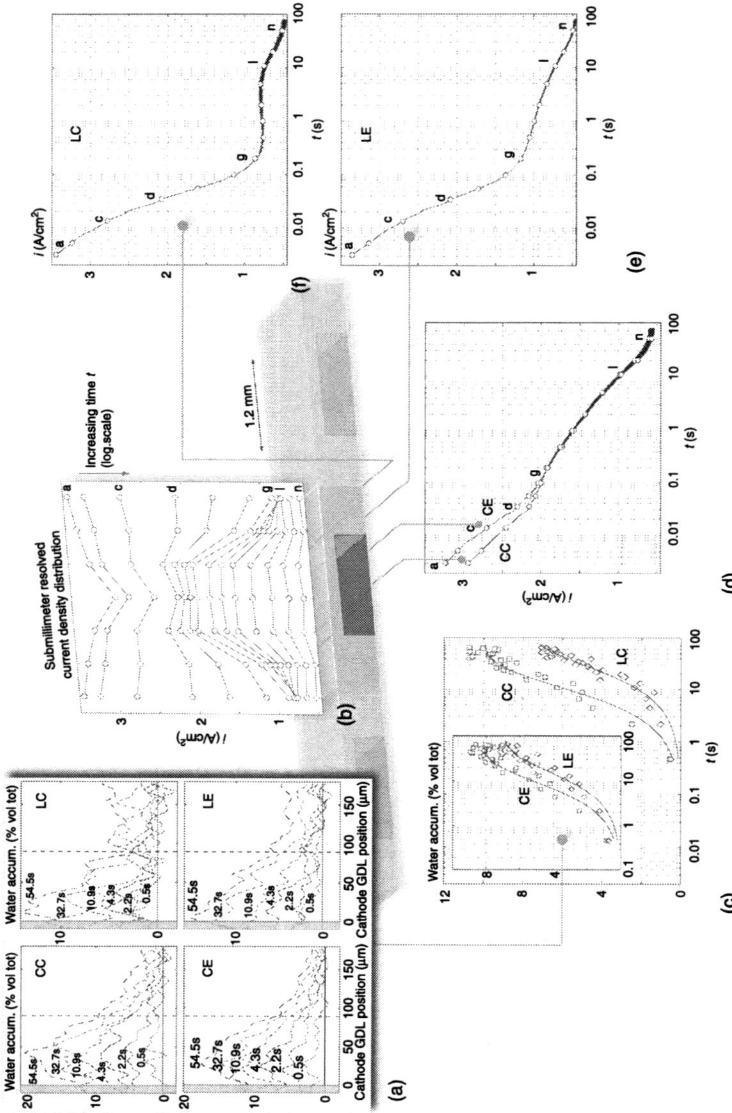

12.13 (a) Liquid water profiles in the four characteristic areas CC, CE, LE, and LC (Fig. 12.1b (iii)) for distinct points in time, (b) current density distribution in channel and land areas for distinct points in time (c) transients of liquid water accumulation on a logarithmic time scale (the full line is a guide for the eye (fit $y(t) = k*(1 - e^{-at})$)), and (d–f) local current transients in the four characteristic areas CC, CE, LE, and LC on a logarithmic time scale. (Source: Schneider *et al.*[94] Copyright 2010 The American Chemical Society.)

Figure 12.13b shows the current density distribution in channel and land areas for distinct points in time. The characteristics are governed by the ohmic resistance, double layer charging, oxygen transport and liquid water accumulation. Upon cell perturbation, the local cell current is limited initially by the local ohmic resistance. Under the fully humidified conditions used here, the resistance will depend mainly on electronic GDL properties and therefore, the current density distribution passes a maximum under the ribs. This characteristic governs the current distribution profile for the first 5–10 ms after the step (Fig. 12.13b (a–c)).

The current decay within this time frame must be attributed to double layer charging and to a decreasing oxygen concentration at the air electrode. The oxygen concentration profiles initially develop within both channel and land areas primarily in the through-plane direction and therefore the observed characteristic decay in both areas is similar within this time frame (Fig. 12.13b (a–c)). With the onset of the development of the oxygen concentration profile in the in-plane direction, the current density starts to drop severely under the ribs, as a result of oxygen depletion (Fig. 12.13f (c–g)). This effect is a direct consequence of the inherent longer diffusion path length for oxygen underneath the ribs and becomes most pronounced in the rib center. The performance loss occurs mainly within the first 100 ms after the step, already before notable liquid water accumulation in the GDL (Fig. 12.13c) and leads temporarily to a pronounced maximum in current density in the channel area (Fig. 12.13b (g)). In the channel region, the short diffusion path length keeps the oxygen concentration at the air electrode at a high level.[94]

However, any reactant gas either consumed in the channel or in the land area must pass through the GDL of the channel region. It is therefore surprising that the strong performance loss, which appears in the channel region within the first 5–10 s of liquid water accumulation (Fig. 12.13c (g–l)) is not observed in the rib center (Fig. 12.13e (g–l)), although the local current is limited here by the rate of oxygen supply through the porous GDL. Evidently, the experimentally obtained cross-sectional water profiles (Fig. 12.13a) show liquid water build-up near to the air electrode and, therefore, it is reasonable that liquid water accumulation primarily affects the cell performance in channel and land areas, locally. Furthermore, the liquid water accumulation shows lowest values of all in the rib center (Fig. 12.13a, 13c (LC)). Nevertheless, the GDL must still provide a sufficient number of reactant pathways to sustain a quasi-constant current in the center of the rib within this time frame (Fig. 12.13f (g–l)). It is likely that the performance loss during liquid water accumulation is not primarily the result of a decreasing effective diffusivity for oxygen in the gas diffusion layer, but basically the result of an ongoing performance loss of the air electrode.[94]

This hypothesis provides a basis to explain the characteristics of the local current transients during the flooding period. An electrode performance loss has the strongest impact in the channel area. Here, the oxygen concentration is kept at a high level. The current initially shows the highest values, but the total performance loss, due to electrode flooding, is the highest too (Fig. 12.13d (g–l)). In contrast, an ongoing loss of electrode performance has no impact at all on the local current density under limiting current conditions, as long as the limiting current, which is governed by the maximum oxygen transport rate, can be sustained by the air electrode. This effect might explain the behavior of the rib center (Fig. 12.13f (g–l)). Here, a quasi-constant current is observed, until the gradually increasing liquid water content (Fig. 12.13a, 13c) degrades electrode performance to a level where even the small limiting current density cannot be maintained, despite the high cell polarization ($U = 0.1$ V). As the electrode performance loss is no longer masked by the oxygen transport limitation, the local current starts to drop after 5–10 s (Fig. 12.13e (g–l)). Finally, the performance in both channel and land areas is governed more or less by the poor electrode performance and a quasi-homogenous current distribution is observed (Fig. 12.13b (n)).

It is important to note here that the strong performance loss of the cell on liquid water accumulation is primarily related to the specific operation conditions ($T_{cell} =$ 40 °C, non-PTFE-treated GDL).[96] Low temperatures result in small saturation pressures and should also favor water condensation in front of the electrode.[97] The cell performance improves significantly at higher temperature ($T_{cell} = T_{hum\_air} = T_{hum\_H_2} = 70$ °C). Under these more practical conditions, the pronounced maximum in current production in the channel region is still observed at a high cell polarization during steady-state operation,[24] as we have also demonstrated in our earlier work.[94]

The results demonstrate that the limitations of the local performance, which are inherent to the system and which must be attributed to the cell design, can be separated from losses due to liquid water accumulation by means of the time constant. By taking advantage of the high temporal resolution of the measurement a current density distribution is obtained, which is temporarily practically unaffected by the effects of liquid water accumulation. This distribution shows a pronounced maximum in the channel area at higher cell polarization and fully humidified conditions, while the performance drops strongly in the land area and shows a minimum in the rib center. This performance loss is primarily the result of the increasing diffusion path length for oxygen and its depletion in diluting nitrogen underneath the ribs. The effect is enforced by the compression of the GDL. Under practical conditions, a pronounced maximum of current generation was still observed in the channel region at steady state.

In the light of this result, a missing clear correlation between the integral cell performance and the liquid water content underneath the ribs is not an unexpected finding.[39] At an already small contribution of the land areas to the total cell current at higher cell polarization, the effect of liquid water accumulation in the GDL on total mass transport related losses is expected to be small. The discussion has also shown that any effect of water accumulation on the electrode performance in the land area is almost irrelevant under the limiting current conditions, as long as the already low limiting current density can be sustained by the air electrode. The performance of the land areas and their contribution to the total cell current is therefore more likely governed primarily by the cell design, e.g. by the flow field structure or the clamping pressure,[98] rather than just by the amount of liquid water. This conclusion is further supported by the result of triangular voltage sweep experiments.

12.5.3 Triangular voltage sweep experiments

The application of a triangular voltage sweep for the *in situ* investigation of two-phase flow phenomena in PEFCs has been reported by Ziegler *et al.*[87] They proposed the use of a variety of voltage perturbation waveforms as a lean and quick alternative to neutron imaging, including voltage step and sinusoidal cell perturbation.[99] In their triangular voltage sweep experiments, they observed a hysteresis in the integral i/U characteristics of a quasi-two-dimensional H_2/air PEFC at higher cell polarization. In the mass transport controlled region, the current density first passed a maximum in the forward sweep and then, after sweep reversal, lower values for the cell current were obtained at a given cell polarization (during the back sweep). In this context, note that the appearance of the maximum in the local i/U curve of the rib center as demonstrated in Section 12.4.2 (Fig. 12.9c) is a different phenomenon. It is a local effect, which occurs under steady-state conditions, while the maximum in the integral i/U characteristics observed by Ziegler *et al.*, is a direct consequence of the high sweep rate ($v = 10$ mV/s). This hysteresis disappears if the i/U curve is taken at a lower sweep rate under quasi-steady-state conditions.

The one-dimensional model of Gerteisen *et al.*[100] attributes the appearance of the hysteresis to pore flooding due to insufficient water removal. The cumulative saturation during the forward sweep, causes increasing mass transport limitations, which in turn force the cell current to decrease even before the maximum cell polarization is reached, at the vertex voltage. In fact, the model actually indicates further increasing liquid water saturation even after the triangular voltage sweep passed the minimum.

The lag between maximum cell polarization and maximum water content as predicted by the model is clearly seen in our experimental data (Fig. 12.14). However, significant differences become apparent in the characteristic transient evolution of the local water content in the four characteristic areas of the cathode flow field (three channels, each channel 1.2 mm wide) (Fig. 12.14b). The lag and the associated saturation hysteresis loop are most strongly pronounced in the rib center, most probably as a consequence of hindered water removal from the GDL in a lateral direction underneath the ribs. Here, the maximum saturation level is not reached until the cell voltage passes a value of $U = 0.4$ V in the back sweep, while the maximum in the channel region is reached earlier, already at a cell voltage of $U = 0.2$ V. At the same time, liquid water accumulates faster in the channel region and, interestingly, both the local water content (Fig. 12.14b) and the local current

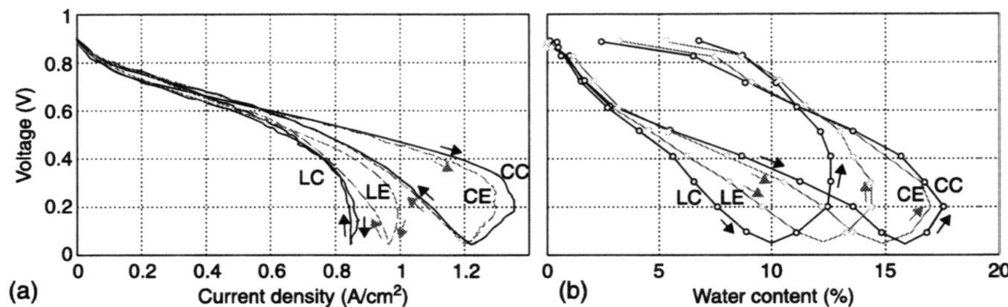

12.14 Triangular voltage sweep experiment (v = 10mV/s) in a H_2/air PEFC: (a) Local i/U characteristics and (b) local water content (electrode section of the GDL) vs. U in the channel center (CC), channel edge (CE), land edge (LE) and land center (LC) region. A_{cell} = 0.6 cm², T_{hum_air} = $T_{hum\_H_2}$ = 40 °C, T_{cell} = 40 °C, V_{air} = V_{H2} = 200 ml/min, ambient pressure, potential range E = 50 mV – OCV, 3 × 1.2 mm channel cathode flow field (channel:land ratio 1:1), GDL: untreated Toray TGP-H-090 (78% porosity, 280 μm thickness), 200 μm thick PTFE gasket, CCM: Gore Primea FCM (0.1 mg Pt/cm² (anode) and 0.4 mg Pt/cm² (cathode)).

density (Fig. 12.14a) reach highest peak values in this region. A hysteresis in the local i/U characteristics is pronounced only in the channel region, while the current in the rib center approaches limiting current density already at a low value of 0.8 A/cm². Here, no notable hysteresis is observed. This indicates that the current in the land center is primarily limited by oxygen mass transport underneath the ribs. This limitation outweighs flooding effects, so that here the water content plays a subordinate role. This corroborates the result of the voltage step experiment (Fig. 12.13). The performance of the land areas is more likely governed primarily by the cell design rather than just by the amount of liquid water.[94]

12.6   Start/stop phenomena in channel and land areas

Cost reduction and durability enhancement are major technical obstacles for the commercialization of PEFCs. A part of PEFCs susceptible to degradation is the catalyst layer. Dissolution and migration of the platinum particles as well as the oxidation of the carbon catalyst support are critical issues.

A well-known phenomenon, which causes carbon corrosion, is the inhomogeneous anode gas composition, due to cathode-to-anode air leakage. This effect was reported for phosphoric acid fuel cells (PAFCs) in 1990.[42] More recently, similar corrosion processes have been identified in PEFCs as a consequence of localized fuel starvation[16, 43, 45, 101] and during the start-up and shutdown processes.[43, 102] Here, the reverse current phenomenon[43] plays an important role.

This phenomenon is illustrated in Fig. 12.15. During a start-up process (Fig. 12.15a), air present at the fuel electrode of a PEFC in off-state is replaced by

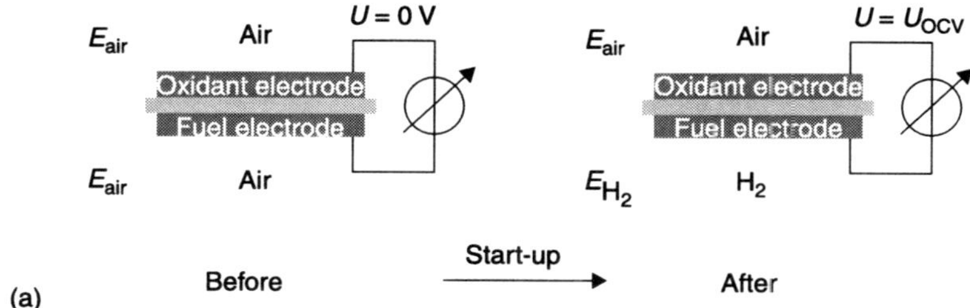

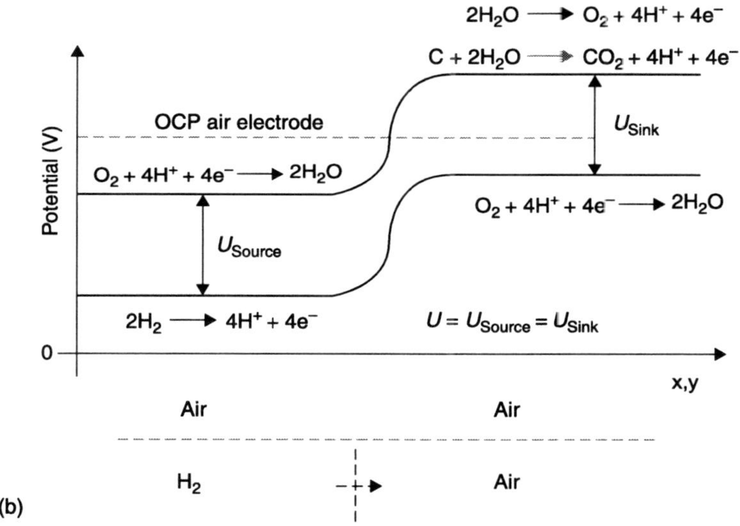

12.15 (a) Schematic to illustrate the fuel and oxidant electrode potential E as well as the cell voltage U during off-state and after the start-up process. (b) Simplified sketch of the reverse-current phenomenon[102,103]. (Source: Schneider *et al.*[114] Copyright 2010 The Electrochemical Society, Inc.)

hydrogen. Thereby, the cell voltage U approaches open circuit voltage U_{OCV} as the potential E of the respective fuel electrode changes from the air electrode potential E_{air} to the hydrogen electrode potential E_{H2} (Fig. 12.15a). Similar considerations held for the shutdown process. However, due to the finite mass transport rate in both the along-the-channel direction and the perpendicular-to-the-channel direction of PEFC flow field structures, the gas is temporarily inhomogeneously distributed over the fuel electrode area (Fig. 12.15b) during start-up and shutdown of a PEFC. As a result, significant in-plane potential gradients exist temporarily over the electrode area (high ionic in-plane resistivity), while the cell voltage is virtually equal in all parts of the cell (high electronic in-plane conductivity of the bipolar plates) as illustrated in Fig. 12.15b.[102, 103] Consequently, the inhomogeneous gas distribution over the fuel electrode leads to electronic in-plane current transients, where hydrogen rich parts of the cell act as a voltage source and drive a reverse-current through hydrogen starved parts, which act as a current sink (Fig. 12.15b). Thereby, the oxidant electrode is temporarily exposed to high potentials[103] and, as a consequence, prone to carbon corrosion and Pt dissolution.

Despite the existence of several proposed mitigation strategies, start-stop phenomena have not completely been understood. Several models try to capture start/stop processes and associated degradation phenomena at elevated potentials.[43–46] In-plane current transients are expected to occur in both the along-the-channel and the perpendicular-to-the-channel (lateral) directions. High local current densities of more than 1 A/cm^2 have been observed experimentally during the start-up process of a fuel cell, which was segmented in the down-the-channel direction.[17, 18, 46] Here a shorter fuel gas introduction time results in both higher and shorter current transients. This correlation between peak height and width is reasonable, if one takes into account that a major part of these in-plane currents is ascribed to capacitive charging currents.[17, 18, 44–46]

Surprisingly, to the best of our knowledge, no attention has been paid to monitor lateral current transients in channel and land areas of PEFC flow field structures, although it has been proposed that the reverse-current phenomenon can occur even within small regions of local hydrogen starvation of a size of fractions of a millimeter, e.g. due to local hydrogen starvation.[43, 101, 104] An assessment of the reverse current phenomenon on the small scale of channel and land areas is therefore crucial.

Here, only those methods (Fig. 12.2–12.4) that employ a fully catalyzed CCM can be used. The partly catalyzed electrode approach (Fig. 12.2) does not allow the monitoring of in-plane phenomena such as reactant starvation underneath the ribs or reverse-currents during start-up and shutdown.

12.6.1 Monitoring of in-plane current transients

Here, the segmented microstructured flow field approach (Fig. 12.4) was used to monitor local current transients in channel and land areas during start-up and shutdown processes of a PEFC. The in-plane currents during the PEFC start-up and shutdown process are the result of a changing gas composition at the fuel electrode (Fig. 12.15b). For an investigation of the effect of the fuel flow field in start/stop experiments, the cell must be operated in a configuration where the microstructured flow field is employed at the oxidant electrode. This arrangement, which is just vice versa to the original setup described in Section 2.3, is acceptable in the experiments, considering the short duration of the current transients and operation of the cell at open circuit voltage U_{OCV}.

During the start/stop experiments, the oxidant electrode was operated on humidified air, while the fuel electrode gas feed was switched from humidified air to humidified hydrogen and vice versa by a magnetic valve. In some of the experiments, mixtures of O_2 and X = He (4 g/mol) or X = SF_6 (146 g/mol) were employed as the fuel gas (O_2 (21%), X (79%)) instead of air (X = N_2 (28 g/mol)) in an investigation on the effect of the diluting inert gas X on mass transport. The flow controllers (Bronkhorst High-Tec BV) were set to a value of V_{air} = V_{H2} = 0.2 l/min. The fuel mass flow controller is calibrated for hydrogen. As a consequence, the gas flow rate increases by a factor of 4/3 if the fuel gas is switched to He/O_2 mixtures. It decreases to 1/3 of the setpoint value in the case of SF_6/O_2 mixtures. However, the high gas flow rate ensures a short residence time of the gas in the fuel flow channels and, even in the case of humidified SF_6/O_2 mixtures this time is well below 25 ms and the gas supply is not limiting.

Two different types of fuel flow fields were employed in the measurements: (i) three channels, each channel 1.2 mm wide, and (ii) two channels, each 2 mm wide, (channel:land ratio 1:1). The pitch of the micro-structured flow field at the air cathode is up to an order of magnitude finer than the channel/land structure of the fuel flow field. The local current transients were averaged in repetitive units of the flow field area; the edge segments were treated as rib center segments. Note that the local current transients add up to zero if no external load is applied.

Local current transients as observed in channel and land areas during a start-up and a shutdown process are shown in Fig. 12.16 for two different types of flow fields. During start-up (Fig. 12.16a, 16c), air is replaced quasi-instantaneously by

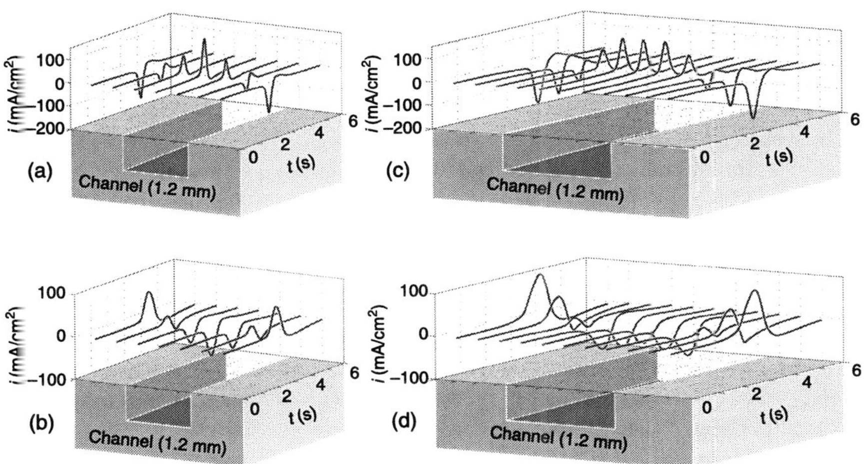

12.16 In-plane current transients during (a, c) start-up and (b, d) shutdown of a H_2/air PEFC (fully humidified gases). The local current transients are shown for a (a, b) 1.2 mm channel and a (c, d) 2 mm channel fuel flow field (channel:land ratio 1:1). CCM: N115, 0.5 mg Pt/cm^2 (Paxitech), Toray TGP-H-090 GDL (10% PTFE), 200 μm thick PTFE gasket. (T_{cell} = 70 °C, fully humidified gases). (Source: Schneider et al.[115] Copyright 2010 The Electrochemical Society, Inc.)

hydrogen in the fuel flow channels. The hydrogen-rich channel areas already operate as a fuel cell and drive a reverse current through the hydrogen-starved land areas. On closer inspection, segments of the land/channel areas act first as a sink/source and subsequently as a source/sink during the start-up period. Inverse characteristics are observed during a shutdown process, when hydrogen is replaced by air in the fuel flow channels (Fig. 12.16b, 16d). The results underline that high gas flow rates and a short residence time in the fuel flow channels cannot avoid mass transport limitations underneath the ribs in the perpendicular-to-the-fuel-channel direction. These limitations are inherent to the system and they can, in principle, only be affected by changing the flow field design or gas diffusion layer (GDL) properties.

The results shown in Fig. 12.16 demonstrate that the local current transients for the 2 mm channel flow field last longer and exhibit higher peak values compared to the respective transients as obtained for the 1.2 mm channel flow field. This effect can, at least partially, be attributed to the increasing diffusion path length for the gases to penetrate underneath the ribs of the flow field. Since higher and longer lasting local peak currents could result in higher and longer lasting local oxidant electrode potentials, the choice of a fuel flow field with narrow land widths should mitigate high cathode potentials and associated corrosion phenomena.

Observations have been made as to whether the start-up or the shutdown process has a higher impact on carbon corrosion, but until now this could not completely be resolved.[45, 105] In experiments we have generally observed a faster change of the cell voltage in the start-up process of a H_2/air PEFC (Fig. 12.17a) compared to the shutdown procedure (Fig. 12.17b). Moreover, experimentally obtained peak currents are, by a factor of two, higher for the start-up process (Fig. 12.17a, 17b).

The different characteristics of start-up and shutdown transients are much less pronounced if helox (21% O_2 in helium) is used instead of air (Fig. 12.17c, 17d). If we take into account that the characteristics of the start-up transient is comparable for a H_2/air and a H_2/helox cell (Fig. 12.17a, 17c), we might conclude that gas-phase mass transport limitations are less important for the transport of hydrogen (through diluted oxygen) toward the fuel electrode during start-up. In contrast, the different characteristics of the respective shutdown transients for a H_2/air cell and a H_2/helox cell (Fig. 12.17b, 17d) indicates that the transport of molecular oxygen to the fuel electrode during shutdown is strongly dependent on the diluting inert gas. This conclusion is further corroborated if a O_2/SF$_6$ mixture is used to purge the fuel electrode (Fig. 12.17e). The shutdown transients become both shorter and higher with decreasing diffusion resistance for oxygen in the order H_2/(O_2/SF$_6$) > H_2/air > H_2/helox (Fig. 12.17e, 17b, 17d). This effect, which occurs in the perpendicular-to-the-flow channel direction, can be regarded as a counterpart to the observation of higher and shorter transients in the along-the-channel direction toward shorter fuel gas introduction time.[18]

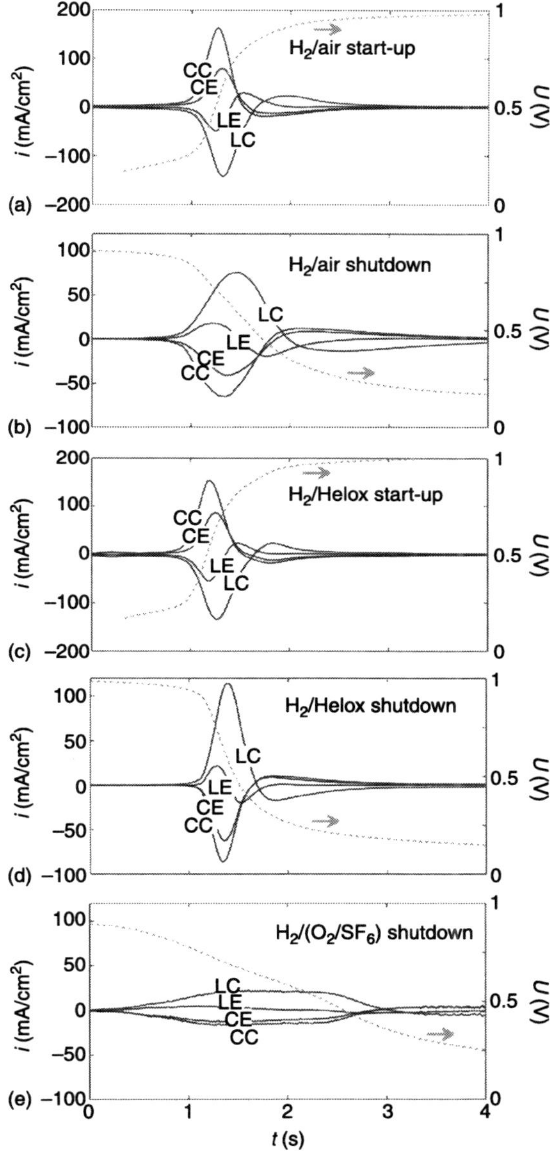

12.17 Local current transients in channel and land areas for different fuel electrode purge gases. (a, b) air. (c, d) Helox, (e) SF_6 (fully humidified gases). The transients are shown during start-up (a, c) and shutdown (b, d, e) of the PEFC in four characteristic areas of the 1.2 mm channel fuel flow field CC: channel center, CE: channel edge, LC: land center, LE: land edge. Experimental conditions as stated in Fig. 12.16. (Source: Schneider *et al.*[115] Copyright 2010 The Electrochemical Society, Inc.)

The results (Fig. 12.17) therefore indicate that the slower voltage change and the smaller in-plane current transients during shutdown of a H_2/air cell (Fig. 12.17a, 17b) are the result of a slower change of the fuel electrode potential from hydrogen to air potential due to a notable gas-phase mass transport limitation for oxygen. This limitation is much less pronounced for hydrogen during the start-up process when the fuel electrode potential changes from the air to the hydrogen potential (Fig. 12.17a). Note that the hydrogen concentration in the channel region is kept high during start-up by the abundant supply of hydrogen to the gas channel. In contrast, during shutdown (channel region is in 'air/air mode') only the hydrogen, which is stored underneath the ribs, is available as fuel for the driving voltage source (land region).

12.6.2 Effect of operation conditions and mitigation strategies

It has been proposed that operation under fully humidified conditions during the start-up and shutdown processes of a PEFC has a more detrimental effect on the catalyst layer[106] than start-up under dry conditions. The local current transients during a start-up process are compared for fully humidified and sub-saturated gases (40% RH) in Fig. 12.18a and 12.18b. Interestingly, the voltage transients show virtually the same characteristics in both cases, yet the local peak currents are remarkably reduced under dry conditions (Fig. 12.18b). This significant effect might be ascribed to the higher ionic membrane resistance. The lower degradation levels, which have been reported for dry operation conditions[106] were attributed to the mitigating effect of the reduced water partial pressure on the carbon corrosion reaction.[106, 107]

Several strategies have been suggested to mitigate the detrimental cell-reversal conditions emerging during start-up and shutdown processes. A comprehensive review of proposed mitigation strategies is given by Perry *et al.*[108, 109] One mitigation strategy addresses the hydrogen introduction time.[45, 46] Here, a lower residence time of the gas front in the cell is supposed to yield a reduced carbon corrosion rate. However, this approach primarily affects phenomena in the along-the-channel direction. In the perpendicular-to-the-channel direction, gas transport is primarily diffusion controlled. Here, both the flow field design and GDL properties play an important role. These properties can, however, only be improved by a change of the cell design.

Another proposed strategy for mitigation of the detrimental reverse-currents is a purge of the fuel gas compartment with inert nitrogen gas during the start-up and the shutdown process.[44] With this approach, the maximum oxidation rate at the oxidant electrode is limited by the rate of oxygen crossover from the oxidant electrode to the nitrogen-purged fraction of the fuel electrode.[44] Figure 12.18c shows the experimentally obtained local current transients and the voltage transient. Very small peak currents and a much slower voltage increase are

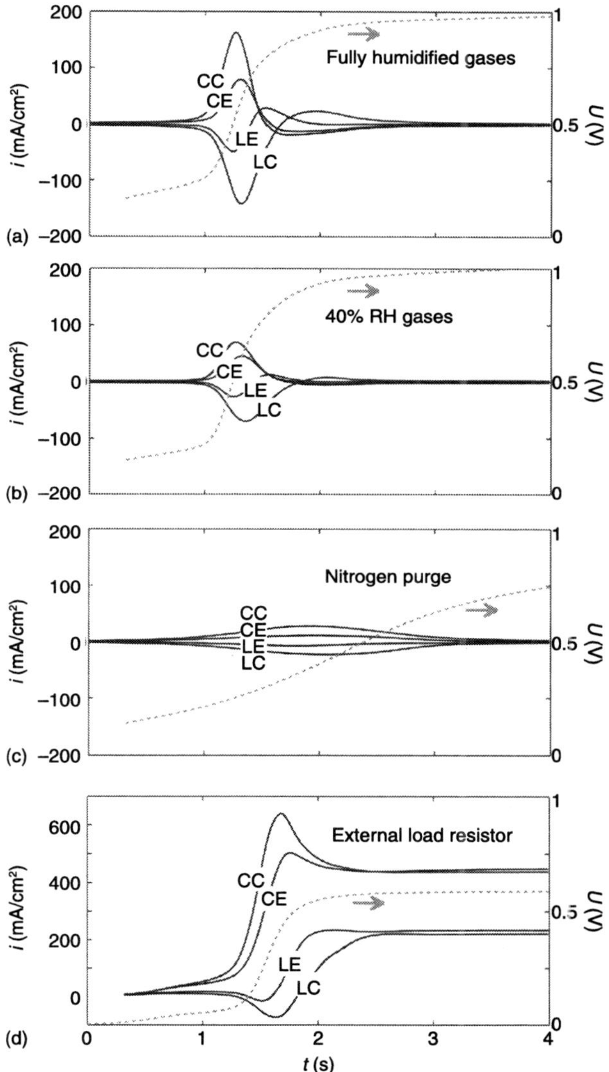

12.18 Effect of operation conditions and mitigation strategies on local current transients in channel and land areas of a hydrogen/air PEFC during start-up: (a) fully humidified gases, (b) 40% RH gases, (c) effect of a nitrogen purge of the fuel gas compartment (fully humidified gases), and (d) effect of an external load resistor (R_{load} = 1.8 Ω cm², fully humidified gases). The transients are shown in four characteristic areas of the 1.2 mm channel fuel flow field: channel center (CC), channel edge (CE), land center (LC), land edge (LE). Experimental conditions as stated in Fig. 12.16. (Source: Schneider *et al.*[15] Copyright 2010. The Electochemical Society, Inc.)

observed. Unfortunately, this approach requires the installation of an external nitrogen supply to the fuel cell system.

Another possible mitigation strategy is a potential-controlled start-up.[44] Here, an external auxiliary load is connected to the cell during start-up and shutdown to prevent the electrode from being exposed to high potentials.[44, 105, 110] A start-up process with an external auxiliary load of 1.8 Ω cm^2 is shown in Fig. 12.18d. Here, the observed negative peak currents are smaller compared to operation without external load. This is consistent with the observed lower degradation rates.[44, 105, 110] However, this approach primarily addresses the cell voltage, which reaches a significantly lower value of approximately 0.55 V. Application of an external auxiliary load resistance substantially reduces the cathode potential during start-up and shutdown processes. High local potentials and their detrimental effect can be mitigated.

The results demonstrate that a short fuel gas introduction time during start-up and shutdown of a PEFC cannot eliminate the reverse-current phenomenon. In general, gas transport in the GDL underneath the ribs is primarily diffusion controlled. In this context, mitigation strategies must focus on the flow field design, GDL properties, and the type of purge gas used at the fuel electrode in the off state. The flow field design and the chosen GDL properties have to be balanced with the requirements of cell operation, while mitigation strategies, which are based on the type of purge gas, the gas humidity or the use of an external load will affect the cell characteristics primarily during the start-up and shutdown period.

12.7   Concluding remarks

The experimental evaluation of the lateral current distribution in channel and land areas of PEFC flow field structures is not straightforward. Only in recent years, three different methods for an experimental assessment have been published. All methods have certain drawbacks. The partly catalyzed electrode approach entails most restrictions, although this method provides the advantage of direct current measurement. In-plane phenomena such as reactant starvation underneath the ribs or reverse-currents during start/stop are not captured by this method, since no fully catalyzed CCM can be used.

All methods have been used for an evaluation of the steady-state current distribution in channel and land areas of PEFCs under various operation conditions. The results were implemented in the design of technical flow field structures.[111–113] Above all, the segmented microstructured flow field approach enabled the application of spatially resolved transient techniques and the monitoring of local transient phenomena in channel and land areas of PEFC flow field structures with submillimeter resolution.

The results demonstrate that phenomena that occur as a consequence of inhomogeneous operation of PEFCs in the down-the-channel direction, typically on a length scale of tens of centimeters, have their counterpart on the millimeter scale of channels and ribs in the direction that is perpendicular to the flow channels. Both should be treated with the same attention. However, there is an important difference. Inhomogeneities, which occur during operation of a PEFC in the down-the-channel direction, depend primarily on the gas stoichiometry. Phenomena such as the appearance of negative low-frequency resistance values in the local spectra of the outlet region due to oxygen depletion in a H_2/air PEFC[9, 62, 64] will therefore appear at low air stoichiometry, regardless of the total cell current. However, they can simply be excluded by using high gas stoichiometries. In contrast, lateral inhomogeneities in the perpendicular-to-the-flow channel direction become important, primarily at higher cell polarization and higher total cell current. These inhomogeneities are a direct consequence of a mass transport limitation underneath the ribs. They are, therefore, inherent to the system and they can only be mitigated by changing the cell design or by operation of the cell on pure fully humidified gases at moderate current densities.

12.8   Acknowledgments

This work was partially funded by the Swiss Federal Office of Energy (BFE). Special thanks go to Thomas Mayer and Andreas Stuker (ETH Zürich) for technical support and Diana Turgi for inspiring discussions.

12.9   References

1. Cleghorn, S. J. C., Derouin, C. R., Wilson, M. S. and Gottesfeld, S. (1998) 'A printed circuit board approach to measuring current distribution in a fuel cell', *J. Appl. Electrochem.*, **28**, 663–672.
2. Stumper, J., Campbell, S. A., Wilkinson, D. P., Johnson, M. C. and Davis, M. (1998) '*In-situ* methods for the determination of current distributions in PEM fuel cells', *Electrochim. Acta*, **43**(24), 3773–3783.
3. Brett, D. J. L., Atkins, S., Brandon, N. P., Vescovic, V., Vasileiadis, N. and Kucernak, A. R. (2001) 'Measurement of the current distribution along a single flow channel of a solid polymer fuel cell', *Electrochem. Commun.*, **3**, 628–632.
4. Noponen, M., Hottinen, T., Mennola, T., Mikkola, M. and Lund, P. (2002) 'Determination of mass diffusion overpotential distribution with flow pulse method from current distribution measurements in a PEMFC', *J. Appl. Electrochem.*, **32**, 1081–1089.
5. Mench, M. M. and Wang, C. Y. (2003) 'An *in situ* method for determination of current distribution in PEM fuel cells applied to a direct methanol fuel cell', *J. Electrochem. Soc.*, **150**(1), A79–A85.
6. Bender, G., Wilson, M. S. and Zawodzinski, T. A. (2003) 'Further refinements in the segmented cell approach to diagnosing performance in polymer electrolyte fuel cells', *J. Power Sources*, **123**, 163–171.
7. Geiger, A. B., Eckl, R., Wokaun, A. and Scherer, G. G. (2004) 'An approach to measuring locally resolved currents in polymer electrolyte fuel cells', *J. Electrochem. Soc.*, **151**(3), A394–A398.

8. Dong, Q., Mench, M. M., Cleghorn, S. and Beuscher, U. (2005) 'Distributed performance of polymer electrolyte fuel cells under low-humidity conditions', *J. Electrochem. Soc.*, **152**(11), A2114–A2122.

9. Schneider, I. A. and Scherer, G. G. (2009) 'Local transient techniques in polymer electrolyte fuel cell (PEFC) diagnostics', in *Handbook of Fuel Cells*, Vol. 6, Vielstich, W., Gasteiger, H. A. and Yokokawa, H., Editors, John Wiley & Sons, New York.

10. Brett, D. J. L., Atkins, S., Brandon, N. P., Vesovic, V., Vasileiadis, N. and Kucernak, A. (2003) 'Localized impedance measurements along a single channel of a solid polymer fuel cell', *Electrochem. Solid-State Lett.*, **6**(4), A63–A66.

11. Brett, D. J. L., Atkins, S., Brandon, N. P., Vesovic, V., Vasileiadis, N. and Kucernak, A. R. (2004) 'Investigation of reactant transport within a polymer electrolyte fuel cell using localised CO stripping voltammetry and adsorption transients', *J. Power Sources*, **133**, 205–213.

12. Hakenjos, A. and Hebling, C. (2005) 'Spatially resolved measurement of PEM fuel cells', *J. Power Sources*, **145**, 307–311.

13. Schneider, I. A., Kuhn, H., Wokaun, A. and Scherer, G. G. (2005) 'Fast Locally Resolved Electrochemical Impedance Spectroscopy in Polymer Electrolyte Fuel Cells', *J. Electrochem. Soc.*, **152**(10), A2092–A2103.

14. Schneider, I. A., Kramer, D., Wokaun, A. and Scherer, G. G. (2007) 'Effect of inert gas flow on hydrogen underpotential deposition measurements in polymer electrolyte fuel cells', *Electrochem. Commun.*, **9**, 1607–1612.

15. Freunberger, S. A., Schneider, I. A., Sui, P.-C., Wokaun, A., Djilali, N. and Büchi, F. N. (2008) 'Cell interaction phenomena in polymer electrolyte fuel cell stacks', *J. Electrochem. Soc.*, **155**(7), B704–B714.

16. Carter, R. N., Gu, W., Brady, B., Yu, P. T., Subramanian, K. and Gasteiger, H. A. (2009) 'Electrode degradation mechanisms studies by current distribution measurements', in *Handbook of Fuel Cells*, Vol. 6, Vielstich, W., Gasteiger, H. A. and Yokokawa, H., Editors, John Wiley & Sons, New York.

17. Siroma, Z., Fujiwara, N., Ioroi, T., Yamazaki, S., Senoh, H. *et al.* (2007) 'Transient phenomena in a PEMFC during the start-up of gas feeding observed with a 97-fold segmented cell', *J. Power Sources*, **172**, 155–162.

18. Maranzana, G., Moyne, C., Dillet, J., Didierjean, S. and Lottin, O. (2010) 'About internal currents during start-up in proton exchange membrane fuel cell', *J. Power Sources*, **195**, 5990–5995.

19. Freunberger, S. A., Santis, M., Schneider, I. A., Wokaun, A. and Büchi, F. N. (2006) 'In-plane effects in large-scale PEMFCs', *J. Electrochem. Soc.*, **153**(2), A396–A405.

20. Gu, W., Baker, D. R., Liu, Y. and Gasteiger, H. A. (2009) 'Proton exchange membrane fuel cell (PEMFC) down-the-channel performance model', in *Handbook of Fuel Cells*, Vol. 6, Vielstich, W., Gasteiger, H. A. and Yokokawa, H., Editors, John Wiley & Sons, New York.

21. Mathias, M., Baker, D., Zhang, J., Liu, Y. and Gu, W. (2008) 'Frontiers in application of impedance diagnostics to H_2-fed polymer electrolyte fuel cells', *ECS Trans.*, **13**(13), 129–152.

22. Makharia, R., Mathias, M. F. and Baker, D. R. (2005) 'Measurement of catalyst layer electrolyte resistance in PEFCs using electrochemical impedance spectroscopy', *J. Electrochem. Soc.*, **152**(5), A970–A977.

23. Freunberger, S. A., Reum, M., Evertz, J., Wokaun, A. and Büchi, F. N. (2006) 'Measuring the current distribution in PEFCs with sub-millimeter resolution', *J. Electrochem. Soc.*, **153**(11), A2158–A2165.

24 Schneider, I. A., von Dahlen, S., Wokaun, A. and Scherer, G. G. (2010) 'A segmented microstructured flow field approach for submillimeter resolved local current measurement in channel and land areas of a PEFC', *J. Electrochem. Soc.*, **157**(3), B338–B341.

25 Kocha, S. S. (2003) 'Principles of MEA preparation', in *Handbook of Fuel Cells*, Vol. 3, Vielstich, W., Gasteiger, H. A. and Lamm, A., Editors, John Wiley & Sons, Hoboken, NJ.

26 Jerkiewicz, G. (1998), 'Hydrogen sorption at/in electrodes', *Progress in Surface Science*, **57**(2), 137–186.

27 Biegler, T., Rand, D. A. J. and Woods, R. (1971) 'Limiting oxygen coverage on platinized platinum; relevance to determination of real platinum area by hydrogen adsorption', *J. Electroanal. Chem.*, **29**, 269–277.

28 Gilman, S., (1964) 'Measurement of hydrogen adsorption by the MPP method', *J. Electroanal. Chem.*, **7**, 382–391.

29 Bett, J., Kinoshita, K., Routsis, K. and Stonehart, P. (1973) 'A comparison of gas-phase and electrochemical measurements for chemisorbed carbon monoxide and hydrogen on platinum crystallites', *J. Catal.*, **29**, 160–168.

30 MacDonald, D. D. (1977) *Transient Techniques in Electrochemistry*, Plenum Press, New York.

31 Macdonald, J. R. and Barsoukov, E. (2005) *Impedance Spectroscopy: Theory, Experiment, and Applications*, 2nd Ed., Wiley, New York.

32 Orazem, M. E. and Tribollet, B. (2008) *Electrochemical Impedance Spectroscopy*, John Wiley & Sons, Hoboken, NJ.

33 Gomadam, P. M. and Weidner, J. W. (2005) 'Analysis of electrochemical impedance spectroscopy in proton exchange membrane fuel cells', *Int. J. of Energy Res.*, **29**, 1133–1151.

34 Berg, P., Promislow, K., St. Pierre, J., Stumper, J. and Wetton, B. (2004) 'Water management in PEM fuel cells', *J. Electrochem. Soc.*, **151**(3), A341–A353.

35 Weber, A. Z., Balliet, R., Gunterman, H. P. and Newman, J. (2008) 'Modeling water management in polymer-electrolyte fuel cells', *LBNL Paper*, 273–415.

36 Wang, C.-Y. (2004) 'Fundamental models for fuel cell engineering', *Chem. Rev.*, **104**, 4727–4766.

37 Hickner, M. A. and Hussey, D. S. (2010) 'Neutron Radioscopy: Industrial and Scientific Applications', in *Encyclopedia of Analytical Chemistry*, Meyers, R. A., Editor, John Wiley & Sons, Chichester, UK.

38 Mukundan, R. and Borup, R. L. (2009) 'Visualising liquid water in PEM fuel cells using neutron imaging', *Fuel Cells*, **9**(5), 499–505.

39 Zhang, J., Kramer, D., Shimoi, R., Ono, Y., Lehmann, E. *et al.* (2006) '*In situ* diagnostic of two-phase flow phenomena in polymer electrolyte fuel cells by neutron imaging Part B. Material variations', *Electrochim. Acta*, **51**, 2715–2727.

40 Hartnig, C., Manke, I., Kardjilov, N., Hilger, A., Grünerbel *et al.* (2008) 'Combined neutron radiography and locally resolved current density measurements of operating PEM fuel cells', *J. Power Sources*, **176**, 452–459.

41 Hickner, M. A., Siegel, N. P., Chen, K. S., Hussey, D. S., Jacobson, D. L. *et al.* (2008) '*In situ* high-resolution neutron radiography of cross-sectional liquid water profiles in proton exchange membrane fuel cells', *J. Electrochem. Soc.*, **155**(4), B427–B434.

42 Farooque, M., Kush, A. and Christner, L. (1990) 'Novel explanation of unusual localized corrosion in energy conversion devices', *J. Electrochem. Soc.*, **137**(7), 2025–2028.

43. Reiser, C. A., Bregoli, L., Patterson, T. W., Yi, J. S., Yang *et al.* (2005) 'A reverse-current decay mechanism for fuel cells', *Electrochem. Solid-State Lett.*, **8**, A273–A276.

44. Meyers, J. P. and Darling, R. M. (2006) 'Model of carbon corrosion in PEM fuel cells', *J. Electrochem. Soc.*, **153**(8), A1432–A1442.

45. Gu, W., Carter, R. N., Yu, P. T. and Gasteiger, H. A. (2007) 'Start/stop and local H_2 starvation mechanisms of carbon corrosion: Model vs. experiment', *ECS Trans.*, **11**(1) 963–973.

46. Wetton, B., Bradean, R. and Eggen, K. (2009) 'A model of PEM fuel cell start-up including capacitance and electrical coupling effects', *Proceedings of the 20th International Symposium on Transport Phenomena*, Victoria BC.

47. Toray Carbon Fiber Paper 'TGP-H' Property Sheet; Toray Industries Inc., Tokyo.

48. Wang, L. and Liu, H. (2008) 'Separate measurement of current density under the channel and the shoulder in PEM fuel cells', *J. Power Sources*, **180**, 365–372.

49. Higier, A. and Liu, H. (2009) 'Direct measurement of current density under the land and channel in a PEM fuel cell with serpentine flow fields', *J. Power Sources*, **193**, 639–648.

50. Reum, M., Freunberger, S. A., Wokaun, A. and Büchi, F. N. (2009) 'Measuring the current distribution with sub-millimeter resolution in PEFCs', *J. Electrochem. Soc.*, **156**(3), B301–B310.

51. Reum, M. (2008) 'Sub-Millimeter Resolved Measurement of Current Density and Membrane Resistance in Polymer Electrolyte Fuel Cells (PEFC)', Ph.D. Thesis 17979, Swiss Federal Institute of Technology, Zürich, Switzerland.

52. Seyfang, B. C., Kuhnke, M., Lippert, T., Scherer, G. G. and Wokaun, A. (2007) 'A novel, simplified micro-PEFC concept employing glassy carbon micro-structures', *Electrochem. Commun.*, **9**, 1956–1962.

53. Romero-Castanon, T., Arriaga, L. G. and Cano-Castillo, U. (2003) 'Impedance spectroscopy as a tool in the evaluation of MEAs', *J. Power Sources*, **118**, 179–182.

54. Springer, T. E., Zawodzinski, T. A., Wilson, M. S. and Gottesfeld, S. (1996) 'Characterization of Polymer Electrolyte Fuel Cells Using AC Impedance Spectroscopy', *J. Electrochem. Soc.*, **143**(2), 587–599.

55. Paganin, V. A., Oliveira, C. L. F., Ticianelli, E. A., Springer, T. E. and Gonzalez, E. R. (1998) 'Modelistic interpretation of the impedance response of a polymer electrolyte fuel cell', *Electrochim. Acta*, **43**(24), 3761–3766.

56. Freire, T. J. P. and Gonzalez, E. R. (2001) 'Effect of membrane characteristics and humidification conditions on the impedance response of polymer electrolyte fuel cells', *J. Electroanal. Chem.*, **503**, 57–68.

57. Kuhn, H., Andreaus, B., Wokaun, A. and Scherer, G. G. (2006) 'Electrochemical impedance spectroscopy applied to polymer electrolyte fuel cells with a pseudo reference electrode arrangement', *Electrochim. Acta*, **51**, 1622–1628.

58. Roy, S. K., Orazem, M. E. and Tribollet, B. (2007) 'Interpretation of low-frequency inductive loops in PEM fuel cells', *J. Electrochem. Soc.*, **154**(12), B1378–B1388.

59. Schneider, I. A., Kramer, D., Wokaun, A. and Scherer, G. G. (2006) 'Oscillations in the gas channels – The forgotten player in impedance spectroscopy in polymer electrolyte fuel cells, A. Exploring the wave', *ECS Trans.*, **3**(1) 1001–1010.

60. Kramer, D., Schneider, I. A., Wokaun, A. and Scherer, G. G. (2006) 'Oscillations in the gas channels – The forgotten player in impedance spectroscopy in polymer electrolyte fuel cells, B. Modeling the wave', *ECS Trans.*, **3**(1) 1249–1258.

61. Schneider, I. A., Freunberger, S. A., Kramer, D., Wokaun, A. and Scherer, G. G. (2007) 'Oscillations in gas channels part I. The forgotten player in impedance spectroscopy in PEFCs', *J. Electrochem. Soc.*, **154**(4), B383–B388.

62. Schneider, I. A., Kramer, D., Wokaun, A. and Scherer, G. G. (2007) 'Oscillations in gas channels II. Unraveling the characteristics of the low frequency loop in air-fed PEFC impedance spectra', *J. Electrochem. Soc.*, **154**(8), B770–B782.

63. Schneider, I. A., Bayer, M. H., Boillat, P., Wokaun, A. and Scherer, G. G. (2007) 'Formation of low frequency inductive loops in polymer electrolyte fuel cell impedance spectra under sub-saturated conditions', *ECS Trans.*, **11**(1) 461–472.

64. Schneider, I. A., Bayer, M. H., Wokaun, A. and Scherer, G. G. (2009) 'Negative resistance values in locally resolved impedance spectra of polymer electrolyte fuel cells', *ECS Trans.* **25**(1) 937–948.

65. Bayer, M. H., Wokaun, A., Scherer, G. G. and Schneider, I. A. (2009) 'Dynamic measurement and modeling of the water vapor concentration during ac impedance measurements in polymer electrolyte fuel cells (PEFCs)', *ECS Trans.*, **25**(1) 949–960.

66. Schneider, I. A., Fuchs, H., Gubler, L., Wokaun, A. and Scherer, G. G. (2005) 'Effect of gas flow rate on local voltammograms in PEFCs', *Annual Report*, Electrochemistry Laboratory, Paul Scherrer Institut, pp. 8–9, ISSN 1661-5379.

67. Eikerling, M. and Kornyshev, A. A. (1999) 'Electrochemical impedance of the cathode catalyst layer in polymer electrolyte fuel cells', *J. Electroanal. Chem.*, **475**, 107–123.

68. Guo, Q. and White, R. E. (2004) 'A steady-state impedance model for a PEMFC cathode', *J. Electrochem. Soc.*, **151**(4), E133-E149.

69. Ciureanu, M. and Roberge, R. (2001) 'Electrochemical impedance study of PEM fuel cells. Experimental diagnostics and modeling of air cathodes', *J. Phys. Chem. B*, **105**, 3531–3539.

70. Bultel, Y., Wiezell, K., Jaouen, F., Ozil, P. and Lindbergh, G. (2005) 'Investigation of mass transport in gas diffusion layer at the air cathode of a PEMFC', *Electrochim. Acta*, **51**, 474–488.

71. Boillot, M., Bonnet, C., Jatroudakis, N., Carre, P., Didierjean, S. *et al.* (2006) 'Effect of gas dilution on PEM fuel cell performance and impedance response', *Fuel Cells*, **6**(1), 31–37.

72. Boillot, M., Bonnet, C., Didierjean, S. and Lapicque, F. (2007) 'Investigation of the response of separate electrodes in a polymer electrolyte membrane fuel cell without reference electrode', *J. Appl. Electrochem.*, **37**, 103–110.

73. Fouquet, N., Doulet, C., Noillant, C., Dauphin-Tanguy, G. and Ould-Bouamama, B. (2006) 'Model based PEM fuel cell state-of-health monitoring via ac impedance measurements', *J. Power Sources*, **159**, 905–913.

74. Kulikovsky, A. A., Kucernak, A. and Kornyshev, A. A. (2005) 'Feeding PEM fuel cells', *Electrochim. Acta*, **50**, 1323–1333.

75. Flückiger, R., Freunberger, S. A., Kramer, D., Wokaun, A., Scherer, G. G. *et al.* (2008) 'Anisotropic, effective diffusivity of porous gas diffusion layer materials for PEFC', *Electrochim. Acta*, **54**, 551–559.

76. Neyerlin, K. C., Gu, W., Jorne, J. and Gasteiger, H. A. (2006) 'Determination of catalyst unique parameters for the oxygen reduction reaction in a PEMFC', *J. Electrochem. Soc.*, **153**(10), A1955–A1963.

77. Roy, S. K. and Orazem, M. E. (2009) 'Graphical estimation of interfacial capacitance of PEM fuel cells from impedance measurements', *J. Electrochem. Soc.*, **156**(2), B203–B209.

78. Basu, S., Renfro, M. W., Gorgun, H. and Cetegen, B. M. (2006) '*In situ* simultaneous measurements of temperature and water partial pressure in a PEM fuel cell under steady state and dynamic cycling', *J. Power Sources*, **159**, 987–994.

79. Basu, S., Renfro, M. W. and Cetegen, B. M. (2006) 'Spatially resolved optical measurements of water partial pressure and temperature in a PEM fuel cell under dynamic operating conditions', *J. Power Sources*, **162**, 286–293.

80. Sur, R., Boucher, T. J., Renfro, M. W. and Cetegen, B. M. (2010) '*In situ* measurements of water vapor partial pressure and temperature dynamics in a PEM fuel cell', *J. Electrochem. Soc.*, **157**(1), B45–B53.

81. Bayer, M. H., Wokaun, A., Scherer, G. G. and Schneider, I. A. (2010) '*In situ* membrane parameter determination using simultaneously locally resolved impedance spectroscopy and dynamic water partial pressure measurement in PEFCs', *Proceedings of the ASME 2010 Eighth International Fuel Cell Science, Engineering and Technology Conference FuelCell2010*, Brooklyn, New York, USA.

82. Schneider, I. A., Kramer, D., Wokaun, A. and Scherer, G. G. (2005) 'Spatially resolved characterization of PEFCs using simultaneously neutron radiography and locally resolved impedance spectroscopy', *Electrochem. Commun.*, **7**, 1393–1397.

83. Bazylak, A. (2009) 'Liquid water visualization in PEM fuel cells: A review', *Int. J. Hydrogen Energy*, **34**, 3845–3857.

84. Boillat, P., Kramer, D., Seyfang, B. C., Frei, G., Lehmann, E. *et al.* (2008) '*In situ* observation of the water distribution across a PEFC using high resolution neutron radiography', *Electrochem. Commun.*, **10**, 546–550.

85. Wang, Y. and Wang, C.-Y. (2005) 'Transient analysis of polymer electrolyte fuel cells', *Electrochim. Acta*, **50**, 1307–1315.

86. Wang, Y. and Wang, C.-Y. (2006) 'Dynamics of polymer electrolyte fuel cells undergoing load changes', *Electrochim. Acta*, **51**, 3924–3933.

87. Yu, H. and Ziegler, C. (2006) 'Transient behavior of a proton exchange membrane fuel cell under dry operation', *J. Electrochem. Soc.* **153**(3), A570–A575.

88. Wang, Y. and Wang, C.-Y. (2007) 'Two-phase transients of polymer electrolyte fuel cells', *J. Electrochem. Soc.*, **154**(7), B636–B643.

89. Wu, H., Berg, P. and Li, X. (2010) 'Modeling of PEMFC transients with finite-rate phase-transfer processes', *J. Electrochem. Soc.*, **157**(1), B1–B12.

90. Hickner, M. A., Siegel, N. P., Chen, K. S., Hussey, D. S. and Jacobson, D. L. (2010) 'Observations of transient flooding in a proton exchange membrane fuel cell using time-resolved neutron radiography', *J. Electrochem. Soc.*, **157**(1), B32–B38.

91. Schneider, I. A., Bayer, M. H., Wokaun, A. and Scherer, G. G. (2008) 'Impedance response of the proton exchange membrane in polymer electrolyte fuel cells', *J. Electrochem. Soc.*, **155**(8), B783–B792.

92. Lehmann, E. H., Frei, G., Kühne, G. and Boillat, P. (2007) 'The micro-setup for neutron imaging: A major step forward to improve the spatial resolution', *Nucl. Instrum. Methods Phys. Res., Sect. A*, **576**, 389–396.

93. Boillat, P., Frei, G., Lehmann, E. H., Scherer, G. G. and Wokaun, A. (2010) 'Neutron imaging resolution improvements optimized for fuel cell applications', *Electrochem. Solid-State Lett.*, **13**, B25–B27.

94. Schneider, I. A., von Dahlen, S., Bayer, M. H., Boillat, P., Hildebrandt *et al.* (2010) 'Local transients of flooding and current in channel and land areas of a polymer electrolyte fuel cell', *J. Phys. Chem. C*, **114**(27), 11998–12002.

95. Ahmed, D. H., Sung, H. J. and Bae, J. (2008) 'Effect of GDL permeability on water and thermal management in PEMFCs—II. Clamping force', *J. Int. J. Hydrogen Energy*, **33**, 3786–3800.

96. Ziegler, C., Heilmann, T. and Gerteisen, D. (2008) 'Experimental study of two-phase transients in PEMFCs', *J. Electrochem. Soc.*, **155**(4), B349–B355.

97. Weber, A. Z. and Newman, J. (2006) 'Coupled thermal and water management in polymer electrolyte fuel cells', *J. Electrochem. Soc.*, **153**(12), A2205–A2214.

98. Ihonen, J., Mikkola, M. and Lindbergh, G. (2004) 'Flooding of gas diffusion backing in PEFCs: physical and electrochemical characterization', *J. Electrochem. Soc.*, **151**(8), 1152–1161.

99. Ziegler, C. and Gerteisen, D. (2009) 'Validity of two-phase polymer electrolyte membrane fuel cell models with respect to the gas diffusion layer', *J. Power Sources*, **188**, 184–191.

100. Gerteisen, D., Heilmann, T. and Ziegler, C. (2009) 'Modeling the phenomena of dehydration and flooding of a polymer electrolyte membrane fuel cell', *J. Power Sources*, **187**, 165–181.

101. Patterson, T. W. and Darling, R. M. (2006) 'Damage to the cathode catalyst of a PEM fuel cell caused by localized fuel starvation', *Electrochem. Solid-State Lett.*, **9**(4), A183–A185.

102. Young, A. P., Stumper, J. and Gyenge, E. (2009) 'Characterizing the structural degradation in a PEMFC cathode catalyst layer: carbon corrosion', *J. Electrochem. Soc.*, **156**(8), B913–B922.

103. Lauritzen, M. V., He, P., Young, A. P., Knights, S., Colbow, V. *et al.* (2007) 'Study of fuel cell corrosion processes using dynamic hydrogen reference electrodes', *J. New Mat. Electrochem. Systems*, **10**, 143–145.

104. Gallagher, K. G., Wong, D. T. and Fuller, T. F. (2008) 'The effect of transient potential exposure on the electrochemical oxidation of carbon black in low-temperature fuel cells', *J. Electrochem. Soc.*, **155**(5), B488–B493.

105. Kim, J. H., Cho, E. A., Jang, J. H., Kim, H. J., Lim, T. H. *et al.* (2010) 'Development of a durable PEMFC start-up process by applying a dummy load: II. Diagnostic study', *J. Electrochem. Soc.*, **157**(1), B118–B124.

106. Kim, J. H., Cho, E. A., Jang, J. H., Kim, H. J., Lim, T. H. *et al.* (2010) 'Effects of cathode inlet relative humidity on PEMFC durability during startup–shutdown cycling: I. Electrochemical study', *J. Electrochem. Soc.*, **157**(1), B104–B112.

107. Schmidt, T. J. (2009) 'High-temperature polymer electrolyte fuel cells: Durability insights', in *Polymer Electrolyte Fuel Cell Durability*, Buechi, F. N., Inaba, M., Schmidt, T. J., Editors, Springer, New York, USA.

108. Perry, M. L., Patterson, T. W. and Reiser, C. (2006) 'Systems strategies to mitigate carbon corrosion in fuel cells', *ECS Transactions*, **3**(1), 783–795.

109. Perry, M. L., Darling, R. M., Kandoi, S., Patterson, T. W. and Reiser, C. A. (2009) 'Operating requirements for durable polymer-electrolyte fuel cell stacks', in *Polymer Electrolyte Fuel Cell Durability*, Buechi, F. N., Inaba, M., Schmidt, T. J., Editors, Springer, New York, USA.

110. Kim, J. H., Cho, E. A., Jang, J. H., Kim, H. J., Lim, T.-H. *et al.* (2009) 'Development of a durable PEMFC startup process by applying a dummy load: I. Electrochemical study', *J. Electrochem. Soc.*, **156**(8), B955–B961.

111. Reum, M., Wokaun, A. and Büchi, F. N. (2009) 'Measuring the current distribution with submillimeter resolution in PEFCs: III. Influence of the flow field geometry', *J. Electrochem. Soc.*, **156**(10), B1225–B1231.

112. Reum, M., Wokaun, A. and Büchi, F. N. (2010) 'Adapted flow field structures for PEFC', *J. Electrochem. Soc.*, **157**(5), B673–B679.

113. Higier, A. and Liu, H. (2010) 'Optimization of PEM fuel cell flow field via local current density measurement', *Int. J. Hydrogen Energy*, **35**, 2144–2150.

114. Schneider, I. A., Bayer, M. H. and von Dahlen, S. (2011) 'Locally resolved electrochemical impedance spectroscopy in channel and land areas of a differential polymer electrolyte fuel cell', *J. Electrochem. Soc.*, **158**(3), B343–B348.

115. Schneider, I. A. and von Dahlen, S. (2011) 'Start-stop phenomena in channel and land areas of a polymer electrolyte fuel cell', *Electrochem. Solid-State Lett.*, **14**(2), B30–B33.

13

Scanning electrochemical microscopy (SECM) in proton exchange membrane fuel cell research and development

W. SCHUHMANN, Ruhr-Universität Bochum, Germany and
M. BRON, Martin-Luther-Universität Halle-Wittenberg, Germany

Abstract: Scanning electrochemical microscopy (SECM) has been established as a powerful technique in fuel cell catalysis research and development. This chapter presents the principles of SECM with a focus on the application of the various SECM modes to the investigation of fuel cell electrocatalysts. For the two most important reactions, namely hydrogen oxidation and oxygen reduction, an overview is given of kinetic studies as well as efforts in catalyst development and testing. Other reactions such as methanol oxidation and hydrogen peroxide formation and the application of SECM to fuel cell electrodes are also discussed.

Key words: scanning electrochemical microscopy, SECM, fuel cell catalysis, catalyst development, high throughput, oxygen reduction, hydrogen oxidation, electrode kinetics.

13.1 Introduction

Scanning electrochemical microscopy (SECM), although employed (Engstrom *et al.*, 1986, Engstrom *et al.*, 1987) and described (Bard *et al.*, 1989, Kwak and Bard, 1989, Engstrom and Pharr, 1989) already in the late 1980s, can still be considered as a relatively young electrochemical technique which, however, successively finds its way into the various areas of electrochemical and materials science research. The potential of SECM to investigate electrochemical processes on a local scale with high spatial resolution, something that cannot be achieved with traditional electrochemical methods, has triggered intensive research in such diverse areas as among others corrosion science (Pust *et al.*, 2008), bioelectrochemistry (Fernández *et al.*, 2004, Karnicka *et al.*, 2007, Nogala *et al.*, 2008), photocatalysis (Lee *et al.*, 2008) and the release and uptake of compounds by biological specimens such as single living cells (Bauermann *et al.*, 2004). Principal considerations of SECM, instrument design and microelectrode fabrication, as well as the vast amount of applications to gather localized knowledge about the electrochemical activity of a large variety of different samples using SECM, have been subject to a considerable number of reviews (e.g. Pust *et al.*, 2008, Mirkin, 1999, Wittstock *et al.*, 2007, Anemiya *et al.*, 2008) and a textbook (Bard and Mirkin, 2001).

The dissemination of this technique was further facilitated by the development of a solid theoretical background, as reviewed in (Pust *et al.*, 2008) allowing, e.g., the determination of reaction rates. Moreover, the instrumental progress such as the possibility of preparing sub-micrometer electrodes combined with high-resolution positioning systems (Amemiya *et al.*, 2008, Katemann and Schuhmann, 2002), including various techniques to precisely control the distance between microelectrode and substrate (Hengstenberg *et al.*, 2000, Katemann *et al.*, 2004), the development of specific measuring modes (Pust *et al.*, 2008, Mirkin, 1999, Wittstock *et al.*, 2007, Amemiya *et al.*, 2008) and, last but not least, the availability of commercial instruments, contributed to the comparatively fast spreading of SECM into a large variety of applications. It is hence no surprise that there is also an increasing use of SECM for answering questions arising in electrocatalysis and fuel cell research with the aim of investigating local properties of catalysts and electrode materials. Numerous studies have been carried out that aim at the elucidation of the detailed kinetics and mechanistic pathways of electrocatalytic processes as well as their local changes. These studies form the basis for more applied investigations, in which SECM is used as a fast and reliable tool to compare catalyst activities for a given reaction in a high-throughput experiment while varying the determining parameters such as pH-value, temperature and applied potential. Finally, as an ultimate goal for using SECM in practical fuel cell research, several attempts have been recently made in which SECM was applied to evaluate the local electrocatalytic properties of porous catalyst layers and gas diffusion electrodes. This chapter presents a compilation of the various approaches for using SECM in proton exchange membrane fuel cell research, discusses advantages and challenges, and highlights results obtained in this field during the past 20 years.

13.2   Basics of scanning electrochemical microscopy (SECM)

Scanning electrochemical microscopy is a technique that provides spatially resolved (electro)chemical information about a sample under investigation. In addition, topographic information may be achieved as well. It has to be pointed out that SECM is a true chemical scanning probe microscopy in which, in most cases, reaction rates of redox processes at the sample surface are used for acquiring information about sample properties of interest.

13.2.1 Measurement principle

In a typical SECM experiment an ultra-microelectrode (UME, i.e. an electrode with typical dimensions of < 30 μm in diameter) is rastered stepwise at a small distance (i.e. typically at about 50–80% of the UME diameter) across the surface of the sample under investigation. The measurement is performed at a timescale such that a hemispherical diffusion zone is established in front of the electrode surface. At each point of the scanning grid an (electro)chemical experiment is performed, either at the tip or at the sample or at both, which provides the anticipated information about the underlying local electrochemical properties of the sample. Experiments with the UME resting at the same place above the substrate surface or with the UME approaching the surface (or being retracted) are possible as well.

UMEs are typically (but not necessarily) disk-shaped and are surrounded by an insulating sheath. The ratio of the sheath diameter to that of the active electrode area, r_s/r_t, is called the RG value. The RG value is important because of its influence on the diffusion of reactants or products into or out of the gap formed between the positioned UME and the sample. This configuration can be treated like a leaking thin-layer electrochemical cell with the overall dimensions given by the insulating sheath and the active UME surface in the centre. Due to the specific properties of an UME, the hemispherical diffusion profiles establish very fast, leading to an efficient mass transfer and high diffusion-limited currents, which may even exceed those obtained in rotating-disk electrode experiments. In the following, the UME will typically be called the 'tip', a term that is generally used in scanning probe microscopy techniques.

13.2.2 SECM modes

SECM measurements can be performed employing various modes, which differ in the detailed experimental procedure. In the following, the general principles of some of the most often used SECM modes will be described, with the emphasis on those that already found applications in electrocatalysis and fuel cell research. The different applications for acquiring local electrochemical information in fuel cell-related research are described in detail in the following section. An in-depth description of the mathematical background of the various SECM modes is beyond the scope of this chapter and can be found elsewhere (Kwak and Bard, 1989, Pust *et al.*, 2008, Cornut and Lefrou, 2007a, 2007b, Lefrou, 2006, Fulian *et al.*, 1999, Martin and Unwin, 1997, 1998).

Feedback mode

The feedback mode employs a free-diffusing redox mediator, which is usually added in either reduced (R) or oxidized (O) form to the electrolyte solution. Assuming that the oxidized form was added, the tip is polarized to a potential at which O is reduced to R under diffusion limitation. With the SECM tip positioned far enough away from the sample surface in the bulk of the redox mediator containing electrolyte solution, an undisturbed hemispherical diffusion profile of O in front of the tip is established that gives rise to a constant diffusion-limited current. The strength of the limited current is dependent on the electrode size and the mediator concentration. This current at infinite distance is usually called $i_{t,\infty}$. If we now consider the tip approaching the surface, then two limiting cases can be distinguished. If the surface is electrochemically inactive, i.e. the reaction rate of

R to O is very small or zero, then the diffusion of O to the tip is hindered as soon as the surface interferes with the hemispherical diffusion profile in front of the tip. This effect is stronger the closer the tip approaches the sample surface, with diffusion being completely blocked when the tip touches the surface. Thus, while approaching the surface, the diffusion-limited tip current will decrease (negative feedback effect). Due to a small tilting between the plane of the UME and the sample surface, which is typically found in practical setups, the insulating sheath touches the sample surface first and a zero current is not achieved. At the other extreme, if the tip was approaching an electrochemically active surface area that is substantially larger than the overall tip dimensions, R that was formed at the tip can be oxidized back to O. In this way, the amount of O available within the diffusion zone in front of the tip is increased as compared to the diffusional mass transport from the bulk solution, leading to an enhanced tip current exceeding $i_{t,\infty}$. This so-called positive feedback effect is again stronger the closer the tip approaches the surface, and will theoretically be infinite as the tip touches the surface. Moreover, it is highly dependent on the reaction rate of R to O at the site of the sample surface that is located opposite to the tip electrode. Since the extent of positive feedback for electrocatalytic reactions depends on the electrochemical or electrocatalytic activity of the sample under investigation, it is possible under certain conditions to distinguish between more active and less active surface sites, as will be discussed further below.

While approaching the substrate, a plot of $i_t/i_{t,\infty}$ with i_t being the actual tip current, over the tip-to-sample distance d, which is typically given in its normalized form as d/r_t, is called the approach curve and can be used to position the tip electrode in close proximity to the sample surface. Furthermore, a fit of the approach curve employing a suitable theoretical model can be used to determine kinetic parameters for the redox reaction under investigation at a specific sample location. This feedback behaviour is shown in Fig. 13.1 where (a) is the microelectrode in the bulk of the solution and (b) and (c) illustrate negative and positive feedback as indicated. Figure 13.1 (d) shows calculated approach curves both for positive and negative feedback.

Generation-collection modes

In contrast to the feedback modes, in the generator-collector modes (GC modes) no redox mediator is added to the electrolyte solution but a redox-active compound is formed either at the tip or at the sample. In the sample generation-tip collection mode (SG/TC-mode), redox active species are formed at the sample by the reaction under investigation. The generated redox species diffuse to the tip across the gap between tip and sample where they are either oxidized or reduced. If the rate of generation of the redox species at the sample surface is related to its (electro)chemical activity, the local differences in (electro)chemical activity of the sample can be identified by monitoring the tip current. Vice versa, redox active

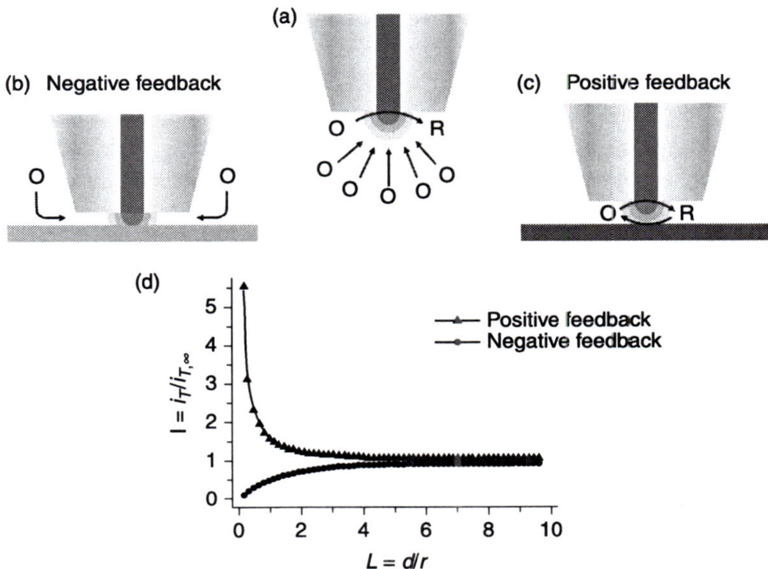

13.1 Illustration of positive and negative feedback in SECM (a–c). The approach curves in (d) are calculated for an R/G value of 10.2.

species formed at the tip diffuse to the sample and are there further converted in the so-called tip generation-sample collection mode (TG/SC-mode). In this case, the sample current is a measure for the local (electro)chemical activity of the sample at the location of the tip. However, the spatial resolution in the TG/SC-mode is inherently limited by the often large area of the sample and the concomitantly observed electrical noise and background current, which makes the detection of small currents with too-small tip sizes demanding, if not impossible.

A main advantage of the TG/SC mode as compared with the feedback mode is seen in the fact that the current is measured without a faradaic background current from redox species present in the bulk solution. Thus, the flux of redox species as generated at the tip or sample can be directly measured. The local changes in activity of a compound that is released from the sample can also be detected potentiometrically (Denuault *et al.*, 1992, Etienne *et al.*, 2007, Gray and Unwin, 2000, Horrocks *et al.*, 1993b). In this case the potentiometric tip is just a spectator and does not modulate the local concentration of the compound. Therefore, despite being often used, the term 'SG/TC mode' is not fully appropriate in this case.

Redox-competition mode (RC-SECM mode)

The redox-competition SECM (RC-SECM) mode was developed with the aim of overcoming the limitations and drawbacks of the GC modes that were previously used for the determination of the local catalytic activity on surfaces (Eckhard *et al.*, 2006). In the RC-SECM mode, both tip and substrate compete for the same analyte that is present in the gap between tip and sample. The analytical signal in this case is the current at the SECM tip, which depends on the concentration of the analyte in the gap and thus on the activity for a reaction occurring at the sample that is consuming the analyte. In this manner, a continuous increase in background current as in the SG/TC mode or a limited resolution as in the TG/SC mode can be avoided. A mode where both tip and substrate compete for the same analyte has also been investigated by Zoski *et al.* and the effect has been labelled as shielding (Zoski *et al.*, 2003).

Further modes of SECM have been developed and applied to various electrochemical questions. A detailed description of these, however, is beyond the scope of this chapter. Details can be found in the reviews cited above.

13.2.3 Instrumentation

A SECM instrument consists of a suitable x-, y-, z-positioning system (typically step motor-driven positioning tables with a nominal resolution of about 100 nm/microstep and piezo elements with integrated encoders with a nominal resolution of a few nm) that allows the positioning of the SECM tip in close proximity to the sample under investigation, an electrochemical cell of which the sample usually forms the bottom, the SECM tip itself and a (bi)potentiostat/galvanostat or another measuring device, depending on the signal that is acquired at the tip (e.g. a highly sensitive potentiometer, a lock-in amplifier, etc.). Figure 13.2 illustrates the

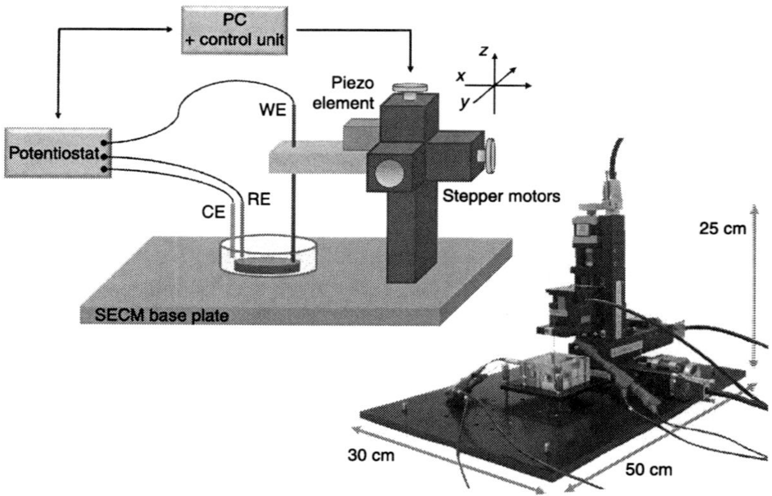

13.2 Principles of SECM instrumentation including a potentiostat, a positioning system and the electrochemical cell. The photo shows a commercial SECM. (Source: Sensolytics, Bochum, Germany.)

principles of the setup of a SECM. The spatial resolution of a SECM is usually not limited by the accuracy of the positioning system but by the size of the tip, which in turn has to be adapted to the problem under investigation, i.e. to the expected resolution that is necessary to visualize the local variation in the electrochemical activity of the sample. However, due to the fact that in general a Faraday current is measured, the resolution is not only limited by the size down to which tips may be fabricated but also by the current resolution of the amplifier used. The best systems have sub-pA resolution and the smallest tips used in SECM are Pt nanoelectrodes with a tip diameter as small as 100 nm. If nanoelectrodes are used as SECM tips the positioning and especially lateral scanning is far more demanding, and additional techniques have to be integrated to keep the tip-to-sample distance constant during scanning (Hengstenberg *et al.*, 2000, Katemann *et al.*, 2004). The resolution in SECM, however, is generally lower than with other scanning probe microscopies owing to the fact that free-diffusing redox species are involved in signal generation and the lowest detectable current still requires about 1000 electrons/s to be transferred. On the other hand, as already pointed out above, SECM is a true 'chemical microscopy' and provides chemical information about the sample under investigation.

13.3   SECM in fuel cell catalyst development and investigation

One main area in fuel cell research in which SECM is predominantly used is catalyst development, catalyst optimization and investigation of lateral differences, as well as the evaluation of important reactions such as the oxygen reduction reaction (ORR), the hydrogen oxidation reaction (HOR) and the hydrogen evolution reaction (HER). Here we review these efforts with a special focus on the advantages of SECM in catalyst screening (parallel catalyst testing/'high-throughput experimentation'). While SECM might be considered to be a slow screening technique compared to other catalyst screening approaches, it offers the possibility of a quantitative kinetic data analysis within the same experiment. Thus, efforts to investigate the detailed kinetics of fuel cell-relevant reactions will be presented as well.

13.3.1 Oxygen reduction reaction (ORR)

Electrochemical reduction of molecular oxygen (oxygen reduction reaction (ORR)) is one of the most important electrocatalytic reactions with a strong impact on fuel cells. It is known that due to the complex four-electron/four-proton transfer the ORR is the predominant limiting factor in fuel cells, but also in modern HCl recycling, electrolytic chlorine production, and corrosion science. However, the ORR is still not well understood despite many years of intensive research. Together with the specific features of suitable catalysts and the fabrication of catalyst-modified electrodes, the

ORR needs to be further investigated both in basic research as well as in more applied studies dedicated to optimizing industrial processes. Thus, it is evident that SECM was used in a variety of measuring modes to address the ORR at a variety of different materials especially with respect to the kinetics of the ORR, for comparing the activity of different materials and to evaluate the selectivity of the reaction at different catalyst materials. SECM was used in the feedback mode, the generation-collection mode, and more recently in the redox competition mode to visualize the local electrocatalytic activity of catalyst samples with respect to the ORR. The different modes have advantages and drawbacks when applied to the ORR. The specific features of the different SECM modes in the elucidation of the ORR will be discussed in the following section.

SECM modes for the visualization of the ORR

In the feedback mode (see above), the redox pair under consideration is the O_2/OH^- pair. OH^- is oxidised at the SECM tip and the O_2 formed diffuses across the gap to the sample where it is reduced back to OH^- that may diffuse to the tip again and be re-oxidised to O_2. The tip current serves as the analytical signal. This mode is possible in alkaline media (pH 9–12), where the reduction product of oxygen is OH^-. In acidic media, in which oxygen is reduced to water or H_2O_2, or in highly concentrated alkaline media in which the OH^- concentration is too high to detect the relatively minor changes in the feedback loop (Liu and Bard, 2002), this mode cannot be applied. In contrast to the conventional feedback mode, in which a redox mediator has to be added on purpose, the modulation of the local concentration ratio of the O_2/OH^- redox pair is used for the visualization of the local catalytic activity of the sample. This approach obviously directly involves reactants of the catalytic reaction and does not rely on a non-specific substitute such as classical redox mediators such as ferrous/ferricyanide.

To overcome the pH limitations of the O_2/OH^- feedback mode and especially to study the ORR also in acidic solutions with SECM, Bard and co-workers have employed the tip generation/sample collection mode (Fernández and Bard, 2003). Here, oxygen is continuously produced at the tip by water oxidation in an initially oxygen-free electrolyte. (The tip in a TG-SC experiment may be either controlled by the bipotentiostat/galvanostat also controlling the substrate, or by an external circuit (e.g. a 9 V battery). The latter approach has also been labeled modified TG-SC.) When the tip is positioned close to an active spot of the sample, the oxygen formed at the tip will diffuse to the sample where it is reduced. The sample current can hence be taken as a measure for the activity of the sample at the position of the tip. The drawback is related to the size of the sample that is investigated, and which is in any case much larger than the dimensions of the SECM tip. The noise in the sample current is rather high, leading to a minimum tip size for the generation of a signal that substantially exceeds the background current noise. Several advantages of the TG-SC mode as compared to the feedback mode have been described by Lu *et al.*, including its applicability in higher electrolyte concentrations, the smaller dependence on the tip-to-sample distance and its higher tolerance to contaminations at the tip (Lu *et al.*, 2007).

To overcome the drawback of higher noise in the TG-SC mode, the so-called redox competition mode (RC mode) was developed, which was initially specifically designed to investigate local electrocatalytic oxygen reduction (Eckhard *et al.*, 2006). Here, the analyte under consideration is oxygen, which is consumed at both the tip and the sample. The oxygen reduction current at the tip, which is taken as the analytical signal, is decreased the more oxygen is consumed by the sample below the tip, i.e. the higher the electrocatalytic activity of the sample is at this location. The principle of RC-SECM in comparison to the GC modes as applied to ORR is illustrated in Fig. 13.3. To ensure constant and unchanged conditions throughout the whole experiment, a pulse potential profile is applied to the tip where an initial short positive potential pulse leads to water oxidation (i.e., oxygen formation) within the gap between tip and sample in an initially oxygen-free electrolyte solution. The potential of the sample can be adjusted to alter the driving force for the electrocatalytic reaction; however, it remains constant throughout the experiment. After the oxygen formation pulse, the tip is immediately polarized to a potential at which oxygen reduction becomes possible. Thus, the tip competes for the oxygen in the solution, which is modulated by the catalytic activity for the ORR at the sample opposite to the tip position. During this so-called measuring potential, a fast data acquisition is invoked and

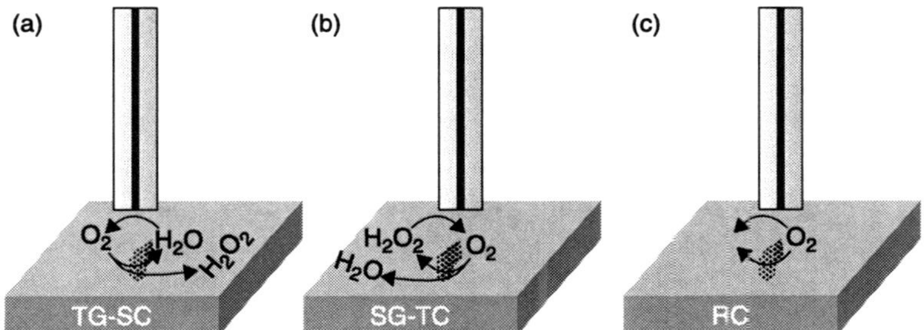

13.3 Schematic representations of different detection modes in SECM. (a) Tip generation, sample collection (TG–SC) mode. The scanning microtip oxidises water to produce O_2. The O_2 reduction current at the sample is the analytical signal. (b) Sample generation, tip collection (SG–TC) mode. The sample is polarized to reduce O_2. The tip oxidises H_2O_2 that may be formed as a by-product at the sample. The analytical signal is the peroxide oxidation current detected at the tip. (c) Redox competition mode (RC-SECM). The O_2 reduction current is detected at the tip. Depletion of this current is appointed to locally diminished oxygen concentration caused by a competing O_2 reducing catalyst site at the sample. (Source: Eckhard and Schuhmann, 2007, with kind permission from Elsevier.)

100 current values are typically recorded within 0.2–0.5 s. Thus, the tip current is monitoring the decrease of the oxygen concentration within the gap between tip and sample. Finally, the tip is polarised to a no-effect potential, moved to the next grid point, and the pulse potential can be applied again to the tip. A further extension of the pulse program also made possible the sequential detection of hydrogen peroxide, as will be discussed in Section 13.3.2 (Eckhard and Schuhmann, 2007).

Studies on oxygen electroreduction kinetics

The kinetics of oxygen reduction at Pt in a diluted alkaline solution have been studied using the feedback mode employing the O_2/OH^--redox pair (see above). Liu and Bard (2002) used a gold tip and applied a potential pulse program to the tip to avoid electrode fouling; a continuous production of O_2 by water/OH^--oxidation could thus be obtained. By fitting approach curves, the rate constants for the heterogeneous electron transfer as well as Tafel slopes could be derived, which agreed well with data obtained from rotating-disc electrode measurements. This approach to the study of oxygen reduction kinetics was later extended to more extreme conditions such as 10 M NaOH (Zhang *et al.*, 2009), which are of relevance, e.g. for NaCl electrolysis. In the latter study, the O_2/O_2^*- redox pair was used to establish the feedback loop. In a follow-up study, Fernandez and Bard also employed the TG-SC mode for kinetic studies for reaction conditions, which are inaccessible by the feedback mode. The theoretical background as well as the example of ORR on Pt in phosphoric acid are described by Fernández and Bard (2004). Simulated steady-state responses correlated well with experimental data, however only within a certain parameter range.

Oxygen reduction electrocatalyst development and testing

One major advantage of the application of SECM in the development and optimization of electrocatalysts as compared to more traditional methods such as the rotating-disc electrode is the possibility to rapidly and, as discussed in the previous chapter, at the same time quantitatively compare, in a single experiment, different catalysts with respect to their activity under a wide range of experimental conditions. The challenge here is not only the catalyst testing itself but, as with all approaches to high-throughput catalyst testing, the selection and fast preparation of suitable arrays of individual catalyst compositions ('catalyst libraries'). Various approaches to this topic have been demonstrated in the literature as will be presented in the following, and further below in the section on hydrogen oxidation/ proton reduction electrocatalysts.

A large number of studies on the ORR has been done by Bard and co-workers, who compared the ORR activities of different bi- and ternary combinations of Pd, Au, Ag, Co, Ti, V and Mn (Fernández *et al.*, 2005a, 2005c, 2006, Walsh *et al.*,

2006) using TG-SC-mode SECM. The choice of catalyst compositions was based on thermodynamic guidelines, and catalyst preparation has been carried out in a high-throughput fashion using mixtures of metal salts deposited as small spots on a conducting substrate by a piezo spotter. The metal composites were obtained by reduction with hydrogen at elevated temperatures and/or $NaBH_4$. SECM proved to be a fast method with which a large number of catalyst compositions could be screened and compared with respect to their oxygen reduction activity in a short time. The visualisation of oxygen reduction activity over such catalyst arrays is shown in Fig. 13.4 (Fernández et al., 2005a). Compositions with comparably high activity such as PdCo (Fernández et al., 2005a, Fernández et al., 2006) and PdV (Walsh et al., 2006) were identified and have been synthesised at a larger scale as carbon-supported materials to confirm their high activity using the rotating-disc electrode. Levels of activity nearly reaching that of Pt were found.

While these studies have been carried out with model samples such as thin-film materials deposited on glassy carbon, several recent attempts aimed at the characterisation of more practical catalysts such as carbon-supported nanoparticles, etc. Schwamborn et al. (2010) prepared patterned carbon nanotubes (CNT) arrays and demonstrated their possible use as a high-surface-area electrode material for combinatorial SECM investigations by electrodepositing Pt onto CNT spots and monitoring oxygen reduction activity by means of RC-SECM. Chen et al. (2009) have used an electrochemical droplet cell to deposit arrays of small spots (ca. 500 μm in diameter) of mono- and bimetallic catalysts (Pt, Au, Ru, and Rh) onto a thin layer of carbon nanotubes on a glassy carbon surface. The resulting catalyst arrays were investigated for ORR activity using the RC mode of SECM. Semi-quantitative results were obtained by relating the tip-current decrease in SECM

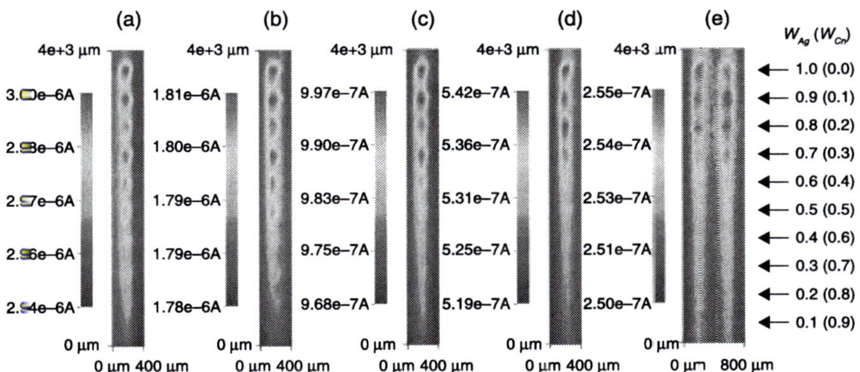

13.4 SECM TG-SC images of oxygen reduction activity measured on binary arrays of Ag-Cu in 0.5 M H_2SO_4. (a) $E_S = 0.0$. (b) $E_S = 0.05$. (c) $E_S = 0.1$. (d) $E_S = 0.15$. (e) $E_S = 0.2$. V vs. HRE. W is the atomic ratio of metal (Ag, Cu) in the spot. (Source: Fernández et al., 2005a. Copyright 2005 American Chemical Society.)

experiments to the deposited amount of the metal under investigation. A similar quantification approach has been used to compare Pt, Ag and PtAg catalysts electrodeposited onto glassy carbon surfaces for their ORR activity. Semi-quantitative SECM results agreed well with RDE experiments, and a PtAg deposit with an atomic ratio of 59:40 has been found to be more active than Pt in neutral electrolyte (Nagaiah *et al.*, 2009). Further studies in this field include the investigation of ORR activity of differently shaped Pt nanoparticles (Sánchez-Sánchez *et al.*, 2010).

Besides noble metals and other metal catalysts, SECM has also been used to investigate non-noble metal oxygen reduction catalysts such as nitrogen-modified CNTs (Kundu *et al.*, 2009) and electrodeposited porphyrins (Okunola *et al.*, 2009). In these studies, which were carried out using the RC mode, the intention was to compare individual catalyst samples that have been prepared or deposited as small spots ($<<$ 1 mm) and to complement activity measurements obtained with rotating-disc electrode (RDE) measurements. Usually good agreement between RDE results and RC-SECM was found. However, differences occurred in the case of electrodeposited porphyrines, where a high amount of H_2O_2 produced at the sample and trapped in the gap between tip and sample most likely led to a degradation of the porphyrin (Okunola *et al.*, 2009). Besides activity, catalyst samples were also screened with regard to their selectivity in oxygen reduction, which will be discussed in the next section.

13.3.2 Hydrogen peroxide formation and reduction

As mentioned above, the ORR is considered as one of the major limiting factors in obtaining high efficiency in PEM fuel cells. This, however, is not only related to the catalyst activity but also to selectivity of the oxygen reduction reaction: Oxygen may either be reduced in a direct four-electron reaction to water or in a two-electron reaction to hydrogen peroxide; the latter may be decomposed or further reduced in a two-electron reaction to water. To determine the extent of H_2O_2 production and the average number of electrons transferred per oxygen molecule (n), measurements employing the rotating ring-disc electrode are usually performed. Several reports have recently demonstrated the applicability of SECM for the detection of the amount of H_2O_2 formed during the oxygen reduction, both qualitatively and quantitatively. First approaches used wired enzymes at the tip (Horrocks *et al.*, 1993a); however, this was restricted to the pH range of the enzyme stability.

A combined RC-SECM/GC-SECM mode to visualize both the local oxygen reduction activity and the H_2O_2 production of a catalyst modified sample in a single SECM scan has been suggested (Eckhard and Schuhmann, 2007). In this mode, an even more complex potential pulse profile is applied to the SECM tip, while the catalyst-modified sample is kept at a predefined constant potential. At each grid point, first the pulse sequence for the RC mode is performed and the decrease in tip current over time is recorded to evaluate the activity of the catalyst with respect to the ORR. Keeping the tip positioned at the same coordinates, a subsequent generation/collection experiment is initiated by applying an oxygen injection pulse. However, instead of jumping to the oxygen reduction potential, a H_2O_2 oxidation potential is applied and the H_2O_2 that is generated at the catalyst during the ORR diffuses to the tip, where it is detected. Thus, the catalytic activity for the ORR at a predefined potential of the sample can be correlated with the selectivity of the electrocatalytic reaction with respect to a direct or consecutive formation of H_2O in a four-electron transfer or 2 + 2 electron transfer reaction or to the predominant formation of H_2O_2 in a two-electron transfer step. It was possible to distinguish between a spot of electrodeposited Au nanoparticles, which exhibit nearly exclusive reduction of O_2 to H_2O_2, and a spot of electrodeposited Pt nanoparticles, which show a negligible amount of formed H_2O_2, as shown in Fig. 13.5 (Eckhard and Schuhmann, 2007). Later, this combined mode was extended to the investigation of a variety of different electrochemically deposited metalloporphyrins (Okunola *et al.*, 2009). Different amounts of H_2O_2 formed over Mn, Fe and Co porphyrins have been detected with the Mn porphyrin producing the lowest amount of H_2O_2 suggesting that the primarily generated H_2O_2 was further reduced to H_2O in a sequential 2+2-electron transfer reaction. However, it also was shown that care must be taken selecting a suitable inert substrate for the experiment, since H_2O_2 formed at the glassy carbon used as a substrate for depositing porphyrins led to a large background current. The principal reactions occurring in such a thin layer are shown in Fig. 13.6.

Efforts to quantify H_2O_2 formation during oxygen reduction in an SECM experiment have been carried out by Sánchez-Sánchez *et al.* (2008) using the SG/TC mode and employing small sample surfaces (diameter $\leq$ 200 μm) to avoid a continuous increase in the background current due to the continuous formation of H_2O_2 at large samples. Calibration curves were derived to predict the tip collection efficiency for various tip geometries (diameter of sample and tip, thickness of insulating sheath, tip-to-sample distance) by applying a combined experimental/simulation approach. The prediction could be verified evaluating the oxygen reduction to H_2O_2 at Hg/Au-electrodes, as an example. This mode might provide advantages over the classical rotating ring disk electrode (RRDE) approach such as the comparably high collection efficiency of 43–49% (Sánchez-Sánchez and Bard, 2009). Furthermore, the H_2O_2 transport within the gap between the positioned tip and the sample surface occurs only via diffusion processes and not via convection as in the RRDE technique. Thus, it is not influenced by distorted convection profiles in front of a possibly non-flat electrode. After demonstration of its feasibility in principle, this mode was applied to determining the potential dependent average number of electrons transferred per molecule during oxygen reduction at a variety of catalytically active metals and alloys (Hg, Au, Ag, Cu, Pt, Pd, $Pd_{80}Co_{20}$, and $Au_{60}Cu_{40}$) in 0.5 M H_2SO_4 (Sánchez-Sánchez and Bard, 2009). The number of electrons transferred during the ORR is ranging from two for Hg

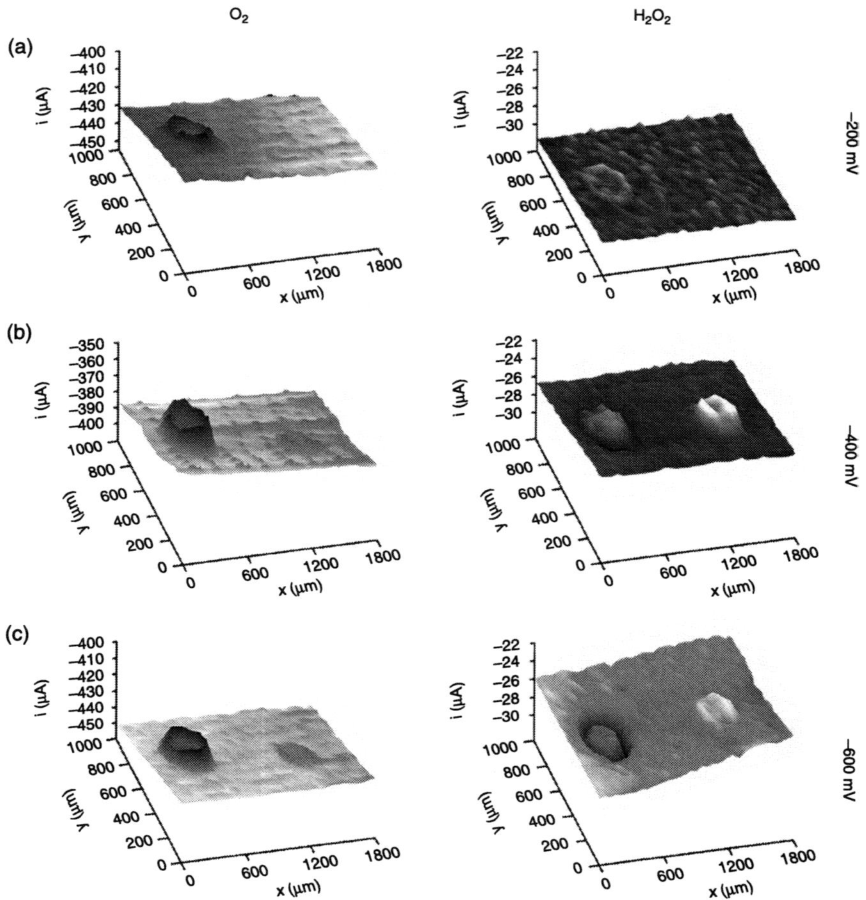

13.5 Visualization of oxygen reduction activity and hydrogen peroxide formation over spots of gold (right-hand spot in each image) and Pt (left-hand spot) nanoparticles, as investigated with RC-SECM. Rows are sample polarizations of E_S vs. Ag/AgCl (a) −200 mV. (b) −400 mV, (c) −600 mV. The first column represents the oxygen competition data. Plotted is the reduction current at the tip as a function of x- and y-position. Darker shaded areas correspond to diminished current flow due to competition for the available O_2 with actively reducing sample sites. The higher ORR activity of Pt is clearly visible. The second column represents the oxidation current of H_2O_2 at the tip simultaneously acquired at the same indicated sample polarization. Despite the oxidation the absolute values are negative since the Faradaic current is overlaid by a capacitive contribution by the potential pulse step. Local differences compared to the image background can be detected, and Au is shown to produce more H_2O_2 than Pt. (Source: Eckhard and Schuhmann, 2007, with kind permission from Elsevier.)

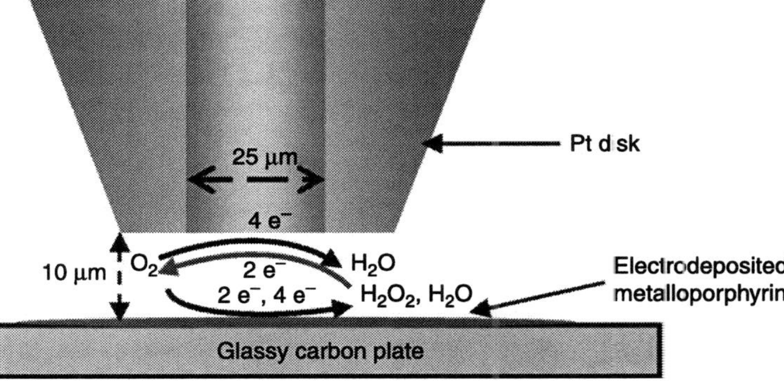

13.6 Schematic representation of the reactions taking place in the gap between the SECM tip and the metalloporphyrin-modified sample during the redox competition mode of SECM in which the SECM tip and the sample compete for the available O_2. Note that the tip current during oxygen reduction is also influenced by the amount of H_2O_2 available. (Source: Okunola *et al.*, 2009, with kind permission from Elsevier.)

to four for Pt and $Pd_{80}Co_{20}$, in agreement with what was known about the ORR of these materials. A further quantitative analysis of H_2O_2 decomposition using SECM has been carried out by the Bard group (Fernández *et al.*, 2005b).

A different procedure to use the SG/TC mode for quantitative H_2O_2 detection has been suggested by the group of Wittstock (Shen *et al.*, 2008a). Instead of the above restriction to small sample surfaces, a sample size of at least one order of magnitude larger than that of the tip was deliberately chosen. However, different to the above approach the sample potential is not kept constant but is pulsed to the ORR potential intermittently in order to avoid the extension of the H_2O_2 diffusion layer in front of the sample surface. The UME is held at a fixed potential and the transient amperometric tip current caused by the H_2O_2 oxidation is recorded after pulsing of the substrate. By fitting these amperometric current transients to an analytical model, effective rate constants for the direct four-electron reduction and the two sequential two-electron reduction steps could be derived for the ORR at Au and a PdCo alloy (8.75 atom % Co). For the ORR at Pt, however, a fast decay in the transient H_2O_2 reduction at the tip was observed, ascribed to a H_2O_2 formation only at the beginning of the pulse (probably on an initially oxidized Pt surface) and a subsequent H_2O_2 reduction also at the sample surface. This technique was also applied to more complex electrodes, namely Pt and Pd nanoparticles electrodeposited into polyelectrolyte films. Here, the effective rate constants of oxygen reduction at Pd could be derived, while again Pt showed a qualitatively different behaviour (Shen *et al.*, 2008b). It has to be pointed out that the rate constants were assuming a certain reaction pathway and rate limiting steps. Although the numbers might be questioned, the trend is clear and supports the hypotheses.

Using similar SG/TC-experiments as those described above, a semi-quantitative analysis of H_2O_2 production on a Nafion-covered Pt microdisc has been carried out by Kishi *et al.* (2010). The authors found that thicker Nafion films led to a decrease in H_2O_2 formation, while oxygen consumption remained unchanged.

13.3.3 Hydrogen oxidation and evolution

SECM modes and principles

Although less challenging than the ORR due to the inherently much less complicated reaction pathway, hydrogen oxidation as the major anode reaction in fuel cells was extensively studied by means of SECM, as discussed in the following sections. Differing from the oxygen electrode, which exhibits large over-potentials for the ORR as well as for oxygen evolution, the hydrogen electrode is quasi-reversible. The H_2/H^+ redox pair can be used as a mediator employing the feedback mode (Zhou *et al.*, 2000, Shah and Hillier, 2000). To study local activity on catalyst samples for hydrogen oxidation, protons produced at the substrate are reduced at the tip at a potential where diffusion limitation occurs. Thus, this mode is restricted to slightly acid solutions; typically 0.01 M acids are employed together with an electrolyte of the same anion. Again, as in the case of the O_2/OH^- redox pair, differences in the local chemical activity of the sample surface influence the feedback, thus making this technique suitable for studying the activity distribution. Other modes that have been employed to study hydrogen oxidation/proton reduction are the generation-collection modes.

Kinetic studies

The kinetics of hydrogen evolution and oxidation have been studied with SECM using the feedback mode, where the approach curves, as in the case of ORR kinetics, are fitted based on a suitable theoretical model. The basic approaches are demonstrated by the groups of Bard and Hillier (Zhou *et al.*, 2000, Shah and Hillier, 2000). Zhou *et al.* derived kinetic parameters of the hydrogen oxidation at Pt at different potentials. A Tafel slope of 117 mV dec^{-1} has been derived, while oxide formation on Pt hindered hydrogen oxidation. The influence of adsorbing species such as Cl^- and Br^- has also been investigated (Zhou *et al.*, 2000). An application to characterize a patterned Au/Pt electrode can be found in Shah and Hillier (2000).

A kinetic analysis of the hydrogen oxidation on clean and CO-covered polycrystalline Pt in H_2SO_4 has been carried out by Jambunathan *et al.* (2001, 2002) using the feedback mode. Clean Pt showed a Tafel slope of 28 mV dec^{-1} consistent with other published data. Expectedly, the CO-covered surface was nearly inactive towards hydrogen oxidation at $\theta > 0.8$ monolayer (ML). However, high activity was observed at $\theta < 0.6$ ML. Using tip–sample voltammetry (TSV), where the sample potential was varied at a fixed tip-to-sample distance, it was shown that subtle dynamic changes in the CO adlayer with the applied potential have strong influences on the transient hydrogen oxidation kinetics. High hydrogen oxidation activity could be sustained on a CO-covered surface by pulsing the potential between 0.1 V and 0.4–0.7 V.

Contrary to the 28 mV dec^{-1} as the Tafel slope for hydrogen oxidation on polycrystalline Pt as reported above, Zoski (2003) found a value of 117 mV dec^{-1}, derived from a similar experimental approach, i.e. using the feedback mode with proton reduction at the tip and hydrogen oxidation at the sample. A similar value has also been found by Zhou *et al.* (2000). Furthermore, polycrystalline Ir and Rh have been investigated with Ir showing hydrogen oxidation activity similar to Pt, which was considerably higher than that of Rh. The Volmer or Heyrovsky reactions are considered to be the rate-determining step of hydrogen oxidation at these metals (Zoski, 2003).

Apart from bulk materials, work has been done to derive kinetic parameters for hydrogen oxidation at nanoparticles. Ahmad *et al.* (2004) investigated the hydrogen reactions at Pt nanoparticles incorporated into polyaniline. Comparably high rate constants as well as a Tafel slope of 217 mV dec^{-1} for H$_2$ oxidation might indicate that diffusion limitation may possibly influence the results.

Catalyst development and testing

As for the evaluation of oxygen reduction catalysts, SECM offers possibilities in the parallel and comparably fast screening of catalyst libraries for hydrogen oxidation catalysts. Furthermore, it is also easily possible to rapidly evaluate influences such as substrate potential and CO poisoning on hydrogen oxidation activity for a wide range of catalyst compositions in a single SECM scan.

McGinn and colleagues (Black *et al.*, 2004, 2005, Cooper *et al.*, 2005) have prepared PtRu thin-film electrodes with varying composition by sputter deposition through masks onto Si wafers. The composition spreads were investigated with SECM towards the hydrogen oxidation activity using the feedback mode. Clear differences in the electrocatalytic activity for hydrogen oxidation were detected and trends in the reactivity with changes in the Pt/Ru composition were observed. Addition of WC or Co to form binary or ternary catalysts can significantly increase activity and CO tolerance (Lu *et al.*, 2006).

Jayaraman and Hillier employed an array of eight individually addressable band electrodes and modified them with particles of varying PtRu or PtRuMo composition by pulsed electrodeposition. The activity of these electrodes was evaluated with respect to hydrogen oxidation in the presence of an adsorbed monolayer of CO using the SECM feedback mode (Jayaraman and Hillier, 2003). The observed onset potentials were consistent with those of CO stripping experiments employing cyclic voltammetry.

Similar to approaches reported above for catalyst screening for oxygen reduction, Weng and Hsieh (2011) deposited chloro-salts of Pt, Ru, Ir and Pd and their bimetallic mixtures in the form of an array onto a glassy carbon substrate from water/glycerol solutions by means of a piezo spotter. After reduction of the deposited metal ions to form the active metal catalyst, the library was characterized by XRD, SEM and EDX. A modified TG-SC mode with constant hydrogen production at the tip has been employed to map the hydrogen oxidation activity of the individual compositions. PtPd catalysts with high Pd loading turned out to be more active than Pt. The most active compositions of each binary mixture (Pt_4Pd_6, Pt_9Ru_1, Pt_3Ir_7) have been investigated in more detail, including tip substrate-voltammetry as well as kinetic analysis by fitting approach curves. The derived kinetic parameters were similar to those reported in other studies (Zhou *et al.*, 2000, Zoski, 2003).

The above approaches lead to a distinct set of samples/catalyst compositions, and the number of catalysts compared in a single experiment is typically far below 100. SECM, however, would in principle allow for the screening of a much larger number of catalyst compositions. In order to increase the sample number in a high-throughput approach, Jayaraman and Hillier (2004) extended their studies to PtRu catalysts with a gradient in the Ru content on an indium-doped tin oxide (ITO) substrate. The gradient has been achieved by diffusion of the metal ions through a gel and subsequent electrodeposition. Such gradients theoretically lead to an infinite number of catalyst compositions. The authors scanned the layer towards activity in hydrogen oxidation in the presence of adsorbed CO employing the feedback mode. By varying the substrate potential, the composition-dependent extent of CO poisoning could be visualized, as shown in Fig. 13.7.

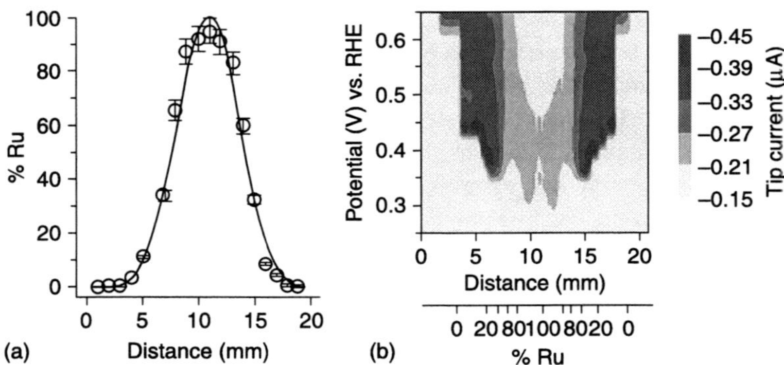

13.7 Composition and reactivity maps of Pt_xRu_y gradient. (a) Composition map as a function of distance along the surface as determined by X-ray energy dispersive spectroscopy (EDS). (b) SECM reactivity map of the CO-coated Pt_xRu_y gradient at a tip–substrate separation of 10 µm in a nitrogen-purged aqueous solution containing 0.01 M H_2SO_4 and 0.1 M Na_2SO_4. This image was constructed by combining SECM line scan data acquired at several different substrate potentials. (Source: Jayaraman and Hillier, 2004. Copyright 2004 American Chemical Society.)

Not only gradients in catalyst composition have been investigated by Hillier and coworkers. The group also employed gradients in the applied potential to quickly obtain a large data set (Jayaraman *et al.*, 2006). A linear gradient in potential was applied to a Pt layer patterned in thin bands and deposited onto an ITO substrate with a suitable electrical resistance. The gradient could be achieved by applying two different potentials at both edges of the ITO substrate using a bi-potentiostat. By applying a suitable potential gradient, potential-dependent phenomena in hydrogen oxidation on a CO-covered Pt electrode could be resolved with SECM in a single experiment. Regions of CO poisoning, hydrogen oxidation and Pt oxide formation could be distinguished.

The same approach of potential gradients has been used by Jayaraman and Hillier (2001) to prepare gradients in the amount of deposited catalysts. Again, SECM in the feedback mode was employed to monitor the local hydrogen oxidation activity in relation to the amount of deposited catalyst.

13.3.4 Other approaches and reactions

Processes at an electrode surface such as hydrogen adsorption/desorption, oxide formation/reduction, etc. are often associated with local pH changes. If these pH changes are comparably large, which can be easily obtained by generating a very small gap between the positioned SECM tip and the sample simultaneously avoiding a too-high buffer capacity, SECM may be able to resolve them, e.g. amperometrically by continuously reducing protons at the tip or using a positioned pH-sensitive microelectrode. For example, Yang and Denuault (1996a, 1996b) used proton reduction at the SECM tip to monitor hydrogen adsorption/desorption and oxide formation/reduction at polycrystalline Pt electrodes.

Besides hydrogen, other fuels are also considered for PEM fuel cells such as methanol, ethanol or formic acid, which may be employed in so-called direct fuel cells. While it might be difficult to devise a generator-collector or feedback experiment in which a tip can differentiate between the fuel and its oxidation products, the oxidation of organic fuels is typically associated with the liberation of protons. Shah and Hillier took advantage of this effect and demonstrated the applicability of SECM to monitor local activity for methanol oxidation (Shah and Hillier, 2000). The protons liberated during methanol oxidation can, in a SG/TC approach, diffuse to the tip, where they can be reduced to hydrogen. The reduction current at the tip is thus a measure for the local methanol oxidation activity of the substrate. Care, however, must be taken in analyzing the results, since changes in local pH, positive feedback due to generated hydrogen, as well as substrate poisoning by intermediates may influence the observed currents (Shah and Hillier, 2000).

13.4 Towards the characterization of fuel cell electrodes with SECM

The characterization of real-world fuel cell electrodes with SECM constitutes a particular challenge. The topography of such electrodes is often far from flat, and ill-defined porosity, e.g. in gas diffusion electrodes for fuel cell applications, provides diffusion paths for species formed or consumed at the substrate, which may limit the collection efficiency or resolution. Furthermore, for example in the feedback mode, not only the outermost layer of the porous electrode, but probably the whole porous layer may contribute to the generation of species, which are back converted at the tip. If the porous layer is inhomogeneous in thickness, then the tip current may not only depend on the local chemical reactivity and tip-to-sample distance, but also on the thickness of the layer, complicating a qualitative understanding of the observed phenomena and making a sound mathematical treatment extremely demanding. However, due to the practical importance of such structures, efforts are undertaken to apply SECM to real-world fuel cell electrodes for obtaining a deeper understanding of electrode processes, e.g. transport and degradation processes, on a mesoscopic level.

A first step towards more applied problems was made in the above-mentioned studies using supported catalysts or polyaniline-incorporated Pt nanoparticles. A further step has been taken by Nicholson *et al.* (2009) using model fuel cell electrodes, which were fabricated by spin-coating several layers of a carbon-supported Pt catalyst with different Pt loadings onto highly ordered pyrolytic graphite (HOPG). The H_2/H^+ redox couple in the feedback mode was employed for the determination of the lateral activity distribution, while topographic effects have been monitored using ferrocenemethanol (FcMeOH) in the feedback mode converting the height of the feedback current into topographic information. FcMeOH turned out to be insensitive towards the surface chemical composition (Pt vs. C). Thus, a deconvolution of topographic effects and activity distribution was suggested. Using this approach the authors were able to identify areas where Pt-rich particles agglomerated, thus demonstrating an inhomogeneous film composition, and could show differences in the behaviour of their two catalyst materials. However, the authors critically discussed the above mentioned challenges in SECM measurements over porous structures. Moreover, despite being used more frequently, a determination of the topography using the feedback mode and then normalizing GC experiments seems to be a risky approach especially if the tip or sample reaction is changing, e.g. the pH value or the concentration of reactants within the gap between the tip and sample. In this case, the assumption that the generator-collector experiment is not modulated by the gap size does not hold and the assumed normalization using a feedback-mode based topography image may be misleading.

The effect of Nafion on the oxygen reduction has been studied on recessed electrodes, where the cavity has been filled with a mixture of Pt/C catalyst and

Nafion (Kishi *et al.*, 2010). The characterization of practical direct methanol fuel cell (DMFC) anodes prepared by brushing or airbrushing of PtRu catalysts onto diffusion layers has also been attempted using the SG/TC mode and proton detection at the tip. Differences in activity distribution could be related to differences in the elemental distribution (Wang *et al.*, 2006).

Besides electrodes, proton-conducting membranes have also been studied using SECM (Kallio *et al.*, 2003), a subject, however, outside the scope of this chapter.

13.5   Future trends

As shown above, SECM has become an indispensible and continuously emerging tool in high-throughput electrocatalyst testing having the advantage that, as well as being able to compare local catalytic activity, quantitative kinetic information may also be obtained in a single experiment. However, the application of SECM to practical fuel cell electrodes is only in its infancy and much work has to be done in this field in the future. Besides the application to practical electrodes, future trends in SECM will include combination of methods to provide an even more complete picture of the specimen under investigation, including SECM in combination with ATR-IR (Wang *et al.*, 2010) or other spectroscopy techniques.

One well-established combination is AFM and STM with SECM. An early demonstration of such a combination to investigate single fuel cell catalyst particles has been published by Kucernak *et al.* (2000). Such studies might, in the future, help to increase our understanding of electrocatalytic processes at the micro and nano scales. Recently, AFM has been combined with alternating current SECM (Eckhard *et al.*, 2007) providing a basis for a more widespread application of topography-independent experiments in fuel cell catalyst research. With the development of specific detection modes and improved availability of SECM instrumentation including commercialized systems, SECM will undoubtedly find its place in catalyst development, high-throughput catalyst comparison, and in the investigation of local properties of fuel cell electrodes including post-mortem analysis of catalyst layers after prolonged fuel cell tests for gaining knowledge about degradation mechanism on a local scale.

13.6   References

Ahmed S, Ji S, Petrik L and Linkov V M (2004), 'Scanning electrochemical microscopic study of hydrogen oxidation and evolution at electrochemically deposited Pt nanoparticulate electrode incorporated in polyaniline', *Anal Sci*, **20**, 1283–1287. DOI:10.2116/analsci.20.1283.

Anemiya S, Bard A J, Fan F-R F, Mirkin M V and Unwin P R (2008), 'Scanning electrochemical microscopy', *Annu Rev Anal Chem*, **1**, 95–131. DOI: 10.1146/annurev.anchem.1.031207.112938.

Bard A J, Fan F-R F, Kwak J and Lev O, (1989), 'Scanning electrochemical microscopy. Introduction and principles', *Anal Chem*, **61**, 132–138. DOI: 10.1021/ac00177a011.

Bard A J and Mirkin M V (2001), Eds., *Scanning Electrochemical Microscopy*, Marcel Dekker, New York, Basel.

Bauermann L P, Schuhmann W and Schulte A (2004), 'An advanced biological scanning electrochemical microscope (Bio-SECM) for studying individual living cells', *Phys Chem Chem Phys*, 6, 4003–4008. DOI: 10.1039/B405233A.

Black M, Cooper J and McGinn P (2004), 'Scanning electrochemical microscope characterization of thin film Pt–Ru alloys for fuel cell applications', *Chem Eng Sci*, 59, 4839–4845. DOI:10.1016/j.ces.2004.08.047.

Black M, Cooper J and McGinn P (2005), 'Scanning electrochemical microscope characterization of thin film combinatorial libraries for fuel cell electrode applications', *Meas Sci Technol*, 16, 174–182. DOI:10.1088/0957–0233/16/1/023.

Chen X, Eckhard K, Zhou M, Bron M and Schuhmann W (2009), 'Electrocatalytic activity of spots of electrodeposited noble-metal catalysts on carbon nanotubes modified glassy carbon', *Anal Chem*, 81, 7597–7603. DOI: 10.1021/ac900937k.

Cooper J S, Zhang G and McGinn P J (2005), 'Plasma sputtering system for deposition of thin film combinatorial libraries', *Rev Sci Instr*, 76, 062221. DOI:10.1063/1.1921552.

Cornut R and Lefrou C (2007a), 'A unified new analytical approximation for negative feedback currents with a microdisk SECM tip', *J Electroanal Chem*, 608, 59–66. DOI: 10.1016/j.jelechem.2007.05.007.

Cornut R and Lefrou C (2007b), 'New analytical approximations for negative feedback currents with a microdisk SECM tip', *J Electroanal Chem*, 604, 91–100. DOI: 10.1016/j.jelechem.2007.02.029.

Denuault G, Troise Frank M H and Peter L M (1992), 'Scanning electrochemical microscopy: potentiometric probing of ion fluxes', *Faraday Discuss*, 94, 23–35 DOI: 10.1039/FD9929400023.

Eckhard K, Chen X, Turcu F and Schuhmann W (2006), 'Redox competition mode of scanning electrochemical microscopy (RC-SECM) for visualisation of local catalytic activity', *Phys Chem Chem Phys*, 8, 5359–5365. DOI:10.1039/B609511A.

Eckhard K and Schuhmann W, (2007), 'Localised visualisation of O_2 consumption and H_2O_2 formation by means of SECM for the characterisation of fuel cell catalyst activity', *Electrochim Acta*, 53, 1164–1169. DOI: 10.1016/j.electacta.2007.02.028.

Eckhard K, Shin H, Mizaikoff B, Schuhmann W and Kranz C (2007), 'Alternating current (AC) impedance imaging with combined atomic force scanning electrochemical microscopy (AFM-SECM)', *Electrochem Commun*, 9, 1311–1315. DOI:10.1016/j.elecom.2007.01.027.

Engstrom R C, Weber M, Wunder D J, Burgess R and Winquist S (1986), 'Measurements within the diffusion layer using a microelectrode probe', *Anal Chem*, 58, 844–848. DOI: 10.1021/ac00295a044.

Engstrom R E, Meaney T, Tople R and Wightman R M, (1987), 'Spatiotemporal Description of the Diffusion Layer with a Microelectrode Probe', *Anal Chem*, 59, 2005–2010. DOI: 10.1021/ac00142a024.

Engstrom R E and Pharr C M (1989), 'Scanning electrochemical microscopy', *Anal Chem*, 61, 1099A–1104A. DOI: 10.1021/ac00194a002.

Etienne M, Dierkes P, Erichsen T, Schuhmann and Fritsch I (2007), 'Constant-distance mode scanning potentiometry. High resolution pH measurements in three-dimensions', *Electroanal*, 19, 318–323. DOI: 10.1002/elan.200603735.

Fernández J L and Bard A J (2003), 'Scanning electrochemical microscopy. 47. Imaging electrocatalytic activity for oxygen reduction in an acidic medium by the tip generation-substrate collection mode', *Anal Chem*, 75, 2967–2974. DOI: 10.1021/ac0340354.

Fernández J L and Bard A J (2004), 'Scanning electrochemical microscopy 50. Kinetic study of electrode reactions by the tip generation-substrate collection mode', *Anal Chem*, **76**, 2281–2289. DOI: 10.1021/ac035518a.

Fernández J L, Mano N, Heller A and Bard A J (2004), 'Optimization Of "wired" enzyme O_2-electroreduction catalyst compositions by scanning electrochemical microscopy', *Angew Chem*, **116**, 6515 –6517. DOI: 10.1002/anie.200461528.

Fernández J L, Walsh D A and Bard A J (2005a), 'Thermodynamic guidelines for the design of bimetallic catalysts for oxygen electroreduction and rapid screening by scanning electrochemical microscopy. M-Co (M: Pd, Ag, Au)', *J Am Chem Soc*, **127**, 357–365. DOI: 10.1021/ja0449729.

Fernández J L, Hurth C and Bard A J (2005b), 'Scanning electrochemical microscopy #54. Application to the study of heterogeneous catalytic reactions – hydrogen peroxide decomposition', *J Phys Chem B*, **109**, 9532–9539 DOI: 10.1021/jp050340c.

Fernández J L, Raghuveer V, Manthiram A and Bard A J (2005c), 'Pd-Ti and Pd-Co-Au electrocatalysts as a replacement for platinum for oxygen reduction in proton exchange membrane fuel cells', *J Am Chem Soc*, **127**, 13100–13101. DOI: 10.1021/ja0534710.

Fernández J L, White J M, Sun Y, Tang W, Henkelman G *et al.* (2006), 'Characterization and theory of electrocatalysts based on scanning electrochemical microscopy screening methods', *Langmuir*, **22**, 10426–10431. DOI: 10.1021/la061164h.

Fulian Q, Fisher A C and Denuault G (1999), 'Applications of the boundary element method in electrochemistry: Scanning electrochemical microscopy', *J Phys Chem B*, **103**, 4387–4392. DOI: 10.1021/jp983732a.

Gray N J and Unwin P R (2000), 'Simple procedure for the fabrication of silver/silver chloride potentiometric electrodes with micrometre and smaller dimensions: application to scanning electrochemical microscopy', *Analyst*, **125**, 889–893. DOI: 10.1039/B001659O.

Hengstenberg A, Kranz C and Schuhmann W (2000), 'Facilitated tip-positioning and applications of non-electrode tips in scanning electrochemical microscopy using a shear force based constant-distance mode', *Chem Eur J*, **6**, 1547–1554. DOI: 10.1002/ (SICI)1521–3765(20000502).

Horrocks B R, Schmidtke D, Heller A and Bard A J (1993a), 'Scanning electrochemical microscopy. 24. Enzyme ultramicroelectrodes for the measurement of hydrogen peroxide at surfaces', *Anal Chem*, **65**, 3605–3614. DOI: 10.1021/ac00072a013.

Horrocks B R, Mirkin M V, Pierce D T, Bard A J, Nagy G *et al.* (1993b), 'Scanning electrochemical microscopy. 19. Ion-selective potentiometric microscopy', *Anal Chem*, **65**, 1213–1224. DOI: 10.1021/ac00057a019.

Jambunathan K, Shah B C, Hudson J L and Hillier A C (2001), 'Scanning electrochemical microscopy of hydrogen electro-oxidation. Rate constant measurements and carbon monoxide poisoning on platinum', *J Electroanal Chem*, **500**, 279–289. DOI:10.1016/S0022–0728(00)00344–2.

Jambunathan K and Hillier A C (2002), 'Scanning electrochemical microscopy of hydrogen electro-oxidation: Part II. Coverage and potential dependence of platinum deactivation by carbon monoxide', *J Electroanal Chem*, **524–525**, 144–156. DOI:10.1016/S0022–0728(02)00675–7.

Jayaraman S and Hillier A C (2001), 'Construction and reactivity mapping of a platinum catalyst gradient using the scanning electrochemical microscope', *Langmuir*, **17**, 7857–7864. DOI: 10.1021/la010930b.

Jayaraman S and Hillier A C (2003), 'Screening the reactivity of Pt_xRu_y and $Pt_xRu_yMo_z$ catalysts toward the hydrogen oxidation reaction with the scanning electrochemical microscope', *J Phys Chem B*, **107**, 5221–5230. DOI: 10.1021/jp0274886.

Jayaraman S and Hillier A C (2004), 'Construction and reactivity screening of a surface composition gradient for combinatorial discovery of electro-oxidation catalysts', *J Comb Chem*, **6**, 27–31. DOI: 10.1021/cc034030k.

Jayaraman S, May E L and Hillier A C (2006), 'Scanning electrochemical mapping of spatially localized electrochemical reactions induced by surface potential gradients', *Langmuir*, **22**, 10322–10328. DOI: 10.1021/la0607048.

Kallio T, Slevin C, Sundholm G, Holmlund P and Kontturi K (2003), 'Proton transport in radiation-grafted membranes for fuel cells as detected by SECM', *Electrochem Commun*, **5**, 561–565. DOI:10.1016/S1388–2481(03)00117–6.

Karnicka K, Eckhard K, Guschin D A, Stoica L, Kulesza P J *et al.* (2007), 'Visualisation of the local bio-electrocatalytic activity in biofuel cell cathodes by means of redox competition scanning electrochemical microscopy (RC-SECM)', *Electrochem Commun*, **9**, 1998–2002. (DOI:10.1016/j.elecom.2007.05.015.

Katemann B B and Schuhmann W (2002), 'Fabrication and characterization of needle-type Pt-disk nanoelectrodes', *Electroanal*, **14**, 22–28. DOI: 10.1002/1521–4109(200201)14:1 <22::AID-ELAN22>3.0.CO;2-F).

Katemann B B, Schulte A and Schuhmann W (2004), 'Constant-distance mode scanning electrochemical microscopy. Part II: High-resolution secm imaging employing Pt nanoelectrodes as miniaturized scanning probes', *Electroanal*, **16**, 60–65. DOI: 10.1002/elan.200302918.

Kishi A, Inoue M and Umeda M (2010), 'Scanning electrochemical microscopy study of H_2O_2 byproduct during O_2 reduction at Pt/C-Nafion composite cathode', *J Phys Chem C*, **114**, 1110–1116. DOI: 10.1021/jp909010q.

Kucernak A R, Chowdhury P B, Wilde C P, Kelsall G H, Zhu Y Y *et al.* (2000), 'Scanning electrochemical microscopy of a fuel-cell electrocatalyst deposited onto highly oriented pyrolytic graphite', *Electrochim Acta*, **45**, 4483–4491. DOI:10.1016/S0013–4686(00)00504–1.

Kundu S, Nagaiah T C, Xia W, Wang Y, van Dommele S *et al.* (2009), 'Electrocatalytic activity and stability of nitrogen-containing carbon nanotubes in the oxygen reduction reaction', *J Phys Chem C*, **113**, 14302–14310. DOI: 10.1021/jp811320d.

Kwak J and Bard A J (1989), 'Scanning electrochemical microscopy. Theory of the feedback mode', *Anal Chem*, **61**, 1221–1227. DOI: 10.1021/ac00186a009.

Lee J, Ye H, Pan S and Bard A J (2008), 'Screening of photocatalysts by scanning electrochemical microscopy', *Anal Chem*, **80**, 7445–7450. DOI: 10.1021/ac801142g.

Lefrou C (2006), 'A unified new analytical approximation for positive feedback currents with a microdisk SECM tip', *J Electroanal Chem*, **592**, 103–112. DOI: 10.1016/j.jelechem.2006.05.003.

Liu B and Bard A J (2002), 'Scanning electrochemical microscopy. 45. Study of the kinetics of oxygen reduction on platinum with potential programming of the tip', *J Phys Chem B*, **106**, 12801–12806. DOI: 10.1021/jp026824f.

Lu G, Cooper J S and McGinn P J (2006), 'SECM characterization of Pt–Ru–WC and Pt–Ru–Co ternary thin film combinatorial libraries as anode electrocatalysts for PEMFC', *J Power Sources*, **161**, 106–114. DOI:10.1016/j.jpowsour.2006.04.089.

Lu G, Cooper J S, and McGinn P J (2007), 'SECM imaging of electrocatalytic activity for oxygen reduction reaction on thin film materials', *Electrochim Acta*, **52**, 5172–5181. DOI:10.1016/j.electacta.2007.02.022.

Martin R D and Unwin P R (1997), 'Scanning electrochemical microscopy – theory and experiment for the positive feedback mode with unequal diffusion coefficients of the redox mediator couple', *J Electroanal Chem*, **439**, 123–136. DOI: 10.1016/S0022–0728(97)00377-X.

Martin R D and Unwin P R (1998), 'Theory and experiment for the substrate generation/tip collection mode of the scanning electrochemical microscope: application as an approach for measuring the diffusion coefficient ratio of a redox couple', *Anal Chem*, **70**, 276–284. DOI: 10.1021/ac970681p.

Mirkin M V (1999), 'High resolution studies of heterogeneous processes with the scanning electrochemical microscope', *Microchim Acta*, 130, 127–153. DOI: 10.1007/BF01244921.

Nagaiah T C, Maljusch A, Chen X, Bron M and Schuhmann W (2009), 'Visualization of the local catalytic activity of electrodeposited Pt–Ag catalysts for oxygen reduction by means of SECM', *ChemPhysChem*, **10**, 2711–2718. DOI: 10.1002/cphc.200900496.

Nicholson P G, Zhou S, Hinds G, Wain A J and Turnbull A (2009), 'Electrocatalytic activity mapping of model fuel cell catalyst films using scanning electrochemical microscopy', *Electrochim Acta*, **54**, 4525–4533. DOI:10.1016/j.electacta.2009.03.021.

Nogala W, Burchardt M, Opallo M, Rogalski J and Wittstock G (2008), 'Scanning electrochemical microscopy study of laccase within a sol–gel processed silicate film', *Bioelectrochem*, **72**, 174–182. DOI:10.1016/j.bioelechem.2008.01.010.

Okunola A O, Nagaiah T C, Chen X, Eckhard K, Schuhmann W *et al.*, (2009), 'Visualization of local electrocatalytic activity of metalloporphyrins towards oxygen reduction by means of redox competition scanning electrochemical microscopy (RC-SECM)', *Electrochim Acta*, **54**, 4971–4978. DOI: 10.1016/j.electacta.2009.02.047.

Pust S E, Maier W and Wittstock G (2008), 'Investigation of localized catalytic and electrocatalytic processes and corrosion reactions with scanning electrochemical microscopy (SECM)', *Z Phys Chem*, **222**, 1463–1517. DOI: 10.1524/zpch.2008.5426.

Sánchez-Sánchez C M, Rodríguez-López J and Bard A J (2008), 'Scanning electrochemical microscopy. 60. Quantitative calibration of the SECM substrate generation/tip collection mode and its use for the study of the oxygen reduction mechanism', *Anal Chem*, **80**, 3254–3260. DOI: 10.1021/ac702453n.

Sánchez-Sánchez C M and Bard A J (2009), 'Hydrogen peroxide production in the oxygen reduction reaction at different electrocatalysts as quantified by scanning electrochemical microscopy', *Anal Chem*, **81**, 8094–8100. DOI: 10.1021/ac901291v.

Sánchez-Sánchez C M, Solla-Gullón J, Vidal-Iglesias F J, Aldaz A, Montiel V *et al.* (2010), 'Imaging structure sensitive catalysis on different shape-controlled platinum nanoparticles', *J Am Chem Soc*, **132**, 5622–5624. DOI:10.1021/ja100922h.

Schwamborn S, Stoica L, Chen X, Xia W, Kundu S *et al.* (2010), 'Patterned CNT arrays for the evaluation of oxygen reduction activity by SECM', *ChemPhysChem* **11**, 74–78. DOI: 10.1002/cphc.200900744.

Shah B C and Hillier A C (2000), 'Imaging the reactivity of electro-oxidation catalysts with the scanning electrochemical microscope', *Journal of the Electrochemical Society*, **147**, 3043–3048. DOI: 10.1149/1.1393645.

Shen Y, Träuble M and Wittstock G (2008a), 'Detection of hydrogen peroxide produced during electrochemical oxygen reduction using scanning electrochemical microscopy', *Anal Chem*, **80**, 750–759. DOI: 10.1021/ac0711889.

Shen Y, Träuble M and Wittstock G (2008b), 'Electrodeposited noble metal particles in polyelectrolyte multilayer matrix as electrocatalyst for oxygen reduction studied using SECM', *PhysChemChemPhys*, **10**, 3635–3644. DOI: 10.1039/B802688B.

Walsh D A, Fernández J L, and Bard A J (2006), 'Rapid screening of bimetallic electrocatalysts for oxygen reduction in acidic media by scanning electrochemical microscopy', *J Electrochem Soc*, **153**, E99–E103. DOI: 10.1149/1.2186208.

Wang Z, Liu Y and Linkov V M (2006), 'The influence of catalyst layer morphology on the electrochemical performance of DMFC anode', *J Power Sources*, **160**, 326–333. DOI:10.1016/j.jpowsour.2006.01.056.

Wang L, Kowalik J, Mizaikoff B and Kranz C (2010), 'Combining scanning electrochemical microscopy with infrared attenuated total reflection spectroscopy for *in situ* studies of electrochemically induced processes', *Anal Chem*, **82**, 3139–3145. DOI: 10.1021/ac9027802.

Weng Y-C, and Hsieh C-T (2011), 'Scanning electrochemical microscopy characterization of bimetallic Pt–M (M = Pd, Ru, Ir) catalysts for hydrogen oxidation', *Electrochim Acta*, **56**, 1932–1940. DOI:10.1016/j.electacta.2010.12.029.

Wittstock G, Burchardt M, Pust S E, Shen Y and Zhao C (2007), 'Scanning electrochemical microscopy for direct imaging of reaction rates', *Angew Chem Int Ed*, **46**, 1584–1617. DOI: 10.1002/anie.200602750.

Yang Y-F and Denuault G (1996a), 'Scanning electrochemical microscopy (SECM) study of pH changes at Pt electrode surfaces in Na_2So_4 solution (pH 4) under potential cycling conditions', *J Chem Soc, Faraday Trans*, **92**, 3791–3798. DOI: 10.1039/FT9969203791.

Yang Y-F and Denuault G, (1996b), 'Scanning electrochemical microscopy (SECM) study of the adsorption and desorption of hydrogen on platinum electrodes in Na_2SO_4 solution (pH = 7)', *J Electroanal Chem*, **418**, 99–107. DOI: 10.1016/S0022–0728(96)04768–7.

Zhang C, Fan F-R F and Bard A J (2009), 'Electrochemistry of oxygen in concentrated NaOH solutions: solubility, diffusion coefficients, and superoxide formation', *J Am Chem Soc*, **131**, 177–181. DOI: 10.1021/ja8064254.

Zhou J, Zu Y and Bard A J (2000), 'Scanning electrochemical microscopy: Part 39. The proton/hydrogen mediator system and its application to the study of the electrocatalysis of hydrogen oxidation' *J Electroanal Chem*, **491**, 22–29. DOI:10.1016/S0022–0728(00)00100–5.

Zoski C G, Aguilar J C, and Bard A J (2003), 'Scanning electrochemical microscopy. 46. Shielding effects on reversible and quasireversible reactions', *Anal Chem*, **75**, 2959–2966. DOI: 10.1021/ac034011x.

Zoski C G (2003), 'Scanning electrochemical microscopy: Investigation of hydrogen oxidation at polycrystalline noble metal electrodes', *J Phys Chem B*, **107**, 6401–6405. DOI: 10.1021/jp027436g.

14

Laser-optical methods for transport studies in low temperature fuel cells

R. LINDKEN and S. BURGMANN, Zentrum für
Brennstoffzellen Technik ZBT GmbH, Germany

Abstract: The flow distribution and structure in a polymer electrolyte membrane fuel cell (PEMFC) or a direct methanol fuel cell (DMFC) strongly affects the performance of the cell. The main functions of the micro-channel flow field are the homogeneous distribution of the reactants, an enhancement of an effective mixing and the expulsion of the reaction products such as liquid water. To properly design the flow field of a cell and the manifold of a stack, detailed experimental velocity measurements are essential. Non-intrusive Laser-optical measurement techniques such as particle-image velocimetry, micro-particle image velocimetry, laser Doppler velocimetry, and molecular-tagging velocimetry qualify for such investigations and are more and more used in fuel cell research. The basic measurement principles of these measurement techniques are explained and the relevant applications in fuel cell research are depicted and critically reviewed. A forecast to the future use of these measurement techniques in fuel cell research is risked and additional sources of information concerning the measurement techniques and fluid mechanical problems are presented.

Key words: particle-image velocimetry, micro-particle image velocimetry, laser Doppler velocimetry, molecular-tagging velocimetry, flow distribution, mixing, two-phase flow, convection, diffusion.

14.1 Introduction

The performance of fuel cells is inherently governed by fluid mechanics. Since the structure of a polymer electrolyte membrane fuel cell (PEMFC) and direct methanol fuel cell (DMFC) are very similar, they both encounter similar fluid mechanical problems. The reactants hydrogen or, in the case of the DMFC, methanol and water, on the anode side, and either oxygen or air on the cathode side are fed to the reaction site by channel structures on the bipolar plates, the so-called flow field. Several requirements must be fulfilled by a properly designed flow field: The reactants must be uniformly distributed, such that the electrochemical reactions, and therefore the current density, are equal over the reactive area of a fuel cell. Mixing across the channel width must be enhanced to reduce the influence of depletion zones. Reactants must be forced into, and reaction products, namely water or carbon dioxide, must be removed from the porous electrodes of the membrane electrode assembly (MEA). Finally, water droplets (PEMFC cathode) or carbon dioxide bubbles (DMFC anode) that may

425

appear and accumulate in the flow channels must be efficiently removed, as they hinder or even block the reactant supply.

This introduction summarizes the characteristics of flows in fuel cells and the most common problems encountered during the fluid dynamical optimization of fuel cells according to the above-mentioned criteria.

The characteristics of the flow in a PEMFC or a DMFC depends on the performance of the electro-chemical reaction, i.e. the normally achieved current densities. A typical current density of a PEMFC is in the order of 0.5 A/cm^2, resulting in volumetric flow rates in the order of ~1 l min^{-1} cm^{-2}. In contrast, a DMFC achieves typical current densities in the order of 0.1 A/cm^2, resulting in volumetric flow rates in the order of ~10 μl min^{-1} cm^{-2}. Taking these values and using a flow field plate with, for example, an area of 50 cm^2 and six parallel channels with a cross section of 1 mm^2 each, we can estimate the relevant dimensionless numbers. The Reynolds number is the ratio of convective to diffusive momentum transport. The Reynolds number can be expressed as the ratio of inertial forces $F_i = \rho\, u^2 d^2$ and viscous forces $F_v = \mu\, u\, d$ (ρ is the fluid density, u is the characteristic velocity, d is the characteristic length, μ is the dynamic fluid viscosity):

$$Re = \frac{\rho\, u\, d}{\mu} \qquad [14.1]$$

In the case of PEMFCs, typical Reynolds numbers are in the order of $Re = 100$; in the case of DMFCs the Reynolds numbers are much lower, generally in the order of $Re = 1$, indicating for both cases that the flow is laminar. Two conclusions arise from these low Reynolds numbers. Firstly, the pressure loss over the channel length is relatively high due to the wall friction. Secondly, turbulence does not exist, so that transverse mixing within the channel occurs by diffusion only. In the manifolds that are supplying the cells of a stack with fuel, the flow conditions can vary widely, depending on the number of cells in the stack, their active area and the diameter of the manifold. Conditions in the manifold can be laminar or turbulent.

Since fuel is consumed only on the electrode side of a flow channel, the fuel concentration decreases on that side. If diffusion is not sufficiently fast, and no measures are taken, a depletion zone along the gas diffusion layer (GDL) may be generated. Looking at the ratio of convective to diffusive mass transport, described by the Péclet number

$$Pe = \frac{u\, d}{D} \qquad [14.2]$$

where u is the characteristic velocity, d is the characteristic length, D is the diffusion coefficient, we find that in the case of the cathode side of PEMFCs the Péclet number is $Pe = 100$, so that diffusion within the channel is sufficiently fast, mostly due to the fact that oxygen easily diffuses through air. On the anode side of DMFCs, the Péclet number is higher by an order of magnitude, being in the order of $Pe = 2000$. As reasoned above, mixing within the channel flow will be governed by diffusion alone, which is comparatively slow. Estimating the diffusion time of a methanol molecule across the water-filled channel in the DMFC, we get $t = x^2/2D \approx 10^{-6}/2 \cdot 5 \cdot 10^{-10} = 1000$ s. Statistically, on average, molecules in the channel at the opposite side of the gas diffusion layer will not reach the reaction sides before exiting the fuel cell. Large quantities of fuel are wasted if no measures are taken to enhance the transverse mixing. For the fuel supply in PEMFCs with a hydrogen self-diffusion constant of the order of 10^{-4} this effect can be ignored.

If droplets and bubbles appear in the channels, the Weber number also has to be taken into account. The Weber number is defined as the ratio of inertial force $F_i = \rho\, u^2 d^2$ to surface tension force $F_\sigma = \sigma\, d$

$$We = \frac{\rho\, u^2 d}{\sigma} \qquad\qquad [14.3]$$

where ρ is the fluid density, u is the characteristic velocity, d is the characteristic length, σ is the surface tension. Concerning this dimensionless parameter, the two-phase flows of the two types of fuel cells exhibit differences. In the case of PEMFC cathodes, the Weber number is in the order of $We = 100$, meaning that inertia has a certain influence, and droplets can be transported by the gaseous main flow, even if they adhere to the wall. In contrast, in DMFC anodes the Weber number is in the order of $We = 10^{-3}$. Inertia is too weak to efficiently remove bubbles if they adhere to the channel wall or the gas diffusion layer. Hence, additional effects have to be induced, e.g. capillary effects, to ensure that the carbon dioxide bubbles produced by the fuel cell are removed.

Supplying the fuel cells stack or, more precisely, a number of fuel cells, with reactants at a first glance seems an easy enough task. But evenly distributing a given volumetric flow rate across several tens of parallel cells, and again splitting that flow into several parallel channels is not trivial. Often this problem is treated in analytical flow network analyses. However, numerical and/or experimental validation is often inevitable, since it is not the channel length that has to be equal or every single possible route the flow can take, but the pressure drop.

As mentioned above, the distribution of the reactants within the channel geometry, including the GDL, is one of the fluid mechanical problems of fuel cells. Effective mixing may be a key feature of a high-performance fuel cell. However, mixing in small channel geometries is still one of the major fluid mechanical challenges. Mixing can be enhanced by secondary fluid motion, e.g. by Dean vortices, evolving due to channel curvature or other vortices generated by micro-structures at the channel wall. Additionally, the thickness of the depletion zone close to the GDL may be reduced simply by tapering the channels. Closely related to mixing effects to redistribute the reactants is the flow in and over the porous GDL of a fuel cell. The transport processes within the GDL and into the GDL are the focus of ongoing numerical and experimental investigations. In general, the flow through a porous medium is described by Darcy's law, whereas for the flow over a porous medium the Brinckman equation is frequently used:

$$\frac{\mu\, U}{k} = -\nabla p + \mu' \nabla^2 U \qquad\qquad [14.4]$$

where μ is the fluid viscosity, U is the surface velocity, k is Darcy's permeability, p is the pressure and μ' is the Brinckman viscosity. The empirical parameters k and μ' have to be determined for each porous material and fluid and combinations of these. Applying fluid mechanical measurement techniques, e.g. laser-optical techniques, enable the measurement of the surface velocity U. The flow through a porous medium becomes even more complex if two different fluids are involved. As mentioned above, water droplets may appear on the cathode or, in DMFCs, carbon dioxide bubbles may appear on the anode. Hence, the behaviour of a two-phase fluid is a major fluid mechanical problem of fuel cells, not only in the porous layer but also in the channels, where droplets and bubbles, respectively, may alter and hinder the flow. In this latter case laser-optical flow measurement techniques may be successfully applied to analyse the fluid mechanical effects in detail.

In summary, in fuel cells highly complex fluid mechanical problems arise that cannot be easily predicted. Computational fluid mechanics (CFD) requires high spatial resolution, complex models and therefore long and expensive computational time. Furthermore, empirical parameters often need to be determined prior to the simulations. Additionally, any simulation results must be validated with appropriate measurements. Experimental investigations such as laser-optical measurement techniques allow the local detection of fluid velocities in gas and in liquid flows, and can also give insight into multiphase-flow phenomena. Hence, they are necessary tools for accurate examinations in fuel cells.

14.2   Basic principles, methods and technology

In this section the basic principles of several different laser-optical measurement techniques are elucidated. Particle image velocimetry (PIV) as well as its further development at micrometer scales, i.e. micro-particle image velocimetry (μPIV), laser Doppler velocimetry (LDV) and molecular tagging velocimetry (MTV) are presented. The methods are described and the relevant technology is explained. In each case relevant applications are presented demonstrating the potential of each measurement technique. In contrast to, e.g., hotwire anemometry (HWA), these three measurement techniques have the main advantages of being non-intrusive. No sensors have to be inserted into the flow; therefore the flow field is not disturbed or altered. These laser-based measurement techniques only require an optical access to the flow field under investigation; e.g. to measure the flow in a channel, a transparent window needs to be installed for the laser illumination and the recording device. The proper modification of an existing device concerning optical access and high quality of the recorded images is one of the key features of a successful velocity measurement.

14.2.1 Particle image velocimetry

Particle image velocimetry (PIV) is a measurement technique that enables the determination of planar velocity fields. PIV can be used for qualitative flow visualization as well as for quantitative analysis of the velocity distribution of a flow. Introduced in the late 1970s, it since has become established as a standard measurement technique in fluid mechanics. This technique is based on the visualization of small tracer particles that in most cases need to be added to the flow. These tracer particles should be homogeneously distributed within the measurement region. The measurement principle is shown in Fig. 14.1.

The flow is illuminated in a plane using specific light-sheet optics and a strong light source. The illuminated tracer particles within this light-sheet plane are recorded by a camera. The recording of two consecutive images enables the determination of the particle displacement. The displacement Δs of the tracer particles in the second image (t_2) with respect to their position in the first image (t_1) can be transformed into a velocity taking into account the magnification of the recording optics and the time interval between the first and second image (Eq. 14.5).

$$v = \frac{\Delta s}{t_2 - t_1}$$

[14.5]

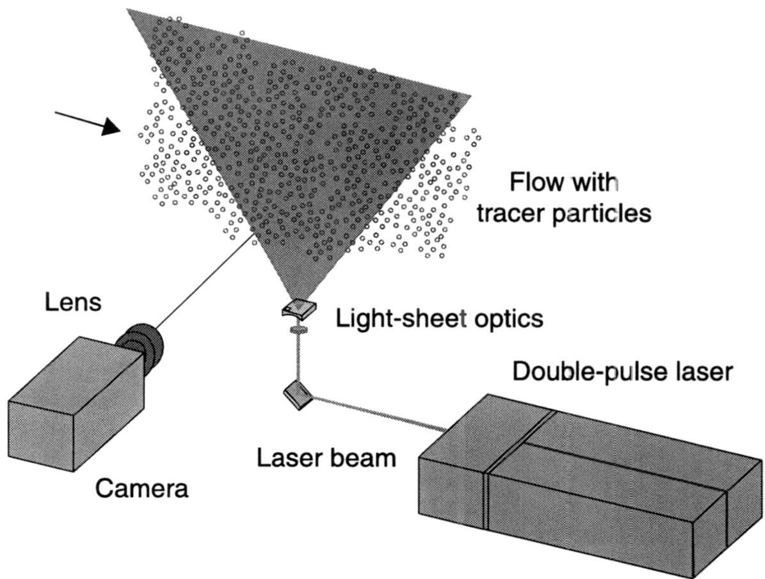

14.1 A typical PIV measurement set-up.

It should be mentioned that particle image velocimetry does not detect the specific velocity of single tracer particles. Instead, the local velocity of a small area of the measurement plane is calculated by statistical means based on the identification of particle patterns. In general, two-dimensional (2D) cross-correlation schemes are applied (Keane and Adrian 1992). The two recorded images that represent the illuminated tracer particles in the flow as a grey-value pattern are divided into small interrogation windows. The pattern of an interrogation window in the first image is correlated with a region of equal size in the second image that is shifted pixel-wise in the vicinity of the location of the interrogation window of the first image. The result of this operation is a local correlation map of a specific particle or grey-value pattern as shown in Fig. 14.2. The position of the maximum correlation value within this map is the most probable displacement of the particle pattern of this specific interrogation window. Further information concerning the cross-correlation procedure for PIV can be found in, e.g., Keane and Adrian (1990, 1991) or Raffel *et al.* (2007). Normally, a PIV analysis leads to 10^3–10^5 velocity measurements per image depending on image format and resolution. Typical measurement uncertainties are in the order of about 1% (McKeon *et al.* 2007).

Complete PIV-systems are now commercially available. These systems consist of a digital camera, a laser, and light-sheet optics. In general, charge-coupled

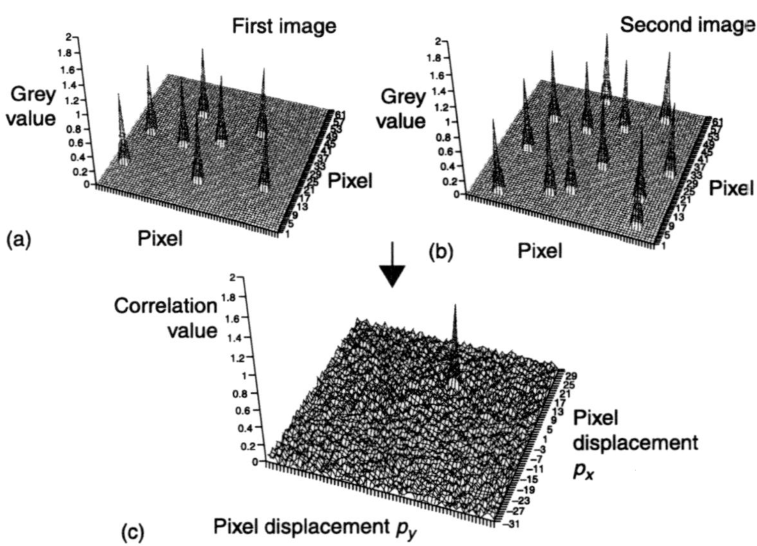

14.2 Illustration of the cross-correlation scheme. (a) First image (peaks indicate particle pattern). (b) Second image (peaks indicate particle pattern). (c) Cross-correlation of the first and second image, highest correlation value indicates the most probable displacement.

device (CCD) cameras are used that allow the recording of a double image within microseconds due to the structure of the CCD-chip. Complementary metal-oxide-semiconductor (CMOS) cameras are currently being introduced. The preferred light sources are pulsed lasers, as they provide sufficiently high energy densities per pulse to illuminate the particles in the flow. The laser pulses are in the order of several nanoseconds. In most cases frequency-doubled Nd:YAG or Nd:YLF lasers emitting green light are used. The particles that are added to the flow need to follow the flow. Furthermore, the particles need to scatter sufficient light to be detectable. Both properties depend on the particle diameter. Smaller particles are likely to follow the flow but the intensity of the scattered light is decreased. The light scattering occurs in the so-called Mie regime. The light intensity depends on the particle diameter, the wavelength of the light and the refraction index ratio between the particle material and the fluid. Typically, for gas flows, liquid particles of a size of a few micrometers are applied, whereas in liquid flows solid particles of tens of micrometers are used. The choice of suitable particles for a specific PIV application is one of the main challenges for a successful experiment.

Several advanced modifications of the PIV principle have been developed to gain further insight into complex flow fields. The classical PIV technique enables the determination of the in-plane velocity distribution of the measurement plane as explained above. However, in some cases the out-of-plane velocity component is of interest. This third velocity component of the velocity vector may be determined by applying a stereo configuration with two cameras (Prasad 2000). In this case complex mapping operations are necessary to be able to combine the information recorded by the two cameras. If the complete three-dimensional (3D) flow structure is of interest, a stereo-configuration can be combined with a moving light-sheet such that the flow field can be scanned in the direction perpendicular to the light-sheet plane as demonstrated by Hori and Sakakibara (2004) and Burgmann and Schröder (2008). Another method for the detection of three-dimensional flow structures has recently been developed, that uses four cameras and a volume illumination instead of a classical light-sheet technique (Elsinga *et al.* 2006). This PIV modification is called tomographic PIV (Tomo-PIV) and is based on the reconstruction of the particle position in space and three-dimensional cross-correlation. So far these volumetric PIV techniques are only applicable for macroscopic flow problems. An in-depth description of the PIV measurement technique is provided by Adrian and Westerweel (2010).

The previously described conventional PIV approach is used to investigate macroscopic flow configurations, i.e. the typical length scales of such PIV experiments are in the order of 10^{-2}–10^{0} m. In these cases the measurement plane is defined by the light-sheet plane. The light-sheet plane has a typical thickness of 0.5–1 mm. This is by far too large for the investigation of flows in microscopic devices by a conventional PIV approach.

14.2.2 Micro-particle image velocimetry

The investigation of flows in microscopic devices by PIV uses a volume illumination instead of a light-sheet plane as explained below. The characteristic length scales of such devices are in the order of 10^{-3}–10^{-7} m. The light sheets of the classical PIV approach are far too large. Hence, the micro-particle image velocimetry (μPIV) technique uses the strongly limited depth of focus of the optical system to define the measurement plane. The measurement principle is the same as for the macroscopic PIV technique, i.e. particles within the flow are illuminated, their displacement is optically detected and the velocity distribution is reconstructed by cross-correlation schemes (Santiago *et al.* 1998, Meinhart *et al.* 1999). A typical set-up of a μPIV measurement is depicted in Fig. 14.3.

As can be seen, μPIV is typically being performed using a special microscope. The laser light is coupled into the optics of the microscope such that the volume illumination and the light of the illuminated particles within the flow pass the same lens. The microscope contains a dichroic mirror to deflect the light of the tracer particles to the camera. This optical principle is called epi-fluorescence and it requires a wavelength shift between the illuminating laser light and the light originating from the tracer particles. Normally this is achieved by using special tracer particles that are coated with or contain a fluorescent dye, e.g. rhodamine-B (peak excitation wavelength λ_{ex} = 544 nm, peak emission wavelength λ_{em} = 600 nm), rhodamine-6G (λ_{ex} = 524 nm, λ_{em} = 580 nm), nile blue A (λ_{ex} = 636 nm, λ_{em} = 680 nm) or ethidium bromide (λ_{ex} = 505 nm, λ_{em} = 602 nm). As discussed in the previous paragraph, the lasers established for PIV applications

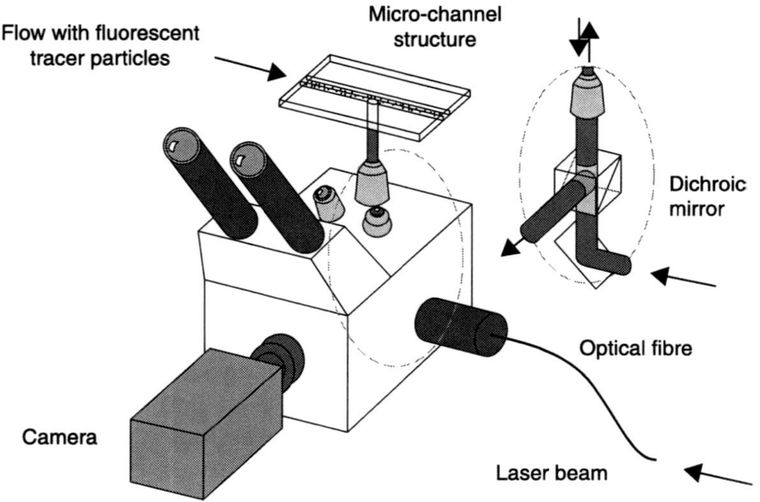

14.3 A typical μPIV measurement set-up.

use green light. The wavelengths are $\lambda_{Nd\text{-}YAG}$ = 532 nm for a Nd:YAG laser or $\lambda_{Nd\text{-}YLF}$ = 527 nm for Nd:YLF lasers. Rhodamine-B is a good choice for excitation with green laser light, since the quantum efficiency of the fluorescence is relatively high (Bindhu and Harilal 2001) and photo-bleaching is negligible. Without using fluorescent particles and a wavelength-specific optical filter inside the beam path of the microscope a high background noise would occur in the recorded images since the complete depth of the microscopic device is illuminated by the laser light. In opposition to the conventional PIV, it is not Mie scattering of the illuminated particles that is now detected in μPIV-configurations but the fluorescence of a dye attached to the particles. For fluorescent particles the brightness decreases with d_p^3 whereas for Mie scattering as in conventional PIV the particle brightness decreases with d_p^2 (McKeon *et al.* 2007). In contrast to conventional PIV the recorded particle images need post-processing before cross-correlation schemes can be applied. Although the incident laser light is filtered out, the background noise is usually still high due to fluorescent particles not in focus. These out-of-focus particles lead to background patterns that may also be displaced from the first to the second recorded image. Hence, this pattern may also lead to a correlation and may overlay and distort the detection of the particle displacement of the measurement plane. In conclusion, the depth of the measurement plane is not determined by the depth of focus of the lens system but mainly by the depth of correlation (Olsen and Adrian 2000).

The μPIV technique has also been developed toward a three-component measurement technique by applying a stereo configuration. Again, complex mapping techniques and a specially modified stereo-microscope are necessary (Lindken *et al.* 2006).

A review of the μPIV method is given by Lindken *et al.* (2009) and Wereley and Meinhart (2010).

14.2.3 Laser Doppler velocimetry

Laser Doppler velocimetry (LDV) is a frequently used point-wise velocity measurement technique. It was first introduced in the 1960s and has been further developed toward commercially available easy-to-use measurement devices. Conventional LDVs work in the dual-beam mode with two crossing laser beams. A laser beam is divided into two beams with equal properties by a beam splitter. Both beams pass through the lens system, where they are refracted such that they cross at a certain point in front of the lens. This region of the superimposing laser beams is the measurement volume. The interference of the laser beams forms an interference fringe-pattern. The distance Δx between the fringes depends on the angle ϕ between the beams and the wavelength of the laser light λ (Eq. 14.6).

$$\Delta x = \frac{\lambda}{2 \, \sin \phi}$$

[14.6]

If a particle crosses this measurement volume, i.e. the fringe pattern, a periodic intensity signal is generated by the light scattering of the particle. This intensity signal is captured by a photo detector. The so-called Doppler frequency f_D of this intensity signal detected by the receiving sensor is directly proportional to the velocity of the particle (Eq. 14.7) scaled by the fringe distance Δx.

$$f_D = \frac{2 \sin\phi}{\lambda} v \qquad [14.7]$$

The laser Doppler velocimetry allows a local time-resolved velocity measurement. The temporal resolution is only limited by the amount of particles in the flow and the intensity of the scattered light, i.e. the detection of particles. A high spatial resolution can be achieved by traversing the sensor in the flow field. Typically, the measurement volume is 1–2 mm long and 100 μm wide. The extent of the measurement volume governs the spatial resolution. However, in contrast to PIV, no planar or volumetric velocity distributions may be recorded at the same time, i.e. instantaneous velocity distributions in space cannot be measured, but mean velocity distributions may be reconstructed with a high precision. Typically, the measurement uncertainty of the velocity of a single particle crossing the fringe pattern is about 0.2%.

LDVs are commercially available in a forward or backward configuration, i.e. the receiving optics are located opposite to the laser optics or on the same side as the laser optics. For a backward configuration, usually the receiving optics are integrated into the laser optics. A typical set-up of a LDV system is depicted in Fig. 14.4.

The configuration described above allows the measurement of only the magnitude of the particle velocity. With a frequency modulation of one of the two

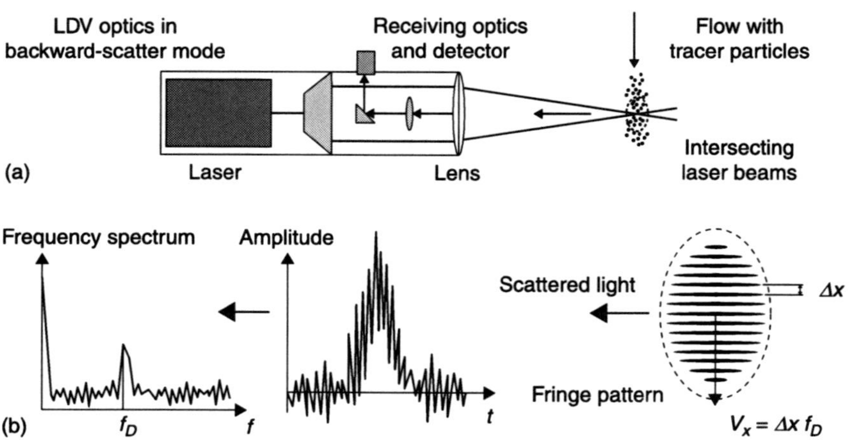

14.4 A typical LDV measurement set-up. (a) Experimental setup. (b) Analysis of results.

intersecting laser beams the direction of the flow is detected. Opto-acoustic modulators such as Bragg cells can be used such that a moving interference pattern occurs due to the slightly shifted wavelength of the modulated beam. With respect to the shift frequency f_S the Doppler frequency f_D is changed depending on the direction of the velocity vector of the particle. With two LDV systems with different wavelengths, e.g. a blue and a green laser, in combination with optical filters for the receiving optics, two overlying measurement volumes can be generated that are oriented perpendicularly to each other. Two velocity components are measured with such a configuration. The addition of a third LDV system offers the measurement of all three velocity components of the flow at one point. In comparison to PIV the LDV technique needs less laser power. Therefore an increasing number of modern small diode lasers are installed in LDV systems, promoting the miniaturization of such systems. A review of this technique can be found in Tropea (1995).

14.2.4 Molecular tagging velocimetry

Molecular tagging velocimetry (MTV) is a velocity measurement principle that is based on the optical detection of molecular tracers instead of particles (Fig. 14.5). The molecules are tagged by changing their properties by laser illumination. In the illuminated region the molecules show a temporal or permanent change of properties that are observed with temporal or local resolution (Koochesfahani 2000). Property changes that are exploited for molecular tagging methods are fluorescence (Lempert *et al.* 1995, Gendrich and Koochesfahani 1996), phosphorescence (Gendrich *et al.* 1997), photo-bleaching (Mosier *et al.* 2002), colour-change (Gharib *et al.* 2002) or temperature change (Chung *et al* 2003). Flow tagging can in principle be performed in two ways. A species for detection by laser-induced fluorescence is either locally generated by optical methods or added to the flow, e.g. by a pulsed valve. The latter approach is also referred to as molecular flow seeding velocimetry. The transport of the generated feature then allows for the determination of velocity information as well as information about mixing by diffusion and local dispersion in transport velocities.

Examples for optical generation of an observable change of properties are vibrational excitation of O_2 (Miles *et al.* 1987), photolysis of hot water forming OH (Boedecker 1989) and photolytic reactions in *tert*-butyl nitrite (Krüger *et al.* 2000), nitric dioxide (Orlemann *et al.* 1999), or N_2/O_2 mixtures (van der Laan *et al.* 2003) forming NO. To be able to detect a velocity distribution with MTV a species needs to be made observable in a spatially structured pattern, e.g. in lines perpendicular to the flow direction. While the pattern convects and diffuses, it is detected in a subsequent step via UV-laser-induced fluorescence imaging.

Alternatively, the change of properties of the molecules by sudden addition of a fluorescent species to the flow, e.g. acetone (Lozano *et al.*, 1992), NO (Grünefeld *et al.* 1998), or Argon (Hecht *et al.* 2010, 2011), results in a step function of

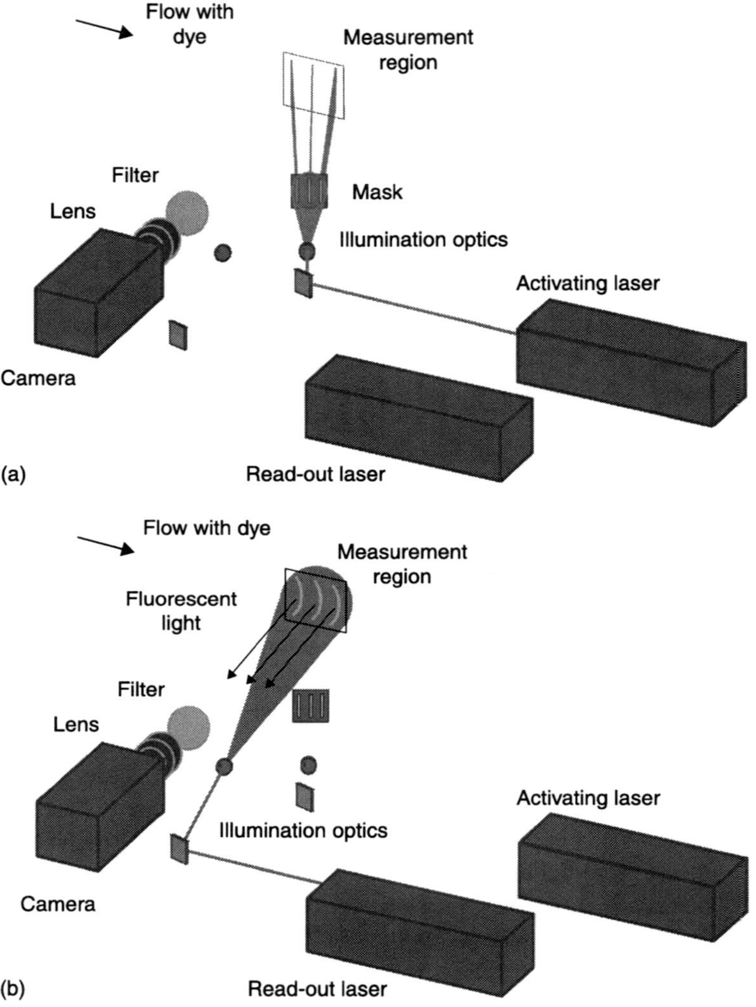

14.5 A typical MTV measurement set-up as described in Roetmann *et al.* (2008). (a) Tagging. (b) Observation.

properties and the change in the slope of the concentration increase as well as the change of the position of the step function can be used to derive velocities.

The velocity information can be retrieved by means of post-processing algorithms such as correlation (Gendrich and Koochesfahani 1996) or optical flow techniques (Garbe *et al.* 2008). MTV is not very often used in fluid mechanical research since marking the fluid with a specific pattern may be more elaborate than simply illuminating tracer particles. However, in the last decade molecular tagging velocimetry has been successfully applied for the diagnostics of micro-fluid systems, e.g. in Koochesfahani and Nocera (2001), Roetmann *et al.* (2008), and Hecht *et al.* (2011). Different tagging techniques and applications are listed in Koochesfahani and Nocera (2001) and Koochesfahani *et al.* (2000). An in-depth description of MTV is given in McKeon *et al.* (2007).

14.3 Development of techniques and application to fuel cell inspection

In this section the development of the previously presented measurement techniques toward the application in fuel cells or fuel-cell-related fluid mechanical problems is depicted. In each case relevant work is presented and critically reviewed.

Several researchers and research-groups applied laser-optical measurement techniques to understand the fluid mechanics of fuel cells or fuel-cell-like configurations in the last six years. However, it has to be stated that in most of these investigations abstractions and simplifications have been made to make these specific measurement techniques applicable. One of the main modifications is the design and construction of an optical transparent fuel cell or fuel-cell-like apparatus. Only a few attempts to directly measure the flow field in an operating fuel cell or even the gas flow have been made.

14.3.1 Development and application of particle image velocimetry to fuel cells

Martin *et al.* (2005) were interested in flow phenomena occurring at the cathode side of a serpentine channel design of a PEMFC. Their goal was improvement of fuel cell performance due to effective mixing in the channels. Due to the electrochemical reactions the oxygen distribution at the cathode side may become inhomogeneous, i.e. the oxygen concentration may decrease close to the electrodes. Effective mixing of the cathode flow may avoid locally insufficient oxygen concentrations. In this experiment a scaled-up U-shaped channel with sharp 90° corners and square cross-section was constructed. The width of the land section between the channels was equal to the channel width. The flow medium was water, applying Reynolds' analogy. The investigated Reynolds numbers ranged from $Re = 0–872$ based on the average velocity and the hydraulic diameter. These Reynolds numbers correspond to flow rates of a real fuel cell operated at low and medium current densities. The flow was investigated using a classical PIV set-up (Fig. 14.6). The flow was recorded in both a streamwise and a cross-flow plane. The authors detected Dean vortices generated by the U-turns in all investigated cases. The corresponding Dean number is larger than 30. However, for larger Reynolds numbers they found some instabilities to evolve that led to a decomposition of the Dean vortices into smaller vortices. Furthermore, the authors observed some kind of vortex shedding stemming from the separation region at

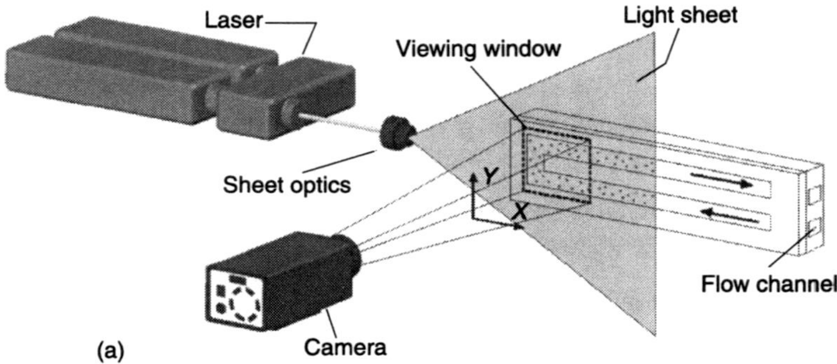

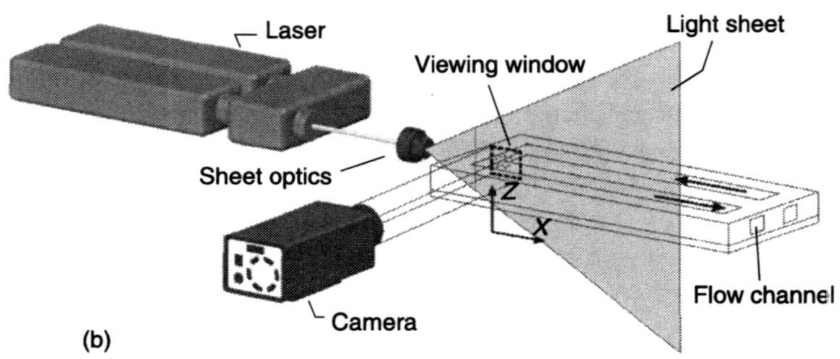

14.6 PIV measurement set-up for the investigation of the velocity distribution in a U-turn. (a) Streamwise measurement. (b) Crossflow measurement. (Source: adapted from Martin *et al.* (2005) by courtesy of Prof. N. Djiali.)

the centre part, i.e. the region of the land, of the U-turn channel. The authors found a critical Reynolds number for the onset of these instabilities of $Re = 381–436$. They deduce that due to the observed instabilities and secondary flow structures, an enhanced mixing can be achieved if the channel design leads to Reynolds numbers close to the critical number. Although the authors do not comment on details concerning their PIV system, this investigation is a good example of a simple and quick PIV measurement that highlights some fundamental fluid mechanical properties of a fuel-cell-like flow. It may directly lead to an improvement of fuel cell performance if taken into account for fuel cell design. However, the effect of the flow through the porous GDL was not investigated or taken into account.

Feser *et al.* (2007) studied the effect of the flow over a porous medium integrated as the lower wall of fuel-cell-like channels. They applied the PIV-technique to a serpentine and interdigitated flow field of the cathode side of a PEMFC to understand the effect of convective bypass into the gas diffusion layer (GDL). In this study, scaled-up channel structures and water as flow medium were again used taking Reynolds' analogy into account. The porous medium was modelled by a randomly oriented e-glass fibre layer with an in-plane permeability of 10^{-9} m^2. Such a material was chosen to cover upper values of real GDLs since the in-plane permeability scales with L^2 and the channel model is scaled up by a factor of 10. The channel-based Reynolds number was 1400 for the serpentine design and 350 for the interdigitated design. Both configurations were fed with the same total mass flow. The light-sheet plane, i.e. the measurement plane, was located to cover streamwise and crossflow planes, respectively. Although a classical PIV set-up with a light-sheet was used, 20–40 μm fluorescent particles were fed to the flow. The authors installed a longpass filter in front of the camera to eliminate background noise. The results reveal significant bypass-flow or crossflow through the GDL, indicating that for the flow under the land sections, i.e. in the GDL, convective transport dominates over molecular diffusion. For the serpentine design, the flow velocities leading to secondary flow structures due to the crossflow are of the same order of magnitude as those due to the U-turn flow, i.e. the Dean vortices. The magnitude of the secondary flow structures is about 1% of the maximum velocity of the channel flow. Again, it was shown that a relatively simple experimental configuration provides significant fluid mechanical information concerning one of the open questions of fuel cell operation, i.e. the supply of reactants to the land region and the transport of product water.

It has to be noted that in these previously described investigations water was used since particles can be easily submerged with the consequence that the effects of the electro-chemical processes cannot be measured. If, however, the reactant fluid is a liquid, as in DMFCs, the PIV technique is directly applicable to fuel-cell-like arrangements without restrictions concerning electro-chemistry. Simple PIV measurements in a single channel, operating DMFC have been performed by van de Schoot *et al.* (2008). This work will be presented below.

Concerning gas-fed fuel cells, using a PIV set-up that works with gas flows is inevitable if the electro-chemistry is of interest. One of the major challenges for the application of PIV in the gas flow is feeding particles into the small channels of the model fuel cell. Huang *et al.* (2010) used the PIV technique in the gas flow through a straight scaled-up channel of an anode supported solid oxide fuel cell (SOFC). The transparent channel was approximately 8 mm high and equipped with three different porous media at one side with a porosity varying from 0.04–0.13 (Fig. 14.7). The aim of this study was the determination of the effective viscosity μ' of the Brinkman equation (see Eq. 14.4). Using a c_w-laser instead of a double pulse laser, a CMOS high-speed camera was necessary to get suitable time intervals between the recorded images. The channel was seeded with 3 μm water droplets generated by a pressure-driven air-fed aerosol generator. The flow was investigated at $Re = 100$. The gradient of the velocity profile at the interface of the free-channel flow and the porous wall was reconstructed based on the PIV results. The effective viscosity μ' can be deduced from the slip velocity at this interface.

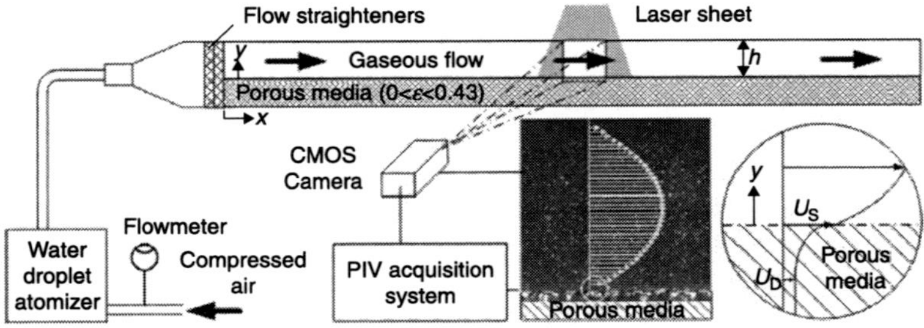

14.7 PIV measurement set-up for the investigation of the velocity profile over a porous layer. (Source: Huang *et al.* (2010).)

The authors found the effective viscosity to be unequal to the fluid viscosity, i.e. μ' is several orders of magnitude smaller than μ_f. However, it has to be stated that the velocity determination close to a surface is a major challenge to the PIV technique. Since the flow field is segmented into interrogation windows, the spatial resolution is limited. Using highly overlapping interrogation windows, as has been done in this work, may give velocity measurements that may lead to wrong results or explanations. Hence, velocity gradient calculations based on PIV results usually need higher spatial resolutions than those obtained (approximately 2.5% of the channel height). Unfortunately, the authors do not comment on particle concentrations, which is another limiting factor for the application of PIV. In general, a single interrogation window should contain at least four particles (Raffel *et al.* 2007). It is questionable whether such a high seeding can be achieved using a pressurized, air-driven aerosol generator and the low flow rates that correspond to the given Reynolds number of $Re = 100$. The problem of seeding small channel structures with particles at low flow rates will be discussed in detail below.

The flow through the single channels of a fuel cell highly depends on the flow distribution and flow structure in the manifold. In other words, effective mixing of the flow in a single channel or effective product water transport may only work for some of the channels (for a parallel, multi-serpentine or interdigitated channel design) if the flow in the manifold is not distributed in an appropriate way. Due to the low viscosity and low density of hydrogen, the Reynolds number of the anode flow of a PEMFC is so low that a homogeneous distribution and volume flow through the manifold and cells can be easily achieved. The cathode flow is more critical concerning the reactant distribution. Again, PIV may be a suitable tool to investigate the flow distribution in a cathode manifold. For example, Grega *et al.* (2007) performed PIV measurements of the flow in the manifold and in several parallel channels of a scaled-up and highly simplified fuel cell model. The aim of the study was the generation of benchmark data for numerical simulations of flow maldistributions. A fuel cell stack of 21 cells was modelled by an arrangement of 21 parallel slot-type channels with an aspect ratio of 25:1. This cell-model arrangement was connected to an inlet and an outlet manifold, each with a cross-section of 13.4 cm × 10.0 cm covering the complete slot-type cell model. The flow is guided in a Z-type configuration through the model. The complete apparatus was placed in a wind tunnel that was seeded with small 1 μm oil droplets. The inlet volume flow of the model depends on the wind tunnel speed. Inlet Reynolds numbers of $Re = 3140$, 6280 and 12 580 were investigated leading to Reynolds numbers (based on the width of the 'cells' of 0.4 cm) of the cell flow of $Re = 210$–850. A light-sheet plane PIV set-up was used to measure the flow at the centreline of the manifold over the complete length of the manifold as well as inside the cells. The authors found a highly unsteady behaviour of the flow within the manifold, including larger vortical structures, especially at the rear end of the manifold. This unsteadiness of the flow field within the manifold was found to affect the flow inside the channels, which also exhibit unsteady flow behaviour. There are several aspects to be critically reviewed in this context. The authors state the channel Reynolds number to be between 210–850. At these Reynolds numbers the flow should be fully developed midway through the channel length where the measurements were performed. However, the authors found some velocity fluctuations and ascribe that finding to a turbulent channel flow. Given the investigated Reynolds numbers, the 2D-flow in the channel (due to the channel geometry) and the measurement position, the flow will be laminar. All velocity fluctuations measured within the channel must stem from some unsteady inflow condition of the manifold. Unsteady flow structures do not indicate a turbulent flow condition under all circumstances. The authors indicate some PIV images were not taken into account during analysis, since the seeding density was poor and varied over time. This clearly hints that the inflow condition was not steady in this work. Furthermore, it has to be stated that the chosen channel geometry does not necessarily provide a pressure drop that would be similar to that of a real cell. The flow distribution within the manifold and the volume flow distribution over the cells and their channels severely depend on the pressure drop. PIV may provide a detailed impression of a flow. However, the interpretation of the detected flow field needs to take into account the design of the PIV experiment. Much care has to be taken regarding the assignment of the measurement data to real configurations and toward design rules of flow structures.

The flow within the cells, or to be more precise, within the channels of the cells, highly depends on the flow into the (cathode) manifold. The flow into the manifold depends on the inlet conditions, i.e. the tubing and the connection of the tubing to the manifold. This aspect was investigated, for example, by Lebaek *et al.* (2010) in a PIV experiment of a full-scale 70-cell stack with a flat-channel type manifold. In this case the fuel cell model with the inlet and outlet manifolds was designed in a U-type flow configuration. The cells were also modelled by slot type channels as in the work of Grega *et al.* (2007) but the pressure drop of real cells was simulated by a perforated plate mounted in the test rig. The flow into the stack model was controlled by a flow meter and seeding was provided by an aerosol generator that produced 1.55 µm paraffin oil droplets. Circular and diffuser-type inlet configurations were experimentally investigated. The circular type inlet was found to lead to the formation of a highly turbulent jet ($u_{max} \approx 150$ m/s), followed by a severe maldistribution of the flow within the manifold. Additionally, such a configuration leads to a large pressure drop in the manifold. The diffuser-type inlet revealed a smoother velocity distribution in the manifold leading to a more homogeneous flow distribution and volume flow into the channels of the model cells (Fig. 14.8). Unfortunately, the corresponding Reynolds number of the flow is not presented. The maximum velocity at the inlet of the manifold was $u_{max} \approx 30$ m/s leading to Reynolds numbers of the order of $Re = 10^4$. The documentation of fluid mechanical dimensionless numbers should be one of the key features of fluid mechanical investigation in general. Otherwise, a comparison of velocity data and flow distributions with CFD or other experimental data is not possible. Most flow distributions are highly Reynolds number-dependent, especially when geometric changes play a role. The combined PIV and CFD investigation by Lebaek *et al.* (2010) clearly shows that CFD does not necessarily provide correct results for a given configuration since the calculated and measured velocity distributions significantly deviate. However, the authors were aware of some measurement errors of the PIV investigation, e.g. they were not able to determine the velocity distribution close to the wall. This was caused by laser light reflections at the wall and the formation of a thin oil film due to the agglomeration of particles. The agglomeration of particles may be a critical aspect for PIV investigations, especially for conditions with sharp bends and high velocities. Due to inertial effects the particles may not exactly follow the flow. At some critical condition the particles may collide with the channel walls. Obviously,

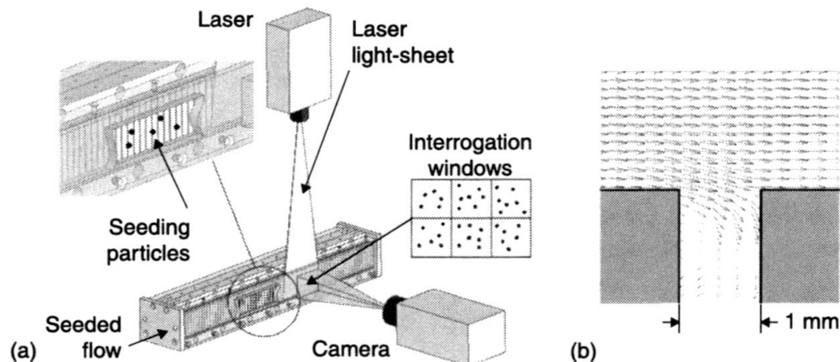

14.8 (a) PIV measurement set-up for the investigation of the velocity distribution within the manifold. (b) Flow into the single channels of a model stack. (Source: adapted from Lebaek *et al.* (2010) by courtesy of Dr J. Lebaek.)

thi happened in the measurements of Lebaek *et al.* (2010). The authors calculated the particle-related Stokes number and found it to be much smaller than 1. However, this is only a very rough estimate whether particles may follow the flow or not. This aspect will be explained in detail in the following section, since this is a key aspect of μPIV in gas-phase flows.

Another experimental investigation regarding the volume flow distribution over a fuel cell stack was performed by Klinner *et al.* (2010, 2011). Cell-to-cell variations of the volume flow may arise due to a local intrusion of the GDL into the channels or uneven pressure distributions due to manufacturing tolerances or due to the specific tubing of the inlet and exit manifolds. This is why the authors measured the velocity distribution from the exit of the cells into the exit manifold by PIV (Fig. 14.9). A specially designed tubular light-sheet optical instrument was constructed that could be introduced through a small hole into the exit manifold of the stack. The illuminated flow was recorded by a camera through a small window. Only small modifications of the real stack were necessary for the application of the PIV technique. The exit velocity distribution from each cell of

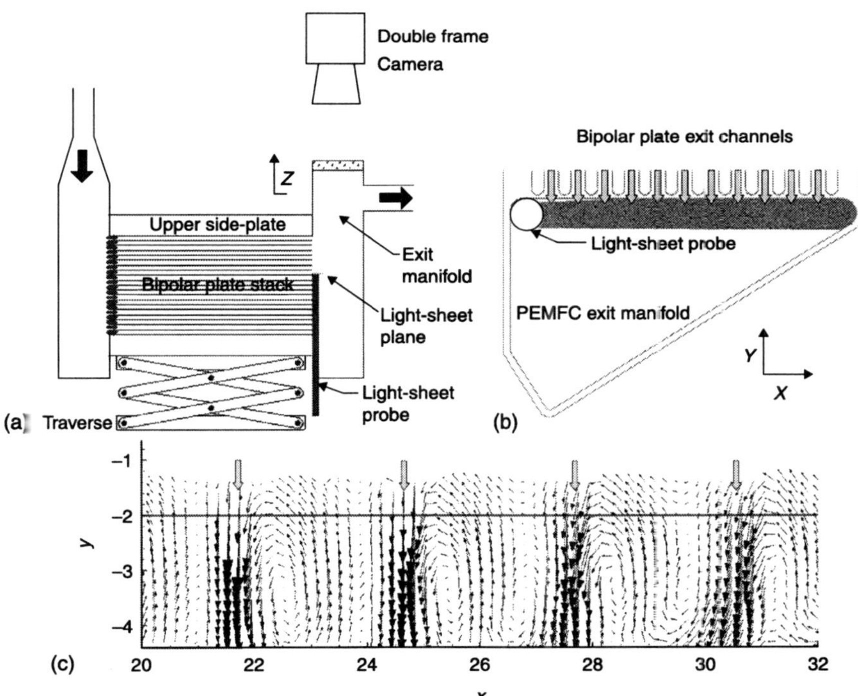

14.9 (a) PIV-set-up for the investigation of the flow distribution at the exit of the cells of a real stack. (b) Light-sheet position. (c) Results of the velocity distribution measured at the maximum flow rate. (Source: Klinner *et al.* (2010) by courtesy of J. Klinner.)

the stack was measured in 150 planes, i.e. five planes per bipolar plate, covering the complete height of the stack. The flow was seeded with small droplets of a mixture of water and 5% polyethylene glycol. Polyethylene glycol was used since this material reduces the evaporation rate of the small water droplets. A separator and a large reservoir upstream of the stack were used to remove larger droplets. The velocity distribution was measured for a flow rate of approximately 7 l/min ($Re = 10$–20 at the cell exit channels) and 210 l/min ($Re = 2000$ at the cell exit channels). In the first case the whole volume flow was led through the particle seeder, i.e. an appropriate seeding density could be reached. In the latter case, only 38% of the flow rate could be directed through the particle seeder. In this case, the particle density was rather poor and PIV post-processing was hampered. The turbulent nature of the exit flow in this case also prevented a reliable validation of the velocity results in this case. However, in the low-volume flow case laminar exit jets could be found, whereas in the high-volume flow case turbulent exit jets were identified, leading to strong neighbouring recirculation regions with an unsteady character. Especially for the high-volume flow case, water droplet agglomeration was likely to occur inside the fuel cell stack due to the high velocities. This aspect was verified by the observation of a pressure-drop increase over time that could be removed by purging the stack with clean, dry air. Again, this experimental investigation shows that particle generation and particle size are critical issues concerning PIV in the gas phase, which becomes even more severe in μPIV applications.

14.3.2 Development and application of micro-particle image velocimetry to fuel cells

The PIV technique may not only be applied in the macroscopic regime, e.g. for the investigation of the manifold flow, but also in the microscopic range, i.e. for the investigation of the flow structure within the channels of a single fuel cell. As depicted in Section 14.2.2 the PIV technique needs to be modified to be able to capture the flow in tiny structures, applying microscopes or microscope optics. In general, μPIV is performed in liquids since in this case the particles can be easily pre-mixed into the fluid and the deviation of the particle traces from the streamlines due to inertial effects is negligible. This effect is mainly based on the density ratio of the order of 1 for the particles and the surrounding fluid for μPIV in liquids alone.

An example of a fuel-cell related μPIV measurement in liquids was performed by van der Schoot *et al.* (2008). Furthermore, this investigation is one of the few concerning laser-optical measurement of an operating fuel cell. In this case, the anode-side flow of a single channel, operating DMFC was investigated. The particles were premixed to the water-methanol mixture of the anode. The challenge of this experiment was the construction of an optical access to the anode side of the fuel cell without restraining the operation of the fuel cell. Hence, the

1 mm × 1 mm channel was milled into 1 mm thick graphite material. To close the channel a 1 mm-thick glass plate was fused to the graphite material by an isolating foi . Using a commercially available membrane electrode assembly (MEA) and a graphite plate containing the cathode channel, an operational DMFC was constructed. A methanol-water mixture with 1 μm-diameter fluorescent polystyrene latex (PSL) particles was fed to the system through a heated pipe by a syringe pump. The electrical circuit was closed by several different load resistances. An inverted microscope with a magnification of 2.5 was used for these measurements. Due to the electrochemical processes CO_2 is generated on the anode side, leading to the formation of gas bubbles and a two-phase flow (Fig. 14.10). Since PIV is a laser-optical method it may serve not only to quantitatively determine the velocity distribution of a flow but can be used to visualize flow effects such as the motion of gas bubbles at the same time. Gas bubble growth and transport was investigated for different volume flow-rates of the methanol-water mixture. The bubbles were found to be continuously generated in a spatially constant location. At lower flow rates the bubbles stick to the channel wall, whereas at higher flow rates the gas bubbles are transported out of the channel. Furthermore, the velocity within the channel and the influence of the transported gas bubbles was determined based on the μPIV measurements. If bubbles are present a local increase of the velocity was measured just before

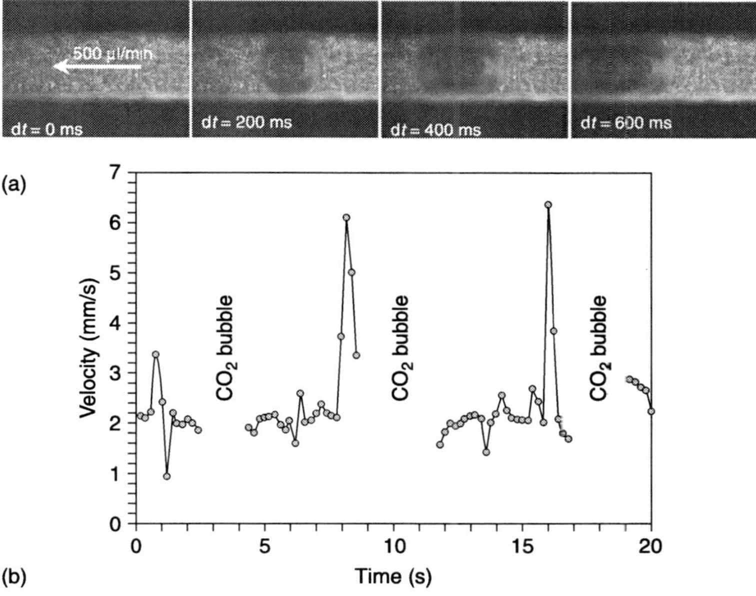

14.10 (a) Visualization of CO_2-bubble movement in the channel of an operated DMFC. (b) Corresponding velocities. (Source: adapted from van der Schoot *et al.* (2008).)

the bubble passes the observation location. This velocity increase was assumed to be related to the interacting effects of channel blockage and flow through the porous layer.

Another example of a μPIV measurement in liquids related to fuel cells is the work of Hidrovo *et al.* (2006). Their measurements focus on the formation of a two-phase flow through a typical PEMFC micro-channel configuration. Hence, a heated micro-channel was constructed equipped with several water-injection slots in line. Water is injected into the channel forming a film at the lower wall. The formation and the thickness of the water film in the micro-channel measured against the inlet flow velocity were investigated by white light and fluorescence imaging. This aspect was indicated to be important for the performance of a fuel cell since a liquid water film may reduce the effective area. Fluorescence imaging is closely related to μPIV. The authors investigated the flow around a gas bubble in water by μPIV. They were able to calculate drag forces around a bubble as well as the bubble growth and detachment regimes, thereby demonstrating the possibilities of μPIV. These aspects may help in understanding the behaviour of CO_2 bubbles in DMFCs.

However, in gas-fed PEMFCs, the flow around a water droplet surrounded by air is of major importance in fuel cell research. Minor *et al.* (2008) analysed the behaviour of a water droplet in a micro-channel by μPIV measurements. The water droplet was artificially injected into a 1 mm × 3 mm wide channel with a real GDL. The corresponding investigated Reynolds numbers were $Re = 212$ and $Re = 500$. The authors intended to mimic the appearance of a water droplet through the pores of a GDL at operating conditions in a PEMFC. This water droplet was seeded with fluorescent particles and had a diameter of approximately 0.7 mm. The measurements revealed a significant circular motion within the droplet (Fig. 14.11). Furthermore, at higher flow rates, droplet deformation including a contact-angle hysteresis was observed. The work included a CFD analysis of the

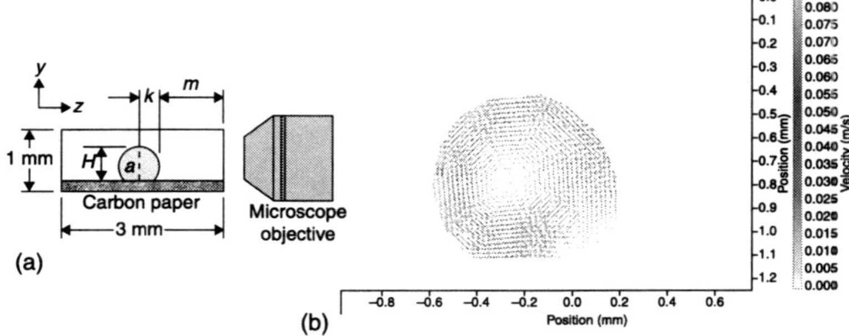

14.11 (a) Experimental set-up for the investigation of a water droplet in a PEM fuel-cell-like micro-channel using μPIV. (b) Result of the measured velocity distribution within the bubble at $Re = 212$. (Source: Minor *et al.* (2008) by courtesy of Prof. N. Djiali.)

droplet dynamics on hydrophobic surfaces taking the experimental results into account. Since transport of liquid water is still one of the main topics concerning the design and operation procedures of PEMFCs, this work demonstrates the manifold possibilities using μPIV for the analysis of liquid droplet behaviour.

μPIV measurements in the gas phase still need to be established. For gas flows the inertial effects on the particles play a larger role since the density ratio of particle and fluid is usually of the order of 10^3. One of the first attempts to apply μPIV measurements in the gas phase in fuel-cell-like micro-channels was performed by Sugii and Okamoto (2006). An optically transparent parallel-channel design was constructed. This model was placed on a simple microscope with a magnification of 10. The channel was illuminated from the side by laser light (Fig. 14.12). Nitrogen was fed to the fuel cell model. Reynolds numbers of 26 and 130 were investigated in the parallel channels of the transparent model. Nitrogen was used since pressurized nitrogen was applied to generate the tracer particles. A custom-made atomizer was used to generate an aerosol of olive oil droplets. The size of the droplets was adjusted by heating the olive oil, i.e. changing the viscosity, and changing the operational pressure of the atomizer.

To perform μPIV measurements it is usually inevitable that fluorescent particles must be generated. This was done by mixing the fluorescent dye nile red into the olive oil (Sugii 2011). The particle size was determined to be 0.5–2 μm. Since the Reynolds number is relatively low and the flow is only investigated in the straight channel part of the fuel cell model the particles were supposed to follow the flow in this region. However, the achieved seeding density was very poor, due to the fact that the volume flow for appropriate working conditions of the atomizer was much higher than that needed to feed the fuel cell model. Hence, only a part of the total volume flow was used. As mentioned before, the appropriate seeding of the flow is one of the main challenges for μPIV in micro-channel gas flows. The authors state that only 10–20 particles could be recorded in a single image. Hence, cross-correlation PIV algorithms cannot be directly applied in such a case. The

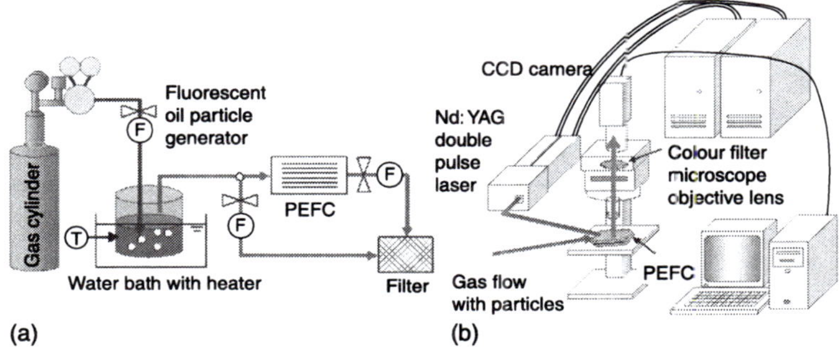

14.12 Investigation of the flow within a parallel-channel PEM-type model. (a) Generation of fluorescent oil particles (b) μPIV set-up. (Source: adapted from Sugii and Okamoto (2006) by courtesy of Prof. Y. Sugii.)

authors chose to apply particle tracking velocimetry (PTV) methods. This method can be regarded as a kind of low-image-density PIV. This method only works if the distance between the particles is much larger than the particle displacement. In this case, particle pairs can be easily identified and their displacement can be detected. However, only very limited information about the velocities can be derived from such measurements. If the flow field does not contain a severe unsteadiness, then the correlation averaging method can be applied (Meinhart *et al.* 2000) instead of PTV methods. This method adds up the correlation results of several PIV double images. If no discrete particle displacement is detected by the cross-correlation evaluation, random noise dominates the correlation result. However, the correlation values in this case are several orders of magnitude smaller than those for a detected particle displacement. For example, summing-up a lot of noise-based correlation results with only a few correlation results due to a real particle displacement will lead to a significant and explicit correlation peak for a given interrogation window. Hence, even with a poor seeding density μPIV measurements are possible. The only prerequisites are a sufficient number of double images and a flow with steady flow behaviour. This method must not be confused with a simple averaging of local velocity vectors over time. The correlation of an interrogation window with no or poor particle images will lead to a spurious vector of random magnitude. Averaging over these vectors with only some 'correct' velocity vectors over time will lead to a false result.

Although Sugii and Okamoto (2006) did not apply the correlation averaging technique and hence the detected velocity distribution of their measurement may be improved, their work provides important information and methods concerning the application of μPIV in gas flows. However, the choice of the particle material is another key mechanism for the successful application of μPIV in operating fuel cells. For example, Sugii and Okamoto (2006) in a preceding test with an operating fuel cell and seeded air observed that olive oil particles were absorbed by the porous layer and therefore the current density decreased. Furthermore, hydrophobic oil-based seeding is likely to irreversibly damage the hydrophilic membrane of a PEMFC (Klinner *et al.* 2010, 2011).

Yoon *et al.* (2006) performed μPIV measurements in a single U-turn micro-channel typical of the channel design of the cathode of a PEMFC, i.e. a similar configuration as in Martin *et al.* (2005) referred to above. Measurements were performed at $Re = 700$, again similar to the conditions reported in Martin *et al.* (2005). However, Yoon *et al.* applied the μPIV-technique in a gas flow in a full-scale fuel-cell-like channel. Establishing the μPIV in wall-bounded gas-phase flows would be the first step toward direct μPIV measurements in operating fuel cells. Hence, Yoon *et al.* tried to establish a flow condition similar to that of a real PEMFC, i.e. the air was humidified and the complete channel and tubing were heated. The micro-channel structure was placed on a standard research microscope and illuminated from the side by a laser. Two types of particles were used in these experiments: smoke particles with a size of 1 μm and water droplets of 1–8 μm.

The measurements with the smoke particles were used to demonstrate the performance of the PIV technique and to identify separation regions within the U-turn channel (Fig. 14.13). However, smoke particles may contaminate the system and are rather inappropriate for application in fuel cells. Hence, the authors applied water droplet seeding. Such a seeding may be used for μPIV investigations in operating fuel cells since no contaminating substance is introduced. In these measurements the humidity level was kept to 100% and the channel was heated to avoid image-obscuring condensation of water at the channel observation window. μPIV measurements were performed in the straight part of the channel, demonstrating the potential of this seeding configuration. However, the water droplet size was determined to be such that the particles could not adequately follow the flow. CFD simulations of the traces of particles of different size within the U-turn section at $Re = 50$ and $Re = 400$ demonstrated that for higher Reynolds numbers the particle size should not exceed 1 µm. For lower Reynolds numbers larger particles can also follow the flow. Again, the Stokes number was used to

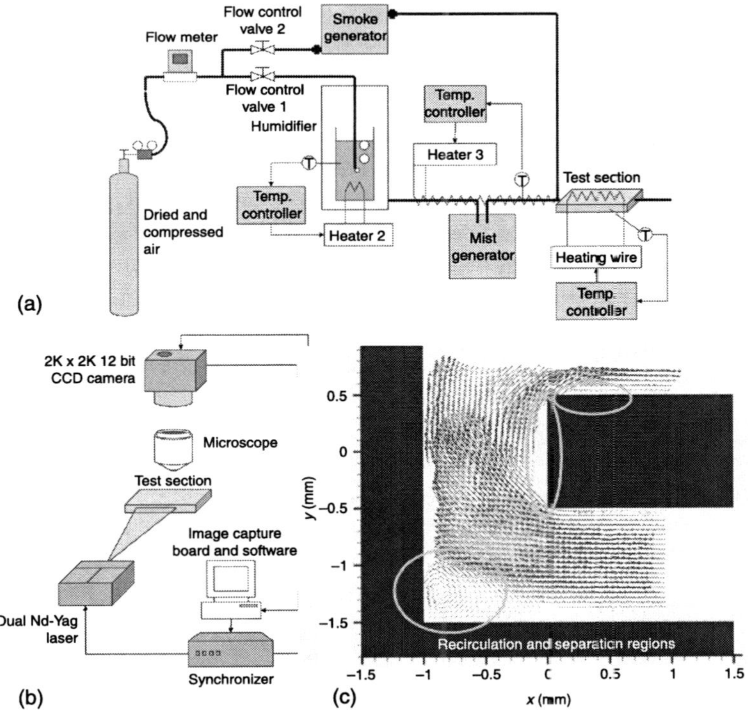

14.13 (a) μPIV concept of the particle generation for the investigation of the flow in a U-turn channel under realistic fuel-cell-like flow conditions. (b) Experimental setup. (c) Typical velocity distribution within the U-turn at $Re = 700$ measured with smoke particles. (Source: Yoon *et al.* (2006).)

assess particles' displacement and their deviation from the streamlines. It is noteworthy that in these measurements no fluorescence was used. It is believed that the particle image quality could be significantly increased by using water droplets with a fluorescent dye. Concerning the seeding with water droplets it has to be stated that water quickly evaporates due to the low vapour pressure. Particle concentration may decrease over time and channel length, as also reported by Yoon *et al.* (2006). Furthermore, water may not be used as a seeding material if high-temperature PEMFCs are to be investigated, since these are operated at approximately 180 °C.

Burgmann *et al.* (2010, 2011) reported results of μPIV measurements of the flow in a 90° bench of 1 mm width at typical fuel cell conditions. They developed μPIV techniques for application in wall-bounded micro-scale gas-phase flows. The papers concentrate on the identification of suitable tracer particle materials and the generation of particle sizes. The particles should be large enough or generate sufficient fluorescence or light intensity that they can be detected. On the other hand, larger particles may not be able to follow the flow. Hence, several tests were performed with different materials and particle seeders. Solid, fluorescence-coated PSL particles were found to be inadequate since they cannot be detected. Tests of different liquids for detectability and for the corresponding particle size distribution generated by an atomizer yielded ethylene glycol as a suitable seeding material. Rhodamine-B, the standard fluorescent dye for μPIV, can easily be dissolved in ethylene glycol, producing a high fluorescence signal. Using a pressurized-air-driven atomizer a mean particle size of approximately 1.4 μm could be generated. With such a seeding μPIV measurements in the gas-phase micro-channel flow at Reynolds numbers up to 138 were successfully performed. The correct mapping of the flow was confirmed by additional μPIV measurements at the same Reynolds number in water. In all measurements a classic μPIV set-up as described in Section 14.2.2 was used. Since an atomizer was used to generate the particles, only a part of the volume flow of the seeder could be used to feed the micro-channel, as already discussed for the experiments of Sugii and Okamoto (2006). Therefore, the particle density was relatively poor. However, the correlation averaging technique was successfully applied. Concerning the ability of the particles to follow the flow the authors found a simple rule to determine the particle drift based on the differential equation of the particle motion. Using this differential equation the relaxation time t of a particle within a moving fluid and the corresponding drift length $s_{p,\tau}$ can be determined (Eq. 14.8).

$$\tau = \frac{(\rho_p - \rho)d_p^2 C_c}{18\eta} \quad \text{and} \quad s_{p,\tau} = \tau\, U_{charact} \qquad [14.8]$$

where ρ_p denotes the density of the particle, ρ is the density of the surrounding fluid, d_p is the particle diameter, C_c is the Cunningham correction factor, and η is the viscosity of the fluid. With a characteristic flow velocity $U_{charact}$ the particle drift is determined.

Using a characteristic length scale the particle drift can be calculated. This has been demonstrated for the particle drift due to centrifugal forces in a 90° bench comparing the result of the analytical calculation and CFD simulations (Burgmann *et al.* 2010). However, since the particles may only be generated with a certain particle size distribution some particle agglomeration of larger particles at the channel wall was observed. Hence, a particle generation principle needs to be used that produces more or less monodisperse particles of a size of approximately 1 μm. Additionally, for the investigation of an operating fuel cell a different tracer material should be used on the cathode side since ethylene gylcol can be easily oxidised at low voltages, i.e. it would work as if fuel was injected on the cathode side.

14.3.3 Development and application of laser Doppler velocimetry to fuel cells

Although laser Doppler velocimetry may provide a detailed view on velocity distributions and temporal devolutions it has not been widely used in fuel cell research. One of the reasons might be the necessary quality of the particle signal. Windows of high optical quality need to be installed in a fuel cell or fuel cell model. Furthermore, the fix angle of the intersecting laser beams may require a relatively large optical access to the flow structure that cannot be provided in all cases.

However, a first attempt to use LDV in fuel cell research was made by Ma *et al.* (2002). The work deals with a network flow analysis and its application to the calculation of flow distribution within fuel cells. The calculated flow distributions were verified by LDV measurements. A design of 21 parallel channels connected by common inlet and outlet manifold was used. The authors applied LDV to measure the velocities in the channels and calculated the flow rates. Unfortunately, the authors do not provide any information concerning the specific LDV system, the measurement performance, or the temporal or spatial resolution. The authors state that the LDV results are strongly affected by measurement errors due to the placement of the laser beams, reflections on the channel surface, bubbles inside the channel, and seeding density. However, they calculated volume flow rates that show similar trends as the measured flow rates.

The second and, to our knowledge, last attempt to apply LDV in fuel cells was performed by Kucernak and co-workers (Kucernak *et al.* 2005, Brett *et al.* 2007). In their experiment a single channel, U-turn type PEMFC was constructed from transparent material and equipped with a Nafion membrane, and with Toray carbon paper for the GDL. The fuel cell was investigated during operation. The channels had a cross-section of 2 mm × 2 mm. The flow was investigated at the cathode side at Reynolds number $Re = 80$. A shifted LDV system in forward

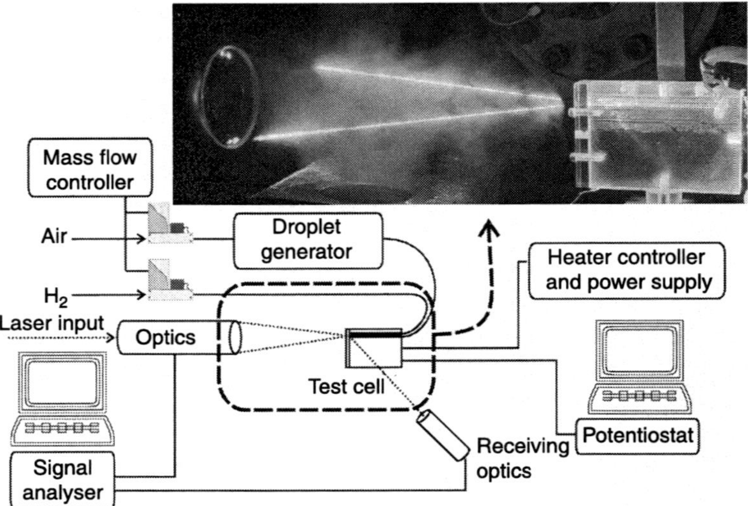

14.14 Measurement set-up for the investigation of the flow in a U-turn channel of an operating PEM fuel-cell with laser Dopper velocimetry. (Source: Kucernak *et al.* (2005) by courtesy of Prof. A. Kucernak.)

scattering mode was used. The receiving optics were placed opposite to the transmitting optics at an angle of 45° (Fig. 14.14). Water droplets of 2–5 μm were used. At each measurement point data were collected and averaged over 5000–10 000 particle detections. The authors were able to detect recirculation regions within the U-turn section. They found a symmetrical vortex pair downstream of the U-turn. Measurements close to the GDL surface revealed the surface-normal velocities to be several orders of magnitude higher than those that can be expected due to pure diffusion. Although this experiment is one of the few laser-optical measurements of an operating fuel cell, the results need to be regarded as preliminary. This work can be regarded as a kind of feasibility study.

14.3.4 Development and application of molecular tagging velocimetry to fuel cells

To our knowledge there is only one publication concerning the application of molecular tagging velocimetry to determine velocity distributions in fuel cells. In Hecht *et al.* (2010, 2011) the authors present a feasibility study. The flow in a typical multiple serpentine channel design of a PEMFC was investigated. The original bipolar-half plate of a PEMFC was covered and sealed by a glass plate. A 100 μm spacer was added between the glass cover and the channel structure to roughly resemble the ability of the flow to cross between the channels through the porous layer. In this case molecular tagging velocimetry was performed by a laser induced fluorescence of nitrogen-monoxide. Argon was used as the continuous fluid medium, and a mixture of argon and nitrogen monoxide was added to the flow for a short period by quickly opening and closing a valve. Using whole-field illumination the flow was recorded several times and the fluorescence signal was averaged. Local variations of the fluorescence signal due to background noise and inhomogeneities in the illumination were corrected. A reference case with channels completely filled with the argon-nitrogen monoxide mixture was recorded. The authors were able to determine relative concentrations. For a qualitative comparison of the detected flow pattern, similar experiments were performed in water at the same Reynolds number of $Re \approx 600$. In this case Rhodamine-B was dissolved in water and served as a fluorescence marker. In both measurements a similar cross-flow over the land between the channels was observed (Fig. 14.15). The velocity of the fluorescence front in the channel was measured and was found to linearly decline over the channel length. Diffusion constants were calculated based on the increase of the intensity signal. A Péclet number calculation for the flow in the channels and the cross-flow over the land ($Pe_{cross} = 400$ and $Pe_{channel} = 10\,000$) led to the assumption that the flow through the porous layer is convection driven. However, it has to be stated that since the porous layer was only roughly modelled by the gap, the transport phenomena through the GDL are not realistically reproduced in these experiments. Nevertheless, this work and measurement technique offers additional insight to

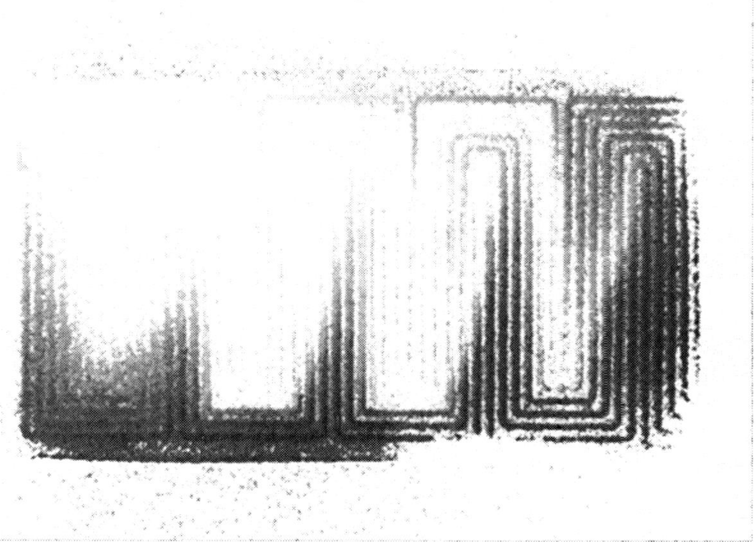

14.15 Image of the nitrogen-monoxide distribution of an MTV measurement performed in a typical multi-serpentine channel design PEM fuel cell. (Source: Hecht et al. (2010) by courtesy of C. Hecht.)

the transport phenomena in fuel cells, since diffusion transport mechanisms can be analysed. In general, these mechanisms cannot be analysed using particle-based measurement techniques as discussed in Section 14.2.4.

14.4   Advantages and limitations

The previous description of the measurement techniques as well as the reviewed practical applications in fuel cell-related problems show that laser-optical velocity measurement techniques have some advantages concerning the analysis of fluid mechanical problems. Common to the three methods, particle image velocimetry, laser Doppler velocimetry as well as molecular tagging velocimetry is their non-intrusive nature. No additional mechanical sensors need to be installed in the flow field that may unacceptably alter the flow. Experimental velocity measurement techniques provide new insight into fluid mechanical problems and may lead to a more detailed understanding of influences and interrelations. The combination of computational fluid mechanics and laser-optical velocity measurement techniques is a powerful tool. As the work of Lebaek *et al.* (2010) shows, CFD does not necessarily provide correct results. CFD results strongly depend on the definition of the boundary conditions. The addition of an experiment may significantly increase the relevance of a fluid mechanical investigation.

However, the depicted laser-optical measurement techniques cannot be applied under all circumstances. They all have in common that they need an optical access. The system needs major modifications or simplifications to make the specific measurement technique applicable. For example, since μPIV in the gas phase still needs to be established, microscopic flows within a fuel cell often need to be simulated by the substitution of the gases by a liquid following Reynolds' analogy.

In the following the advantages and limitations of each previously described measurement technique will be briefly discussed.

The main advantage of PIV is the ability to measure planar velocity distributions with a relatively large spatial resolution. Depending on image format and resolution up to 10^3–10^5 velocity measurements per image can be recorded at the same time. Fluid structures can be reconstructed in time and space. The measurement technique is relatively robust and may be used with only little previous knowledge if simple flow fields are to be investigated. As for all measurement techniques, complicated flow structures and configurations need an expert to perform and evaluate the PIV measurements. However, ready-to-use, complete PIV systems are commercially available. Note that since PIV is a planar optical measurement technique it may not only serve as a quantitative measurement tool but also as a mean to visualize flows, e.g. van der Schoot *et al.* (2008). Good examples for simple and quick PIV measurements that highlight some fundamental fluid mechanical properties of a fuel-cell related flow field are presented in Martin *et al.* (2005) and Feser *et al.* (2007). PIV measurements have been performed in liquid flow in an operating direct methanol fuel cell (van der Schoot *et al.* 2008). The application of PIV in gas flows of operating fuel cells is still in development. Contamination of the fuel cell, i.e. the poisoning of the catalyst, is a current research topic. The integration of an optical access to a PEMFC with high power density is a challenge, when optically transparent and electrically conducting indium tin oxide (ITO) coatings are not able to conduct the high electric charge. Additionally, laser-light reflections and particle agglomerations leading to a film at the optical transparent surface may hinder PIV measurement. The agglomeration of particles is still one of the main problems of PIV in gas-phase flows. Since the density ratio between the seeding particles and the surrounding fluid usually is in the order of 1000, inertial effects have a strong influence on particle motion. The particles need to be sufficiently small. Furthermore, the seeding material must not change the operating conditions of a fuel cell, i.e. must not block the pores of the GDL, change the electro-chemical reaction, etc. Hence, it is obvious that the added particles need to adequately follow the flow. Seeding of micro-channels with suitable small particles with a sufficiently high particle density is still an unsolved problem (Yoon *et al.* 2006, Burgmann *et al.* 2011). Note that PIV may provide a detailed impression of the velocity distribution of a flow. An estimate of the flow conditions by means of vector plots can be obtained even in cases where the necessary particle density is not reached or the image quality is poor. This is due to the robust nature of the cross-correlation processes applied in the PIV post-processing. PIV experiments need to be carefully designed to obtain reliable velocity results. The measurement error of a properly designed PIV experiment is about 1%.

Velocity measurements with higher measurement accuracy can be performed by LDV. Usually the accuracy is about 0.2%. LDV may provide pointwise velocity information with a high temporal resolution. If highly unstable flow structures are to be investigated LDV may be a better choice than PIV. Applying a temporal averaging and using a traversing mechanism a high spatial resolution can be achieved. LDV probes are commercially available from easy-to-use unshifted measurement devices to more complex shifted LDV-probes that allow the detection of the velocity direction. Furthermore, 2D and 3D-probes are also available. Kucernak *et al.* (2005) demonstrated the applicability of the LDV technique in operating fuel cells. However, the application of LDV may be complicated since reflections at interfaces and walls may hamper the detection of the scattered light of the particles. The signal quality needs to be relatively high, requiring optical transparent windows of high quality. Additionally, the fixed angle between the intersecting beams may also provide some limitations concerning the access to the flow field. As for PIV, particles need to be added to the flow. These particles need to fulfil the same requirements concerning agglomeration at walls and the ability to follow the flow. As pointed out in Yoon *et al.* (2006) and Burgmann *et al.* (2011) the ability of particles to adequately follow a gaseous flow is one of the critical issues of particle-based velocity measurement techniques. The ratio of the particle density to the fluid density is the main source of error, i.e. the deviation of the particle path from the streamline. Since the density of hydrogen is 10 000 times smaller than that of the commonly used particles there is actually no way to reliably measure the flow in hydrogen with a particle-based velocity measurement technique, although in most cases the Reynolds number is very low.

In contrast to these particle-based velocity measurement techniques, MTV allows the direct measurement of the fluid velocity. The major advantage of MTV compared to PIV and LDV is the absence of flow-tracing particles. The risk of contamination of the fuel cell is reduced. The only prerequisite is the generation of a (laser-induced) fluorescence signal by locally changing the molecular structure of the fluid or the addition of a molecular marker to the flow. MTV allows the detection of a motion due to convection and/or diffusion by the detection of pattern displacement and change of fluorescence gradients. However, the illumination of the fluid and the generation of a fluorescence pattern in the flow may be complex. For the investigation of the flow an O_2-fluorescence method based on two-photon excitation exists, but its high cost of implementation limits the application to fuel cells. The generation of a fluorescence signal may afford the addition of a specific species. For example, to our knowledge there is no method to perform molecular tagging in hydrogen. For that reason fuel cell flows are investigated with substitute gases. Nitrogen monoxide (NO) in Argon (Ar) is a suitable system with a similar diffusion. Furthermore, MTV is an integral method since the fluorescence pattern is generated over the complete depth of the flow field. The spatial resolution is low, since the velocity detection is only possible due to an intensity change and not within a fluorescing pattern. Diffusion mechanisms may smear out the fluorescence pattern over time which, for example, may hinder the determination of a locally changing velocity profile of a convection driven flow. Unfortunately, MTV systems are not commercially available. However, Hecht *et al.* (2011) demonstrated possible flow measurement by MTV in fuel cell applications.

14.5　Future trends

Since fluid mechanical problems come more and more into the focus of fuel cell research, velocity measurements are becoming essential. Non-intrusive measurement techniques such as PIV, μPIV, LDV and MTV are commonly used in fluid mechanical investigations. The design and construction of optically transparent fuel cells or partially transparent fuel cells is part of ongoing research (Rosli *et al.* 2009). In the near future the number of publications concerning laser-optical measurements in those transparent fuel cells will increase. Techniques to construct operating fuel cells with optical access are available and the measurement techniques already have been successfully applied in fuel cell research, partially even under realistic operational conditions. The classical PIV technique is developing into an established measurement technique in fuel cell research, since it s relatively easy to use and can quickly provide detailed insight into flow problems. Concerning μPIV in the gas phase some more steps need to be taken. However, the problem of seeding the flow with suitable tracers and at an adequate concentration is subject to ongoing research. Successful μPIV measurements in the gas phase allow significant new insights in fluid mechanical problems such as mixing and flow over porous media. In the liquid phase, μPIV is a widely-used technique. Flows in DMFCs can be directly measured with currently available configurations and materials. Even two-phase flows may be investigated, as it is widely done in lab-on-a-chip structures (Günther and Jensen 2006). The LDV technique will be applied here since it allows for the local control and monitoring of a flow. The flow distribution of a fuel cell stack may be monitored at the exit of the different cells using several LDV systems. Such a system would allow for the detection of local blockage effects during fuel cell operation. Since miniaturization of LDV systems is currently in the focus of LDV development (Modarress *et al.* 2010) the application of several LDVs in a restricted space at the same time will be possible. However, similarly to μPIV applications in fuel cells, suitable seeding for the application in small scales still needs to be found. The MTV technique will be a measurement principle that may cover a niche in fuel cell research. MTV allows for the determination of diffusion effects.

14.6   Sources of further information

General information concerning experimental techniques in fluid mechanics can be found in Yarin *et al.* (2007). This handbook covers the most fundamental concepts and most frequently applied measurement techniques and fluid behaviours. The basics of PIV, LDV and MTV are depicted and example results are shown.

Comprehensive reviews of the PIV technique can be found in Raffel *et al.* (2007) and Adrian and Westerweel (2010). Details of the μPIV technique and its applications are given in Wereley and Meinhart (2009) and Lindken *et al.* (2009).

Concerning the LDV technique detailed reviews are given by Durst *et al.* (1976) and Tropea (1995).

The MTV technique is explained in detail and example applications are presented in Koochesfahani (1999), Lempert and Harris (2000) and Koochesfahani (2000).

Details about the design and construction of transparent fuel cells can be found in Rosli *et al.* (2009).

Comprehensive reviews of the fluid mechanical aspects in fuel cells are lacking. Fluid mechanics are partly reviewed concerning the water management and water transportation phenomena, e.g. in Li *et al.* (2008). The review by Bazylak (2009) deals with the visualization of liquid water in PEMFCs. Kejang *et al.* (2009) give an overview about microfluidic fuel cells, also reviewing mixing aspects.

A review of fluid mechanics in micro- and minichannels is presented in Stone *et al.* (2004). Additionally, the review of multi-phase microfluidics by Günther and Jensen (2006) is recommended.

14.7   References

Adrian RJ, Westerweel J, *Particle Image Velocimetry*, Cambridge University Press, Cambridge, UK, 2010.

Bazylak A, 'Liquid water visualization in PEM fuel cells: A review', *International Journal of Hydrogen Energy*, 2009, **34**, 3845–3857.

Bindhu CV, Harilal SS, 'Effect of the excitation source on the quantum-yield measurements of rhodamine-B laser dye studied using thermal-lens technique', *Analytical Sciences*, 2001, **17**, 141–144.

Boedeker LR, 'Velocity measurements by H_2O photolysis and laser-induced fluorescence of OH', *Opt Lett*, 1989, 14, 473–475.

Brett DJL, Kucernak AR, Atkins S, Blewitt R, Brandon NP *et al.* 'Developing an Experimental Functional Map of a Polymer Electrolyte Fuel Cell', in Alemo PV (Ed.) *Progress in Fuel Cell Research*, Nova, New York, 1–53, 2007.

Burgmann S, Schröder W, 'Investigation of the vortex induced unsteadiness of a separation bubble via time-resolved and scanning PIV measurements', *Exp Fluids*, 2008, **45**, 675–691.

Burgmann S, van der Schoot N, Asbach C, Wartmann J, Lindken R, 'Analysis of tracer particle characteristics for micro-PIV in wall-bounded gas flows', *Proceedings of the 2nd European Conference on Microfluidics*, 2010, Toulouse, France.

Burgmann S, van der Schoot N, Wartmann J, Lindken R, 'Micro particle-image-velocimetry für gasströmungen in mikrokanälen', *tm – Technisches Messen*, 2011, **78**, 243–259.

Chung JW, Grigoropoulos CP, Greif R, 'Infrared thermal velocimetry for nonintrusive flow measurement in silicon microfluidic devices', *Review of Scientific Instruments* 2003, **74**, 2911–2917.

Durst F, Melling A, Whitelaw JH, *Principles and Practice of Laser-Doppler Anemometry*, Academic Press, London, 1976.

Elsinga GE, Scarano F, Wienecke B, van Oudheusen BW, 'Tomographic particle image velocimetry', *Exp Fluids*, 2006, **41**, 933–947.

Feser JP, Prasad AK, Advani SG, 'Particle image velocimetry measurements in a model proton exchange membrane fuel cell', *J. Fuel Cell Sci Technol*, 2007, **4**, 328–335.

Garbe CS, Roetmann K, Beushausen V, Jaehne B, 'An optical flow MTV based technique for measuring microfluidic flow in the presence of diffusion and Taylor dispersion', *Exp Fluids*, 2008, **44**, 439–450.

Gendrich CP, Koochesfahani MM, 'A spatial correlation technique for estimating velocity fields using molecular tagging velocimetry (MTV)', *Exp Fluids*, 1996, 22, 67–77.

Gendrich CP, Koochesfahani MM, Nocera DG, 'Molecular tagging velocimetry and other novel applications of a new phosphorescent supramolecule', *Exp Fluids*, 1997, **23**, 361–372.

Gharib M, Kremers D, Koochesfahani MM, Kemp M, 'Leonardo's vision of flow visualization', *Exp Fluids*, 2002, **33**, 219–223.

Grega L, McGarry M, Begum M, Abruzzo B, 'Flow characterization of a polymer electronic membrane fuel cell manifold and individual cells using particle image velocimetry', *J. Fuel Cell Sci Technol*, 2007, **4**, 272–279.

Grünefeld G, Graber A, Diekmann A Kruger S, Andresen P, 'Measurement system for simultaneous species densities, temperature, and velocity double-pulse measurements in turbulent hydrogen flames', *Combustion Science and Technology*, 1998, **135**, 135–152.

Günther A, Jensen KF, 'Multiphase microfluidics: from flow characteristics to chemical and materials synthesis', *Lab Chip*, 2006, **6**, 1487–1503.

Hecht C, van der Schoot N, Kronemayer H, Lindken R, Schulz C, 'Visualization of the gas flow in fuel cell bipolar plates using molecular tagging velocimetry and micro PIV', *Proceedings of the 15th Int. Symp. on Applications of Laser Techniques to Fluid Mechanics*, 5–8 July 2010, Lisbon, Portugal.

Hecht C, van der Schoot N, Kronemayer H, Wlokas I, Lindken R, Schulz C, 'Visualization of the gas flow in fuel cell bipolar plates using molecular flow seeding and micro-particle image velocimetry', *Exp Fluids*, 2011, online first, DOI 10.1007/s00348-011-1112-4.

Hidrovo CH, Kramer TA, Wang EN, Vigneron S, Steinbrenner JE *et al.* 'Two-phase microfluidics for semiconductor circuits and fuel cells', *Heat Transfer Engineering*, 2006, **27**, 53–63.

Hori T, Sakakibara J, 'High-speed scanning stereoscopic PIV for 3D vorticity measurement in liquids', *Meas Sci Technol*, 2004, **15**, 1067–1078.

Huang C, Shy S, Chien C, Lee C, 'Parametric study of anodic microstructures to cell performance of planar solid oxide fuel cell using measured porous transport properties', *J Power Sources*, 2010, **195**, 2260–2265.

Keane RD, Adrian RJ, 'Optimization of particle image velocimeters. Part I: Double pulsed systems', *Meas Sci Technol*, 1990, **1**, 1202–1215.

Keane RD, Adrian RJ, 'Optimization of particle image velocimeters. Part II: Multiple pulsed systems', *Meas Sci Technol*, 1991, **2**, 963–974.

Keane RD, Adrian RJ, 'Theory of cross-correlation analysis of PIV images', *Appl Sci Res*, 1992, **49**, 191–215.

Kjeang E, Djilali N, Sinton D, 'Microfluidic fuel cells: A review', *J Power Sources*, 2009, **186**, 353–369.

Kinner J, Willert C, Schneider A, Mack-Gardner A, 'Flow distribution measurements at the exit of bipolar plates in a PEM fuel cell stack by a scanning light sheet method', *Fachtagung 'Lasermethoden in der Strömungsmesstechnik'*, 2010, Cottbus, Germany.

Kinner J, Willert C, Schneider A, Mack-Gardner A, 'Messung der strömungsverteilung am austritt der bipolarplatten eines brennstoffzellen-stacks', *tm – Technisches Messen*, 2011, **78**, 246–252.

Koochesfahani MM, 'Molecular tagging velocimetry (MTV): Progress and applications', *AIAA paper*, 1999, 3786.

Koochesfahani M, 'Special feature: Molecular tagging velocimetry', *Meas Sci Technol*, 2000, **11**(9).

Koochesfahani MM, Nocera DG, 'Molecular tagging velocimetry maps fluid flows', *Laser Focus World*, 2001, 103–108.

Krüger S, Grünefeld G, 'Gas-phase velocity field measurements in dense sprays by laser-based flow tagging', *Appl Phys B*, 2000, **70**, 463–466.

Kucernak A, Ladewig B, Blewitt R, Shrimpton J, 'Laser Doppler anemometry study of reactant flow in fuel cell channels', *3rd European PEFC Forum*, 2005, Lucerne, Poster 21.

Lebaek J, Andreasen MB, Andresen HA, Bang M, Kaer SK, 'Particle image velocimetry and computational fluid dynamics analysis of fuel cell manifold', *J Fuel Cell Sci Technol*, 2010, **7**, 031001.

Lempert WR, Magee K, Ronney P, Gee KR, Haugland RP, 'Flow tagging velocimetry in incompressible-flow using photo-activated nonintrusive tracking of molecular-motion (PHANTOMM)', *Exp Fluids*, 1995, **18**, 249–257.

Lempert WR, Harris SR, 'Molecular tagging velocimetry', in Smits AJ, Lim TT (Eds.) *Flow visualization – techniques and applications*, Imperial College Press, London, 2000.

Li H, Tang Y, Wang Z, Shi Z, Wu S *et al.* 'A review of water flooding issues in the proton exchange membrane fuel cell', *J Power Sources*, 2008, **178**, 103–117.

Lindken R, Rossi M, Grosse S, Westerweel J, 'Micro-particle image velocimetry (μPIV): Recent developments, applications, and guidelines', *Lab Chip*, 2009, **9**, 2551–2567.

Lindken R, Westerweel J, Wieneke B, 'Stereoscopic micro particle image velocimetry', *Exp Fluids*, 2006, 41, 161–171.

Lozano A, Yip B, Hanson RK, 'Acetone: A tracer for concentration measurements in gaseous flows by planar laser-induced fluorescence', *Exp Fluids*, 1992, **13**, 369–376.

Ma Z, Jeter SM, Abdel-Khalik SI, 'Flow network analysis application in fuel cells', *J Power Sources*, 2002, **108**(1–2), 106–112.

Martin J, Oshkai P, Djilali N, 'Flow structures in a u-shaped fuel cell flow channel: quantitative visualization using PIV', *J Fuel Cell Sci Technol*, 2005, **2**, 70–80.

McKeon BJ, Comte-Bellot G, Foss JF, Westerweel J, Scarano F *et al.* 'Velocity, vorticity and Mach number', in: Yarin A, Tropea C, Foss JF (Eds.) *Handbook of Experimental Fluid Mechanics*, Springer, 83–616, 2007.

Meinhart CD, Wereley ST, Santiago JG, 'PIV measurements of a microchannel flow', *Exp Fluids*, 1999, **27**, 414–419.

Meinhart CD, Wereley ST, Santiago JG, 'A PIV algorithm for estimating time-averaged velocity fields' *Journal of Fluids Engineering–Transaction of the ASME*, 2000, **122**, 285–289.

Miles R, Cohen C, Conners J, Howard P, Huang S *et al.* 'Velocity measurements by vibrational tagging and fluorescent probing of oxygen', *Opt Lett*, 1987, **12**, 861–863.

Minor G, Zhu X, Oshkai P, Sui P, Djilali N, 'Water transport dynamics in fuel cell micro-channels', in: Kakac S, Pramuanjaroenkij A and Vasiliev L (Eds.) *Mini-Micro Fuel Cells: Fundamentals and applications*, Springer, 153–170, 2008.

Modarress D, Fourguette D, Tuagwalder F, Gharib M, Forouhar S *et al.* 'Miniature and micro-Doppler sensors', *Proceedings of the 15th Int Symp on Applications of Laser Techniques to Fluid Mechanics*, 2010, Lisbon, Portugal.

Mosier BP, Molho JI, Santiago JG, 'Photobleached-fluorescence imaging of microflows', *Exp Fluids*, 2002, **33**, 545–554.

Olsen MG, Adrian, RJ, 'Out-of-focus effects on particle image visibility and correlation in microscopic particle image velocimetry', *Exp Fluids*, 2000, **29**, 166–174.

Orlemann C, Schulz C, Wolfrum J, 'NO-flow tagging by photodissociation of NO_2: A new approach for measuring small-scale flow structures', *Chem Phys Lett*, 1999, **307**, 15–20.

Prasad AK, 'Stereoscopic particle image velocimetry', *Exp Fluids*, 2000, **29**, 103–116.

Raffel M, Willert C, Wereley ST, Kompenhans J, *Particle image velocimetry: a practical guide.* Springer, Heidelberg, 2007.

Roetmann K, Schmunk W, Garbe CS, Beushausen V, 'Micro-flow analysis by molecular tagging velocimetry and planar Raman-scattering', *Exp. Fluids*, 2008, **44**, 419–430.

Rosli MI, Pourkashanian M, Ingham DB, Ma L, Ismail MS, 'Transparent PEM fuel cells for direct visualisation experiments', *Proceedings of FuelCell2009, 7th International Fuel Cell Science, Engineering and Technology Conference*, 2009 Newport Beach, California, USA.

Santiago JG, Wereley ST, Meinhart CD, Beebe DJ, Adrian RJ, 'A particle image velocimetry system for microfluidics', *Exp Fluids*, 1998, **25**, 316–319.

Stone, HA Stroock AD, Ajdari A, 'Engineering flows in small devices: Microfluidics toward a lab-on-a-chip', *Annu Rev Fluid Mech*, 2004, **36**, 381–411.

Sugii Y, personal communication, *2011*.

Sugii Y, Okamoto K, 'Velocity measurement of gas flow using micro PIV technique in polymer electrolyte fuel cell', *Proceedings of ICNMM2006 Fourth International Conference on Nanochannels, Microchannels and Minichannels*, 2006, Limerick, Ireland.

Tropea C, 'Laser Doppler anemometry: recent developments and future challenges', *Meas Sci Technol*, 1995, **6**, 605–619.

van der Laan WPN, Tolboom RAL, Dam NJ, ter Meulen JJ, 'Molecular tagging velocimetry in the wake of an object in supersonic flow', *Exp Fluids*, 2003, **34**, 531–534.

van der Schoot N, Lindken R, Sharp KV, Peil S, Wartmann J, Westerweel J, 'PIV-messungen zur untersuchung in brennstoffzellen', *Fachtagung 'Lasermethoden in der Strömungsmesstechnik'*, 2008, Karlsruhe, Germany.

Wereley ST, Meinhart CD, 'Recent advances in micro-particle image velocimetry', *Annu Rev Fluid Mech*, 2009, **42**, 557–576.

Yarin A, Tropea C, Foss JF, *Handbook of Experimental Fluid Mechanics*, Springer, Berlin-Heidelberg, 2007.

Yoon S, Ross J, Mench M, Sharp KV, 'Gas-phase particle image velocimetry (PIV) for application to the design of fuel cell reactant flow channels', *J Power Sources*, 2006, **160**, 1017–1025.

15

Synchrotron radiography for high resolution transport and materials studies of low temperature fuel cells

C. HARTNIG, Chemetall GmbH, Germany and I. MANKE, Helmholtz Centre Berlin for Materials and Energy, Germany

Abstract: The evolution and transport of liquid water in operating low temperature polymer electrolyte membrane (PEM) fuel cells can be observed on a microscopic level by means of synchrotron radiography; this method allows for spatial resolutions down to 1 μm with time resolutions of just a few seconds. In direct methanol fuel cells (DMFCs), this method has been successfully applied to study the evolution of carbon dioxide. As an extension to radiographic imaging, synchrotron tomography provides insight into three-dimensional structures and water distributions in operating fuel cells. Application examples are shown for both PEMFCs and DMFCs; in addition, ex-situ results are provided which can be used to study three-dimensional effects of degradation and aging phenomena on fuel cell components.

Key words: PEMFC, DMFC, high resolution, in-situ method.

15.1   Introduction

Several tools are nowadays commonly applied to study the water distribution and transport in fuel cells and fuel cell components *ex situ* and *in situ*. X-ray based methods usually suffer from the fact that most of the fuel cell components such as the end plates are an optical obstacle for the incident beam due to their metallic nature. However, X-ray-based imaging methods represent interesting tools, as the achievable resolution in space and time is usually several orders of magnitude higher compared to methods as e.g. neutron-based methods as described in Chapter 9.

The following sections will start with the theoretical background of synchrotron-based imaging as well as a short introduction to different synchrotron sources. After an introduction to *ex situ* applications, prerequisites for *in situ* radiography and tomography as well as research examples are shown and discussed.

15.1.1 Theoretical background

An important prerequisite for imaging approaches are X-ray beams of high brilliance in terms of intensity and collimation; usually these beam qualities are provided by third-generation synchrotron sources. Technical details on beam sources and the technical principles can be found in Banhart[1] and Reimers *et al.*[2] X-ray imaging is based on the fact that the different materials, from

462

which fuel cell components are manufactured, have different linear attenuation coefficients μ. The value of μ is influenced by a number of effects such as the photoelectric effect, Rayleigh scattering or Compton scattering. The contribution of the different effects is a function of the energy of the incident beam. For fuel cell materials, X-ray energies in the range of 5 keV to 100 keV are normally employed. This is below the limit where electron-positron pair production (1022 keV) is possible. For energies below 100 keV, the photoelectric effect and Compton scattering are the dominating interaction mechanisms.

15.1.2 X-ray tubes and synchrotron X-rays sources

Common X-ray tubes work in the standard manner of radiation sources: accelerated electrons hit a metallic target (e.g. copper or tungsten) and produce an energy spectrum of X-rays that depends on the maximum acceleration potential, typically ranging from a few 10 keV to some 100 keV. A different kind of radiation source is found in synchrotron facilities; X-ray beams from these sources provide several orders of magnitude higher fluxes, i.e. beam intensities. These high fluxes allow for very fast measurements with a typical time resolution of down to several milliseconds and even a few microseconds in some cases.

In addition, the high flux allows the utilization of a monochromator to generate monochromatic beams, which means that only a specific energy is selected. The monochromaticity offers several advantages, especially quantification of radiographic and tomographic measurements is simplified, as only one specific attenuation coefficient has to be considered when calculating the respective images. No integration over a broad spectrum of coefficients has to be taken into account and the fluxes of the monochromatic beams are usually high enough to get excellent signal-to-noise-ratios at time resolutions of typically a few seconds to some milliseconds at concurrently high local resolutions (spatial resolution in the $x/y/z$ dimensions) of less than 1 μm.

15.1.3 *In situ* X-ray radiography

Images of the dry (water-free) cell are taken and serve as reference images. At the aimed point of operation, images of the water-containing setup are recorded. All images are corrected for two contributions, both of which might lead to errors in the final image, the so-called dark fields and flat fields. Dark fields describe the noise per pixel caused by the detector of the camera (the CCD chip) and contributions from the ambient light (which are also minor error sources) that reach the camera from the outside. These images are recorded with the radiation beam switched off. Flat fields, on the contrary, are recorded with the radiation beam switched on to identify the shape of the unperturbed beam without a sample.

Typically achieved imaging frequencies (image-to-image ratios) are in the range of 0.1–5 seconds and are in many cases only limited by the reading speed of the sensor of the camera. Shorter exposure times are required for more intense beams, which always bear the risk of local accelerated aging, as has been reported in the literature.

15.1.4 X-ray tomography

A tomographic measurement provides a three-dimensional (3D) image of the whole interior structure of a fuel cell. The final tomogram is a product of numerous individual images, where the sample is rotated around one defined axis for a defined angle, resulting in a set of images, which is the input to a reconstruction algorithm.[3]

Depending on the setup and the focus of the research, tomographic imaging can be used for a variety of questions in fuel cell research, which will be summarized below and discussed in more detail in the following sections. Synchrotron radiography has so far been almost exclusively used for *in situ* studies focusing on the observation of water transport phenomena, while synchrotron tomography is used for both *ex situ* and *in situ* applications. *Ex situ* studies take advantage of the high resolution and the complete absence of a time restriction. These studies are employed to gain insight, e.g. into water distribution in the two-phase system comprising water/carbon fiber. As a non-destructive tool (after sample preparation hardly anything is modified at the sample) the method is applied to study, e.g., degradation phenomena, for which any artificial modification should be excluded. *In situ* tomographic studies require fuel cells that have been adapted to the boundary conditions of the beam setup, taking into account the limited region of interest (the examined part has to be completely in the beam) and careful choice of the employed materials (Table. 15.1).

Table 15.1 Comparison of radiographic and tomographic imaging and achievable resolutions

	Radiography	Tomography
Typical spatial resolution	0.6–5 µm	
Ex situ		0.6–5 µm
In situ		2–10 µm
Typical time resolution[†]	0.1–5 s	
Ex situ		10–180 min
In situ		5–60 min
Addressed questions		
Ex situ		Material studies
In situ	Water transport/dynamics	3D water distribution

[†]Adequate beam intensities provided

15.2   *Ex situ* studies

In fuel cell research most of the *ex situ* studies focus on stationary conditions and systems composed of a fuel cell component and water. High time resolution is (compared to *in situ* studies, see below) not as important as local resolution. The main application of synchrotron radiation for *ex situ* studies is synchrotron tomography, which allows for the best use of the mentioned advantages. Typical fields of application are studies of water distribution in diffusion media (the commonly used term 'gas diffusion media' is replaced by 'diffusion media', because the materials considered are also employed in liquid-fed DMFCs), the composition and distribution of the micro layer or just the imaging of the fibers of diffusion media, which in turn forms the basis for simulations of the pore distribution or transport pathways for water and reactant gases.

15.2.1 Tomographic imaging of diffusion media

After sample preparation, which is required to fit the sample to beam size dimensions, this method can be considered non-destructive, allowing for an unperturbed view of different compounds in the diffusion media and the complete membrane electrode assembly (MEA) structure. Depending on the object, the setup is comparatively easy apart from the required synchrotron facility. The preparation includes the careful cutting of the sample from a larger piece. The sample must be small enough to fit in the X-ray beam, but large enough so that an unperturbed center without artifacts resulting from the preparation is obtained. In Fig. 15.1, the setup for *ex situ* synchrotron tomography is displayed. The sample with typical diameters of around 2–3 mm is glued on top of a holding device, so that in a 360° rotation all parts are within the beam area. The sample position is as close as possible to the scintillator and the lens of the camera while avoiding additional blurring artifacts due to X-ray beam refraction. The sample size has to be chosen to counterbalance the desired resolution and the camera aperture. As an example, a 2000 pixel-wide sensor and a sample size of 2 mm allow for a spatial resolution of 1 μm, which is sufficient to image structures of the microporous layer, the catalyst layer or the membrane as well as the interfaces formed by these components.[4]

Generally speaking, two main systems have been studied so far for fuel cell applications. The two-phase system composed of water and the gas diffusion media has been investigated with regard to water distribution and surface wetting. Also, artificially or accelerated aged MEA samples have been imaged to study degradation effects. Two examples are described below to highlight the capabilities of this imaging method.

In Fig. 15.2a, the tomogram taken from a system consisting of a GDL sample and water is shown. Samples of the porous material have been put into a sample holder, which is filled with the required amount of liquid water to achieve a certain filling level at a given porosity. In this case, a hydrophilic material has been

15.1 Experimental setup for synchrotron tomography; the sample (white arrow) is mounted on a screw in close proximity to the camera lens.

chosen as a proof of principle for water visualization. The different absorption coefficients allow for a clear distinction between the different phases: fibers, water, and void volume. At the same time, this image enhances the advantage of non-destructive methods, as it is impossible to gain an insight into small pores or different humidification conditions by means of any other microscopic method.

In Fig. 15.2b–d, a separated view of the different phases has been achieved. Due to the almost identical atomic composition, carbon fibers and binder material cannot be distinguished and are displayed together (b). Because of the aforementioned hydrophilic nature of the base material, the water phase is homogeneously spread across the open pore volume (c).

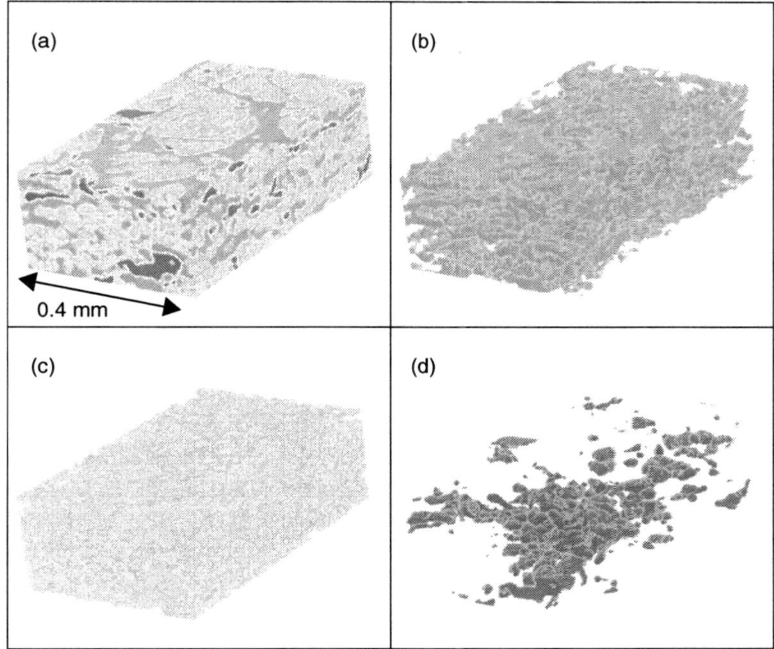

15.2 3D imaging of a water-filled GDL sample. (a) Three-phase representation. (b) Carbon fibers. (c) Water-filled pores. (d) Empty pores.

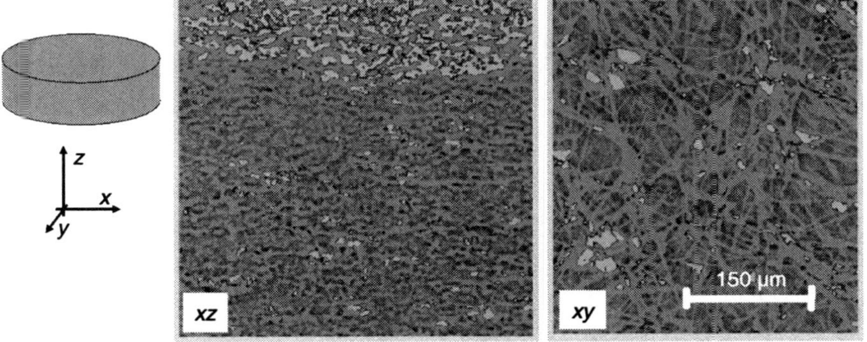

15.3 2D imaging of a water-filled GDL sample.

Based on a set of 3D information subsets, the complete tomogram can be examined by means of single slices as demonstrated in Fig. 15.3. A sectional plane in the *xy* direction reveals the paper-type structure of the fiber material, reflecting the anisotropic properties of this material. In between the lengthy fibers, liquid water is visualized as dark areas with just a few empty spots marked as bright grey blurs. A sectional plane in the *xz* direction elucidates the distribution of liquid water in more detail. Due to gravity, empty pores can be found preferentially in the upper part and liquid-filled pores are present in the lower section. In conclusion, based on the different absorption coefficients of water and carbon (fibers and binder), the structure as well as a static water distribution can be studied un-perturbed in a very detailed manner. Besides the insight into the distribution pattern, the structural information serves as a basis for a stochastic description of the pore space[5] or input for modeling tools.[6]

The experimental setup of water-containing two-phase systems can be employed to study and visualize the effect of hydrophobic treatments on the distribution of liquid water in the pores of the diffusion media. In operating fuel cells, an optimum hydrophobicity has been determined empirically; the PTFE content of the samples taken for *ex situ* studies can be varied and the size of the water cluster determined. In Fig. 15.4 slices taken from tomograms in the *xy* cross-sectional plane are displayed (compare to Fig. 15.3). The three materials employed differ in their PTFE content, decreasing from left to right, which means the left-hand sample can be considered very hydrophobic and the right-hand sample is hydrophilic, with the middle intermediate.

At a first glance, the water clusters align preferentially in the fine structure of the diffusion media, which is a result of the production process. Within the porous media, liquid water preferentially aligns in channel-like templates. The effect of the surface treatment is evident: With increasing PTFE content the total amount of liquid water inside the material decreases. At the same time, the average size of the water clusters decreases. This in turn means that, employed in a fuel cell, the intrusion rate of water is lowered, possibly leading to enhanced flooding effects in the catalyst layer. On the other hand, hydrophilic gas diffusion media tend to soak up too much water, leading to large water clusters. A magnified view of the water structure in a hydrophilic

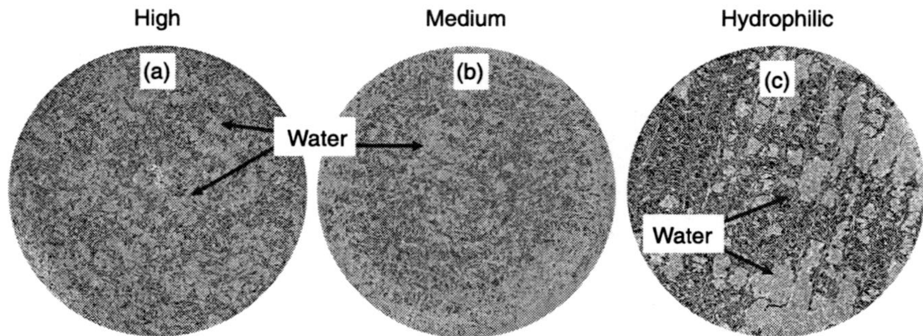

15.4 Distribution of liquid water cluster (bright grey spots, see arrows) in diffusion media with different degrees of hydrophobicity.

diffusion medium is shown in Fig. 15.5. The water clusters have a diameter of close to 1 mm; for fuel cell applications the size is large enough to block a complete cross-section of a gas channel. For applications in fuel cells, hydrophilic materials are not chosen, since flooding effects dominate the water transport behavior, and a drastic performance drop is observed once liquid water is formed.

Another example to demonstrate the advantage of synchrotron-based methods is described in Fig. 15.6. In Fig. 15.6a a tomogram of a cut-out sample taken from an operated DMFC MEA is displayed. The remarkably different absorption coefficients can be used to clearly differentiate the fibers from the woven diffusion media and the adjacent catalyst layer, which is dark-colored (the membrane has been rendered invisible).

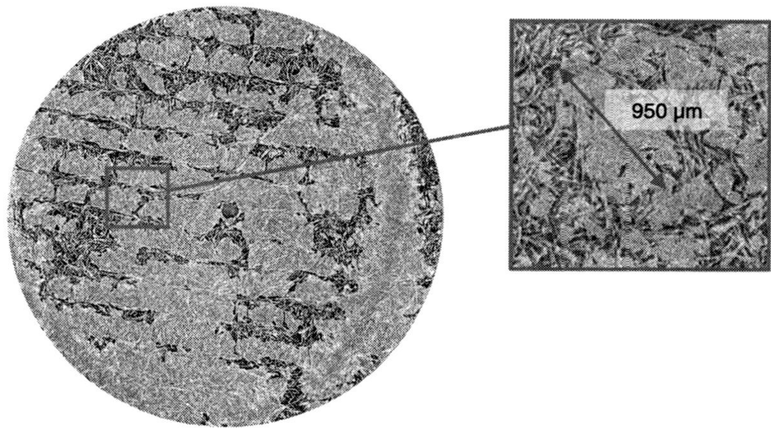

15.5 Water cluster formation in a hydrophilic GDL. The cut-out demonstrates the average size of the water clusters.

15.6 Tomogram of an operated MEA sample. (a) Complete sample. (b) Woven diffusion media turned colorless. Diameter of sample is approximately 3.5mm.[7] (Source: Henning Markötter, Helmholtz-Zentrum Berlin.)

Once the materials with a low absorption coefficient, i.e. the fibers of the diffusion media and the microporous layer, are rendered transparent, the structure of the catalyst layer can be visualized and studied in more detail (Fig. 15.6b). Enhanced crack formation after operation and repeated drying and wetting cycles can be observed; compared to a new sample these cracks appear deeper and wider. Contrary to direct imaging methods such as SEM, which require a thorough and often tedious sample preparation, the MEA sample is not impacted and any artifacts can be excluded.

15.3 *In situ* studies

Ex situ studies as described in 15.2 are usually more focused on material properties under equilibirium conditions. In situ studies require a more elaborate setup as custom made fuel cell designs have to be designed to fulfill method based requirements. At the same time, the high temporal and spatial resolution allows for investigations of dynamic processes under operating conditions.

15.3.1 General aspects

For *in situ* methods based on synchrotron radiation most commonly applied methods refer to diffraction methods such as, e.g., energy dispersive X-ray diffraction, which has been successfully applied to monitor the hydration state of the membrane.[8] In contrast to diffraction methods, synchrotron imaging allows for a direct visualization of the processes taking place in a fuel cell. Depending on the

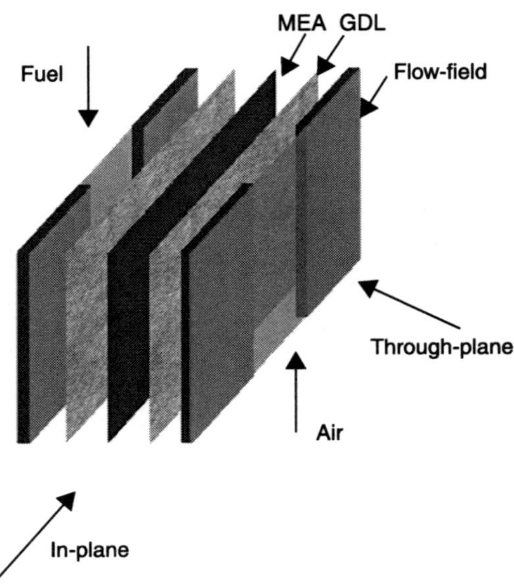

15.7 Viewing directions employed for *in situ* studies.

aspects of interest, the cell setup can be chosen so that almost no changes from state-of-the-art designs are necessary. The focus of these studies is typically shifted away from transport in flow fields, but is more on the evolution of primary liquid water clusters and their transfer from the diffusion media to the flow field channel. The transport of liquid water can also be addressed by means of modified fuel cells that include transparent parts. As these modifications lead to changed surface properties (wetting behavior, contact angle), an unperturbed insight is advantageous. In the following, two examples will be considered, differing in the direction of view (Fig. 15.7). The through-plane viewing direction gives an integral view on water distribution and transport processes across the complete MEA, whereas the in plane view allows for a distinction between water agglomerates in the different components of the fuel cell. Illustrating the importance of these different viewing options, any formation of diffusion barriers can be determined by a combination of the results obtained by viewing from different directions.

In terms of application, only minor changes have to be made to the fuel cell setup in order to achieve optimal contrast ratios. In particular, metallic end-plates have to be modified at the position of interest, and any additional material that is in the direction of the beam has to be minimized. At the same time, only negligible changes to the thermal and electrical setup at this position should be induced to the cell, to ensure realistic conditions at this position. The in-plane viewing direction requires some more preparation efforts. As the integral imaging of processes in more than one channel results in an unusable amount of information, only one channel plus lands can be effectively studied. Therefore, a single channel cell has to be constructed, which at the same time allows for realistic conditions in terms of temperature, current density and gas flow as well as gas stoichiometries.

15.3.2 Through-plane imaging

The combination of high resolution down to 1 μm and short image-to-image frequencies typically in the range of 1–4 s is an excellent prerequisite to the study of water evolution and transport processes on a microscopic level. The focus of through-plane studies reported so far has been on the transport of water from the gas diffusion media to the channel and transport in the channel.

The main advantage compared to acrylic end plates, which have previously been used to study liquid water evolution, becomes obvious when studying the origin of the liquid water. With transparent layers on the back side of the gas channels, the formation of liquid water droplets in the gas channel can be studied easily. However, the origin of the liquid water, which at the end collapses to form a droplet, still remains unclear, as regions inside the gas diffusion media and/or under the lands of the flow field remain hidden. Employing synchrotron-based methods these spots can be rendered visible. Normalizing images with respect to images taken from the water-free cell allows visualization of liquid water formation. Calculating the difference between two consecutive images helps to clearly outline the steps of the water transport process. As mentioned at the beginning of this section, special attention has to be paid to ensure sufficient contrast of the imaging procedure. Endplates and current collectors are typically made from materials such as steel (which does not necessarily influence the local thermal and electronic equilibrium) and are therefore not X-ray transparent and have to be furnished with suitable windows. The thickness of other components, such as flow field plates or additional sealing components, has to be minimized at the region of investigation. Cooling and heating of the cell requires special attention. In a standard setup, a separate flow field is engaged to temper the cell, which in a single cell setup usually requires heating. The liquid water used for this purpose has to be diverted around the studied spot in order not to add additional noise to the images in the regions where the X-ray energy is carefully selected to achieve maximum sensitivity for water.

In Fig. 15.8, a non-normalized image as directly obtained from synchrotron imaging is displayed. Here, the channels and the lands of the flow-field can be clearly distinguished and the overlaying structures of the anodic and cathodic flow-fields are obvious. The slightly tilted angle of the walls of the flow-field channels leads to the impression of different cross-sections of the channel and the land. As the information that can be derived from this raw image is not sufficient to study processes on a very small scale, normalization is usually carried out with an empty object. In the figure, the empty object is the water-free cell.

The derived image (quotient of the two images) removes the surrounding material, which appears in more or less one grey value and the shift of water droplets can be clearly identified. An example is given in Plate VI in the color section between pages 252 and 253. Two consecutive images with an image-to-image period of approximately 2 s have been chosen, where the formation of a water droplet has been observed. The quotient of these two images quite clearly reveals the processes that are taking place during these seconds. 'Water pixels',

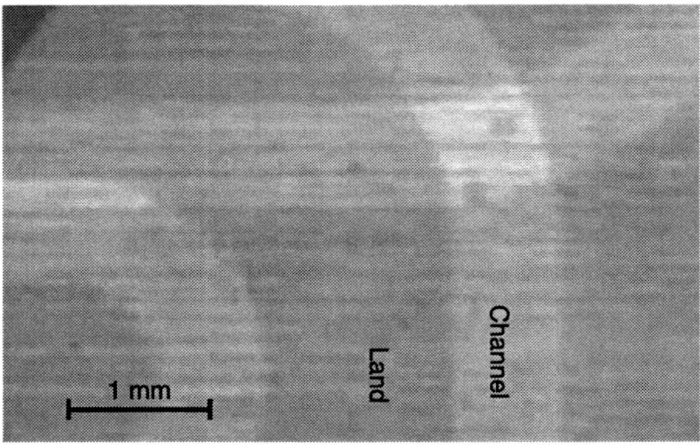

15.8 Non-normalized image as obtained during the experiments.

which are dominated by an absorption coefficient from liquid water, and 'materials pixels', in which the major contribution to the absorption coefficient comes from material components, can be clearly differentiated. Once the images are divided by each other, the changes appear in lower or higher (or unchanged if no changes happened) absorption of the respective pixels and changes can be monitored easily.

The water droplet in the channel is formed by small water agglomerations stemming from the area under the land of the flow-field, which is up to 300 μm × 300 μm². The process here can be described as a sudden eruption of the water volume: a slow increase of the filling level of the pores is followed by an eruptive evolution of the liquid phase to the channel where the smaller volumes in the diffusion medium collapse to form a larger droplet.[9] The overall process takes place in a cyclic mode with consecutive steps of filling and emptying, so-called hydrostatic pumping.

15.3.3 In-plane studies

Compared to the previously described viewing direction, in-plane or cross-sectional investigations are considered to address the water content in the different layers of a fuel cell such as the membrane, diffusion media or the microporous layer. The high local resolution of synchrotron radiography allows, for example, the determination of the water content of the microporous layer with a thickness of less than 30 μm.

In Fig. 15.9, the in-plane view of a fuel cell at different current densities is displayed; apart from the current density the other operating conditions are kept constant. The method can be used to accurately describe the onset of liquid water formation: At the lowest current density observed (250 mA/cm²), no liquid water is formed. With increasing load the onset of liquid water formation can be easily observed (420–500 mA/cm², as marked by arrows). The minimum volumes that can be detected with synchrotron-based methods are as low as a few picoliters.

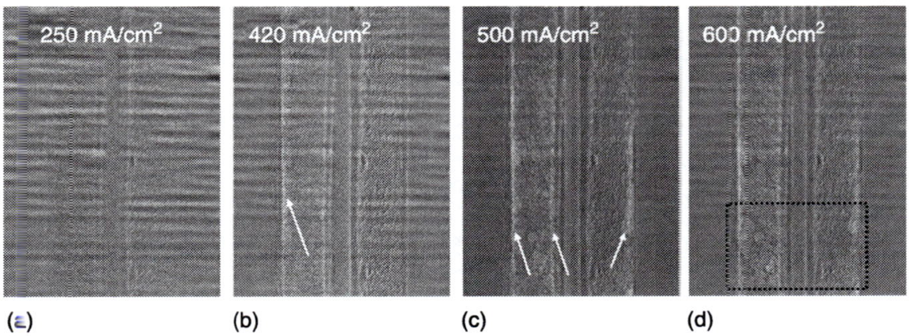

15.9 Liquid water formation as a function of current density. (a) 250 mA/cm². (b) 420 mA/cm². (c) 500 mA/cm². (d) 600 mA/cm². The black dotted rectangle in (d) is the input area for liquid water quantification as displayed in Fig. 15.10.

Based on these images, the relative amount of liquid water as a function of the current density can be estimated. The water thickness is calculated in a standardized way based on the attenuation coefficient of liquid water at the corresponding energy.

As already estimated qualitatively (compare to Fig. 15.9), at low current densities only small amounts of water condense in the vicinity of the microporous layer (MPL) indicating an almost liquid water-free situation at $T = 55$ °C. At higher current densities larger clusters are formed close to the land of the flow field and next to the MPL, which might act as a diffusion barrier for reactant gases. A first maximum of liquid water distribution is located directly in the GDL in close proximity to the MPL. A second maximum resides beneath the rib next to the channel. The combination of the two viewing directions allows for the exact location of liquid water in the diffusion media (Fig. 15.10). As demonstrated above by through-plane imaging and in recent literature,[9] initial water formation starts under the land of the flow field. With the additional information gained from this viewing direction, a pseudo-3D insight can be drawn from these studies: liquid water is formed under the land of the flow field next to the gas channel with a narrow depth profile covering just a few tens of microns.

The dynamics as well as the affected area can be estimated by consecutive images as described above. In Fig. 15.11 two adjacent images and the quotient are displayed. The complete affected volume can be estimated from these images.

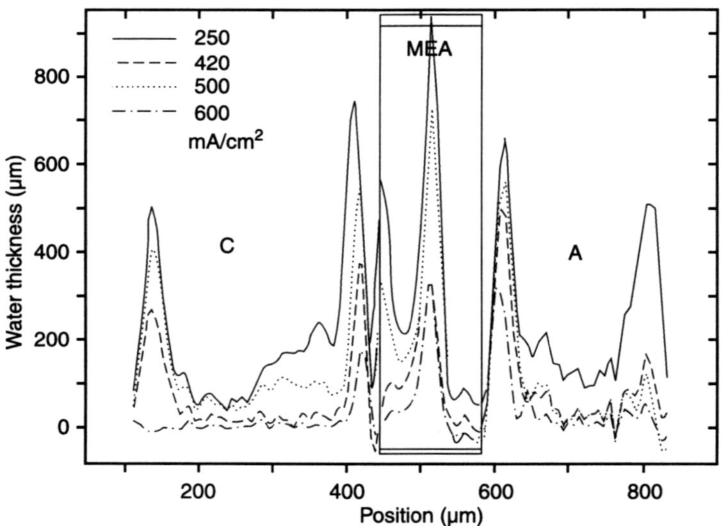

15.10 Quantification and location of liquid water in the cathodic (C) and anodic (A) gas diffusion layer as function of current density; the input for quantification is described in Fig. 15.9.

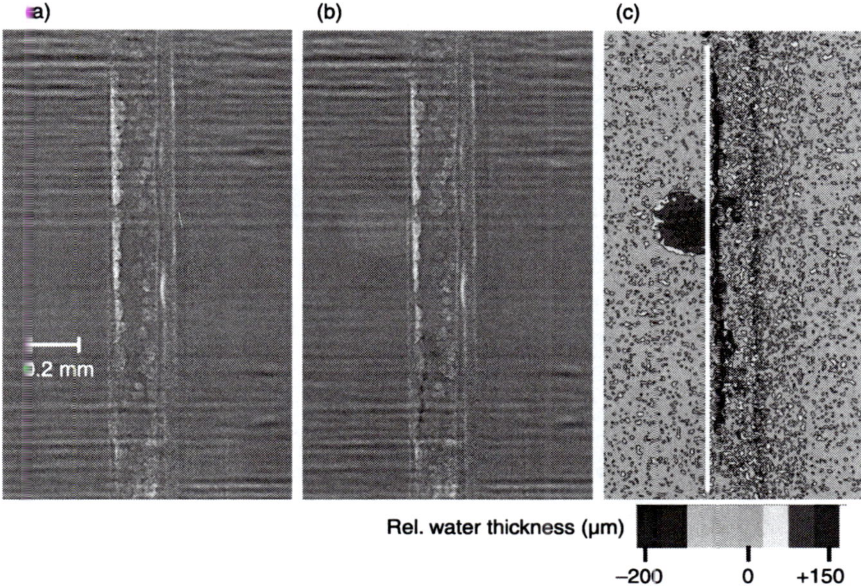

15.11 Eruptive water transport from the gas diffusion media to the flow-field channel. (a, b) white spots denote water agglomerates before and after water eruption. (c) differential image of the images (a) and (b). (Source: Hartnig et al.[10])

15.3.4 *In situ* synchrotron radiography of direct methanol fuel cells (DMFCs)

The lessons learned so far can be directly transferred to liquid-fed systems such as direct methanol fuel cells (DMFCs). As in this class of fuel cells, liquid water evolution is not of main interest, at least at the liquid-fed anode; the focus here is on gas evolution and on flooding effects occurring on the cathode. The compilation below about this aspect is taken from the literature.

Flooding effects and their influence on local performance have been studied in more detail by means of neutron radiography (see Chapter 8). However, the detection of small gas bubbles formed in the electrochemical reaction requires a higher time and local resolution. In general, there are not too many differences in the imaging method when considering PEMFCs and DMFCs. In both cases, the same viewing directions can be distinguished and similar questions can be answered. Also, the same modifications have to be considered as described before, i.e. no metallic parts in the beam direction, and a minimum thickness of absorbing components. In Fig. 15.12, a through-plane image and an in-plane image taken from an operating DMFC are displayed. In the cross-sectional imaging (Fig. 15.12a) dark areas in the anodic diffusion media can be observed. These

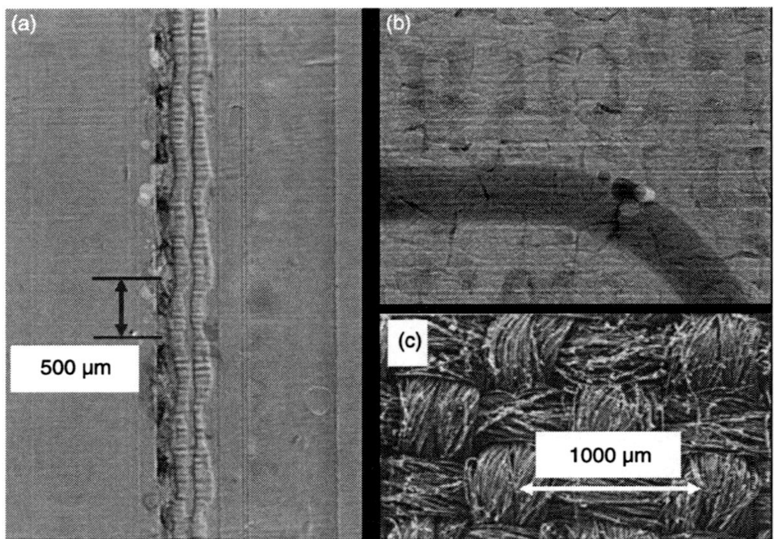

15.12 (a) Visualization of gas agglomerations in the anodic diffusion media by means of cross-sectional imaging. The structure of the diffusion media is reflected in the maxima of the absorption coefficient. (b) Through-plane view of the gas agglomerations which reside in the bends of the carbon cloth and can be identified as a regular dark grid. (c) SEM image of a woven diffusion medium. (Source: Hartnig *et al.*[11])

areas indicate open space in the diffusion media, which are gas bubbles that cannot 'escape' from the carbon cloth structure and are stored in the diffusion media. An identical pattern is observed in a through-plane view (Fig. 15.12b) denoting the formation of gas troughs, which in turn resemble the pattern of the woven structure. Synchrotron imaging proves itself to be a valuable tool to observe these phenomena without interference with the realistic setup. In order to correlate and explain the findings from the imaging approaches, the patterns are compared to an SEM image. The periodicity of the structure of the diffusion media becomes obvious and the maxima of the adsorption distribution correlate well with the bends of the woven material: the maximum-to-maximum distance of 500 µm is identical to the structure as observed in the SEM image (Fig. 15.12c). Additional information can be drawn, as it can be easily observed that no gas bubbles are trapped close to the crossing of the fiber strands. The crossings serve as active spots where carbon dioxide bubbles leave the diffusion media to transfer to the solution.

15.4   *In situ* synchrotron tomography

With radiographic methods a two-dimensional (2D) insight into processes taking place in a fuel cell can be obtained. In addition, combining different viewing directions gives a pseudo-3D impression. However, radiographs suffer from the fact that only integral representations of material distribution are gained, which can be overcome by tomographic methods. These are in principle a series of radiographs that are reconstructed resulting in a tomogram.

Tomographic imaging implies a bunch of limitations with regard to cell design. Similar to what has been explained in Chapter 9 the following pre-requisites have to be fulfilled to achieve an optimum outcome from imaging experiments:

- The fuel cell should fit completely in the beam cross-section.
- No metallic or other strongly absorbing parts should be in th. beam.
- Additional water layers (cooling/heating flow field) in the beam direction should be avoided.

These preconditions require specially designed micro fuel cells. The active area is usually in the range of less than 10 mm × 10 mm, and in order to avoid artifacts stemming from additional water layers, the cells are only indirectly heated, based on the thermal conductivity of the carbon composite flow fields.

Similar to radiographic studies, two different rotational directions can be distinguished, as shown in Fig. 15.13, which will have an impact on the cell design. In the case of in-plane rotation (Fig. 15.13b), metallic parts cannot be employed to ensure a sufficient compression of the fuel cell layers. On the other hand, the perpendicular rotational axis (Fig. 15.13a) requires a careful assembly of the cell so that no external metallic parts, which ensure that the cell assembly is tight, are in the beam direction.

In Fig. 15.14, a tomogram of a micro fuel cell is displayed. The cell was designed according to the requirements mentioned above; at the same time realistic operating conditions with respect to temperature and gas stoichiometry

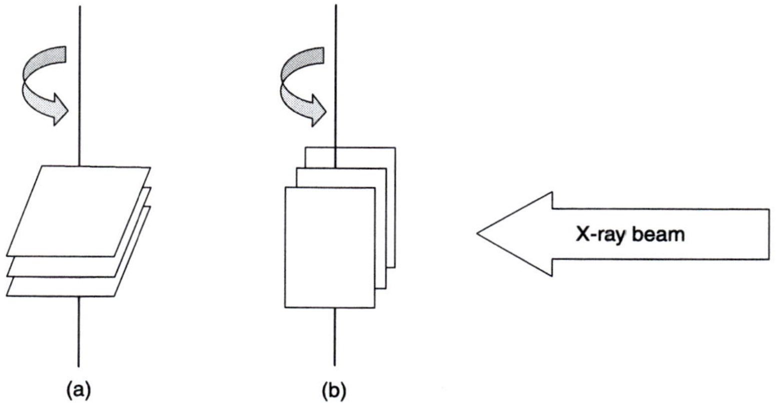

15.13 Rotating directions employed for synchrotron tomograms. (a) Perpendicular rotation (axis) (b) In-plane rotation.

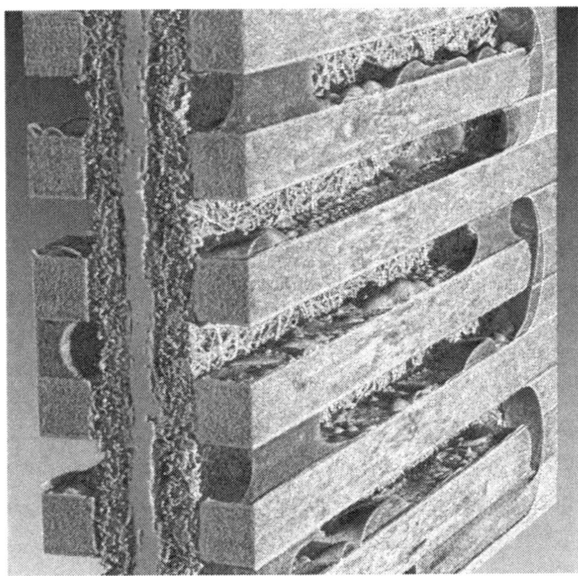

15.14 Tomogram of a small hydrogen-driven fuel cell; back parts of the flow fields are rendered transparent for clarity.

should be achieved. Due to the large number of images that have to be taken for a resulting high-quality tomogram (up to 1800 images/angle positions), the total imaging time is in the range of several hours. By stopping the gas supply, sealing the gas inlets and outlets and at the same time keeping the temperature of the cell constant, artifacts resulting from moving water droplets can almost be excluded. The positive result of this approach is demonstrated in the tomogram, in which no shading or other blurring, which would result from changing positions of the droplets, is observed.

The different components of the fuel cell can be differentiated quite clearly. In addition, the position of the droplets in the gas channel, their shape and contact angle formed with the channel walls can be determined. Compared to initial studies employing acrylic end plates, 'standard materials' can be used to avoid or at least minimize the risk of artificial results.[12] The effect of gravity is obvious and is indicated in the position of the droplets, which reside preferably on the wall oriented towards the bottom of the channel. Especially in the channel bends the adsorption of the water droplets is characterized by the slightly hydrophilic properties of the flow field, which is machined from graphite composite.

Based on the 3D information gained from tomograms, slices in different directions can be taken to study a variety of effects and material properties that are not discussed in more detail here:

- Water-filled and water-free domains with a local resolution of about 5 μm limited only by the extensions of the cell and the available pixels per line in the CCD detector can be detected.
- Transport pathways from the catalyst layer, the place where liquid water is formed, to the gas channel can be highlighted.
- Structural effects of the diffusion media on the droplet formation pattern can be observed, i.e., size of droplets, internal contact angle.

Another example underlining the advantage of a non-destructive method is the determination of the membrane thickness as a function of external parameters. Under operation, the thickness of the membrane is influenced by several factors, and as such is a function of the humidification level, which can be adapted by humidifying the reactant gases, the applied current density or the employed material. A certain humidification level of the membrane is required to achieve a reasonable protonic conductivity. Perfluorinated sulfonic acid-based materials, such as Nafion®, obtain their optimum conductivity at a humidification of $\lambda > 20$, where λ is defined as water molecules per sulfonic acid headgroup. Excessive water content, on the other hand, leads to flooding phenomena in the catalyst layer and/or channels of the flow field, underlining the importance of well-balanced water management.

The current density has a significant influence on swelling behavior and therefore the thickness of the membrane: at zero or very low current densities, the humidification of the membrane is based on diffusion processes, where water molecules formed in the cathodic catalyst layer or stemming from the external humidification have to 'wet' the membrane based on a concentration gradient. Once an external load is applied and a certain protonic current is established, the water molecules in contact with a proton (specifically, hydrated protons) are dragged with these charged particles. Since the protonic current requires the formation of transport channels, the above-mentioned humidification level has to be reached to assure sufficient conductivity. This migration of water is often referred to as the 'electro-osmotic drag effect' and describes the pull-effect of migrating protons in the membrane as the result of the external field. As an effect of this electro-osmotic dragging and the overall increased water content, the humidification level of the membrane is increased up to a certain threshold. From a particular material-specific current density onwards, no further increase of the water content of the membrane can be observed, and quasi-stationary humidification is achieved.

In Fig. 15.15 the membrane thickness as measured from tomographic imaging is given as a function of the current density. On the one hand, the advantage of a non-perturbing method becomes obvious: without impact on the cell and regular performance, the conditions inside the fuel cell can be observed, which would be impossible with plain *ex situ* experiments. At the same time, the 3D-information allows for the determination of the thickness at different spots, which in turn

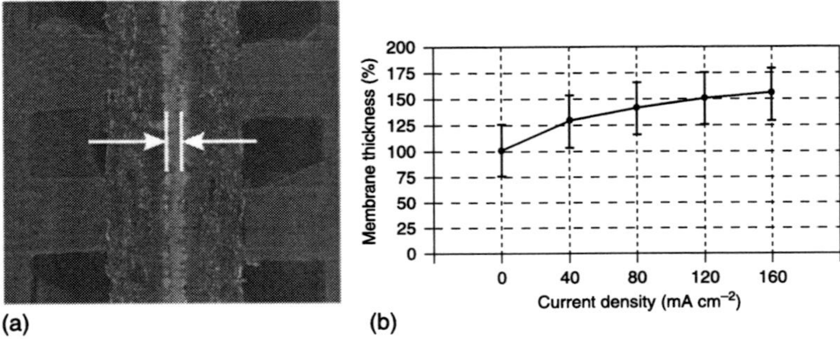

(a)   (b)

15.15 (a) Definition of the membrane thickness as obtained from tomographic imaging. (b) Membrane thickness as a function of the current density. 0 mA/cm² is taken as the reference value (Source: Krüger *et al.*[13])

qualifies the method and allows for the definition of an error bar included in these experiments. The membrane thickness, which is defined as 100% at zero current density, increases steadily with increasing current density up to approximately 200 mA/cm². As this is the maximum value that can be obtained with this setup, any further increase of the membrane thickness is only speculative, but a saturation value can be expected from the slope of the curve.[14]

15.4.1 Input for future studies

A variety of different disciplines can benefit from the findings gained from this comparatively demanding and time-consuming method, ranging from direct results with regard to design improvements up to input for modeling activities, which are summarized and highlighted below.

In *ex situ* studies employing optical microscopy[14, 15] an eruptive mechanism has been detected, which is best described as a successive filling and emptying of pores in the diffusion media. This transport process has been verified in *in situ* studies and, in addition, the source of liquid water and the area affected by this transport mechanism have been determined. A pseudo-3D or real 3D insight into water transport processes can be taken as an input for modeling studies based on a variety of models. Pore network modeling as well as lattice Boltzmann simulations benefit from realistic input data, which is gained from these studies.

Computational fluid dynamics (CFD) modeling activities manage to combine several effects that influence the location of liquid water formation. The primary place of liquid water formation is under the lands of the flow field. By means of synchrotron imaging the very first agglomerates can be detected and theoretical findings can be verified. In the next step, the flow behavior of a single cell can be transferred to model a complete fuel cell stack, taking into account the different filling stages of the GDL and the flow field channels.

15.5   Conclusion and future trends

Compared to neutron-based methods, which are well established in fuel cell research, synchrotron X-ray applications are not too widely used. Methods based on X-rays can be considered complementary to neutron radiography and tomography (see Chapter 9). The observable region is in most cases much smaller, in the range up to 1 cm^2, at least two orders of magnitude smaller than in neutron imaging. Also, the experimental setup is more difficult, as shielding parts, such as metallic end plates or metallic current collectors, have to be prepared so that the optimum resolution can be gained. However, in terms of resolution the advantages of X-ray imaging become obvious. The local as well as time resolution is at least one order of magnitude higher than that which can be achieved with neutron imaging. Depending on the sample diameter and the size of the camera chipset a resolution as low as 0.6 µm can be obtained. The time resolution (image-to-image period) is mainly limited by the beam intensities and can be as low as 0.1 s per image (2010). Due to the development of new, more sensitive scintillator materials and optimized cameras, shorter time resolutions are anticipated. The combination of high local and short time resolution allows not only for the imaging of liquid water transport processes, but also for a fundamental insight into the affected areas. Synchrotron X-ray radiography can therefore be employed to study material and structural effects on droplet formation and the transport mechanism. However, the intensity of the incident beam has to be carefully chosen, as the impact on the catalyst layer might cause enhanced degradation rates, which in turn might be caused by local heating effects.[16] Compared to *ex situ* methods focusing on the migration of water from the GDL to the channel, the origin of liquid water can also be addressed.

The suitability of synchrotron sources to tune the beam energy in very narrow steps allows the detection not only of liquid water, but opens up a wide field of similar applications. The method has been successfully applied to questions regarding the local structure and composition of the MEA under operating conditions, and from these findings deduce the distribution of phosphoric acid in high-temperature PEM fuel cells.[17, 18]

15.6   References

1. J. Banhart (Ed.) (2008) *Advanced Tomographic Methods in Materials Research and Engineering*, Oxford University Press.
2. W. Reimers, A.R. Pyzalla, A.K. Schreyer, H. Clemens (Eds.) (2008) *Neutrons and Synchrotron Radiation in Engineering Materials Science*, Wiley-VCH, Weinheim.
3. K.-W. Harbich, P. Klobes, M. P. Hentschel (2004) *Colloids and Surfaces A: Physicochemical and Engineering Aspects* 241 p. 225–229.
4. Ch. Hartnig, R. Kuhn, Ph. Krüger, I. Manke, N. Kardjilov (2008) 'Water management in fuel cells – a challenge for non-destructive high resolution methods', *Material Testing*, **50**, 609.

5. R. Thiedmann, Ch. Hartnig, I. Manke, V. Schmidt, W. Lehnert (2009) 'Local structural characteristics of pore space in GDLs of PEM fuel cells based on geometric 3D graphs' *J. Electrochem. Soc.*, **156**, B1339–B1347.

6. www.geodict.com.

7. Image courtesy of Henning Markötter, Helmholtz-Zentrum Berlin.

8. A. Isopo, V. Rossi Albertini (2008) 'An original laboratory X-ray diffraction method for in situ investigations on the water dynamics in a fuel cell proton exchange membrane', *J. Power Sources*, **184**, 23–28.

9. I. Manke, Ch. Hartnig, M. Grünerbel, W. Lehnert, N. Kardjilov *et al.* (2007) *Appl. Phys. Lett.* **90**, 174105.

10. Ch. Hartnig, I. Manke, R. Kuhn, S. Kleinau, J. Goebbels *et al.* (2008) 'High resolution in-plane investigation of the water evolution and transport in PEM fuel cells', *Journal of Power Sources*, **188** p. 468–474, Copyright (2008), with permission from Elsevier.

11. Ch. Hartnig, I. Manke, J. Schloesser, P. Krüger, R. Kuhn *et al.* (2009) *Electrochemistry Communications*, **11** pp. 1559–1562, Copyright (2009), with permission from Elsevier.

12. S. J. Lee, N.-Y. Lim, S. Kim, G.-G. Park, C.-S. Kim (2008) 'X-ray imaging of water distribution in a polymer electrolyte fuel cell', *J. Power Sources*, **185**, 867–870.

13. Ph. Krüger, H.Markötter, J.Haußmann, M.Klages, T.Arlt *et al.* (2010) Synchrotron X-ray tomography for investigations of water distribution in polymer electrolyte membrane fuel cells, Copyright (2010), with permission from Elsevier. doi:10.1016/j. jpowsour.2010.09.042.

14. A. Bazylak, D. Sinton, N. Djilali (2008) 'Dynamic water transport and droplet emergence in PEMFC gas diffusion layers', *J. Power Sources*, **176**, 240–246.

15. S. Litster, D. Sinton, N. Djilali (2006) '*Ex situ* visualization of liquid water transport in PEM fuel cell gas diffusion layers', *J. Power Sources*, **154**, 95–105.

16. A. Schneider, C. Weiser, J. Roth, L. Helfen (2010) 'Impact of synchrotron radiation on fuel cell operation in imaging experiments', *J. Power Sources*, **195**, 6349–6355.

17. W. Maier, T. Arlt, C. Wannek, I. Manke, H. Riesemeier *et al.* (2010) '*In-situ* synchrotron X-ray radiography on high temperature polymer electrolyte fuel cells', *Electrochem. Comm.*, **12**, 1436–1438.

18. R. Kuhn, J. Scholta, Ph. Krüger, Ch. Hartnig, W. Lehnert *et al.* (2011), 'Measuring device for synchrotron X-ray imaging and first results of high temperature polymer electrolyte membrane fuel cells', *J. Power Sources*, **196**, 5231–5239.